ENGINEERING AND MANUFACTURING FOR BIOTECHNOLOGY
VOLUME 4

FOCUS ON BIOTECHNOLOGY

Volume 4

Series Editors
MARCEL HOFMAN
Centre for Veterinary and Agrochemical Research, Tervuren, Belgium

JOZEF ANNÉ
Rega Institute, University of Leuven, Belgium

Volume Editors
MARCEL HOFMAN
Société de Chimie Industrielle,
Centre for Veterinary and Agrochemical Research, Tervuren, Belgium

PHILIPPE THONART
Faculté Univ. des Sciences Agronomiques Gembloux,
Centre Wallon de Biologie Industrielle, Belgium

Colophon

Focus on Biotechnology is an open-ended series of reference volumes produced by Kluwer Academic Publishers BV in co-operation with the Branche Belge de la Société de Chimie Industrielle a.s.b.l.

The initiative has been taken in conjunction with the Ninth European Congress on Biotechnology. ECB9 has been supported by the Commission of the European Communities, the General Directorate for Technology, Research and Energy of the Wallonia Region, Belgium and J Chabert, Minister for Economy of the Brussels Capital Region.

The series is edited by Marcel Hofman, Centre for Veterinary and Agrochemical Research, Tervuren, and Jozef Anné, Rega Institute, University of Leuven, Belgium.

Engineering and Manufacturing for Biotechnology Volume 4

Edited by

MARCEL HOFMAN
Société de Chimie Industrielle,
Centre for Veterinary and Agrochemical Research, Tervuren, Belgium

and

PHILIPPE THONART
Faculté Univ. des Sciences Agronomiques Gembloux,
Centre Wallon de Biologie Industrielle, Belgium

SPRINGER-SCIENCE+BUSINESS MEDIA, B.V.

Library of Congress Cataloging-in-Publication Data

Engineering and manufacturing for biotechnology / editors, Marcel Hofman, Philippe Thonart.
p. cm. -- (Focus on biotechnology ; v. 4)
Includes bibliographical references and index.
ISBN 978-90-481-5689-4 ISBN 978-0-306-46889-6 (eBook)
DOI 10.1007/978-0-306-46889-6
1. Biotechnology--Congresses. 2. Biotechnological process control--Congresses. I. Hofman, Marcel. II. Thonart, Philippe. III. Series.

TP248.14 .E545 2001
660.6--dc21

2001022470

ISBN 978-90-481-5689-4

Printed on acid-free paper

EDITORS PREFACE

Early integration of process engineering and biological research is the key for success in industrial biotechnology. This is true as well when a selected wild-type organism is put to work as when an organism is engineered to purpose.

Focus on Biotechnology reports on biotechnology achievements in the recent past, but also provides a strategic view on the evolution in the next decade. The present volume "Engineering and Manufacturing for Biotechnology" took advantage of the 9th European Congress on Biotechnology (Brussels, Belgium, July 11-15, 1999) : by topics handled and by expertise of the contributors the engineering science symposia of this congress offered just what was needed to cover this important topic.

The editors have solicited the authors of a number of outstanding contributions to illustrate the intimate interaction between productive organism and the numerous processing steps running from the initial inoculation to the packaged product. Upstream processing of the feed streams, selection of medium components, product harvesting, downstream processing and product conditioning are just a few major steps. Each step imposes a number of important choices. Every choice is to be balanced against time to market, profitability, safety and ecology.

It should be readily apparent from this volume that the development of a truly effective biotechnology process requires a broad command of leading-edge engineering science, a spark of genius, and last but not least much hard work. That is why the editors wish to express their gratitude to all the authors of this volume, for finding the time after busy hours at the lab, on the pilot floor or in the production plant to share their experience and enthusiasm.

A final word of esteem is due to all those that through their devoted and outstanding secretarial skills have made the edition of this volume possible.

Marcel Hofman   Philippe Thonart

TABLE OF CONTENTS

PART I
UPSTREAM PROCESSES AND FERMENTATION

PRETREATMENT PROCESSES OF MOLASSES FOR THE UTILIZATION IN FERMENTATION PROCESSES

GÜZIDE ÇALIK, MELIHA BERK, FATMA GÜL BOYACI, PINAR ÇALIK, SERPIL TAKAÇ, TUNÇER H. ÖZDAMAR
Ankara University Biotechnology Research Center, Industrial Biotechnology Department, Tandoğan 06100, Ankara, TURKEY

Abstract

The composition of beet molasses was modified by 14 pretreatment processes which were the combination of physical, physicochemical, and chemical processes, i.e. acid hydrolysis of sucrose, precipitation of metal ions, removal of organic acids by ion exchange resins, and entrapment of metal ions in solution. Acid hydrolysis was favourable for glutamic acid fermentation, while the use of diluted and centrifuged molasses was advantageous for alkaline protease fermentation.

1. Introduction

Molasses, which are the by-products of the sugar-beet or sugarcane extraction processes are among the most important raw materials of the fermentation industry; especially for the production of baker's yeast, citric acid, feed yeasts, acetone/butanol, organic acids, amino acids, antibiotics, and enzymes. The suitability of molasses for industrial fermentations cannot be evaluated according to their origin and chemical composition as different criteria determine the productivity and quality for their use in different processes. In processes in which they are the sole carbon source, the molasses should be pretreated and the inhibitors should be removed. For yeast and methanol production molasses are often simply neutralised with calcium carbonate. For many other processes they are only boiled in an acidic or alkaline medium and after setting out separated from the precipitate. For citric acid production the molasses are boiled with potassium ferrocyanide and generally fermented together with the precipitate (Palacios, 1966, Kundu *et al.*, 1984, Cejka, 1985, Sharma *et al.*, 1991).
Although beet or cane molasses has been used in the fermentation media for the production of glutamic acid with *Brevibacterium* or *Corynebacterium* strains in literature the aim were generally to investigate the effect of other operational parameters

M. Hofman and P. Thonart (eds.), Engineering and Manufacturing for Biotechnology, 21–28.

such as the use of surface active agents to increase the membrane permeability in fed-batch operation (Yu-cheng, 1973), shift in operation temperature with temperature sensitive mutants (Momose and Takagi, 1978), estimation of cell growth with on-line CO_2 measurement (Park *et al.*, 1983a,b, 1984, Wu *et al.*, 1989), continuous production with immobilised biocatalyst (Kim and Ryu 1982). The pretreatment processes, if applied any, has not been described in the above-mentioned literature.

Table 1. Composition and properties of the beet molasses.

Dry solids (%)	81.6
Sucrose (%)	51.0
Invert sugar (%)	0.3
Raffinose (%)	1.2
Ash (%)	12.7
Total nitrogen (%)	1.6
Purity quotient	62.5
pH	7.5

The effects of various defined and semi-defined media involving simple carbon sources such as glucose and/or fructose, organic acids, and amino acids, or complex carbon sources such as casein, corn steep liquor, and starch (Çalık *et al.*, 1998, Çalık *et al.*, 2000), in the production of serine alkaline proteases that are the most important group of industrial enzymes using *Bacillus* strains have been investigated, however molasses has not been used for the protease production processes in the literature.

This work investigates the effects of several pretreatment processes (PP), which were the combination of physical, physicochemical, and chemical processes, systematically on the composition of molasses and finally on the efficiency of their use on two bioprocesses, namely, glutamic acid and serine alkaline protease production.

2. Materials and methods

2.1. PRETREATMENT PROCESSES (PP)

Beet molasses, the composition and properties of which was given in Table 1 was modified with 14 pretreatment processes that have been designed as the combination of the following unit processes and abbreviated as PP1 - PP14 as given in Figure 1 (Berk, 1995). Not pretreated, that is, only diluted and centrifuged molasses was abbreviated as PP0. The details of the physical, physicochemical, and chemical pretreatment shown in the Figure are as follows: *Dilution:* 100 g molasses was diluted with water to obtain 200 cm^3 solution. *Centrifugation:* Diluted or insoluble impurity containing molasses was centrifuged for 20 min at 6000 *g* and +4 °C conditions. *Hydrolysis:* Sucrose hydrolysis of molasses was accomplished either with a liquid acid, i.e. H_2SO_4 and H_3PO_4, or a solid acid, i.e. Amberlite IR-120P. Liquid acid hydrolysis was made at pH=3, T=60°C temperature, N=40 rpm agitation rate, and t=1 h conditions with 6M acid. In some of the pretreatments further hydrolysis with liquid at pH=1 was made in order to hydrolyse sucrose completely. Solid hydrolysis reaction was established in packed column including strongly acidic cation exchange resin, Amberlite IR-120P, at T=60°C temperature and 5 cm^3 min^{-1} flow rate conditions by using molasses solution containing 160 g dm^{-3} sucrose. *Precipitation:* Metal ions in molasses were precipitated with $Ca(OH)_2$, NH_4OH, Na_2HPO_4, $CaHPO_4$, $(NH_4)_2SO_4$, and $K_4[Fe(CN)_6]$ solutions. $Ca(OH)_2$ precipitation was carried out either at pH=8 or pH=10, and t=1 h, N=40 rpm, and 70°C temperature conditions with 1.5 M base. NH_4OH precipitation was similar to $Ca(OH)_2$. In the precipitation with Na_2HPO_4 and $CaHPO_4$, 40 cm^3 1.5 M Na_2HPO_4 and 80 cm^3 1.5 M $CaHPO_4$ were added separately to molasses solution to adjust pH to 3 and 8, respectively. The reaction was carried out for 1 h at 70°C temperature. To separate Ca^{+2} ions from the medium $(NH_4)_2SO_4$ was added. $K_4[Fe(CN)_6]$ precipitation was performed with 10 cm^3, 0.436 mol dm^{-3} solution. *EDTA Entrapment:* 5 % (w/v), 60 cm^3 EDTA was used in order to entrap metal ions from the solution. *SACE Pretreatment*: In one of the pretreatment processes metal ions were removed with a strongly acidic cation exchange (SACE) resin, Amberlite IR-120P. For this purpose, molasses solution was fed to the ion exchange column at 10 cm^3 min^{-1} flow rate. *SBAE Pretreatment:* Organic acids were removed with a strongly basic anion exchange (SBAE) resin, Amberlite IRA-400. After the pH of 160 g dm^{-3} molasses solution had adjusted to 7, it was fed to the ion exchange column at 5 cm^3 min^{-1} flow rate. *Filtration:* The solution was filtered with Whatman No:1 filter paper after the hydrolysis and precipitation steps for the separation of the solid impurities.

2.2. BIOPROCESSES

2.2.1. Glutamic acid fermentation

Corynebacterium glutamicum (NRRL B-2784) was grown, inoculated and cultivated as described elsewhere (Berk, 1995). 6 g dm^{-3} urea was added to molasses solution involving 60 g dm^{-3} reduced sugar. pH was adjusted with 25 % NH_3 to 7.3. Batch experiments were conducted in agitation and heating rate controlled orbital shakers, using microbial air filtered 150 cm^3 flasks having 10 cm^3 working volume capacities at T=30°C temperature and N=105 rpm agitation rate conditions. C_p=0.85 U cm^{-3} penicillin G was added to the bioconversion medium at t=8 h cultivation time.

2.2.2. Alkaline protease fermentation

Bacillus licheniformis (DSM 1969) was grown, inoculated and cultivated as described elsewhere (Çalık *et al.*, 1998). The reference medium for the investigation of pretreatment processes in alkaline protease fermentation was (g dm^{-3}): glucose, 6.0; $(NH_4)_2PO_4$, 4.7; KH_2PO_4, 2.0; $CaCl_2$, 0.2 (Çalık *et al.*, 1998). Glucose was replaced by pretreated molasses solution PP0, PP1, and PP13 in order to investigate the effects of the pretreatment. Batch experiments were conducted in agitation and heating rate controlled orbital shakers, using microbial air filtered 150 cm^3 flasks having 30 cm^3 working volume capacities at pH=7.25, T=37°C temperature and N=200 rpm agitation rate conditions.

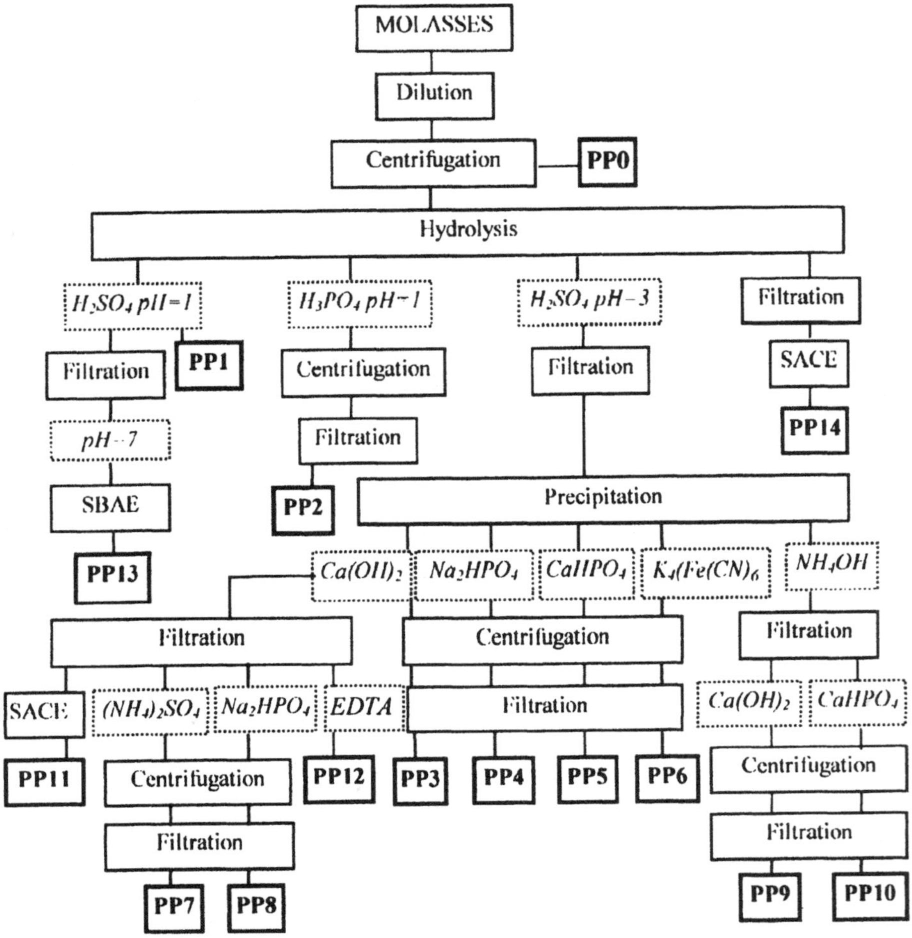

figure 1. The order and nomenclature of pretreatment processes applied to molasses.

3. Results and discussions

3.1. EFFECT OF PP ON METAL ION CONCENTRATIONS

The beet molasses solution contains compounds used as substrates in bioprocesses such as sucrose, invert sugar, amino acids, organic acids, inorganic compounds and vitamins (Schneider,1979; Cejka,1985). However, some compounds can inhibit growth Table 2. The efficiency of metal ion removal with the pretreatment processes for molasses of the microorganism and product formation. Moreover, the stability of the product can be decreased by the compounds involved in molasses. Consequently, the molasses must be modified by some pretreatment processes before using as a substrate in bioprocesses. One of the objectives of this work was to remove the metal ions, Fe^{+2}, Fe^{+3}, Pb^{+2}, Ca^{+2}, Cu^{+1}, Cu^{+2}, Mn^{+2}, Zn^{+2}, Mg^{+2}, K^{+1}, and Na^{+1}, by pretreatment processes and to determine the pretreatment process effects in glutamic acid and alkaline protease fermentations. The order of different pretreatment processes designed in this work, which were the combination of dilution, centrifugation, acid hydrolysis of sucrose, precipitation of metal ions, removal of organic acid, entrapment of metal ions and filtration are given in detail in Figure 1. After the employment of the pretreatment processes from PP1 to PP11, the efficiency of metal ion removal can be seen in Table 2. The metal ions were removed completely by the pretreatment processes PP12 and PP14.

Table 2. Efficiency of metal ion removal

PP	% Removal								
	Fe^{+2}, Fe^{+3}	Pb^{+2}	Ca^{+2}	Cu^{+1}, Cu^{+2}	Mn^{+2}	Zn^{+2}	Mg^{+2}	K^{+1}	Na^{+1}
PP1	2.6	49.0	10.4	40.0	3.1	11.0	7.6	9.4	8.6
PP2	16.0	1.8	6.7	28.6	7.3	10.0	20.0	9.0	9.9
PP3	41.5	50.0	-	100	19.0	11.6	73.0	11.0	16.6
PP4	39.0	62.5	31.0	63.7	64.0	38.0	23.0	12.0	-
PP5	27.0	50.0	-	72.0	12.0	19.0	12.0	3.8	1.7
PP6	-	50.0	60.0	100	100	94.0	28.6	-	2.2
PP7	12.0	91.8	100	100	33.8	45.0	2.6	31.7	35.0
PP8	78.0	95.0	60.0	100	99.0	66.0	34.0	13.6	-
PP9	-	-	100	100	100	100	-	-	8.0
PP10	17.0	40.0	42.5	100	62.0	10.0	-	17.6	9.8
PP11	100	100	100	91.0	100	100	100	99.8	100

3.2. EFFECT OF PP ON GLUTAMIC ACID FERMENTATION

The glutamic acid fermentation was carried out by PP0, PP1, PP2, PP3, PP7, PP11 and PP14 molasses for 37 h. The effect of the pretreatment processes on the relative concentrations of *Corynebacterium glutamicum* and glutamic acid are shown in Figures 2 and 3, respectively. As it is clear in Figure 2, pretreatment processes inhibit cell growth to some extent, where PP14 has the most detrimental effect, because all the metal ions are removed totally. When glutamic acid concentration is considered the best results were achieved with PP1, PP13 and PP2. The comparison of results of fermentations carried out with untreated and pretreated molasses showed that the pretreated molasses increased the glutamic acid yield up to 70 %. The highest glutamic acid concentration was obtained with PP1. In PP1, Ca^{+2} and Pb^{+2} ions were precipitated together with SO_4^{-2} ions as their sulphates; however, in PP2, Ca^{+2}, Mn^{+2}, Mg^{+2} and Zn^{+2} ions precipitate as their phosphates. From the results given in Figure 2 one can conclude that phosphate ion is still a stronger inhibitor than sulphate ion for the microorganism in glutamic acid fermentation. Therefore, its concentration in the media should be very low. Organic acids were removed by strongly basic anion exchange resin with PP13. PP13 was almost as effective as PP1, hence the concentration of organic acids does not produce a negative effect in glutamic acid fermentation.

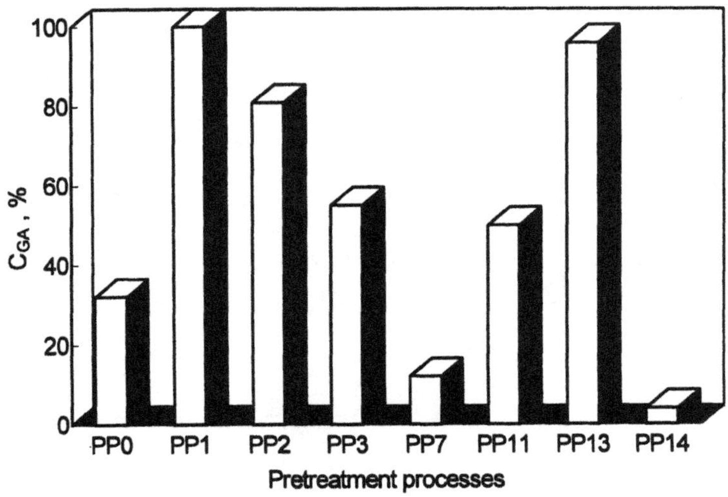

Figure 2. Effect of pretreatment processes on the relative Corynebacterium glutamicum concentration.

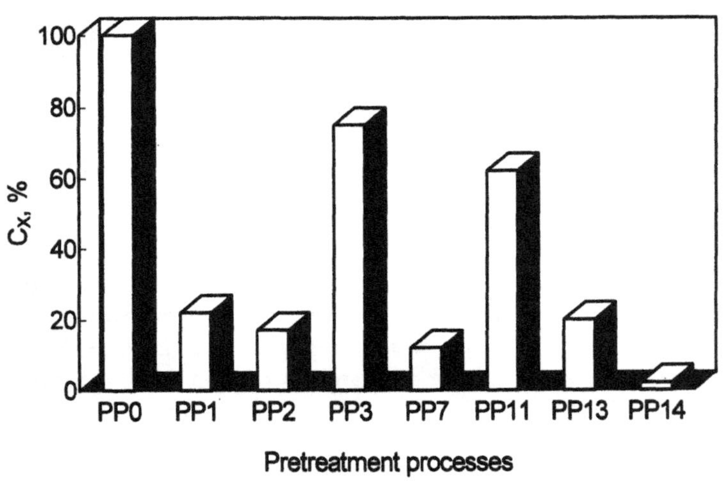

Figure 3. Effect of pretreatment processes on the relative glutamic acid concentration.

3.3. EFFECT OF PP ON SERINE ALKALINE PROTEASE FERMENTATION

The effect of pretreatment processes on alkaline protease fermentation was investigated with PP0, PP1, and PP13 molasses. Alkaline protease productions were carried out for t=46 h. For this purpose, readily accessible carbon source glucose in the fermentation media was replaced with PP0, PP1, and PP13 molasses corresponding to different sucrose or reduced sugar concentrations between 1-50 g dm^{-3}. The variations of cell concentration and enzyme activity were measured throughout the bioprocess. The activity and cell concentrations obtained with PP1 and PP13 molasses did not differ significantly. The results showed that when sucrose/reduced sugar concentration of the PP0 and PP1 molasses increased, the cell concentration first increased and then decreased, giving a maximum at 45 and 15 g dm^{-3}, respectively. With PP0 and PP1 molasses, maximum cell concentrations were 2.5 and 3.4 g dm^{-3}, while for the reference medium involving 6 g dm^{-3} glucose as the carbon source this value was 1.2 g dm^{-3}. Reduced sugar concentration affected the enzyme activity in the same manner and 15 g dm^{-3} gave maximum relative activity with PP1 when compared with the reference solution, as 1.1 and it was obtained at t=43 h. However, only diluted and centrifuged molasses PP0 was more effective in the protease production with the relative activity value 2.1 obtained with 30 g dm^{-3} sucrose containing molasses (Figure 4). These results show that besides the positive effects of amino acids, organic acids, inorganic compounds and vitamins involved in molasses the use of sucrose instead of a readily accessible carbon source, e.g. glucose and fructose, may cause the microorganism to function the bioreaction network under stress that it increases the protease production.

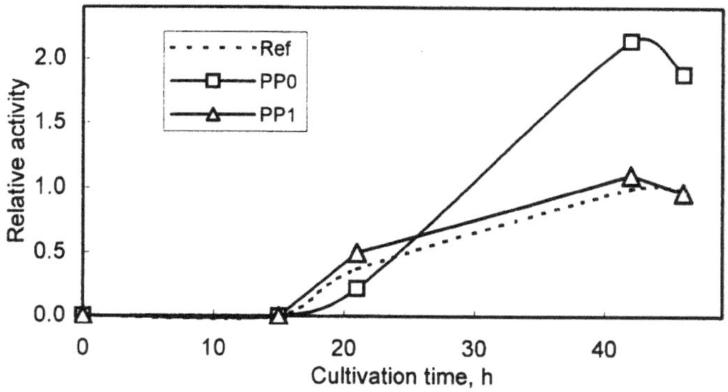

Figure 4. Effect of the hydrolysis of molasses on relative protease activity.

4. Conclusions

14 pretreatment processes were developed for the utilisation of beet molasses in fermentation process, i.e., glutamic acid and serine alkaline protease. The best results were obtained with PP1 for glutamic acid production, however, PP0 molasses was more beneficial for the protease enzyme production. When essential components are removed from the media, the cell growth is strongly inhibited. This consequently causes a decrease in the product formation. However, the presence of some ions may induce the stability of a product, e.g. an enzyme as well. Therefore, the design of the pretreatment

process to be applied to a complex carbon and energy source should be made considering the needs of the cell and its metabolism. The investigation of different pretreatment processes to protease enzyme production as well as other bioprocesses are being continued.

Acknowledgements

The financial support of SPO (Turkey) Grant No. 89K120390 and 95K120290 are gratefully acknowledged.

References

Berk, M. (1995) Adaptation of molasses to the bioprocess for L-glutamic acid, *M.S. Thesis (Turkish)*, Ankara University, Ankara.

Cejka, A. (1985), Preparation of media, in: H.-J.Rehm and G. Reed (eds.), *Biotechnology*, VCH, Weinheim, 2, pp. 629-639.

Cohen, A. (1984) The Pico.Tag system: A new method to analyse primary and secondary amino acids with one picomole sensitivity, *Bio.Tech., September-October*, 273-279.

Çalık, P., Çalık, G. and Özdamar, T.H. (1998) Oxygen transfer effects in serine alkaline protease fermentation by *Bacillus licheniformis*: Use of citric acid as the carbon source, *Enzyme Microb. Technol. 23*, 451-461.

Çalık, P., Çalık, G., and Özdamar, T.H. (in press) Bioprocess development for serine alkaline protease production, *Reviews in Chemical Engineering.*

Kim, H.. and Ryu, D.D.Y. (1982) Continuous glutamate production using an immobilised whole-cell system, *Biotechnol. Bioeng., 24,* 2167-2174.

Kundu, S.T., Panda, S.K., Majumdar, B., and Guha (1984) Pretreatment of Indian cane molasses for increased production of citric acid, *Biotechnol. Bioeng. 26,* 1114-1121.

Momose, H. and Takagi, T. (1978) Glutamic acid production in biotin-rich media by temperature sensitive mutants of *Brevibacterium lactofermentum*, a novel fermentation process, *Agric. Biol. Chem. 42*, 1911-1917.

Palacios, D.R. (1966) Citric acid from ferrocyanide-treated blackstrap molasses, *Biotechnol. Bioeng., 11,* 103-104.

Park, S.H., Hong, K.T., Lee, J.C., and Bea, J.C. (1983a) On-line estimation of cell growth for glutamic acid fermentation system, *Eur. J. Appl. Microb. Biotechnol., 17*, 168-172. Korean J. Food Sci. Technol., 15,

Park, S.H., Hong, K.T., You, S.J., and Lee, J.C. (1983b) Automation of glutamic acid fermentation, *Korean J. Food Sci. Technol., 15,* 202-204.

Hong, K.T., Park, S.H., Lee, J.C., Choi, C.Y. and Bae, J.C. (1984) Control of sugar feeding for glutamic acid fermentation, *J. Ferment Technol.., 62,* 49-54.

Schneider, F. (1979) *Sugar analysis official and tentative methods recommended by the international commission for uniform methods of sugar analysis (ICUMSA),* England.

Sharma, C.B. and Garg, K. (1991) Repeated batch production of citric acid from sugarcane molasses using recycled solid-state surface culture of *Aspergillus niger, Biotechnol. Letters 13,* 913-916.

Wu, W., Tsao, J., and Fan, C. (1989) On-line estimation of cell mass in glutamic acid production, *Journal of Fermentation and Bioengineering 3,* 220-221.

Yu-cheng, C. (1973) MSG production by the fermentation of cane molasses, *Taiwan Sugar 5,* 180-184.

LACTIC ACID FERMENTATION OF HEMICELLULOSE LIQUORS AND THEIR ACTIVATED CARBON PRETREATMENTS

PERTTUNEN, J., MYLLYKOSKI, L. AND KEISKI, R.L.
University of Oulu, Department of Process and Environmental Engineering, PO Box 4300, FIN-90014 University of Oulu, Finland. Fax +358 8 553 2304, E-mail Jyrki.Perttunen@oulu.fi

Summary

In this research the activated carbon pretreatments for reed hemicellulose liquor produced in a MILOX process were studied. The remaining sugar fraction was utilised as a raw material for lactic acid fermentation. Earlier results from birch hemicellulose experiments were used for comparison. Pretreated reed hemicellulose liquor can be used as a substrate for lactic acid fermentation by *Lactobacillus pentosus*. Nearly complete conversion was achieved in 48 h, and the product contained 33 g/l lactic acid and 17 g/l acetic acid.

1. Introduction

The use of lignocellulosic materials, such as wood and grass residues, as a source of chemicals has been studied actively in the recent years. The main organic components of wood and grass are cellulose, hemicellulose, and lignin. Hemicellulose is a heteropolymer that constitutes 20-30% of the lignocellulosic dry weight (Olsson and Hahn-Hägerdal, 1996). Hemicelluloses consist, by definition, of short-branched chain heteropolysaccharides of mixed hexosans and pentosans that are easily hydrolysed. D-xylose and L-arabinose are the major constituents of pentosans, while D-glucose, D-galactose, and D-mannose are the constituents of hexosans (Singh and Mishra, 1995). Hemicellulose acid hydrolysates of reed grass and hardwood contain largely xylose accompanied by smaller quantities of arabinose, glucose, galactose, and mannose. The fermentation of hemicellulose liquor is complicated by the presence of inhibitory compounds. The inhibitors include organic acids, lignin derivatives, and carbohydrate degradation products (Buchert, 1990; Olsson and Hahn-Hägerdal, 1996). Fermentation media made of concentrated hydrolysates are not suitable for fermentation purposes because the concentration of non-volatile inhibitors may be too high. Detoxification of

M. Hofman and P. Thonart (eds.), Engineering and Manufacturing for Biotechnology, 29–38.

hydrolysates can be done by activated charcoal adsorption, ion exchange, ion exclusion or chemical pretreatment (Parajo *et al.*, 1995; Olsson and Hahn-Hägerdal, 1996).

The tolerances of different microbial strains against the potential inhibitors found in hemicellulose hydrolysates also vary significantly. Because the hydrolysate contains both hexoses and pentoses, the bacteria should be capable of utilising all the sugar present, in order to fully utilise the raw material (Olsson and Hahn-Hägerdal, 1993).

Lactic acid occurs widely in nature, being found in man, animals, plants, and microorganisms. It has been produced in biotechnological processes since 1881 (Chahal, 1990). More than half of the total amount consumed (about 40 000 tons/year) is produced by fermentation. It is used in food industry as a pH regulator, flavour enhancer, buffering agent or microbial preservative, and in pharmaceutical industry for various purposes. Lactic acid can also be used in the production of polylactides, which are used as intermediates for biodegradable polymers (Parajo *et al.*, 1996; Padukone *et al.*, 1993). The microorganisms used for lactic acid production belong to the family *Lactobacillaceae* and are differentiated into various genera. Lactic acid bacteria are fastidious organisms, as they require a carbon source, a nitrogen source, several vitamins, growth substances, and minerals for growth. Only a few species of *Lactobacillus* are capable of fermenting xylose to lactic acid (Singh and Mishra, 1995). Most of the lactic acid bacteria are homofermentative, i.e. they produce only lactic acid. Heterofermentative bacteria also produce other products, such as acetic acid, ethanol, CO_2, and formic acid. Lactic acid is found as a racemate (DL) and in two optically active forms (L and D) (Chahal, 1990). Few reports have been published on the utilisation of hemicellulose hydrolysate (Linko *et al.*, 1984), sulphite waste liquor (Leonard *et al.*, 1948) or wood (Parajo *et al.*, 1996 and 1997a; Griffith and Compere, 1976) for lactic acid production.

Common reed grows in a wide area reaching from the Equator to the Arctic Circle, and is very abundant in the estuaries of the great rivers of Eastern Europe, Asia, and Africa. The composition of canary reed is as follows: alpha cellulose 35-37%, hemicellulose 36-38%, lignin 19-20%, extractives 3.6%, and ash 9% (Lindholm *et al.*, 1995). The composition of reed varies largely from one season and region to another. Composition is also different in the different parts of the plant. Normally, the stem is used in pulping processes.

The utilisation of hemicellulose liquor in a biotechnology process was studied, to determine whether reed hemicellulose liquor is a suitable raw material for lactic acid fermentation. Comparisons with the results of previous birch hemicellulose experiments (Perttunen *et al.*, 1996) were made. Granular and powdered activated carbons were tested for pretreatment of hemicellulose liquors.

2. Materials and methods

The raw material was reed (*Phragmites communis*) hemicellulose liquor produced as a by-product by the MILOX pilot at Chempolis Ltd., Oulu, Finland. In this "organosolv" pulping method, wood or grass is delignified with concentrated formic acid and hydrogen peroxide. In the first stage, the wood or grass is cooked with formic acid alone followed by cooking with a mixture of formic acid and hydrogen peroxide. The cooking times and temperatures were 120°C/75 min and 80°C/3 h in the first and second stages, respectively (Seisto *et al.*, 1996). The reed hemicellulose liquor contained about 30% dry matter. Monosaccharides accounted for about 45% of total dry matter. The monosaccharides consisted of xylose (82.5%), glucose (8.5%), galactose (2.5%), arabinose (6%), and mannose (0.5%). The amount of formic acid in reed hemicellulose liquor was 200-400 g/l before any treatment. The composition of birch hemicellulose liquor was 58% of dry matter, of which monosaccharides accounted for 50%. The monosaccharides consisted of xylose (85%), mannose (5%), glucose (4.5%), galactose (4%), and arabinose (1.5%). The activated carbons used in the experiments were the granular carbons Chemviron CAL (carbon A), Norit PK 1-3 (carbon B), and Norit ROW 0.8 SUPRA (carbon C) and the powdered carbons CECA Acticarbone 3S (carbon D) and Acticarbone CXV (carbon E). The granular carbons Chemviron CPG LF 12x40 and Aquasorb BG-09 14x40 were also tested, but they proved to be less good than the other carbons for colour removal. They were therefore omitted from the figures and the discussion.

Lactobacillus pentosus ATCC 8041 and *Lactococcus lactis* IO-1 JCM 7638 were used in the studies of hemicellulose fermentations. Both of these strains are capable of utilising xylose as substrate.

The MRS medium was used for growing the inoculum for *L. pentosus*. For *L. lactis,* the medium consisted of sodium chloride, peptone, and yeast extract. The fermentation medium consisted of reed or birch hemicellulose (diluted with appropriate sugar concentrations) and various nutrients, such as yeast extract and salts. The nutrients were autoclaved before they were introduced into the fermentor. The fermentation medium was filtered using 0.2 µm microfiltration membrane (Millipore). Batch fermentations were carried out in a bioreactor (Biostat E, B. Braun Melsungen AG) with a working volume of 2-5 l. pH was adjusted to 6.0 by adding 5.9 M NaOH. Temperature was maintained at 37°C. Agitation rate was maintained at 150 rpm. The fermentation process was controlled and monitored on-line using the FermExpert (Vinter *et al.*, 1992) program running in a Microsoft Windows environment.

The fermentation samples were analysed using an HPLC (Merck-Hitachi) equipped with an ORH-801 ion exclusion column for organic acids and a CHO-682 carbohydrate column for sugars from InterAction. Bacterial growth was determined by measuring the change in the optical density of the medium at 570 nm.

3. Results and discussion

The first step was to reduce by evaporation the amount of formic acid in hemicellulose liquor to a level that microbes can tolerate. A good indicator was a rise of pH from about one to above three. After evaporation, the liquor was treated with activated carbon. Both granular and powdered activated carbons were tested, to reduce the inhibitors and the dark colour of the reed hemicellulose liquor. In Figure 1, the amount of xylose is presented as a function of the amount of various activated carbon. Figure 2 presents the amount of formic acid before evaporation as a function of the amount of various activated carbon.

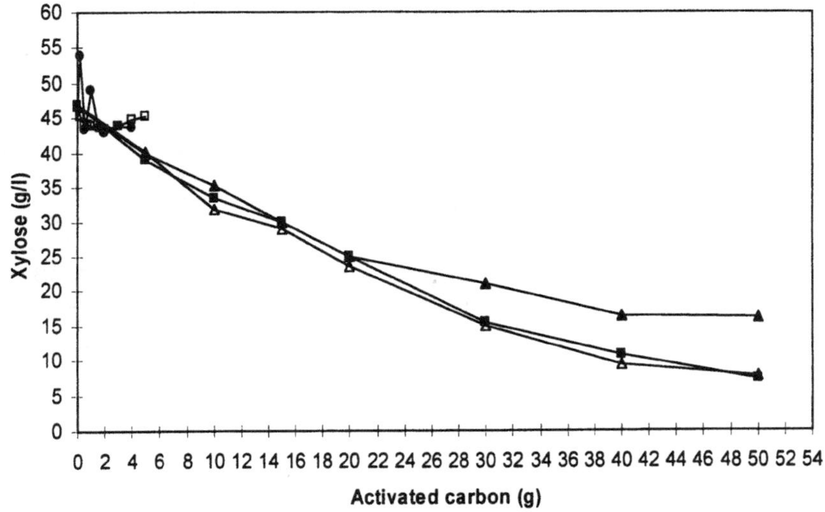

Figure 1. The amount of xylose as a function of the amount of various activated carbons. Symbols: ■ carbon A, ▲ carbon B, Δ carbon C, □ carbon D and • carbon E.

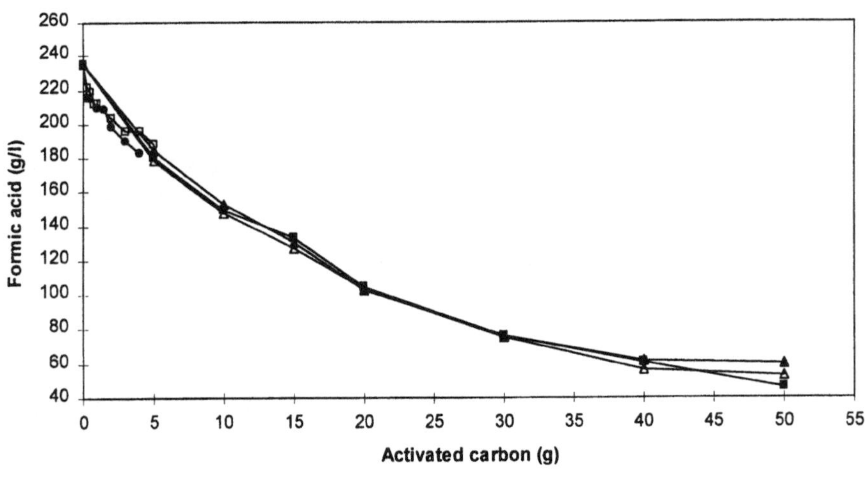

Figure 2. The amount of formic acid before evaporation in hemicellulose liquor as a function of the amount of various activated carbons. Symbols: ■ carbon A, ▲ carbon B, Δ carbon C, □ carbon D and • carbon E.

Granular activated carbons (carbons A, B, and C) proved to be unsuitable for the pretreatment of reed hemicellulose liquor, because they considerably reduced the sugar content of the liquor. This did not happen with powdered activated carbons (carbons D and E, see Figure 1). Davison and Scott (1992) also found granular activated carbons to adsorb glucose. Frazer and McCaskey (1989) further noticed that activated carbon reduced the sugar level in the wood hydrolysate. The particle size distribution of activated carbons could explain the results. The amount of carbon needed for colour removal was 17 times higher in the case of granular compared to powdered carbon. As it can be seen in Figure 2, activated carbons also reduced the formic acid concentration, but treatment with activated carbon alone is not an efficient way to diminish the formic acid concentration. The target level of formic acid concentration is under 0.5% before fermentation. This can be achieved by evaporation and treatment with activated carbon.

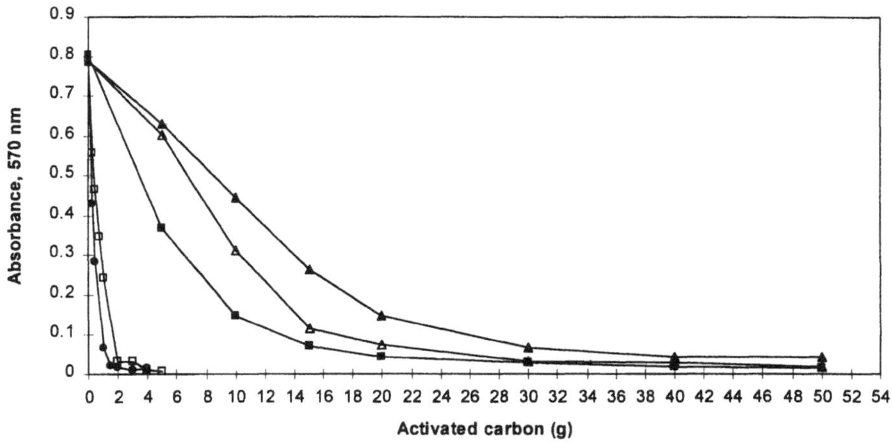

Figure 3. Absorbance as a function of the amount of various activated carbons. Symbols: ■ carbon A, ▲ carbon B, △ carbon C, □ carbon D and ● carbon E.

Figure 3 presents the absorbance curves of various activated carbons at 570 nm. When the absorbance value was under 0.02, the hemicellulose liquid solution was visibly clear. This was not achieved with all the activated carbons tested. The absorbance of hemicellulose liquor before and after activated carbon treatment was measured at 200-800 nm. The absorbance value at 200-400 nm was considerably lower when activated carbon was used. Figures 4a and b represent reed hemicellulose liquor treated with activated carbon. The amount of carbon is 0.4 g/g sugars in hemicellulose in Figure 4a and 1.5 g/g sugars in hemicellulose in Figure 4b. Without treatment, the area of the peak was much larger and the peak diverged from zero absorbance at a higher wavelength. Parajo *et al.* (1997b) reported that reduction of the 279 nm absorbance is indicative of lignin derivatives. Activated carbon treatment removed some of the lignin components from the hemicellulose liquor, which can be seen from the absorbance curves in Figures 4a and b.

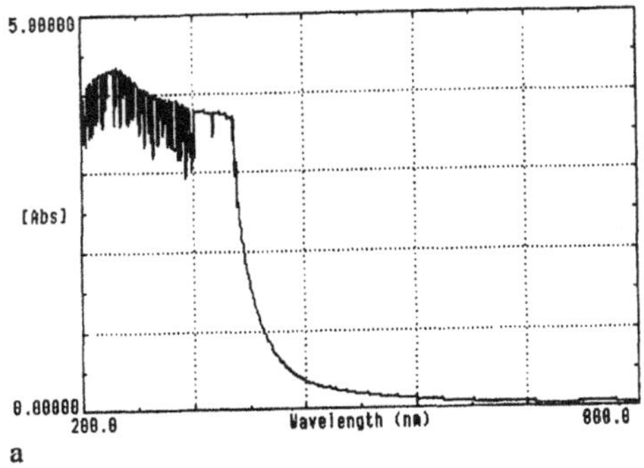

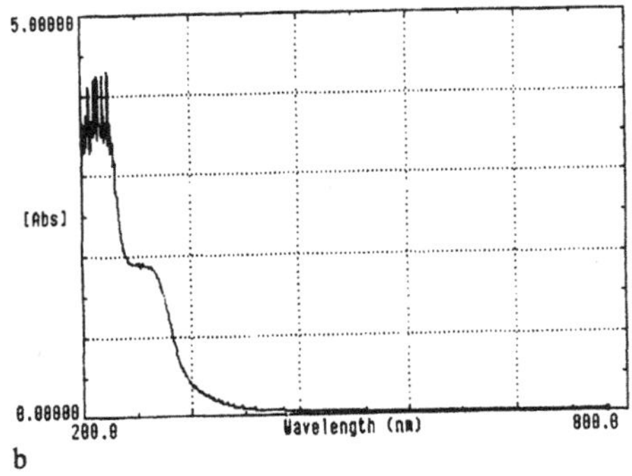

Figure 4. Absorbance curves of reed hemicellulose liquor at 200-800 nm. Figure 4a shows hemicellulose liquor treated with 0.4 g activated carbon/g sugars in hemicellulose. In Figure 4b, the amount is 1.5 g activated carbon/g sugars in hemicellulose.

Table 1. Improvement of the fermentability of hemicellulose liquor by different pretreatments. Symbol (b) indicates birch hemicellulose and symbol (r) reed hemicellulose. LP stands for Lactobacillus pentosus *and LL for* Lactococcus lactis.

Treatment	Cultivation time (h)	Initial sugar content (g/l)	Conversion (%)	Productivity (g/lh) lactic acid	acetic acid
No treatment LP (b)	92	36.1	82.1	0.19	0.05
Activated carbon LP(b)	44	34.2	92.7	0.49	0.19
No treatment LP (r)	93	47.4	68.5	0.31	0.15
Activated carbon LP (r)	53	48.7	97.4	0.57	0.32
Activated carbon LP (r)	45	47.3	94.1	0.59	0.43
Activated carbon LL (r)	70	47.8	47.5	0.27	0.03

Improvements of the fermentability of hemicellulose liquor by different pretreatments are presented in Table 1. Previous results from birch hemicellulose experiments (Perttunen *et al.*, 1996) are given as a reference for the reed hemicellulose experiments. In the experiment with *L. lactis,* both conversion and volumetric productivity were smaller compared with the results obtained with *L. pentosus*, although the literature (Ishizaki and Ueda, 1995) suggests that *L. lactis* might be a useful microorganism for the production of lactic acid from hydrolysed lignocellulose. Activated charcoal removed the dark colour from the hemicellulose liquor. The fermentation time was reduced considerably compared with the untreated liquor substrate. The maximum volumetric productivity (0.59 g/lh) obtained showed the same magnitude as the volumetric productivity presented in the literature (Padukone *et al.*, 1993; Linko *et al.*, 1984). In comparison with the other pretreatments (treatment with CaO or $Ca(OH)_2$) tested (Perttunen *et al.*, 1996), activated charcoal appears to be the most suitable pretreatment when lactic acid bacteria are used. This contrasts with the results obtained with *Gluconobacter oxydans* (Buchert, 1990).

The molecular weight of reed lignin was 650-2500. Because lignin and lignin derivatives play a major role in causing the colour of hemicellulose liquor, one possibility for colour removal from reed hemicellulose liquor could be ultrafiltration or nanofiltration.

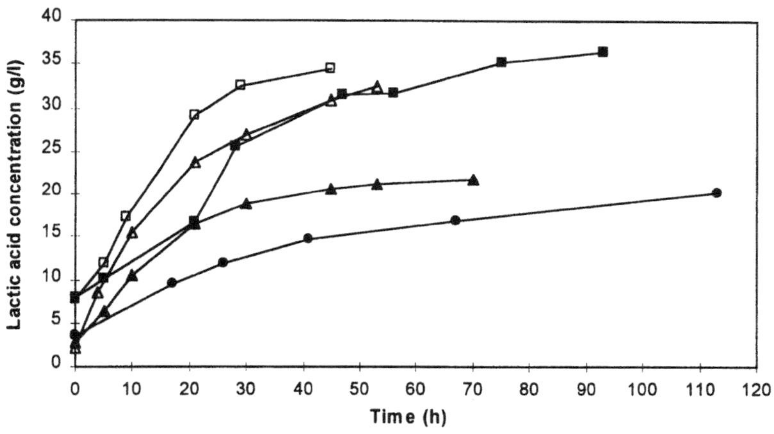

Figure 5. Lactic acid concentration in reed hemicellulose experiments. Symbols: ■ RA (no treatment with activated carbon, L. pentosus), △ RB (activated carbon treatment, L. pentosus), ▲ RC (activated carbon treatment, L. lactis), □ RD (activated carbon treatment, L. pentosus) and ● RE (activated carbon treatment, L. pentosus, no added nutrients).

Figure 5 shows the amount of lactic acid in reed hemicellulose fermentations. Figure 6 presents the amount of acetic acid in reed hemicellulose fermentations. In the RA experiment, reed hemicellulose liquor was not treated with activated carbon. The experiments RB and RD included treatment with activated carbon. Experiment RC also involved treatment with activated carbon, but the lactic acid bacterium was *L. lactis*. In the other experiments, the bacterium was *L. pentosus*. In the RE experiment, no nutrients were added to the activated carbon-treated hemicellulose liquor. The results show that it is necessary to add nutrients in order to get more product within a shorter time. The time needed for complete conversion and product formation is shorter when the hemicellulose liquor is treated with activated carbon compared with untreated hemicellulose liquor

The effect of lactic acid bacterium species can also be seen. *L. pentosus* produced more lactic and acetic acid than *L. lactis*. The reason for this might be some component present in reed hemicellulose and inhibitory to *L. lactis,* because, with xylose as substrate, *L. lactis* produced more lactic acid than *L. pentosus*. Rodewald-Rudescu (1974) also used reed hydrolysate as substrate for lactic acid fermentation. He tested five strains of lactic acid bacteria and found *Lactobacillus pentosus* to be the best acid producer.

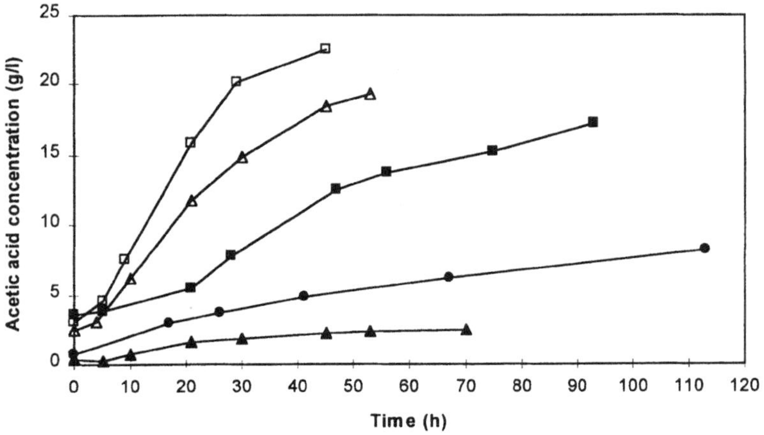

Figure 6. Acetic acid concentration in reed hemicellulose experiments. Symbols: ■ RA (no treatment with activated carbon, L. pentosus), Δ RB (activated carbon treatment, L. pentosus), ▲ RC (activated carbon treatment, L. lactis), □ RD (activated carbon treatment, L. pentosus) and ● RE (activated carbon treatment, L. pentosus, no added nutrients).

4. Conclusions

A potential pretreatment for reed hemicellulose liquor is treatment with powdered activated carbon. A major drawback for granular activated carbons was that they also reduced the sugar content of the liquor. The amount of powdered activated carbon needed to remove the dark colour of reed hemicellulose liquor was 33 g/l when the sugar content of hemicellulose was about 50 g/l. Before treatment with activated carbon, the amount of formic acid has to be reduced to a level that does not inhibit lactic acid fermentation. This can be done by evaporation. One suitable bacterium for the fermentation of reed hemicellulose liquor is *Lactobacillus pentosus*, which produces lactic and acetic acids as the main products. Nearly complete conversion was achieved in 48 h, and the product contained 33 g/l lactic acid and 17 g/l acetic acid. Volumetric productivity was 0.6 g/lh. The amount of formic acid was invariable in most fermentation experiments. Even though *Lactococcus lactis* ferments xylose, it proved to be an unsuitable bacterium for reed hemicellulose liquor. A comparison of fermentations using birch and reed hemicellulose as substrate showed that both produced similar results.

Hemicellulose liquors are very complex in nature, and various treatments are needed before they can be used as substrate for fermentation. Many components that are inhibitory to the microorganisms have to be removed. This can be done with various techniques, but the use of activated carbons is one of the most feasible methods. After fermentation, there is still a need to purify the lactate further, depending on the final grade of the desired product. The product has to be concentrated by evaporation, and after that, treatments with ion exchange resin and activated carbon may be needed. Microorganisms capable of utilising both hexoses and pentoses should be used, but the product can also be other than lactic acid, for example, another organic acid, ethanol or xylitol. If the purpose is to build a commercially feasible process, the costs may be too high if different treatments are used, unless the product is of high enough value. There is a lot of work to be done and many problems to be solved before lactic acid can be produced from hemicellulose liquor economically.

5. References

Buchert, J. (1990) *Biotechnical oxidation of D-xylose and hemicellulose hydrolysates by Gluconobacter oxydans*, Technical Research Centre of Finland, Publications 70, Espoo.

Chahal, S.P. (1990) Lactic acid, in B. Elvers, S. Hawkins and G. Schulz (eds.), *Ullman's Encyclopaedia of Industrial Chemistry* Vol. A 15, VCH Verlagsgesellschaft, Weinheim, pp. 97-105.

Davison, B.H. and Scott, C.D. (1992) A proposed biparticle fluidised-bed for lactic acid fermentation and simultaneous adsorption, *Biotechnology and Bioengineering 39*, 365-368.

Frazer, F.R. and McCaskey, T.A. (1989) Wood hydrolysate treatments for improved fermentation of wood sugars to 2,3-butanediol, *Biomass 18*, 31-42.

Griffith, W.L. and Compere, A.L. (1976) Continuous lactic acid production using a fixed-film system, *Developments in Industrial Microbiology 18*, 723-726.

Ishizaki, A. and Ueda, T. (1995) Growth kinetics and product inhibition of Lactococcus lactis IO-1 culture in xylose medium, *Journal of Fermentation and Bioengineering 80*, 287-290.

Leonard, R.H., Peterson, W.H. and Johnson, M.J. (1948) Lactic acid from fermentation of sulphite waste liquor, *Industrial and Engineering Chemistry 40*, 57-67.

Lindholm, J., Yilmaz, Y., Johansson, A., Gullichsen, J. and Sipilä, K. (1995) A new process of combined energy and pulp production from agricultural plants, The 8th International Symposium on Wood and Pulping Chemistry, The Finnish Pulp and Paper Research Institute (KCL), Jyväskylä, Vol. 2, pp. 455-460.

Linko, P., Stenroos, S.-L., Linko, Y.-Y., Koistinen, T., Harju, M. and Heikonen, M. (1984) Applications of immobilised lactic acid bacteria, in A.I. Laskin, G.T. Tsao and L.B. Wingard (eds.), *Enzymes Engineering VII*, New York Academy of Sciences, New York, pp. 406-417.

Olsson, L. and Hahn-Hägerdal, B. (1993) Fermentative performance of bacteria and yeasts in lignocellulose hydrolysates, *Process Biochemistry 28*, 249-257.

Olsson, L. and Hahn-Hägerdal, B. (1996) Fermentation of lignocellulosic hydrolysates for ethanol production. *Enzyme and Microbial Technology 18*, 312-331.

Padukone, N., Schmidt, S.L., Goodman, B.J. and Wyman, C.E. (1993) Study of lignocellulose components for production of lactic acid, First Biomass Conference of the Americas: Energy, Environment, Agriculture, and Industry, National Renewable Energy Laboratory, Golden, pp. 1311-1319.

Parajo, J.C., Alonso, J.L. and Moldes, A.B. (1997a) Production of lactic acid from lignocellulose in a single stage of hydrolysis and fermentation, *Food Technology 11*, 45-58.

Parajo, J.C., Alonso, J.L. and Santos, V. (1996) Lactic acid from wood. *Process Biochemistry 31*, 271-280.

Parajo, J.C., Dominguez, H. and Dominguez, J.M. (1995) Study of charcoal adsorption for improving the production of xylitol from wood hydrolysates, *Bioprocess Engineering 16*, 39-43.

Parajo, J.C., Dominguez, H. and Dominguez, J.M. (1997b) Improved xylitol production with Debaryomyces hansenii Y-7426 from raw or detoxified wood hydrolysates, *Enzyme and Microbial Technology 21*, 18-24.

Perttunen, J., Sohlo, J. and Mäentausta, O. (1996) The fermentation of birch hemicellulose liquor to lactic acid, in E. Srebotnik and K. Messner (eds.), *Biotechnology in the Pulp and Paper Industry: Recent Advances in Applied and Fundamental Research*, Facultas-Universitätsverlag, Vienna, pp. 295-298.

Rodewald-Rudescu, L. (1974) *Das Schilfrohr Phragmites communis trinius*, E. Schweizerbart'sche Verlagsbuchhandlung (Nägele u. Obermiller), Stuttgart.

Seisto, A., Poppius-Levlin, K. and Jousimaa, T. (1996) *Grass pulp for papermaking by the milox method*, KCL, PSC Communications 87, Espoo.

Singh, A. and Mishra, P. (1995) *Microbial Pentose Utilisation: Current Applications in Biotechnology*, Elsevier Science B. V., Amsterdam.

Vinter, T., Paalme, T. and Vilu, R. (1992) "FermExpert" - an expert system for studies and optimisation of processes of microbial synthesis, in M.N. Karim and G. Stephanopoulos (eds.), *Modelling and Control of Biotechnical Processes*, Pergamon Press, Oxford, pp. 467-470.

ENZYMIC SOLUBILISATION OF PROTEINS FROM TROPICAL TUNA USING ALCALASE AND SOME BIOLOGICAL PROPERTIES OF THE HYDROLYSATES

FABIENNE GUERARD[1], ROZENN RAVALLEC-PLE[2], DENIS DE LA BROISE[1], ADRIEN BINET[1] AND LAURENT DUFOSSE[1]
[1] Université de Bretagne Occidentale, LUMAQ, EA 2651, IUP IIA, Créac'h Gwen, F-29000 Quimper, France - email:guerard@univ-brest.fr
[2] Marine Biology – MNHN – 29182 Concarneau, France

Summary

Tuna protein hydrolysates have been prepared using Alcalase at several concentrations and characterised both by the hydrolysis degree and by the molecular weight distribution of peptides using size exclusion chromatography (SEC). Preliminary results showed that tuna protein hydrolysates performed effectively as cellular growth factors on fibroblastic cells and as nitrogen sources in microbial growth media. The presence of "secretagogue" molecules (gastrin-like peptides) was also detected.

1. Introduction

Only 50% of the total catches of the fishery industries in the European areas are actually eaten by man. The remains (fish viscera and skeletons, crustacean exoskeletons) are rarely upgraded if not rejected into the sea. These wastes constitute a potentially important source of biological molecules, some of them possessing peculiar properties and offering practical application promises in various areas (agriculture, food, medicine, biotechnology and chemistry). Their retrieval and purification is likely to enlarge the range of the presently available biotechnological products (Le Gal and Stenberg, 1998). Moreover, the wastes from marine origin are considered as a safe material and provide proteins with high nutritional properties and a good pattern of essential amino acids (Diniz & Martin, 1998; Shahidi, 1995).

In this context, tuna by-products constitute a biomass of particular interest to upgrade because of the global economic importance of tunas and their international trade for canning. Yellowfin (*Thunnus albacares*) is commercially the second most important species of tuna and, in 1994, accounted for about 1.1 million tons or 35 % of the world-wide total tuna catches (FAO, 1997). During the fish processing, solid wastes

M. Hofman and P. Thonart (eds.), Engineering and Manufacturing for Biotechnology, 39–50.

including viscera, head, skin, bone and some muscle tissue can be as high as 70% of the original material. Traditionally, these wastes have been used as fishmeal or fertiliser (Benkajul & Morrissey, 1997). Another way of upgrading for fish proteins has been the production of Fish Protein Hydrolysates (FPH) in controlled conditions mostly by enzymatic hydrolysis (Shahidi *et al.*, 1995; Martin & Porter, 1995; Hoyle & Merrit, 1994). The use of exogenous commercial enzymes is preferred to autolysis by endogenous enzymes since the hydrolysis and properties of resultant product could be controlled (Diniz and Martin, 1998). From both a technical and economic point of view, enzymes from microbial sources operating at alkaline pH such as Alcalase were shown to be one of the most efficient in the hydrolysis of fish proteins (Guérard *et al.*, 2000; Dufossé *et al.,* 1997). The hydrolysates obtained are characterised using different techniques. One of them is based on the direct estimate of the Degree of Hydrolysis (DH) according to Adler -Nissen (1982). However, for proper definition of the different components resulting from the protein hydrolysis, the peptide molecular weight and size-distribution has to be studied. A method based on size exclusion chromatography (SEC), which is one of the most attractive techniques allowing an accurate study of the peptides generated during hydrolysis, has been developed in our lab.

For a few years, there has been a great interest for finding new applications - with better added value - to the Fish Protein Hydrolysates. The European research project FAIR CT 97-3097 explores the possibility of obtaining biologically active peptides from hydrolysates of marine food processing wastes. As an example, cod hydrolysates prepared from heads, stomach and viscera gave positive results in the calcitonin gene related peptide (CGRP) radioimmunoassay (Fouchereau-Peron *et al.*, 1999) and in gastrin/CCK radioimmunoassay (Cancre *et al.*, 1999). "Secretagogue" molecules (gastrin, cholecystokinin) exhibit a large spectrum of activities ranging from the stimulation of protein synthesis to the secretion of digestive enzymes. The presence of these peptides in fish and shellfish hydrolysates could be of importance because of the development of aquaculture, which requires strict control of feed quality and ingredients sources. Otherwise, the use of FPH as nitrogenous substrates for growth stimulation of micro-organisms is also investigated since growth substrate costs often make up the major part of the production cost of microbial cells and bioproducts from the fermentation industry. A few peptones from marine origin (fish and shellfish hydrolysates) are now being included into some companies' media catalogues.

This work explores the possibility of obtaining biologically active compounds from tuna stomach by controlled enzymatic hydrolysis. Previous results (unpublished data) have shown that the apparent molecular weight of biologically active fractions exhibited a molecular weight ranging from 500 to 10 000 Da. So, the first step was to determine the optimum conditions for use in producing active fragments using a non-specific industrial protease, named Alcalase (Novo Nordisk). The second step was to search for the selected biological activities and to identify them in hydrolysed products. The overall methodology of the work was based on the use of several biological and biochemical tests such as specific radioimmunoassays (gastrin-CCK activities) and cellular growth assayed on cultivating fibroblast cells. A tuna protein hydrolysate was tested as nitrogenous source for microbial growth and compared to other peptones from fish and casein origin using various microbial strains of industrial interest.

2. Materials and methods

2.1. MATERIALS

Stomachs of Yellowfin tunas (*Thunnus albacares*) caught in the Indian Ocean were taken from frozen fish. Heat inactivation of endogenous stomach enzymes (100° C -20 min.) was carried out prior to pH adjustment and addition of the enzyme. The protease Alcalase 2,4 L (a declared activity of 2.4 AU/Kg and a density of 1.18 g/ml) was kindly provided by Novo Nordisk (Denmark). All reagents used were of analytical grade.

2.2. PREPARATION OF THE HYDROLYSATE

Hydrolysis experiments were carried out in a 1-l reactor using the pH-stat method in controlled hydrolysis conditions (pH 8, temperature 50°C and stirring speed 500 rpm). Enzyme concentrations varied in the different tests covering the range from 5.664 to 85 AU/Kg of wet stomach, i.e. Alcalase was added to the sample at enzyme/substrate (E/S) concentration ranging from 0.2 to 3% (wet weight basis). All experiments were carried out in duplicate. During each hydrolysis, pH was maintained constant at the desired value by addition of 2N NaOH. Reactions were terminated by heating the solution to 95°C for 20 min., which assured the inactivation of the enzyme. The resulting slurry was centrifuged at 20000xg for 20 min.

2.3. DETERMINATION OF THE DEGREE OF HYDROLYSIS

$$DH(\%) = \frac{Number\ of\ peptide\ bonds\ cleaved}{Total\ number\ of\ peptide\ bonds} \times 100$$

Reactions were monitored by measuring the extent of proteolytic degradation by means of the DH according to the pH-stat method described by Adler-Nissen (1982). The degree of hydrolysis is defined as follows:
The values for DH can be determined using the following equation:

$$DH(\%) = \frac{B.N_b}{M_p.\alpha.h_{tot}} x100$$

Where DH is the percent ratio between the number of peptide bonds cleaved (h) and the total number of peptide bonds in the substrate studied (h_{tot}). The variable B is the amount of alkali consumed to keep the pH constant during the reaction, N_b is the normality of the alkali, Mp is the mass of the substrate (protein, determined as N x 6.25) in the reaction and α is the average degree of dissociation of α-NH_2 groups released during hydrolysis.

2.4. SIZE EXCLUSION CHROMATOGRAPHY (SEC)

The molecular weight distribution of peptides for each sample was analysed using Fast Protein Liquid Chromatography (FPLC) of gel filtration. The liquid chromatographic system consisted in a Waters 600 automated gradient controller pump and a Waters 996 photodiode array detector. The SEC column was a Superdex Peptide HR 10/30 column from Pharmacia (fractionation range of the column was 7000 to 100 Da). The mobile phase (isocratic elution) consisted of water with TFA 0.1% and acetonitrile (70: 30). The flow rate was 0.5 ml/min. MILLENIUM software was used to collect, plot and process the chromatographic data.

Peptides of known molecular weight (SIGMA) were used to calibrate the column. A relationship between the retention time and the log of the molecular mass of peptides used as standards has been established. Samples injected were dissolved in mobile phase and filtered at 0.2 µm before injection. Absorbance was monitored at 220 nm. For each chromatogram, peptides were sorted out into 3 fractions from 0 to 500 Da (fraction III), 500 to 2000 Da (fraction II) and above 2000 Da (fraction I). The relative areas of each fraction were given in percentage relative to the total area.

2.5. MITOGENIC ACTIVITY

The mitogenic activity was evaluated using MTT (3-[4,5-Dimethylthiazol-2-yl]-2,5-diphenyltetrazolium bromide; Thiazolyl blue), a water tetrazolium salt yielding a yellowish solution when prepared in media or salt solution (Carmichael *et al.*, 1987). Dissolved MTT was converted to an insoluble purple formazan by cleavage of the tetrazolium ring by dehydrogenase enzymes. This water insoluble formazan was solubilised using isopropanol and the dissolved material was measured spectrophotometrically (570 nm) yielding absorbance as a function of concentration of cells (3T3 fibroblasts). Fibroblastic cells were cultured in the presence of tuna hydrolysate for a series of concentrations ranging from 0.05 to 0.5µg /ml of dry weight over 48 hours. Results are expressed in percentage of stimulation reported to the control (100%).

2.6. GASTRIN RADIOIMMUNOASSAY (RIA)

Gastrin radioimmunoassays were carried out using a rabbit antiserum, synthetic ^{125}I gastrin as tracer and synthetic gastrin as standard (GASK-PR, CIS Bio International, France). The rationale of the assay is based on the competition between gastrin radiolabelled with ^{125}iodine and gastrin (or cholecystokinin, CCK) contained in the standards or samples to be assayed for a given limited number of anti-gastrin antibody sites. At the end of the incubation period, the amount of radiolabelled gastrin bound to the antibody is inversely proportional to the amount of non-radiolabelled gastrin (or CCK) originally present in the assay. Several dilutions of each hydrolysate assayed in duplicate were submitted to gastrin radioimmunoassay.

2.7. MICROBIAL CULTIVATIONS

Four fish peptones were compared to a reference peptone from casein (Table 1).

2.7.1. Microorganisms and cultivation media.

The bacteria (*Escherichia coli* ATCC 25922, *Lactobacillus casei* ATCC 7469), the yeasts (*Sporobolomyces odorus* CBS 2636, *Saccharomyces cerevisiae* from IUT Biologie Appliquée, Quimper), and the fungi (*Aspergillus niger* from ESMISAB, Brest, *Penicillium roquefortii* CSL-PV) were grown at 25°C in liquid media previously autoclaved at 121°C for 15 min. The medium consisted in (w/w): 1.5% glucose, 0.5% peptone (except salmon peptone in liquid form, 0.5% v/v), 0.2% KH_2PO_4, 0.013% $CaCl_2$, 2 H_2O, 0.001% $FeSO_4$, 7 H_2O, 0.3% $MgSO_4$, 7 H_2O, pH 6.0. Cultivation was performed for 3 to 5 days in 250 ml culture flasks containing 100 ml of medium.

Table 1. Short presentation of the peptones used in this study.

Trade name	Raw material	Manufacturer	Presentation (aspect)	Dry matter (g DM/100 g of product)	Nitrogen contents (g N/100 g of Dry Matter)*
Bacto-Tryptone	casein	Difco (USA)	powder	95.06	11.5
-	cod viscera	Biotec Maczymal (N)	powder	89.90	9.5
-	tuna stomach	Lumaq (F)	powder	94.84	13.4
Marine Peptone S490	salmon co-products	Primex (N)	liquid	66.53	10.5
Fish Peptone N°1	various fish	Difco (USA)	powder	94.53	10.6

* Kjeldahl method results

2.7.2. Growth kinetics, modelling the growth curve.

Bacterial and yeast growths were followed using optical density measurements (650 nm). Each growth curve for a micro-organism / peptone combination was obtained from four cultures.

A lot of mathematical models can be used to obtain lag phase (λ), maximum growth rate (μ_{max}) and maximum biomass at the stationary phase (A). GOMPERTZ model (1), well suited for such a purpose, was applied to the growth curves obtained on peptones.

$$\log \frac{N}{N_0} = A * \exp\left(- \exp\left(\frac{\mu_{max} * \exp(1) * (\lambda - t)}{A} + 1 \right) \right) \qquad (1)$$

N_0 = initial population; N = population at instant t

Calculations were made on EXCEL (Microsoft) using the least square method to adjust the model to the data, and correlation coefficient to estimate the fitness between the data and the model.

3. Results and discussion

3.1. EFFECT OF THE ENZYME CONCENTRATION ON THE DEGREE OF HYDROLYSIS

The hydrolytic curves obtained with Alcalase at different initial enzyme concentrations are given in Figure 1. A DH up to 23% was observed with the highest enzyme concentration. Significant changes in DH occurred with the enzyme treatment at concentration ranging from 0 to 28.3 AU/Kg. Less significant increases were found with treatment enzyme at concentration above 28.3 AU/Kg. Prolonging the reaction beyond 5.5 hours did not produce any significant improvement in the DH. Similar curves were reported for the enzymatic hydrolysis of sardine (Quaglia & Orban, 1987), capelin (Shahidi et al., 1995) and shark muscle (Diniz & Martin, 1998).

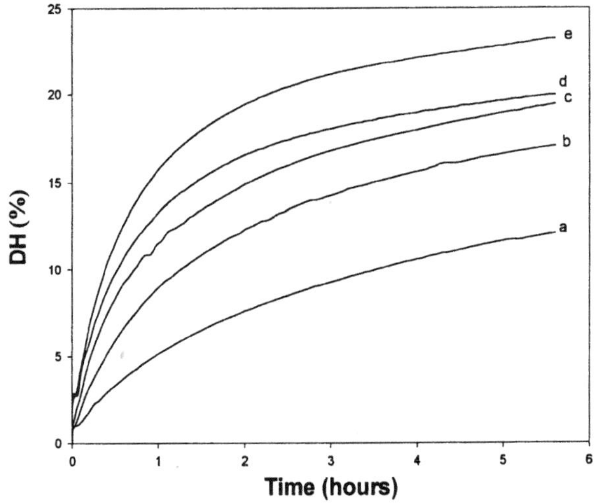

Figure 1. Hydrolysis curves for tuna proteins treated with Alcalase (pH 8; 50°C). Enzyme concentrations (AU/kg): a,5.6 ; b, 14.1 ; c, 28.3 ; d, 45.31 ; e, 85. Reaction conditions: pH = 8; temperature = 50°C.

When log10 (enzyme concentration) versus DH (%) was plotted, a linear relationship was observed (Figure 2). The correlation coefficients were obtained for Alcalase at different enzyme concentrations (R^2 # 0.99). From this relationship, the exact concentration of enzymes required to hydrolyse tuna proteins to a required DH , from half an hour to 5.5 hours, could be calculated.

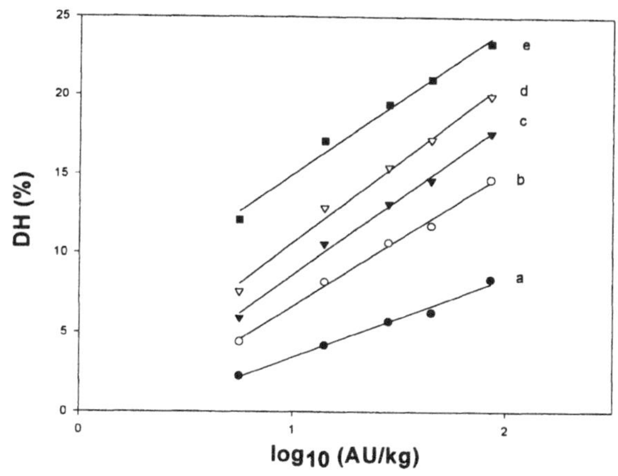

Figure 2. Relationship between $\log_{10}$ (enzyme concentration) and DH for tuna stomachs treated with Alcalase. The hydrolytic reaction was run for 5.5 hours at pH 8 and 50°C. Length of hydrolysis period (hours): a, 0.3; b, 0.8; c, 1.4; d, 2.2; e, 5.5.

Table 2. Area peptide distribution for hydrolysates collected after 5.5 hours hydrolysis. Analyses were performed on Superdex Peptide HR 10/30.

Enzyme concentration (AU/Kg)	**5.6**	**14.1**	**28.3**	**45.3**	**85**
Final DH (%)	12	17	19.5	20	23.3
fraction I (%) [above 2000 Da]	19.3	7.46	4.9	3.52	2.58
fraction II (%) [2000 - 500 Da]	37.07	31.5	27.35	23.98	20.48
fraction III (%) [under 500 Da]	43.63	61.04	67.75	72.5	76.94

3.2. STUDY OF CHROMATOGRAPHIC PROFILES

In order to confirm these results, chromatograms of the hydrolysates were analysed (Figure 3). Samples were collected after 5.5 hours of hydrolysis. A decrease in the high molecular weight fractions is noted as the enzyme/substrate ratio increased. Table 2 shows the values of final DH for each hydrolysate and the molecular weight distribution of peptides sorted in 3 fractions from 0 to 500 Da (fraction III), 500 to 2000 Da (fraction II) and above 2000 Da (fraction I). It can be seen that the controlled hydrolysis of tuna stomach protein through the action of Alcalase 2.4 L gave a high proportion of peptides in the target size range (3000 - 500 Da). The hydrolysate obtained with Alcalase at 5.6 AU/Kg concentration is quite different from the other hydrolysates. The area of fraction

I (19.3%) is very high compared to those of other hydrolysates ranging from 7.5 to 2.6 %. This result is in accordance with the low DH obtained for this hydrolysate.

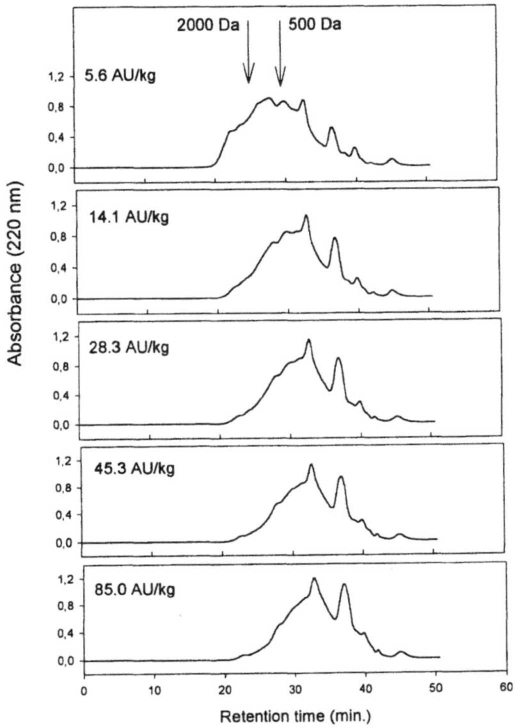

Figure 3. Elution of tuna stomach hydrolysate from Superdex HR 10/30. Samples were prepared using the enzyme Alcalase at different concentrations ranging from 5.6 to 85 AU/Kg at pH 8 and 50°C. Hydrolysis was stopped after 5.5 hours.

The Size Exclusion Chromatography (SEC) method using the SUPERDEX HR10/30 column described in this work allowed a rapid characterisation and analysis of hydrolysates. The separation and the identification of peptide sizes gave a better knowledge about the composition of the hydrolysate . The results obtained were additional to those provided by the degree of hydrolysis. Moreover, this technique was useful for comparing peptidic profiles from different runs and for checking the profile adequacy of identical runs. Bautista *et al.* (1996) outlined the limits of the SEC, e.g. : (i) the approach used could result in underestimation of small peptides and free amino acids and (ii) it could not be applied to absolute determination of molecular weight distribution. However, this technique was a very valuable tool suited to the follow up of proteolysis of protein and for the routine analysis of a large number of samples.

One of the problems encountered with the protein hydrolysate from fish viscera was the lack of reproducibility caused by the presence of endogenous proteases which can act on the hydrolysis process (Guérard *et al.*, 2000). Because of this concern, tuna stomachs were cooked before enzymatic treatment in order to inactivate the endogenous proteases, mainly pepsin. Consequently, protein denaturation might induce loss ability of Alcalase to hydrolyse efficiently the heated proteins because of lower protein flexibility. Nevertheless, in the case of the product we are interested in producing, i.e. FPH as potential source of bioactive peptides, it was desirable to control the size of peptides obtained. This was accomplished with initial standardised material and thus free of by-side enzymatic activities.

3.3. BIOLOGICAL ACTIVITIES OF TUNA HYDROLYSATES

The results presented in this work are given as an example of the most recent development of FPH applications. The biological activity of a tuna hydrolysate obtained using alcalase at the lower concentration (5,6 AU/Kg) is studied.

3.3.1. Mitogenic activity

The tuna hydrolysate exerted stimulation on tritiated thymidine incorporation in fibroblastic cells as shown in Figure 4. A significant stimulation of the 3T3 growth was observed when the cells were incubated with 0,05 or 0,1 µg /ml of tuna hydrolysate. Higher concentrations inhibited the cell proliferation. This effect can be compared with the biological effect of growth factors or other mitotic factors.

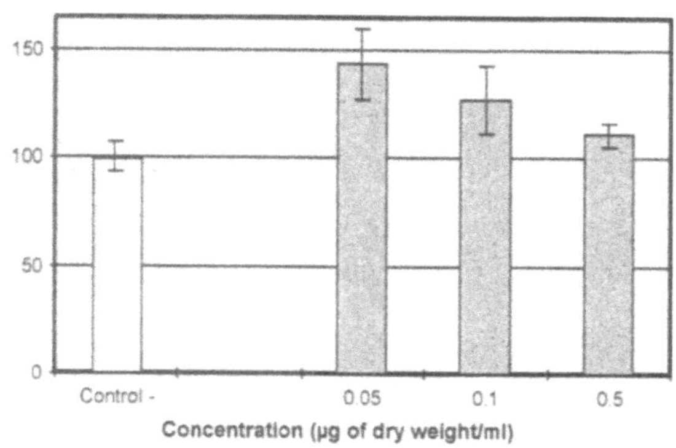

Figure 4. Effect of the tuna hydrolysate on the growth of 3T3 cell line (MTT test). Each value represents the mean of six assays and the bar shows the Standard Deviation.

3.3.2. Gastrin radioimmunoassay

Several amounts of the same tuna hydrolysate were subjected to a Radioimmunoassay (RIA). The slope coefficients obtained indicated if peptides present in these samples were biologically related to gastrin, cholecystokinin (CCK4 or CCK8. First measurements of the gastrin contents showed a good parallelism between the different straight line of tuna and straight line of CCK 8 that could be related to the presence of CCK 8-like peptides in the tuna hydrolysate (Figure 5). Further work will focus on purification and characterisation of these active fractions.

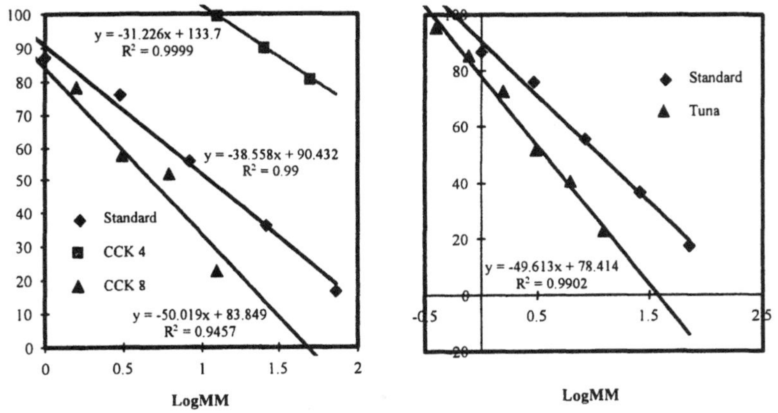

Figure 5. Tuna hydrolysate subjected to a gastrin RIA and compared with native peptides (Standard : gastrin).

3.3.3. Nitrogenous substrate for microbial growth

Tuna peptone produced in our lab was included in the culture media of six micro-organisms belonging to bacteria, yeasts and fungi genera. Among bacteria, one gram negative was chosen, i.e. *Escherichia coli,* as transformed *E. coli* is frequently used in biotechnology ; this micro-organism is rather easy to grow. The gram-positive bacteria was *Lactobacillus casei*, which is harder to grow, is present in dairy starters, and is also used for lactic acid production or post-koji making. Among the yeasts, one Ascomycete, *Saccharomyces cerevisiae* and one Basidiomycete, *Sporobolomyces odorus* were tested. *Saccharomyces cerevisiae* is a very common yeast in biotechnology, in food manufacture (bakery, beer and wine making...) and *Sporobolomyces odorus* is used for aroma production. For the fungi, fish peptones were tested on *Penicillium roquefortii*, which is used for cheese making (Roquefort) or aroma production (methyl ketones), and on *Aspergillus niger* which produces citric acid on an industrial scale.

Growth followed by spectrophotometric measurements allowed us to calculate lag phases, growth rates and maximum biomass at the stationary phase. Results obtained for the yeast Sporobolomyces odorus are given in Figure 6 as an example.

Data from 3 fish peptones (including tuna) out of 4 tested were very close, so tuna peptone could compare with well-established industrial products like casein hydrolysate. As results obtained for the other five micro-organisms were as good as for Sporobolomyces odorus (data not shown), fish peptones - and the tuna one in particular -should have a promising future in biotechnology.

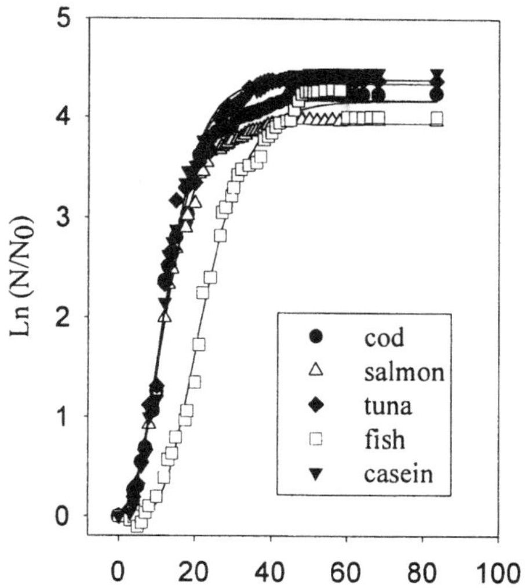

Figure 6. Growth of Sporobolomyces odorus in liquid media containing one of the 5 different peptones as sole nitrogen source (symbols : experimental data ; lines : mathematical model).

4. Conclusion

The controlled hydrolysis of tuna stomach proteins was a good alternative to upgrade marine by-products from the fisheries industry. The reproducibility of the hydrolysates chromatographic profiles between each run was checked. This outlined the importance of the first step of raw material preparation (heat inactivation) and of using exogenous proteases (Alcalase, for example) for ensuring reproducibility of sample preparation and for obtaining the target peptide size.

The fractionation method presented in this work was based on FLPC gel filtration chromatography using the SUPERDEX HR10/30 column. This method gave a good resolution of peptidic fractions and was useful for the follow up of protein proteolysis and the evaluation of the hydrolysis degree.

As far as cellular growth factors are concerned, the tuna hydrolysate tested exerted a stimulatory effect on the protein synthesis in fibroblast cells and the presence of gastrin-like peptides in the tuna hydrolysate has been detected.

The tuna hydrolysates used as nitrogenous substrate are promising with regards to the preliminary results obtained for the strains studied, and fish peptones should be thoroughly investigated to find industrial applications.

References

Adler-Nissen, J. (1982). Limited enzymatic degradation of proteins : a new approach in the industrial application of hydrolases. *J. Chem. Tech. Biotechnol.* 32, 138-156.

Bautista, J., Hernadez-Pinson, I., Alaiz, M., Parrado, J., Millan, F. (1996). Low molecular weight sunflower protein hydrolysate with low concentration in aromatic amino acids. *J. Agric. Food Chem* 44, 967

Benjakul, S. & Morrissey, M. T. (1997). Protein hydrolysates from Pacific Whiting solid wastes. *J. Agric. Food Chem.* 45, 3423-3430.

Cancre, I., Ravallec, R., Van Wormhoudt, A., Stenberg, E., Gildberg, A. & Le Gal, Y. (1999)., Secretagogues and growth factors in fish and crustacean protein hydrolysates. *Mar Biotechnol* 1 :489-494

Carmichael, J., Degraff, W.G., Gazdar, A.F., Minna, J.D. & Mitchell, J.B.(1987). Evaluation of a tetrazolium-based semiautomated colorimetric assay : assessment of chemosensitivity testing. *Cancer Res.* 47(4), 936-942.

Diniz, F. M. & Martin, A. M. (1998). Influence of process variables on the hydrolysis of shark muscle protein. *Food Sci. Technol. Intern.* 4, 91-98.

Dufosse, L., De La Broise, D. *&* Guérard, F. (1997) - Fish protein hydrolysates as nitrogen sources for microbial growth and metabolite production. *In* : « Recent Research Developments in Microbiology, vol.1 », S.G. PANDALAI (Ed), © Research Signpost, ISBN 81-86481-50-8, pp. 365-381.

F.A.O. (1997). Review of the state of world fishery resources : marine fisheries. FAO Fisheries Circular N° 920 FIRM/C920

Fouchereau-Peron, M., Duvail, L., Michel, C., Gildberg, A, Batista, I.& Le Gal, Y. (1999) Isolation of an acid fraction from a fish protein hydrolysate with a CGRP-like biological activity. *Biotechnol.Appl. Biochem.*, 29,87-92.87

Guerard, F., Dufossé, L., De La Broise, D. & Binet, A. (2000). Enzymatic hydrolysis of proteins from yellowfin tuna (*Thunnus albacares*) wastes using Alcalase. *Journal of molecular catalysis B :Enzymatic (in press)*

Hoyle, N. T. & Merritt, J. H. (1994). Quality of fish protein hydrolysates from Herring (Clupea harengus). *J. Food Science* 59, 76-79.

Le Gal Y. & Stenberg E. (1998). Tout est bon dans le poisson, Biofutur, 179

Martin, A. M. & Porter, D. (1995). Studies on the hydrolysis of fish protein by enzymatic treatment. In *Food flavours : generation, analysis and process influence.*, pp. 1395-1404. Edited by G. Charalambous. Amsterdam: Elsevier.

Quaglia, G. B. & Orban, E. (1987). Enzymatic solubilisation of proteins of sardine (Sardina pilchardus) by commercial proteases. *J. Sci. Food Agric.* 38, 263-269.

Shahidi, F., Han, X. Q. & Synowiecki, J. (1995). Production and characteristics of protein hydrolysates from capelin (Mallotus villosus). *Food Chem.* 53, 285-293.

Acknowledgements

This work was supported by the Commission of the European Communities, Directorate-General XII for Research, Technological Development and Demonstration in the Field of Agriculture and Agro-Industry (FAIR) - FAIR contract CT 97-3097. The authors are grateful to the NOVO company, which generously provided the Alcalase preparation.

INFLUENCE OF THE EXPERIMENTAL CONDITIONS ON THE HYDROLYSIS PROCESS IN FISH HYDROLYSATES.

ROZENN RAVALLEC-PLE*, LAURA GILMARTIN, ALAIN VAN WORMHOUDT* AND YVES LE GAL***
** Marine Biology-MNHN- 29182 Concarneau, France*
*** University of Plymouth-Plymouth-Devon PL4 8AA, United Kingdom*

Summary

Protein hydrolysates were prepared from cod muscle (*Gadus morhua*) using commercial Alcalase® in different experimental conditions and were studied in order to determine the influence and the importance of each factors such as pH, temperature and enzyme/substrate ratio on the hydrolysis degree.

1. Introduction

Atlantic cod is an abundant source of waste, particularly due to the filleting process, and has previously been included in Atlantic salmon (*Salmo salar*) diets (Gildberg *et al.*, 1995). The nutritional value of processing discards of cod has been investigated and has been found to be a sustainable and economically attractive protein feed supply for the aquaculture industry (Shahidi *et al.* 1991).

It is well known that a high quality fish protein hydrolysate (FPH) can be commercially produced from fish wastes with simple engineering (Chakraborty and Madhavan , 1977) and can be used as animal feed or fertiliser (Venugopal, 1994). Different processes (hydrolysis, autolysis) permitted to generate molecules larger than individual amino-acids that could be of economical interest in the development of aquaculture. A lack of diets adapted to the larval development in fish or crustacean species encourages the research of adequately alimentation. The level of hydrolysis seems to play an important role for the presence of biological peptides related to growth factors or gastrin (Cancre *et al.*, 1999), calcitonin (Fouchereau-Peron *et al.*, 1999) or opioïds (Piot *et al.*, 1992). Gastrin and cholecystokinins are small peptides but growth factors are bigger molecules and we observed that an extensive process is not necessary to obtain interesting biological activities, on the contrary, a high hydrolysis degree

M. Hofman and P. Thonart (eds.), Engineering and Manufacturing for Biotechnology, 51–58.

correspond to an important reduction in the size of the molecules (Ravallec-Plé *et al.* 2000).

These preliminary results have shown that extensive hydrolysis is not an advantage important but the reduction of size of the original macromolecules is necessary.
One of the potential use of enzymes for the modification and improvement of protein functionality is trough controlled hydrolysis to assure reproducibility of the process (Quaglia and Orban, 1987b). Some studies showed that different conditions of hydrolysis like time, temperature, and the ratio enzyme/substrate will generate different hydrolysis process (Quaglia and Orban, 1987a).

In this context, the purpose of this work was to study the influence of the process parameters on the hydrolysis of cod stomach by a bacterial endopeptidase with low specificity, the Alcalase®.

2. Materials and methods

2.1. SUBSTRATE

Specimens of cod (*Gadus morhua*) were kindly provided by the market of Concarneau. Only the muscle was utilised ; this was boiled 20 minutes, blended and kept frozen at -20°C until it was used.

2.2. ENZYMES

The three enzymes used for the hydrolysis were provided by Novo NORDISK Industri, Denmark. Alcalase® is a serine bacterial endopeptidase (generic name : subtilisin Carlsberg) prepared from a strain of *Bacillus licheniformis* with a specific activity of 2.4 AU/g (Anson units, Anon, 1988). Neutrase® is an endoprotease produced by a selected strain of *Bacillus subtilis* with a specific activity of 0.5 AU/g. Protamex® is a *Bacillus* protease complex developed for hydrolysis of food proteins with a declared activity of 1.5 AU/g. The food-grade enzymes were stored at 5°C until they were used for the hydrolysis experiment. Their optimum activities occur at temperatures between 40°C and 60°C (70°C for Alcalase®) and at pH values between 6 and 10. The deactivation was made at 85°C for 10 minutes (Anon 1988, 1991).

2.3. HYDROLYSIS

Cod muscle (*Gadus morhua*) hydrolysates were produced by hydrolysis in different experimental conditions of pH, temperature, time and enzyme concentration. The hydrolysis was controlled using the pH-stat method (Boyce, 1986) by addition of 2N NaOH for 2h on 50g of raw material in 450 ml of distilled water. After deactivation of the enzyme, the hydrolysate was centrifuged at 20000g during 30 minutes and lyophilised to obtain a powder. The hydrolysis degree was calculated using the pH-stat and the trinitrobenzenesulphonic acid (TNBS) methods (Adler-Nissen, 1982; Adler-

Nissen, 1979). The protein content was determined by the Kjeldahl nitrogen analysis (Lynch *et al.*, 1998).

2.4. STATISTICAL ANALYSIS

Degree of hydrolysis (DH) is generally used as proteolysis factor when the pH-stat method was used. The pH-stat reaction allows the estimation of DH based on the consumption of alkali to maintain a constant pH at desired value. Response Surface methodology (Statgraphic 2.0) was employed to determine the influence and the importance of the different factors such as pH, temperature and ratio enzyme/substrate on the degree of hydrolysis (DH) of the cod muscle by the enzyme Alcalase® and to optimise them. The following table details the experimental design and the average results of three experiments at each point. Using this data contour plots were drawn for each of the factors.

Table 1 : Box-Behnken design matrix and the responses of the dependent variable (DH %). Assays were carried out in triplicate.

Box-Behnken Model

Factors: 3	Blocks: 1		
Runs: 15	Center points: 3		
Factors	**Low**	**High**	**Units**
pH	7,8	8,2	pH-units
Temperature	40	60	°C
E/S ratio	0,5	2	%
Response	Hydrolysis degree %		

pH	Temperature	Ratio E/S	Hydrolysis degree
7,8	50	2	18,92
8	60	2	29,65
8	40	2	22,27
8	40	1	24,3
8,2	40	1	24,6
8	50	1	31,5
8	60	1	33,6
8	40	2	25,7
8	40	0,5	22,3
7,8	40	0,5	16,6
7,8	40	1	18,5
8,2	50	2	30,51
8	40	1	24,32
8	40	1	23,9
8	40	1	24,7

2.5. FPLC CHROMATOGRAPHY

Hydrolysates (1mg/ml) were further analysed by gel filtration on a Superdex Peptide HR 10/30 column (1×30 cm) using acetonitrile (30%) in water with TFA (0.1%) as eluent and a flow rate of 0.5 ml/min according to Guerard *et al.*(2000).

The chromatography was monitored by measuring the absorbance at 220 nm. Column calibration was performed using Ribonuclease A (13700 Da), Aprotinin (6500 Da), Angiotensin I (1296 DA), Bradykinin (1060 Da), Angiotensin III (931 Da), Hexaglycine (360 Da), Tetraglycine (246 Da), Triglycine (189 Da) and Diglycine (132 Da).

3. Results and discussion

3.1. EFFECT OF THE ENZYME ON THE DEGREE OF HYDROLYSIS

The DH values after the 2 hours hydrolysis of cod proteins by the three different enzymes were reported figure 1.

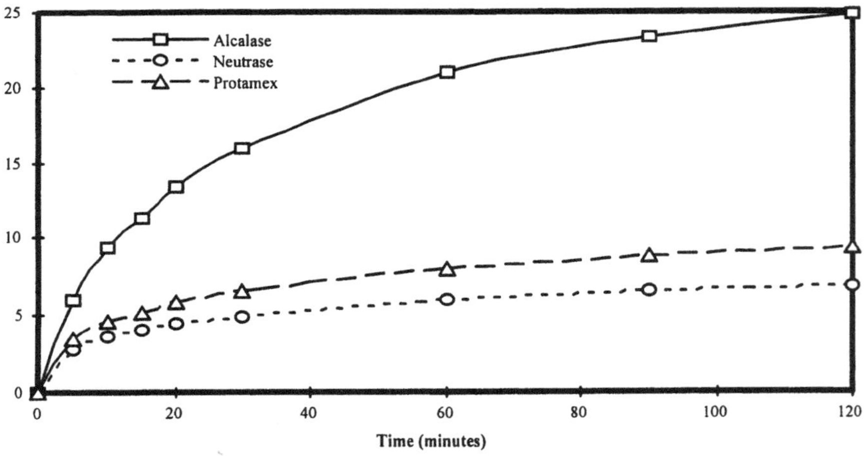

Figure 1 : Effect of the enzyme on the hydrolysis degree of cod muscle (Gadus morhua).

Enzymes were used at a concentration of 1% (volume/weight of raw material) at specific pH and temperature (Alcalase® : 8, 40°C - Neutrase® and Protamex® : 7, 40°C). The highest DH was obtained with Alcalase® (24.85%), the lowest with Protamex® (6.85%), with an intermediate final DH of 9.35% with Neutrase®. Alcalase® showed a higher efficiency than the two others for the hydrolysis of the cod muscle and was choose for the optimisation of the process.

3.2. OPTIMIZATION OF PROCESSING CONDITIONS USING ALCALASE®

In a first time, the hydrolysis factors such as pH, temperature, time, and the amount of enzyme were changed to obtained different fractions with the same raw material, cod muscle. The process is reproducible and the inactivation of endogenous enzymes before the hydrolysis permit to follow each parameter. The hydrolysis degree was calculated with the two methods.

These following figures show the experimental response, in three dimensions under the form of surface plots, of the combined effects of pH and ratio E/S (a), of pH and temperature (b), and of ratio E/S and temperature (c).

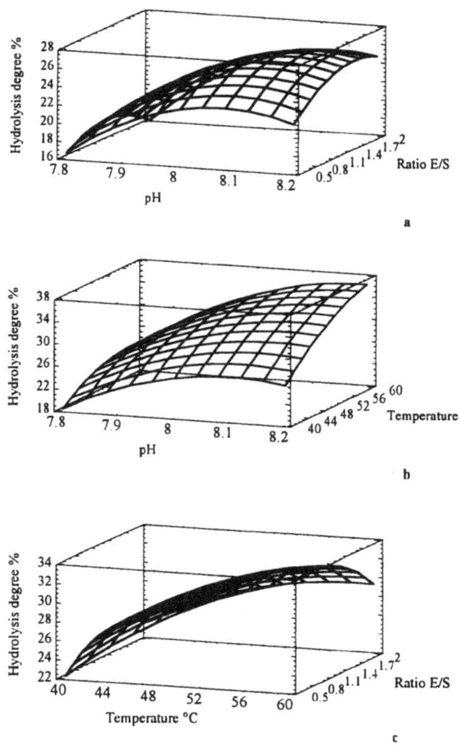

Figure 2 : Influence of processing temperature, pH and enzyme concentration on the degree of Alcalase®-assisted hydrolysis of cod (DH %)

Each graphic is the representation of the evolution of the hydrolysis degree function of the experimental factors. Statistical analysis indicated that within each term all three hydrolysis factors had influence on DH. In fact, Adler-Nissen (1986), investigating the hydrolysis of soy protein by bacterial endoproteases, pointed out the pH, temperature and enzyme-substrate ratio markedly influenced the peptide bond cleavage in the protein substrate.

The highest degree of hydrolysis is obtained with high ratio and pH, even if important temperature reduce the extent of the process with the time of the hydrolysis. The largest surface is obtained with the combined pH and the temperature, what could indicate that these two parameters are more important than the amount of enzyme.

3.2. CHROMATOGRAPHIC PROFILES

The evolution of the peptidic size profile as a function of the time was represented in the figure 3.

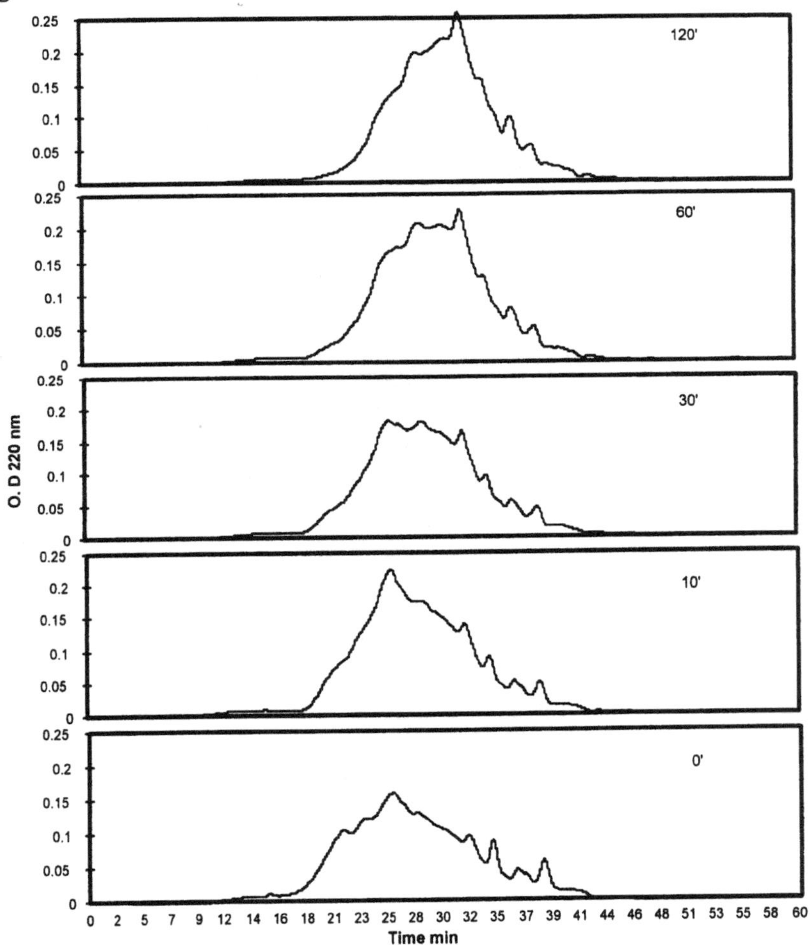

Figure 3 : Elution profiles on the Superdex HR 10/30 exclusion chromatography of hydrolysates removed at different time during the hydrolysis of 50 g of cod muscle by 1% of Alcalase®.

With the increasing time and the extend of hydrolysis, the size distribution is varying and shows an increase of the low molecular weight peptides what is in correlation with the protein degradation and solubilisation during the process.

Figure 4 gives the relative area percentage (corresponding to the area under the curve of the different range of molecular weight) as a function of the time.

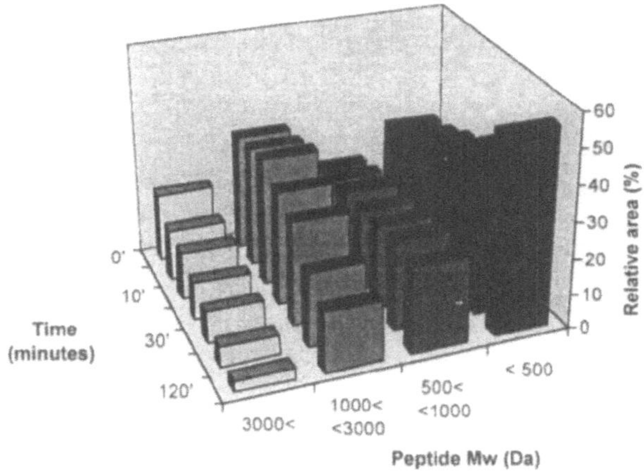

Figure 4 : Relative percentage of molecules as a function of the hydrolysis time after the elution on a Superdex HR 10/30 between 0 minutes and 120 minutes.

The most significant area increase is obtained with molecules of molecular weight under 500 daltons, corresponding to the decrease of molecules with a weight superior at 1000 daltons. Between 500 and 1000 daltons, the amount of peptides seems to stay constant during the process.

4. Conclusion

The degradation depends of the characteristics of the enzyme. Alcalase® is a bacterial endopeptidase with low specificity and is relatively effective for hydrolysis of fish proteins (Mohr, 1978). The combined effect of each pair of variable indicate that in the hydrolysis of cod muscle proteins an increase in DH is achieved by increases in pH and temperature, more than with increases enzyme-substrate ratio, up to certain levels, beyond which DH slightly decreases. Such decrease in the percentage hydrolysis over the higher temperature values is explained by the increasing denaturation of the protease, reducing its biological activity (Diniz and Martin, 1997). The elution profiles on Superdex HR 10/30 gave further indications on the evolution of the hydrolysate composition during the process.

Further *in vitro* and *in vivo* tests will complete these preliminary results. The combination of these tools with biological tests will give more information about the peptides present in the final product and could be of important interest for the determination of the best conditions to obtain high quality fish by-products.

References

Adler-Nissen, J. (1979) Determination of the degree of hydrolysis of food protein hydrolysates by trinitrobenzenesulphonic acid, *J. Agric. Food Chem. 27(6)*, 1256-62.

Adler-Nissen, J. (1982) Limited enzymatic degradation of proteins: a new approach in the industrial application of hydrolases, *J. Chem. Tech. Biotechnol. 32*, 138-156.

Adler-Nissen, J. (1986) *Enzymatic hydrolysis of food proteins*, Elsevier Applied Science Publishers, Barking, UK, 9-24, 57-109, 110-131.

Anonymous (1988) *Alcalase® Food Grade*, B 318b-GB 2000, Bagsvaerd, Denmark: Novo Industry A/S.

Anonymous (1991) *Enzymatic Modification of Proteins using Novo Nordisk Proteases*, B 163g-GB 2500, Bagsvaerd, Denmark: Novo Industri A/S.

Boyce, C. O. L. (1986) *Novo's Handbook of Practical Biotechnology*, Bagsvaerd, Denmark: Novo Industri A/S, 125.

Cancre, I., Ravallec, R., Van Wormhoudt, A., Stenberg, E., Gildberg, A. and Le Gal, Y. (1999) Secretagogues and growth factors in fish and crustacean protein hydrolysates, *Mar Biotechnol 1*, 489-494.

Chakraborty, P. K. and Madhavan, P. (1977) A pilot scale set up for the manufacture of fish hydrolysate. *Fish. Technol. 14(2)*, 159-158.

Diniz, F. M. and Martin, A. M. (1997) Optimisation of nitrogen recovery in the enzymatic hydrolysis of dogfish (*Squalus acanthias*) protein- Composition of the hydrolysates, *Int. J. Food Sci. Nutrit. 48*, 191-200.

Fouchereau-Peron, M., Duvail, L., Michel, C., Gildberg, A., Batista, I. and Le Gal, Y. (1999) Isolation of an acid fraction from a fish protein hydrolysate with a calcitonin-gene-related-peptide-like biological activity, *Biotechnol. Appl. Biochem. 29*, 87-92.

Guérard, F., Dufossé, L., De La Broise, D., Binet, A. (2000) Enzymatic hydrolysis of proteins from yellowfin tuna (*Thunnus albacares*) wastes using alcalase, *Journal of Molecular Catalysis B : Enzymatic*, In Press.

Gildberg, A., Johansen, A. and Bogwald, J. (1995) Growth and survival of Atlantic salmon (*Salmo salar*) fry given diets supplemented with fish protein hydrolysate and lactic acid bacteria during a challenge trial with *Aeromonas salmonicida*., *Aquaculture 138*, 23-24.

Lynch, J. M., Barbano, D. M. and Fleming, J. R. (1998) Indirect and direct determination of the casein content of milk by Kjeldahl nitrogen analysis: collaborative study, *J. AOAC Int. 81(4)*, 763-774.

Mohr, V. (1978) Fish protein concentrate production by enzymatic hydrolysis, in *Biochemical aspects of new protein foods*, J. Adler-Nissen, B. D. Eggum, L. Munck and H. S. Olsen eds, Proc. 11th FEBS Meeting, Pergamon Press, Oxford, 44, 53-62.

Piot, J. M., Zhao, Q., Guillochon, D., Ricart, G. and Thomas, D. (1992) Isolation and characterisation of two opioïd peptides from a bovine haemoglobin peptidic hydrolysate, *Biochem. Biophys. Res. Commun 189*, 101-110.

Quaglia, G. B. and Orban, E.(1987a) Enzymatic solubilisation of proteins of sardines (*Sardina pilchardus*) by commercial proteases, *J. Sci. Food Agric. 38*, 263-269.

Quaglia, G. B. and Orban, E.(1987b) Influence of the degree of hydrolysis on the solubility of the protein hydrolysates from sardines (*Sardina pilchardus*), *J. Sci. Food Agric. 38*, 271-276.

Ravallec-Plé, R., Gilmartin, L., Van Wormhoudt, A. and Le Gal, Y. (2000), Influence of the hydrolysis process on the biological activities of the protein hydrolysates from cod muscle (*Gadus morhua*). J. Sci. Food and Agr. (submitted).

Shahidi, F., Naczk, M., Pegg, R.B. and Synowiecki, J. (1991) Chemical composition and nutritional value of processing discards of cod (*Gadus morhua*), *Food Chem 42*, 145-151.

Venugopal, V. (1994) Production of fish protein hydrolysates by microorganisms. In *Fisheries Processing : Biotechnological applications*, A. M. Martin ed., Chapman & Hall Press, London, 223-243.

PART II
PROCESS MODELLING

MATHEMATICAL MODELLING OF MICROBIAL PROCESSES – MOTIVATION AND MEANS

TEIT AGGER AND JENS NIELSEN
Center for Process Biotechnology, Department of Biotechnology, Technical University of Denmark, Building 223, DK-2800 Lyngby, Denmark.

Abstract

In this paper the motivation for using mathematical models to describe microbial processes is discussed. Mathematical models have a unique ability to extract information from the wealth of experimental data constantly accumulating in the fields of basic and applied microbiology. They allow for detailed investigations of the interactions in complex biological systems that are otherwise practically impossible. Modelling can be applied to optimise the performance of industrial processes, e.g. by use in advanced control algorithms or by simulating different operating conditions. Furthermore, mathematical models used for computer simulations of microbial processes are invaluable educational tools. Mathematical models can be grouped in three classes - whole cell models, segregation models and element models. A whole cell model describes growth and product formation, often in an empirical fashion. A segregation model is used to describe different cell types, and element models are used to give detailed mechanistic descriptions of specific processes. Any level of detail can be included in each of the three classes of models, and the different models may be combined when a fermentation process is to be described. Here a general mathematical framework is given for whole cell models and a few examples of relatively simple, yet very applicable, models are given.

1. Introduction

Modern biotechnology is a rapidly growing discipline encompassing an enormous range of applications. Common to all applications, however, is the involvement of life processes at some stage, directly or indirectly. One such process is the growth of micro-organisms, a process used extensively in many of modern society's vital industries, such as the manufacturing of food, alcoholic beverages, pharmaceuticals, fine chemicals and enzymes. Although micro-organisms are relatively simple life forms their growth is the result of a machinery of immense complexity, involving sophisticated molecular

M. Hofman and P. Thonart (eds.), Engineering and Manufacturing for Biotechnology, 61–75.

processes in which numerous genes, proteins and metabolites take part. This intrinsic complexity of microbial growth, coupled with the fact that far from all aspects of the growth processes are known at present, makes the analysis, development and optimisation of processes that involve the growth of micro-organisms a difficult task. One way of approaching this task in a rational manner is by using mathematical models. These models can help in structuring the wealth of information constantly accumulating in the field, extracting correlations that would otherwise be difficult, if not impossible, to discover. They can increase our understanding of the multitude of processes occurring in the microbial cells and their complex interplay, thereby enabling more rational and efficient experimental strategies to be developed. This paper presents an overview of the purposes and applications of mathematical modelling of microbial growth as well as some of the tools needed, i.e. general mathematical frameworks.

2. Motivation

Mathematical modelling is a powerful scientific tool, but when applying mathematical models to microbial processes it is very important to clearly state the reasoning for setting up the model and using it for simulations. The model should aim at fulfilling a purpose other than just fitting experimental data, which is rarely of any scientific value. In the following a few of the many good reasons for using mathematical models are discussed.

Experimental research involving microbial processes often produces large amounts of data and it can be difficult if not impossible to interpret these data without the aid of mathematical models. The ability of mathematical models to *extract information* is invaluable in the processing and comparison of experimental data. An example of this is the use of simple mathematical models to quantify the morphology of filamentous fungi - Spohr et al. (1997) measured the average total hyphal length and the average number of tips of three different strains of *Aspergillus oryzae* during submerged growth. Using simple mathematical models, parameters such as the maximal tip extension rate and the maximal branching frequency were extracted from the data, enabling the investigators to easily compare the different strains examined. The concepts of metabolic flux analysis and metabolic control analysis can also be used to extract valuable information from experimental data concerning the magnitude of intracellular fluxes and the degree to which different enzymatic steps in the metabolism are rate controlling. Using ^{13}C-labelled substrates metabolic pathways can be analysed in even greater detail, and information on new pathways or the intracellular localisation of reactions can be obtained. This information can then be used to target an experimental effort in order to increase productivity for example. An example of this is the work of Pedersen et al. (1999) who, using metabolic flux analysis, found an increase of 15-26 % in the flux through the pentose phosphate pathway in a recombinant *A. oryzae* producing higher levels of α-amylase than the wild type. Work on recombinant *Bacillus subtilis* producing riboflavin (Sauer et al., 1997) also showed that the pentose phosphate pathway is a major pathway for carbon catabolism.

The complex nature of biological systems often makes it a very hard task to predict the effects of an alteration of the system just by looking at it. The use of mathematical models to *examine the interactions in a complex system* by describing individual parts and their interplay makes such predictions possible. As an example, control of gene expression is a highly complex process, often involving several regulatory proteins, and predicting the effects of performing changes in the genes involved is not an easy task. Lee and Bailey (1984a-c) used a mechanistic approach to model the *lac* operon in *E. coli* and were able to correctly predict the effects of mutations in the genes involved. They also investigated how the efficiency of a cloned *lac* promoter depends on the number of promoters per cloning vector and the cloning vector size. A similar approach was used by Agger and Nielsen (1999) for modelling the *alcA* system in *A. nidulans*, a system with a fairly complex regulatory structure involving a repressor (CreA), an inducer (AlcR) and the structural gene (*alcA*) interacting as shown in Figure 1 (Mathieu and Felenbok, 1994).

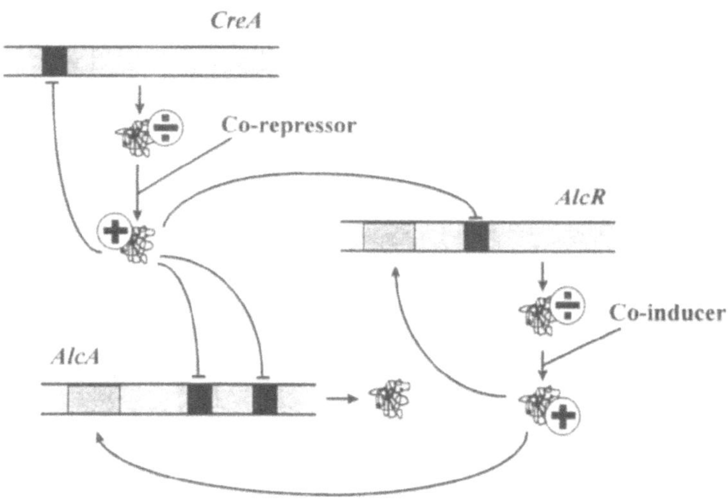

Figure 1. Overview of the regulatory mechanism of the alcA *gene. It is assumed that the CreA and AlcR proteins are synthesised in inactive forms (–) and activated (+) by repressor (e.g. glucose) and inducer (e.g. ethanol) molecules, respectively.*

The general features of this regulatory system are similar to many other systems, e.g. regulation of α-amylase production in *Aspergillus oryzae* and regulation of the *gal*-genes in *Saccharomyces cerevisiae*. Estimation of model parameters such as binding affinities of regulatory proteins to target genes were based on experimental data on intracellular levels of mRNA and protein obtained by Panozzo et al. (1998). By changing only the relevant parameters the model was able to simulate the effects of genetic alterations such as deletion of repressor sites in the structural gene and expression of an extra copy of the activator gene with a constitutive promoter.

Models like these contain systems of non linear differential equations with many parameters, the numerical solution of which pose a heavy computational burden.

However, with the explosive development in computing power this is not a practical problem and simulations using even very complex models can be routinely performed on desk-top computers. These highly mechanistic and extremely detailed models hold the potential for enabling investigators to test hypotheses concerning the interactions in regulatory systems without spending hours in the laboratory. They are able to aid in experimental planning by pointing towards the targets of manipulation that will most likely have the desired impact on the system. And as genomic sequence data continues to build up with ever increasing speed, they will be invaluable in interpreting this wealth of information. In theory, models like these will eventually form the basis for a complete, mechanistic description of all events in the cell.

Another, more practical, use of mathematical models is for *optimising the performance of industrial fermentation processes*. When producing Bakers yeast by fed-batch fermentation with *S. cerevisiae*, control of the feed rate is important. If it is too high, the yeast will produce ethanol, thereby significantly reducing the yield of biomass on substrate. On the other hand, the productivity of the process increases with the feed rate that makes it desirable to operate at a feed rate just below that resulting in ethanol formation. Since a fed-batch fermentation is a dynamic process with a constantly changing biomass concentration the feed rate has to be adjusted throughout the fermentation. A way of optimising the process performance is by applying a control strategy involving information about the metabolism of the micro-organism in the form of a mathematical model (internal model control) which allows for a faster and more precise regulation. Model based regulation can also be used for effective control of the dissolved oxygen tension and the feed rate during production of products such as penicillin. If the process model is sufficiently robust it can be used for designing and optimising fermentation processes as well, e.g. choosing the optimal feeding strategy for a fed-batch fermentation.

Mathematical models are also excellent *educational tools* for teaching everything from basic fermentation technology to advanced metabolic pathway analysis. Interactive illustrations of microbial processes based on mathematical models are fast, comprehensive and inexpensive and are valuable supplements to laboratory training of students. Simulations allow students to explore the dynamics of microbial systems, obtaining instant results and covering many subjects in a short time.

3. Means - General modelling frameworks

A biochemical system involving microbial cells is by nature very complex and heterogeneous. It includes a huge array of chemical compounds ranging from simple metabolites to extremely complex macromolecules, it involves interactions on many levels and exhibits dynamics spanning a wide timescale. This makes a complete mathematical description of such a system virtually impossible and any modelling effort will therefore result in a more or less crude approximation. Keeping this in mind, it is obvious that the modelling should strive to capture the most important elements of the system, these being determined by the purpose of the modelling exercise. As an example, if one wants to be able to quantify the impact of genetic changes on the morphology of a micro-organism it is most likely of little use to focus on a detailed description of primary carbon catabolism. It is thus essential to clearly define the intended use of the model before initiating the mathematical description of the system.

When modelling a microbial process one has to take into account the environment in which the organisms are growing, i.e. the bioreactor. If the dynamics of the processes that the model is supposed to describe are much slower than those of the bioreactor, then the environment can be assumed to be homogenous ('ideal' bioreactor). However, even in small-scale laboratory bioreactors the time of e.g. substrate mixing can be sufficiently long to have a major impact on cellular processes, and then the mass transfer and flow patterns in the bioreactor have to be given special attention. This treatment is out of the scope of this text and hence the models considered here will refer to a homogenous environment.

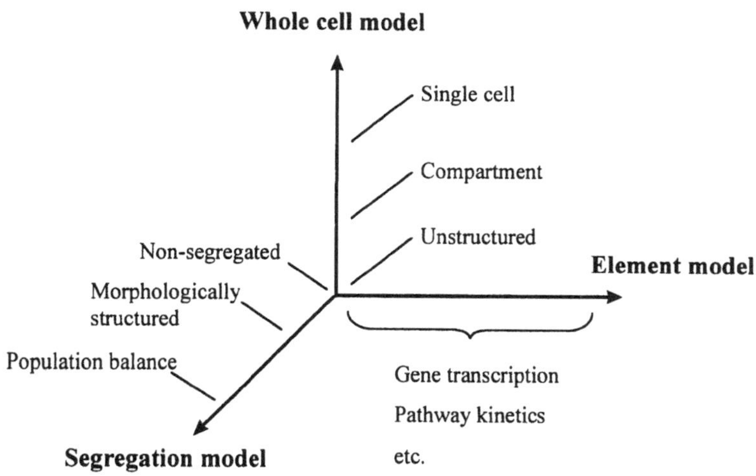

Figure 2. Composition space for mathematical models of microbial systems. Example terms are given for the three dimensions. An element model describes key processes in the cell, e.g. transcription of a certain gene or flux through a single pathway. A whole cell model is a more or less empirical description of cellular processes resulting in cell growth and/or product formation. A segregation model may account for the differences in cellular behaviour within the cell population.

When constructing a mathematical model for a microbial system, three basic questions have to be posed:

- How detailed does the general functions (e.g. substrate uptake, product formation) of the cell have to be described?
- Does the application of the model require any element of the cellular functions to be modelled in detail?
- Are the cells to be treated as average cells or is a description of several different cell types required?

The answers to these three questions will place the model composition somewhere in the space outlined in Figure 2. The element model will often be a highly detailed, mechanistic model of certain cellular functions, whereas the whole cell model gives an empirical description of e.g. substrate uptake and product formation. In theory, a whole cell model can be constructed from a large net of detailed element models describing all the key processes involved in cell growth, e.g. gene transcription, protein synthesis, catabolism and anabolism. However, this is practically impossible and only the elements of cellular function relevant to the specific application are normally included in the element model, the remaining functions being described in a more or less empirical fashion by a whole cell model.

Traditionally, a mathematical model of a microbial system is referred to as being 'unstructured' if only one variable (e.g. biomass concentration) is used to describe the cells, whereas 'structured' models make use of several variables (Fredrickson et al., 1970). Similarly, a 'non-segregated' model describes all cells in the system considered as being equal whereas a 'segregated' model includes a description of the variation between cells (Ramkrishna, 1979). The simplest model is thus the unstructured, non-segregated model given in Eq. (1) - (3) for growth on a single substrate with concentration c_s (g/l), formation of biomass with concentration c_x (g DW/l) and a single product with concentration c_p (g/l):

$$\mu = f(c_s) \tag{1}$$

$$r_s = Y_{xs}\mu + m_s \tag{2}$$

$$r_p = Y_{xp}\mu + m_p \tag{3}$$

Biomass is formed with a specific growth rate of μ (g DW (g DW h)$^{-1}$) which is often described by Monod kinetics, i.e. f(s) = $k \cdot c_s/(K+c_s)$. The specific rates of substrate utilisation r_s (g (g DW h)$^{-1}$) and product formation r_p (g (g DW h)$^{-1}$) are given as linear functions of the specific growth rate μ with the yield (e.g. Y_{xs}) and maintenance (e.g. m_s) coefficients being constants. Although completely empirical, this very simple model performs rather well in a number of situations, especially when substrate is plentiful and balanced growth prevails (all internal cell components grow at the same rate). However, the model does not supply any information about the processes occurring during growth of the micro-organism and it will most likely fail when the growth situation becomes just a little more complex.

If a model is needed that is able to describe microbial growth during a broader range of conditions, the model structure has to incorporate more detailed descriptions of the processes taking place. These descriptions can be included by extending the model complexity along one or more of the three axes shown in Figure 2, e.g. in an empirical fashion by including more detail in the whole cell model or by a description of different cell types or morphology and also in a mechanistic fashion by including an element model employing descriptions based on the vast amount of knowledge available on the biochemistry and genetics of microorganisms. Depending on the purpose of the model, one or more of these subjects can be predominant in the model structure, thus giving rise to e.g. 'morphologically structured' models or 'genetically structured' models. A general framework for the biochemistry of microbial cells, based on that of Nielsen and Villadsen (1992), is given below. It incorporates Q morphological forms each having J intracellular reactions involving L intracellular components (X), N substrates (s) and M metabolic products (p). All biochemical reactions, including substrate uptake and product excretion, can be described by eq. (4) whereas eq. (5) describes K irreversible metamorphosis reactions, converting one morphological form (with mass fraction Z_q) to another:

$$\sum_{i=1}^{N} \alpha_{jiq} s_{iq} + \sum_{i=1}^{L} \gamma_{jiq} X_{iq} + \sum_{i=1}^{M} \beta_{jiq} p_{iq} = 0, \quad j = 1,2,\ldots J \tag{4}$$

$$\sum_{q=1}^{Q} \delta_{jq} Z_q = 0, \quad \mathrm{j} = 1,2,\ldots\mathrm{K} \tag{5}$$

A mass balance for the intracellular components yields eq. (6) in which $\mathbf{R}$ is the net rate of formation of the cell component vector $\mathbf{X}$ (g (g DW h)$^{-1}$):

$$\mathbf{R}_\mathrm{q} = \Gamma_q^T \mathbf{r}_\mathrm{q} + \frac{1}{Z_q}\left[\mathbf{X}(\Delta^-)^T - \mathbf{X}_q \mathbf{e}_K^T\right]\mathbf{u}\Delta_q^+ - \mu_q \mathbf{X}_\mathrm{q} \tag{6}$$

The Γ matrix contains the stoichiometric γ coefficients, $\mathbf{r}_\mathrm{q}$ is the rate vector of the reactions (4) for the qth morphological form, Δ^- and Δ^+ are matrices containing the negative and positive stoichiometric δ coefficients, respectively, Δ_q is the qth column of Δ, $\mathbf{e}_K$ is a vector with K elements, all being unity, $\mathbf{u}$ is a diagonal (K x K) matrix which contains the rates of the metamorphosis reactions and μ_q is the specific growth rate of the qth morphological form. The first term on the right hand side of eq. (6) thus describes the change in the cell component vector $\mathbf{X}$, caused by the reactions given in eq. (4), the second term gives the change caused by the metamorphosis reactions and the third term describes the dilution caused by growth. Mass balances for a bioreactor with sterile feed are given in eqs. (7) - (11):

$$\frac{dc_x}{dt} = (\mu - D)c_x \tag{7}$$

$$\frac{d\mathbf{c}_s}{dt} = -\mathbf{r}_s x + D(\mathbf{c}_s^f - \mathbf{c}_s)c_x \tag{8}$$

$$\frac{d\mathbf{c}_p}{dt} = \mathbf{r}_p x + D(\mathbf{c}_p^f - \mathbf{c}_p)c_x \tag{9}$$

$$\frac{d\mathbf{X}_q}{dt} = \mathbf{R}_q \tag{10}$$

$$\frac{d\mathbf{Z}}{dt} = \Delta^T \mathbf{u}\mathbf{e}_k + \mathbf{MZ} - \mu\mathbf{Z} \tag{11}$$

Here **M** is a (Q x Q) diagonal matrix with μ_q in the diagonal, $\mathbf{c}_s^f$ and $\mathbf{c}_p^f$ are the inlet concentrations of substrates and products, respectively, D is the dilution rate and $\mathbf{r}_s$ and $\mathbf{r}_p$ are given by

$$\mathbf{r}_s = -\sum_{q=1}^{Q} \mathbf{A}_q^T \mathbf{r}_q Z_q \tag{12}$$

$$\mathbf{r}_p = \sum_{q=1}^{Q} \mathbf{B}_q^T \mathbf{r}_q Z_q \tag{13}$$

with the matrices $\mathbf{A}_q$ and $\mathbf{B}_q$ containing the stoichiometric α_q and β_q coefficients, respectively. The specific growth rate μ of the total biomass is given by

$$\mu = \sum_{q=1}^{Q} \mu_q Z_q \tag{14}$$

where μ_q is the specific growth rate of the qth morphological form. With sterile feed, the dilution rate D is zero for a batch cultivation, v/V for a continuous cultivation and

$$D = \frac{1}{V}\frac{dV}{dt} \tag{15}$$

for a fed-batch cultivation. The mass balances (7) - (11) can be used to describe a segregated model to the level of a morphologically structured model containing a finite number Q of cell types and a whole cell model containing any level of detail. If an infinite number of cell types are to be considered the mass balances changes to

$$\frac{\partial \Psi(\mathbf{X},t)}{\partial t} + \sum_{i=1}^{L} \frac{\partial}{\partial X_i}\left[R_i(\mathbf{X},t)\Psi(\mathbf{X},t)\right] = \left[\mu(\mathbf{X},t) - \mu(t)\right]\Psi(\mathbf{X},t) \tag{16}$$

$$\frac{d\mathbf{c}_s}{dt} = -\int_0^\infty \dots \int_0^\infty \mathbf{r}_s(\mathbf{X},t)\Psi(\mathbf{X},t)\prod_{i=1}^{L} dX_i + D(\mathbf{c}_s^f - \mathbf{c}_s) \tag{17}$$

$$\frac{d\mathbf{c}_p}{dt} = \int_0^\infty \dots \int_0^\infty \mathbf{r}_p(\mathbf{X},t)\Psi(\mathbf{X},t)\prod_{i=1}^{L} dX_i + D(\mathbf{c}_p^f - \mathbf{c}_p) \tag{18}$$

$$\mu(t) = \int_0^\infty \dots \int_0^\infty \mathbf{e}_L \Gamma^T \mathbf{r}(\mathbf{X},t)\Psi(\mathbf{X},t)\prod_{i=1}^{L} dX \tag{19}$$

where $\Psi(\mathbf{X},t)$ is a dimensionless discrete distribution function of cell forms, normalised by

$$\int_0^\infty \dots \int_0^\infty \Psi(\mathbf{X},t)\prod_{i=1}^{L} dX = 1 \tag{20}$$

with the average property vector $<\mathbf{X}>$ given by:

$$\int_0^\infty \dots \int_0^\infty \mathbf{X}\Psi(\mathbf{X},t)\prod_{i=1}^{L} dX = <\mathbf{X}> \tag{21}$$

The element models will often need a customised mathematical framework to deal with the processes in question. If a description of gene transcription, translation and regulation is needed, a general framework is given in Agger and Nielsen (1999) based on the work of Lee and Bailey (1984a-c). The model is based on a description of the binding of regulatory proteins to genes by conventional equilibrium kinetics and describes in a highly mechanistic fashion the actions of positive and negative regulatory proteins on their own genes as well as structural genes. Other examples of element models is the modelling framework for protein glycosylation by Shelikoff et al. (1996) and the modelling of the penicillin pathway kinetics in *Penicillin chrysogenum* by Pissara et al. (1996).

A somewhat different modelling approach, the so-called cybernetic modelling, was undertaken by Ramkrishna and co-workers (Dhurjati et al., 1985), based on the idea that cellular processes will function in a way that optimises the resulting outcome. This modelling concept aims at being used in cases where detailed information on regulatory processes is sparse and works by optimising an objective function rather than solving mass balances. Among its successful uses has been the description of microbial growth on multiple substrates (Ramakrishna et al., 1996).

4. Selected applications

An example of a very simple yet rather useful unstructured whole cell model, with the simplest possible segregation model and not containing any element models, is that of Sonnleitner and Käppeli (1986) who developed a model describing the growth of *S. cerevisiae* on glucose, as shown in eqs. (22) - (27). The model is based on an assumption of a limited respiratory capacity and includes oxidative and respirofermentative glucose metabolism [eq. (22) and (23), respectively] and oxidative ethanol metabolism [eq. (24)].

$$s_{glc} + \alpha_{12} s_{O_2} + \alpha_{13} s_N \rightarrow \gamma_1 X + \beta_{11} p_{CO_2}, \quad r_1 = \min\left(r_{glc}, \frac{r_{O_2}}{\alpha_{12}} \right) \tag{22}$$

$$s_{glc} + \alpha_{23} s_N \rightarrow \gamma_2 X + \beta_{21} p_{CO_2} + \beta_{22} p_{EtOH}, \quad r_2 = r_{glc} - r_1 \tag{23}$$

$$p_{EtOH} + \alpha_{32} s_{O_2} + \alpha_{33} s_N \rightarrow \gamma_3 X + \beta_{31} p_{CO_2}, \\ r_3 = \min\left(r_{EtOH}, \frac{r_{O_2} - \alpha_{12} r_{glc}}{\alpha_{32}} \right) \tag{24}$$

$$r_{glc} = k_{s,\max} \frac{s}{s + K_s} \tag{25}$$

$$r_{O_2} = k_{O_2,\max} \frac{O_2}{O_2 + K_{O_2}} \tag{26}$$

$$r_{EtOH} = k_{EtOH,\max} \frac{p}{p + K_p} \frac{1}{1 + s / K_i} \tag{27}$$

Despite its simple empirical structure, the model captures many essential growth characteristics such as the aerobic formation of ethanol at high glucose concentrations and the decrease in the oxidative capacity with decreasing oxygen concentration (the Pasteur effect). A model like this should be used for simulation of growth under relatively stable conditions as it will most likely fail if it is used to describe growth under dynamic operating conditions. This model has been refined a number of times since its first appearance, e.g. to include a description of overexpression of a homologous protein (Carlsen et al., 1997).

The growth of filamentous fungi pose a more difficult modelling objective as their growth pattern results in a continuum of cell types. Agger et al. (1998) approached the problem by using a morphologically structured model in which the hyphal elements are divided into three different regions: the extension zone (the very tips of the hyphal elements), the active region (responsible for all metabolic activity) and the inactive hyphal region. Two metamorphosis reactions are included in the model to describe formation of extension zones and hyphal cells from active cells [eq. (28)]. The whole cell model used to describe growth and substrate uptake is unstructured and uses kinetics based on models of fungal microscopic morphology; equations (29) - (31) give the volumetric rates of formation of the extension zone, the active region and the hyphal region, respectively. The values of almost all of the rather few model parameters can be determined from independent experiments and hence the model can be easily adapted to different fungal species by measuring the key morphological parameters.

$$\begin{bmatrix} 1 & -1 & 0 \\ 0 & 0 & 0 \\ 0 & -1 & 1 \end{bmatrix} \begin{bmatrix} x_e \\ x_a \\ x_h \end{bmatrix} = \begin{bmatrix} 0 \\ 0 \\ 0 \end{bmatrix} \tag{28}$$

$$q_1 = \begin{cases} 0 & ; \quad \dfrac{x_a}{c_n} < \left(\dfrac{x_a}{c_n}\right)_0 \\[2ex] \dfrac{k_1 s}{a(s + K_{s1})} x_a & ; \quad \dfrac{x_a}{c_n} \geq \left(\dfrac{x_a}{c_n}\right)_0 \end{cases} \tag{29}$$

$$q_2 = k_2 x_a \tag{30}$$

$$q_3 = \frac{k_3 s}{s + K_{s3}} \frac{x_a / c_n}{x_a / c_n + K_3} a x_e \tag{31}$$

To validate the model structure, a combination of fluorescence microscopy and digital image analysis was applied. This method allowed a quantification of the active region of the cells, as shown in Figure 3.

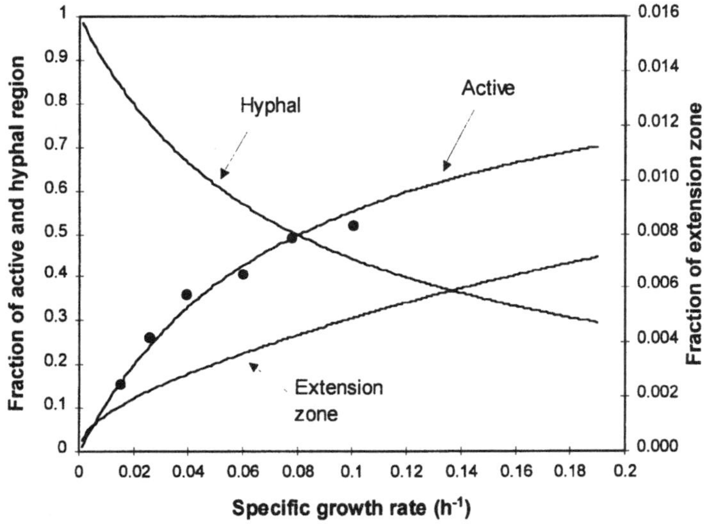

Figure 3. Model simulations of the fractional concentrations of the three morphological regions as described in the model by Agger et al. (1998), given as a function of the specific growth rate. Experimental data are indicated by (●)

Despite the simplicity of this model it performs rather well during simulations of dynamic growth conditions. This model, as well as that of Sonnleitner and Käppeli (1986), should be used for purposes where a relatively simple description of overall process performance is needed, but they can also be combined with element models if a detailed description of certain parts of the metabolism is required.

5. Future prospects

As illustrated with a few examples and discussed here mathematical models are very powerful tools in modern biology. However, as argued it is very important to define the motivation for modelling, and there are roughly four different areas where mathematical models find useful applications:

- Teaching of quantitative aspects of biological systems.
- Design and control of biological processes, *e.g.* simulation and control of fed-batch fermentation processes.
- Extraction of quantitative information from experimental data.
- Integration of information about complex biological systems, *e.g.* complex regulatory networks as seen in some models for gene expression.

Traditionally mathematical models have been used in the two first areas, and there are also many examples where mathematical models have been applied for processing and comparison of experimental data. When it comes to integration of information about complex biological systems there are, however, only a few examples in the literature – best illustrated by the pioneering work of Lee and Bailey (1984a-c) who described gene transcription of the *lac*-operon in *E. coli*. In the future mathematical models will, however, play an increasingly important role in biological research for two reasons:

- There is an explosive increase in the amount of experimental data available, and novel analytical techniques offer the possibility to study individual cellular processes at a level of detail not previously possible.
- The increase in computer power allows for simulation of even very complex mathematical models, and much more information can therefore be included in future models compared with the past.

In order to integrate the wealth of information supplied by advanced experimental studies of cellular function it is necessary to apply mathematical models that in a quantitative fashion allows for an evaluation of the importance of the individual cellular processes. Mathematical models offer the possibility to apply a whole cell view, or a systems approach, in the analysis of experimental data. For this reason they will become essential when information ranging from genomic research to advanced bioimaging is to be evaluated and interpreted.

References

Agger, T., Spohr, A. B., Carlsen, M. and Nielsen, J. (1998) Growth and Product Formation of Aspergillus oryzae during Submerged Cultivations: Verification of a Morphologically Structured Model Using Fluorescent Probes, *Biotechnol. Bioeng.* **57**: 321-329.

Agger, T. and Nielsen, J. (1999) Genetically Structured Modelling of Protein Production in Filamentous Fungi, *Biotechnol. Bioeng.*, **66**: 164-170.

Carlsen M., Jochumsen K.V., Emborg C. and Nielsen, J. (1997) Modelling the growth and proteinase a production in continuous cultures of recombinant *Saccharomyces cerevisiae*, *Biotechnol. Bioeng.* **55**: 447-454.

Dhurjati, P., Ramkrishna, D., Flickinger, M. C. and Tsao, G. T. (1985) A Cybernetic View of Microbial Growth: Modelling Cells as Optimal Strategists, *Biotechnol. Bioeng.* **27**: 1-9.

Fredrickson, A. G., McGee, R. D. III and Tsuchiya, H. M (1970) Mathematical Models in Fermentation Processes, *Adv. Appl. Microbiol.*, **23**, 419.

Lee, S. B. and Bailey, J. E. (1984a) Analysis of Growth Rate Effects on Productivity of Recombinant Escherichia coli Populations Using Molecular Mechanism Models. *Biotechnol. Bioeng.* **26**: 66-73.

Lee, S. B. and Bailey, J. E. (1984b) Genetically Structured Models for lac Promoter-Operator Function in the *Escherichia coli* Chromosome and in Multicopy Plasmids: *lac* Operator Function. *Biotechnol. Bioeng.* **26**: 1372-1382.

Lee, S. B. and Bailey, J. E. (1984c) Genetically Structured Models for lac Promoter-Operator Function in the Chromosome and in Multicopy Plasmids: *lac* Promoter Function. *Biotechnol. Bioeng.* **26**: 1383-1389.

Mathieu, M. and Felenbok, B. (1994) The *Aspergillus nidulans* CreA protein mediates glucose repression of the ethanol regulon at various levels through competition with the ALCR-specific transactivator. *EMBO J.* **13**: 4022-4027.

Nielsen, J. and Villadsen J (1992) Modelling of Microbial Kinetics, *Chem. Eng. Sci.*, **47** (17/18), 4225-4270.

Panozzo, C., Cornillot, E. and Felenbok, B. (1998) The CreA Repressor Is the Sole DNA-binding Protein Responsible for Carbon Catabolite Repression of the *alcA* Gene in *Aspergillus nidulans* via Its Binding to a Couple of Specific Sites. *J. Biol. Chem.*, **273** (11), 6367-6372.

Pedersen, H., Carlsen, M. and Nielsen, J. (1999) Identification of enzymes and quantification of metabolic fluxes in the wild type and in a recombinant Aspergillus oryzae strain. *Appl. Environ. Microbiol.*, **65**, 11-19

Pissara, P.D., Nielsen, J. and Bazin, M.J. (1996) Pathway kinetics and metabolic control analysis of a high-yielding strain of Penicillium chrysogenum during fedbatch cultivations, *Biotechnol. Bioeng.* **51**, 168-176.

Ramakrishna, R., Ramkrishna, D. and Konopka, A.E. (1996) Cybernetic modelling of growth in mixed, substitutable substrate environments: Preferential and simultaneous utilisation. *Biotechnol. Bioeng.* **52**, 141-151.

Ramkrishna, D. (1979) Statistical models for cell populations, *Adv. Biochem. Eng.* **11**, 1-48.

Sauer, U., Hatzimanikatis, V., Bailey, J.E., Hochuli, M., Szyperski, T. and Wuthrich, K. (1997) Metabolic fluxes in riboflavin-producing *Bacillus subtilis*, Nature Biotechnol., **15**, 448-452.

Shelikoff, M., Sinskey, A. J. and Stephanopoulos, G. (1996) A Modelling Framework for the Study of Protein Glycosylation, *Biotechnol. Bioeng.* **50**, 73-90.

Sonnleitner, B. and Käppeli, O. (1986) Growth of *Saccharomyces cerevisiae* is controlled by its limited respiratory capacity: formulation and verification of a hypothesis, *Biotechnol. Bioeng.* **28**, 927-937.

Spohr, A., Carlsen, M., Nielsen, J. and Villadsen, J. (1997) Morphological characterisation of recombinant strains of *Aspergillus oryzae* producing α-amylase during batch cultivations, *Biotechnol. Let.* **19**, 257-261.

Nomenclature

Symbol	Unit	Definition
a	tips (g extension zone DW)$^{-1}$	Constant
c_n	kg^{-1}	Concentration of hyphal elements
$\mathbf{c}_p$		Metabolic product concentration vector
$\mathbf{c}_p^f$	g l^{-1}	Metabolic product concentration vector in feed
$\mathbf{c}_s$	g l^{-1}	Vector of substrate concentrations
$\mathbf{c}_s^f$	g l^{-1}	Vector of substrate concentrations in the feed
c_x	g kg^{-1}	Biomass concentration
D	h^{-1}	Dilution rate
$\mathbf{e}_j$		Column vector of dimension j with all elements being 1
k_1	tips (g active DW h)$^{-1}$	Specific branching frequency (active)
k_2	h^{-1}	Rate constant
k_3	g active DW (tip h)$^{-1}$	Maximal tip extension rate
k_t	h^{-1}	Rate constant

Nomenclature

Symbol	Unit	Definition
K_{s1}	$g\ l^{-1}$	Saturation constant for branching
K_{s3}	$g\ l^{-1}$	Saturation constant for tip extension
K_3	g	Saturation constant for tip extension
K_i		Saturation constant
m_s, m_p	$g\ (g\ DW\ h)^{-1}$	Maintenance coefficients
$\mathbf{r}$	h^{-1}	Rate vector for intracellular reactions
$\mathbf{r}_p$	h^{-1}	Rate vector for product formation
$\mathbf{r}_s$	h^{-1}	Rate vector for substrate uptake
r_m	h^{-1}	Maintenance coefficient
$\mathbf{R}$	$g\ (g\ DW\ h)^{-1}$	Vector of net rates of formation of intracellular components
t	h	Time
$\mathbf{u}$	h^{-1}	Diagonal matrix containing the forward reaction rates of the metamorphosis reactions
v	$l\ h^{-1}$	Volumetric flow to the bioreactor
V	l	Volume of the bioreactor
q_i	$g\ (kg\ h)^{-1}$	Reaction rates
x_e	$g\ kg^{-1}$	Concentration of extension zone
x_a	$g\ kg^{-1}$	Concentration of active region
x_h	$g\ kg^{-1}$	Concentration of hyphal region
$\mathbf{X}$	$g\ (g\ DW)^{-1}$	Vector/matrix of concentrations of intracellular components
$\mathbf{Z}$	$g\ (g\ DW)^{-1}$	Vector of mass fractions of morphological forms

Greek letters		
α_{jiq}		Stoichiometric coefficient for the *i*th substrate in the *j*th reaction running in the *q*th morphological form
$\mathbf{A}$		Matrix containing the stoichiometric coefficients for the substrates
β_{jiq}		Stoichiometric coefficient for the *i*th product in the *j*th reaction running in the *q*th morphological form
$\mathbf{B}$		Matrix containing the stoichiometric coefficients for the products
γ_{jiq}		Stoichiometric coefficient for the *i*th intracellular component in the *j*th reaction running in the *q*th morphological form
Γ		Matrix containing the stoichiometric coefficients for the intracellular components
δ_{jq}		Stoichiometric coefficient for the *q*th morphological form in the *j*th metamorphosis reaction
Δ		Matrix containing the stoichiometric coefficients for the metamorphosis reactions
Δ^+		Matrix containing the positive stoichiometric coefficients for the metamorphosis reactions
Δ^-		Matrix containing the negative stoichiometric coefficients for the metamorphosis reactions
μ	h^{-1}	Specific growth rate of the whole biomass
μ_q	h^{-1}	Specific growth rate of cells in the *q*th morphological form
Ψ		Distribution function for the mass fraction of cells

MACROSCOPIC MODELLING OF BIOPROCESSES WITH A VIEW TO ENGINEERING APPLICATIONS

PH. BOGAERTS AND R. HANUS
Control Engineering and System Analysis Department
Université Libre de Bruxelles
Av. F.-D. Roosevelt, 50 C.P.165/55 B-1050 Brussels (Belgium)
Tel. 32-2-650.26.75 Fax. 32-2-650.26.77
E-mail: pbogaert@labauto.ulb.ac.be

Abstract

Several motivations exist to use macroscopic models for engineering applications and to define a general modelling methodology. In this context, the framework of system of mass balances based on macroscopic reaction schemes is recalled and a new general kinetic model structure is presented and analysed. A general methodology for the parameter identification (kinetic and pseudo-stoichiometric coefficients) is summarised. Necessary conditions of validation of the reaction scheme (based on the identified model parameters) are proposed. The flexibility of the general kinetic model structure and a part of the parameter identification methodology are illustrated on simulated bacteria cultures.

1. Introduction

Engineering applications in biotechnology are concerned with the synthesis of several useful tools for monitoring cell cultures in bioreactors. Among these tools, simulators allow to reproduce the behaviour of cell cultures in bioreactors and this in a cheap and fast way in comparison with real experiments. These virtual experiments can be used to determine optimal experimental conditions (e.g., the dilution rate leading to a maximum amount of biomass in a given time), to train human operators before placing them in the real world, or to test other tools like controllers and software sensors. These latter consist of a second kind of engineering tool. They are able to replace some of the hardware sensors thanks to a combination of the remaining ones, a mathematical model of the bioprocess and a state estimation algorithm. Such a kind of software solution is particularly interesting in the field of bioprocesses where the hardware sensors often exhibit several drawbacks (cost, destruction of the samples, measurements only in discrete time, time delay in the measurements, need for sterilisable materials,

M. Hofman and P. Thonart (eds.), Engineering and Manufacturing for Biotechnology, 77–109.

perturbation of the fluid dynamics, etc.). A third kind of engineering tool is made of controllers which pilot the bioreactors in order to respect the set points (pH, temperature, dissolved oxygen, product concentration, ...) and to reject the disturbances acting on the process.

For each of these engineering tools, it is necessary to build a model of the process. The structure of this model (and especially its complexity) has to be adapted to the final aim (simulation, state estimation, control). This means that, first of all, the model structure must be compatible with this aim. For instance, a software sensor must be based on a model which is observable, i.e. allows to reconstruct the state trajectory (e.g., all the concentrations, measured and non measured) in finite time on the basis of the available measured signals. Note that this definition of observability is just a heuristic one. In the same way, the model structure that will be used in a controller synthesis must involve the evolution of the signals to be controlled. Given a model structure that is supposed to be adapted to the final aim, a second critical point is the possibility to accurately identify the model parameters on the basis of experimental data. Too simple structures would not be able to reproduce the experimental data, whereas too complex structures could lead to overfitting if the number of experimental data is quite limited. Last but not least, it is interesting (at least from the economic point of view) that the model involves a number of measured signals which is quite limited. For example, if one is looking for a software sensor for the biomass concentration, a model involving the main substrate concentration, one product concentration and the biomass concentration will only require two hardware measurements (for the main substrate and for the product) whereas a detailed biological model involving numerous substrates and secondary metabolites would be much more requiring regarding the problem of the necessary hardware measurements.

The previous discussion shows that it is necessary to limit the complexity of the model structure and to limit the number of signals that have to be measured. Therefore, it is quite natural to consider the class of the unstructured and unsegregated models within the classification of Fredrickson (Tsuchiya *et al.*, 1966; Tziampazis and Sambanis, 1994). A model is unstructured if the cell is considered as a unique lumped compartment (the biomass). A model is unsegregated if the cell population is considered as homogeneous (without distinctions based on the cell cycle, the cell size, etc.). In both cases, the structured and segregated features involve more state variables: several interior compartments and their component concentrations for structured models and several subpopulations of the biomass, depending on their position in the cell cycle, for segregated models. Structured models are based on the intracellular metabolism and often lead to the use of large amount of biochemical reactions (up to several hundreds). Techniques of simplification of the model are sometimes used (Barford *et al.*, 1992). In many cases the model parameters are immediately deduced from the literature rather than identified on the basis of experimental data. Structured models have been used with bacteria and yeast cells (Roels, 1983), but also with animal cell cultures (Xie and Wang, 1996a, 1996b) whose behaviour seems to be much more complex. Some authors have also combined structured and segregated features within a model (Martens *et al.*, 1995). Finally, segregated models are interested with the influence of the position within the cell cycle on the cell size (Nielsen *et al.*, 1997) or on the production rate of valuable components like monoclonal antibodies produced by hybridoma cells (Cazzador and Mariani, 1993).

The class of unstructured and unsegregated models which we are interested with consists of quite simple model in comparison with the above mentioned structured and/or segregated models. They are often used in engineering applications like state estimation (Bastin and Dochain, 1990; Gauthier *et al.*, 1992; Ryckaert and Van Impe, 1996; Bogaerts, 1999b) or control of bioreactors (Bastin and Dochain, 1990; Bastin and Van Impe, 1995; Van Impe and Bastin, 1995). This kind of model has generally been introduced within the framework of simple bacteria or yeast cell cultures, but their use has also been extended to the animal cell cultures (Glacken *et al.*, 1988; Goergen *et al.*, 1994). They mainly link the evolution of the biomass to the consumption of the main substrates and to the production of some metabolites and/or products of interest. Most of them are based on or inspired from the Monod law describing the specific growth rate (see Appendix 1 of the book by Bastin and Dochain (1990)). The research in the field of the unstructured and unsegregated models is still ongoing (Zeng and Deckwer, 1995; Tan *et al.*, 1996).

There is a large amount of model structures which are able to describe particular phenomena (like the limitation and/or the inhibition of the specific growth rate by the main substrate) and/or which are specific to particular applications (given cell lines within particular types of bioreactors). However, some authors have proposed more general mathematical modelling frameworks. Among them, the Biochemical System Theory was proposed by Savageau (1969a, 1969b). In this approach, the microorganisms are supposed to be composed by several components (structured model) whose overall production and consumption rates are given by power laws. These models are still used nowadays, for instance to determine the steady state for which the production rate of one (or several) product(s) of interest is maximised in microorganism cultures (Torres *et al.*, 1996, 1997). Another general framework is given by the Metabolic Control Analysis (Westerhoff and Kell, 1987) which is based on a sensitivity analysis of the metabolic systems. This analysis is quantified by control coefficients and elasticity coefficients (Delgado *et al.*, 1993). These coefficients are properties of the metabolic system (in steady state) and allow to determine the most sensitive steps of a reaction scheme. A third general framework consists in the cybernetic modelling developed by Ramkrishna and his co-workers (Dhurjati *et al.*, 1985) which was introduced to model the growth of microorganisms on multiple substrates. This approach considers the cells as optimal controllers that use the different available substrates so as to maximise a performance criterion (like the production of biomass). Finally, we also mention the general modelling methodology for animal cell cultures (Chotteau, 1995) which is based on the concept of macroscopic reaction network (Bastin and Dochain, 1990) and uses polynomial models for the overall consumption and/or production rates of the macroscopic species involved in the reaction scheme.

Although some general modelling frameworks exist and although the class of unstructured and unsegregated models corresponds to the appropriate level of mathematical complexity, several drawbacks are present in these approaches. They are discussed in the sequel and motivate the development of a new general kinetic model structure. Moreover, there also exist some important lacks in the available parameter identification procedures. Therefore, a systematic parameter identification methodology is also proposed in this text.

In Section 2, the basic concept of macroscopic reaction network and its associated mass balances is recalled. Section 3 presents the main drawbacks of the above mentioned general frameworks and of the usual models belonging to the class of the unstructured and unsegregated models. This motivates the proposition of a new general kinetic model structure exhibiting several interesting properties. Section 4 motivates the proposition of a systematic methodology for the parameter identification of the model structure described in Sections 2 and 3. The methodology is described and necessary conditions for reaction scheme validation are provided. Section 5 illustrates the flexibility of the new kinetic model structure and a part of the identification methodology by applying these concepts on simulated bacteria cultures (based on several Monod-type laws for the specific growth rate of the simulator). Finally, Section 6 summarises the material and presents some recent developments and perspectives in this field.

2. Macroscopic reaction network and associated mass balances

As we have seen in the introduction, it is necessary to limit the complexity of the model structures to be used in engineering applications. It is therefore often sufficient (and sometimes necessary) to limit the description of the biological and chemical phenomena to a limited number of essential events. The latter may be contained within a macroscopic reaction network (Bastin and Dochain, 1990)

$$\sum_{i \in R_k} \left(- \nu_{i,k} \right) \xi_i \xrightarrow{\varphi_k} \sum_{j \in P_k} \nu_{j,k} \, \xi_j \qquad k \in [1, M] \tag{1}$$

where

- M is the number of reactions;
- ξ_i the ith component;
- φ_k the reaction rate;
- $\nu_{i,k}$ and $\nu_{j,k}$ the pseudo-stoichiometric (or yield) coefficients (positive when associated to a component which is produced, negative when it is consumed).

It is important to note that this kind of macroscopic reaction network does not respect the elementary mass balances (oxygen, carbon, ...) but quantifies the yields $\nu_{i,k}$ between the consumption and production of the macroscopic species ξ_i. Typically, these networks contain between one and six reactions. Several examples are given in the book of Bastin and Dochain (1990). It is possible to distinguish three main types of reactions. The first one consists of the simple reaction, e.g.

$$(-\nu_A)\xi_A + (-\nu_B)\xi_B \rightarrow \xi_C \tag{2}$$

where ξ_A and ξ_B are consumed and ξ_C is produced, without any catalyse or autocatalyse. A second kind of reaction is the catalytic reaction, e.g.

$$(-\nu_A)\xi_A + (-\nu_B)\xi_B \rightarrow (-\nu_B)\xi_B + \xi_C \tag{3}$$

where ξ_A is consumed, ξ_C is produced and ξ_B catalyses the reaction. This last component is consumed and produced by the reaction, in the same quantity. It activates reaction (3) and is therefore necessary so that the reaction can occur. A third kind of reaction consists of the autocatalytic reaction (denoted by $\rightarrow^{+}$), e.g.

$$(-\nu_A)\xi_A \rightarrow^{+} \nu_B \xi_B + \xi_C \tag{4}$$

where ξ_A is consumed, ξ_C is produced and ξ_B autocatalyses the reaction. This latter component is produced by the reaction. Moreover, it activates the reaction.

On the basis of the reaction scheme (1), the basic mathematical model structure is made of the mass balances for each of the macroscopic species ξ_i:

$$\frac{d\,\xi(t)}{d\,t} = K\,\varphi(\xi,t) - D(t)\,\xi(t) + F(t) - Q(t) \tag{5}$$

where

- $\xi \in \Re^{N}$ is the vector of concentrations;
- $K \in \Re^{N \times M}$ is the pseudo-stoichiometric coefficients matrix;
- $\varphi \in \Re^{M}$ is the vector of reaction rates;
- $D \in \Re$ is the dilution rate;
- $F \in \Re^{N}$ is the vector of external feed rates;
- $Q \in \Re^{N}$ is the vector of gaseous outflow rates.

If the external substrates are diluted in the incoming culture medium, the corresponding components of the vector of external feed rates can be written

$$F_i(t) = D(t)\,\xi_i^{IN}(t) \tag{6}$$

where ξ_i^{IN} is the concentration of the ith component in the incoming stream.

It may also happen that an external substrate is introduced in gaseous form (such as the oxygen in aerobic cultures). Then the corresponding component of the vector of external feed rates is given by

$$F_i(t) = k_l a(F_g)\left(\xi_{is} - \xi_i(t)\right) \tag{7}$$

where

- $k_l a$ is the gas-liquid transfer coefficient (function of the gaseous flow F_g);
- ξ_{is} is the saturation concentration.

By neglecting the dynamics of the liquid-gas transfer, the gaseous outflow rate of some component ξ_i may be assumed to be proportional to the dissolved concentration:

$$Q_i(t) = \beta_i \xi_i(t) \quad \left(0 \le \beta_i \;\; 0 \le \xi_i(t) \le \xi_{is}\right) \tag{8}$$

where

- β_i is the specific rate of liquid-gas transfer;
- ξ_{is} is the saturation concentration.

Taking into account (6) and (8), the system (5) may be rewritten

$$\frac{d\,\xi(t)}{d\,t} = K\,\varphi(\xi,t) - D(t)\,\xi(t) + D(t)\,\xi^{IN}(t) - B\,\xi(t) \tag{9}$$

where

- $\xi^{IN} \in \Re^N$ is the vector of concentrations in the components of the incoming stream;
- $B \in \Re^{N \times N}$ is a diagonal matrix containing the specific rates of liquid-gas transfer.

Sufficient conditions of bounded input - bounded state (BIBS) stability have been proposed in Bogaerts (1999) and in Bogaerts *et al.* (1999). The BIBS stability of a nonlinear differential system

$$\frac{d\,x(t)}{d\,t} = f\,(x(t), u(t)) \tag{10}$$

(where $x(t)$ is the state vector and $u(t)$ is the input vector), guarantees that

$$\begin{cases} u_{\text{MIN}} \le u(t) \le u_{\text{MAX}} \\ x^0_{\text{MIN}} \le x(0) \le x^0_{\text{MAX}} \end{cases} \forall t \ge 0 \;\rightarrow\; x_{\text{MIN}} \le x(t) \le x_{\text{MAX}} \;\; \forall t \ge 0 \tag{11}$$

i.e., that any bounded initial state $x(0)$ and bounded input profile $u(t)$ implies that the state trajectory $x(t)$ remains bounded. Sufficient conditions of BIBS stability for the system of mass balances (9) can be summarised as follows.

2.1. FIRST SUFFICIENT CONDITION OF BIBS STABILITY OF (9)

A reference reactant for a given reaction is defined as a reactant which is neither a catalyst nor an autocatalyst and which is not produced in any other reaction of the scheme.

If

- the dilution rate is non negative (which is obvious from a physical point of view)

$$D(t) \geq 0 \;\; \forall t; \tag{12}$$

- the input concentrations are bounded

$$0 \leq \xi_i^{IN}(t) \leq \xi_{\text{MAX}}^{IN} \;\; \forall i, \forall t; \tag{13}$$

- each reaction of the scheme (1) contains at least one reference reactant

then

the states $\xi(t)$ of the model (9) are positive and upper bounded for any time t.

2.2. SECOND SUFFICIENT CONDITION OF BIBS STABILITY OF (9)

If

- the dilution rate is non negative (which is obvious from a physical point of view)

$$D(t) \geq 0 \;\; \forall t; \tag{14}$$

- the input concentrations are bounded

$$0 \leq \xi_i^{IN}(t) \leq \xi_{\text{MAX}}^{IN} \;\; \forall i, \forall t; \tag{15}$$

- each reaction of the scheme (1), *except reaction* j_l, contains at least one reference reactant;
- *reaction* j_l *contains at least one reactant (neither catalyst nor autocatalyst) which is produced in another reaction* k_l

then

the states $\xi(t)$ of the model (9) are positive and upper bounded for any time t, *provided reaction* k_l *contains at least one reference reactant.*

By taking the italic asserts away, the second condition reduces to the first one. The proofs of these sufficient conditions are sketched in Bogaerts *et al.* (1999) and detailed in Bogaerts (1999).

3. Kinetic model structure

3.1. MOTIVATIONS FOR A NEW KINETIC MODEL STRUCTURE

As shown in the previous section, the model structure consists of the mass balances (5) based on the reaction scheme (1). However, the structure of reaction rates contained in the vector $\varphi(\xi)$ has not yet been given. Each of the reaction rates $\varphi_j(\xi)(j \in [1,M])$ is usually represented by a nonlinear function of the concentrations ξ. The choice of these nonlinear functions is the matter of still ongoing research.

We have already explained in the introduction why structured and/or segregated models will not be used in this formalism. They would lead to a significant increase in the vector ξ dimension, hence requiring several concentrations of different components (or subpopulations of one component) to be measured. It could also lead, in some cases, to the impossibility of building engineering tools like software sensors (due to the unobservable feature of the model).

We have also recall the existence of some general mathematical modelling frameworks. The Biochemical System Theory (Savageau, 1969a, 1969b) uses power laws to describe the overall consumption and the overall production of each component. This kind of model can be linearised w.r.t. (with reference to) the parameters (thanks to a logarithmic transformation). This allows to benefit from the advantages of the linear parameter estimation, namely the existence, uniqueness and complete independence regarding any initial guess of the set of parameters which minimise a quadratic criterion (e.g., the least squares criterion). However, this property of the models in Biochemical System Theory is only available in steady state. Moreover, only the overall rates of consumption and production are given, without distinguishing the rates of a reaction scheme. Finally, BIBS stability and positiveness of the concentrations are not guaranteed *a priori*. Metabolic Control Analysis (Westerhoff and Kell, 1987) is a very structured approach with the drawbacks already mentioned before. Moreover the control coefficients and elasticity coefficients are not easy to determine. Cybernetic models (Dhurjati *et al.*, 1985) seem to be difficult to apply in the (quite usual) case of a simultaneous consumption of several substrates. In order to reproduce the behaviour of such cultures, the structured part of the model has to be more complex (Ramakrishna *et al.*, 1996). Moreover the cybernetic approach seems not to be applicable with animal cell cultures (Tziampazis and Sambanis, 1994). Finally the general modelling methodology for animal cell cultures of Chotteau (1995) exhibits several problems like the computational load (several millions of solutions to be tested), the overall feature of the rate of consumption and/or production, the use of a black box polynomial model for these overall rates (quasi without physical interpretation), the absence of guarantee for the BIBS stability of the model and for the concentration positiveness or, even, the possibility to simulate a spontaneous growth of living microorganisms (without any living cell present at the initial time of the experiment).

For all the above mentioned reasons, the existing general frameworks will not be used in the sequel and we will only focus on the remaining class of unstructured and unsegregated models which contains a large number of different structures. However, there are also several drawbacks exhibited by these solutions. First of all there is a lack of general structures. Indeed, there exist several Monod-type laws (see Appendix 1 of the book by Bastin and Dochain (1990)), each one describing a limited number of physical effects (like the limitation and/or inhibition by substrate and/or biomass, etc.). Even for a given set of physical effects, there exist several solutions, e.g., the Monod, Tessier and Ming laws to describe the limitation of the specific growth rate by the substrate. Therefore a couple of questions arise: What set of physical effects must be chosen *a priori* ? For this set, what kind of law ? There is no systematic way to answer these questions. A second kind of problem, often linked to the solutions which are claimed to be more general, is the lack (or even absence) of physical meaning of the parameters. This is the case, for instance, with the artificial neural networks (Montague and Morris, 1994; Syu and Tsao, 1993) or with the general modelling methodology for animal cell cultures (Chotteau, 1995). A third problem is the absence of guarantee for the concentration positiveness, e.g., when using the famous Pirt law (Pirt, 1965) leading to a constant specific rate for the consumption of substrate linked to the maintenance. A fourth problem, occurring for instance in the methodology of Chotteau (1995), is the absence of guarantee of BIBS stability. Finally, a fifth disadvantage is that the models are nonlinear (which is necessary to reproduce the complex behaviour of bioprocesses) but, in most of the cases, non-linearisable w.r.t. the parameters, although this possibility is of great interest for the parameter identification problem (as stated above concerning Biochemical System Theory).

The overall conclusion of the above discussion is that there does not exist sufficiently general and flexible mathematical model structures for the reaction rates, with interesting (or even necessary) properties like the BIBS stability, the physical meaning of the parameters or the possibility to linearise the structure w.r.t. the parameters. This motivated the development of a new general kinetic model structure (Bogaerts, 1999; Bogaerts *et al.*, 1999).

3.2. GENERAL KINETIC MODEL STRUCTURE

The reaction rate $\varphi_j(\xi)$ ($j \in [1,M]$) may be described by

$$\varphi_j(\xi_1,\ldots,\xi_N,t) \doteq \alpha_j \prod_{k \in R_j^*} \xi_k^{\gamma_{k,j}}(t) \prod_{\ell \in P_j^*} e^{-\beta_{l,j}\xi_\ell(t)} \quad j \in [1,M] \tag{16}$$

where

- $\alpha_j > 0$ is a kinetic constant (function, if necessary, of any physical influence different from the component concentrations, e.g., the temperature dependence according to an Arrhenius law);
- R_j^* the set of indices of the components which activate the reaction *j* (reactants, catalysts and auto-catalysts);
- P_j^* the set of indices of all the components appearing in reaction *j* (or even, if necessary, in other reactions of the scheme);
- $\gamma_{k,j} > 0$ the activation coefficient of component *k* in reaction j;
- $\beta_{\ell,j} \geq 0$ the inhibition coefficient of component *l* in reaction *j*.

This structure has the advantage to be very general in the sense that the activation and/or the inhibition of the reaction by any component can be taken into account. It is obvious that the kinetic parameters have a physical meaning, namely, the activation and/or inhibition .We will see that this property allows to propose necessary conditions to be satisfied by the reaction scheme. A sufficient condition to guarantee the concentration positiveness is that the activation coefficients are such that $\gamma_{k,j} \geq 1 \; \forall \; k, j$. Otherwise, it is necessary to use saturation on the concentration vector ξ such that the lower bound is equalled to zero. Provided these latter precautions to guarantee the concentration positiveness, the BIBS stability of the system of mass balances (9) using the kinetic model structure (16) is guaranteed if one of the sufficient conditions of Section 2 are satisfied. Finally, the structure (16) is nonlinear w.r.t. the kinetic parameters (α_j, $\gamma_{k,j}$, $\beta_{\ell,j}$) but may be linearised thanks to a logarithmic transformation. Indeed, the logarithm of both members of eq. (16) leads to

$$log\varphi_j(\xi_1,\ldots,\xi_N,t)=log\alpha_j+\sum_{k\in R_j^*}\gamma_{k,j}log\xi_k(t)-\sum_{\ell\in P_j^*}\beta_{l,j}\xi_\ell(t) \quad j\in[1,M \qquad (17)$$

which is linear w.r.t. the parameters $log\,\alpha_j$, $\gamma_{k,j}$ and $\beta_{\ell,j}$.

The model structure considered in the sequel is made of the system of mass balances (5) (based upon the reaction scheme (1)) using the general kinetic model structure (16). The question which arises is the identification of the model parameters (i.e., the pseudo-stoichiometric coefficients and the kinetic coefficients) given a set of experimental data containing measurements of the concentration vector and the corresponding input data (dilution rate, external feed rates).

4. Parameter identification

4.1. MOTIVATIONS FOR A SYSTEMATIC PROCEDURE

In the same way as for the kinetic model structure, several motivations exist to propose a complete systematic procedure well suited for the problem. Indeed, in several contributions to the mathematical modelling of bioprocesses, it is the model structure that is mainly discussed and neither the parameter identification procedure nor the model validation. Hereafter, several problems often encountered in the literature are summarised.

In some cases, the parameter identification method is not given. For instance, Ramakrishna *et al.* (1996) just tell that "parameters were estimated from experimental data, literature studies, previous cybernetic modelling efforts, and order-of-magnitude estimations" without any more detail. In other cases, parameter values are just taken from the literature, by using eventually mean values coming from different sources (Martens *et al.*, 1995). Other authors use a trial-and-error method where the model validation criterion is the visual difference between experimental data and simulated values (Goergen *et al.*, 1994; Barford *et al.*, 1992). In most of the studies in the literature, the identification procedure does not take into account the confidence that can be associated with the measurements. For instance, Julien *et al.* (1998) identify a simulation model for an activated sludge process and compare graphically simulated values and experimental data, these latter being presented with their confidence interval. Although these authors analyse the identification problem much more deeply than in lots of other contributions, they do not use this precious information of the confidence intervals in the cost function of the parameter identification procedure. As an obvious consequence of the last point, the confidence intervals for the identified parameters are rarely provided together with the identified values.

Another important problem in the usual literature is that the initial conditions of a simulation model are generally measured values that are supposed to be infinitely accurate. Let us consider a simulation model given by the differential equation

$$\frac{d\,x(t)}{dt} = f(x(t), u(t); \vartheta) \quad x(0) = x_0 \tag{18}$$

where

- $x(t)$ is the simulated state vector;
- x_0 is the initial state vector;
- $u(t)$ is the input vector;
- ϑ is the vector of the model parameters.

For identifying the parameters contained in the vector ϑ, sampled measurements are available:

$$y(t_k) = x(t_k) + \varepsilon(t_k) \tag{19}$$

which may be decomposed into the true state values $x(t_k)$ (corresponding to the solution of system (18) with the true ϑ) and a stochastic noise (generally assumed to be white, with Gaussian distribution of zero mean, see paragraph 4.2). In most of the cases of the literature, the vector ϑ is simply identified by minimising a least squares criterion (which is also based on the assumption of a constant standard deviation for the noise):

$$\hat{\vartheta} = \underset{\vartheta}{ArgMin} \sum_{k=1}^{N} \left(y(t_k) - x(t_k) \right)^T \left(y(t_k) - x(t_k) \right) \tag{20}$$

Given the model (18), the measurements (19) and the above assumptions on the measurement noise, the solution $\hat{\vartheta}$ is the one corresponding to the most likely errors $\varepsilon(t_k)$. But the measurements at the initial time are treated in a different way from all the other ones as the initial conditions of the simulation model are fixed to

$$x_0 = y(0) \tag{21}$$

Hence, the solution (20) is meaningful only if the measurement error $\varepsilon(0) = 0$. As there is no reason *a priori* to satisfy this condition, the only coherent procedure consists in identifying the initial conditions x_0 too:

$$\hat{\vartheta} = \underset{\{\vartheta, x_0\}}{ArgMin} \sum_{k=0}^{N} \left(y(t_k) - x(t_k) \right)^T \left(y(t_k) - x(t_k) \right) \tag{22}$$

We have proved in (Some *et al.*, 1999) the importance of the identification of the initial conditions in the framework of the estimation of drug stability parameters, namely activation energy and shelf-life of acetylsalicylic acid. Of course, this problem arises in any kind of simulation model.

Another frequent problem is that one measured signal is assumed to be corrupted by noise while the others are supposed to be perfectly known. This assumption is often made (but almost never explicitly) with linear regressions where all the regressors are supposed to be perfectly known. This problem is particularly important in the framework of bioprocess modelling as the pseudo-stoichiometric coefficients are usually identified on the basis of linear regressions (Bastin and Dochain, 1990; Chen and Bastin, 1996; Chotteau, 1995). However, it is unacceptable and completely unrealistic to assume that only one concentration measurement is corrupted by noise while the others are not. The way to overcome this problem is to use a maximum likelihood cost function for the identification instead of a least squares cost function (see paragraph 4.2).

Finally, several model identifications are only validated in "simple validation". This means that the final test consists in trying to simulate the data that have been used for the parameter identification. Of course this is just a necessary condition but not a sufficient one. Indeed, if this test fails then it proves that the structure is not able to reproduce the experimental field. But, the greater the number of parameters to be identified the greater the facility to reproduce the available measurements (together with their noise). If the number of parameters becomes too large, the model is then able to reproduce the specific experiments used for the identification rather than the macroscopic behaviour of the system. The "cross validation" is able to detect this kind of problem. It consists in trying to simulate experimental data which have not been used for the parameter identification, hence verifying the ability of the model to reproduce the system behaviour (or even to extrapolate the system behaviour, which means to reproduce experimental data outside the field delimited by the measurements used for the identification). This test is of crucial importance but is rarely made. Of course, other tests are able to provide valuable information about the model validation, see in Murray-Smith (1998) or in Walter and Pronzato (1997) for a more detailed discussion. Generally, any test allowing to verify *a posteriori* what has been assumed *a priori* in the identification procedure (especially concerning the noise properties) is really worthy.

In order to tackle the main problems highlighted above, a systematic procedure is proposed in the following paragraph.

4.2. SYSTEMATIC PROCEDURE FOR THE PARAMETER IDENTIFICATION

A three-step procedure has been proposed in Bogaerts (1999). The basics of this methodology are given hereafter.

4.2.1. First step: estimation of the pseudo-stoichiometric coefficients (independently of the kinetic coefficients)

This first step uses the decoupling method proposed in Bastin and Dochain (1990) and in Chen and Bastin (1996). It is always possible to find a full row rank submatrix $K_a \in \Re^{pxM}$ (where $p = rank\ K$) of a partition $K^T = \begin{bmatrix} K_a^T & K_b^T \end{bmatrix}$. Hence, there exists a unique solution $C \in \Re^{(N-p)\,x\,p}$ to the matrix equation

$$CK_a + K_b = O_{N-p,M} \qquad (23)$$

(Where $O_{N-p,M} \in \Re^{(N-p)\,x\,M}$ is a null matrix).

It is then possible to define an auxiliary vector

$$z = C\,\xi_a + \xi_b \qquad (24)$$

whose dynamics are independent of the reaction rates $\varphi(\xi)$:

$$\frac{d\,z(t)}{dt} = -\,D(t)\,z(t) + Cu_a + u_b \tag{25}$$

(where $u^T = \begin{bmatrix} u_a^T & u_b^T \end{bmatrix}$ corresponds to the partition $K^T = \begin{bmatrix} K_a^T & K_b^T \end{bmatrix}$). C can be estimated on the basis of relation (24) where $z(t)$ is obtained by integration of (25):

$$z(t) = \left(z(0) + \int_0^t \left(C\,u_a(\tau) + u_b(\tau) \right) \left(e^{\int_0^\tau D(\kappa)\,d\kappa} \right) d\tau \right) \left(e^{-\int_0^t D(\kappa)\,d\kappa} \right) \tag{26}$$

Considering the particular (but very usual case) where $p = rank\ K = M$, a necessary and sufficient condition in order to be able to univocally determine K_a and K_b from relation (23) and the knowledge of C, is that there exists a partition $K^T = \begin{bmatrix} K_{a'}^T & K_{b'}^T \end{bmatrix}$ (with $K_{a'} \in \Re^{MxM}$ invertible) such that $K_{a'}$ does not contain any unknown coefficient of K (Chen and Bastin, 1996). When this necessary and sufficient condition is satisfied, the model is called C-identifiable. Several solutions are proposed in Bogaerts (1999) in order to estimate the matrix K on the basis of this decoupling method. One of the solutions consists in using a maximum likelihood criterion allowing to take into account all the measurement errors (for each signal and each sample time, including the initial one) and is summarised hereafter.

Let us first inject the solution (26) into the relation (24), which leads to

$$\eta_b(t) = -\,C\,\eta_a(t) + z(0) \left(e^{-\int_0^t D(\kappa)\,d\kappa} \right) \tag{27}$$

where

$$\eta_a(t) = \xi_a(t) - \left(e^{-\int_0^t D(\kappa)\,d\kappa} \right) \int_0^t u_a(\tau) \left(e^{\int_0^\tau D(\kappa)\,d\kappa} \right) d\tau \tag{28}$$

$$\eta_b(t) = \xi_b(t) - \left(e^{-\int_0^t D(\kappa)\,d\kappa} \right) \int_0^t u_b(\tau) \left(e^{\int_0^\tau D(\kappa)\,d\kappa} \right) d\tau \tag{29}$$

Defining a vector ϑ_C containing all the unknown parameters of C together with all the initial conditions $z_s(0)$ ($s \in [1,S]$, indices of the experiment), its maximum likelihood estimation is given by

$$\hat{\vartheta}_C = \underset{\vartheta_C}{ArgMin} \frac{1}{2} \sum_{s=1}^{S} \sum_{k=1}^{N_s} \left(Y_{m,s,k} - \vartheta_s^T(\vartheta_C)\, \varphi_{m,s,k} \right)^T \left(Q_{Y,s,k} + \vartheta_s^T(\vartheta_C)\, Q_{\varphi,s,k}\, \vartheta_s(\vartheta_C) \right)^{-1} \left(Y_{m,s,k} - \vartheta_s^T(\vartheta_C)\, \varphi_{m,s,k} \right) \tag{30}$$

where

- $$Y_{s,k} = \vartheta_s^T(\vartheta_C)\, \varphi_{s,k} \tag{31}$$

- $$Y_{s,k} = \eta_{b,s}(t_{s,k}) \in \Re^{N-p} \tag{32}$$

- $$\varphi_{s,k}^T = \begin{bmatrix} -\eta_{a,s}^T(t_{s,k}) & e^{-\int_0^{t_{s,k}} D(\kappa)\, d\kappa} \end{bmatrix} \in \Re^{1 x (p+1)} \tag{33}$$

- $$\vartheta_s^T(\vartheta_C) = \begin{bmatrix} C^{(1,:)}(\vartheta_C) & z_s^{(1)}(0) \\ \vdots & \vdots \\ C^{(N-p,:)}(\vartheta_C) & z_s^{(N-p)}(0) \end{bmatrix} \in \Re^{(N-p)x(p+1)} \tag{34}$$

($t_{s,k}$ being the kth sample time of the sth experiment and $C^{(i,:)}$ being the ith row of matrix C)

- $$Y_{m,s,k} = Y_{s,k} + \varepsilon_{Y,s,k} \text{ (measured values)} \tag{35}$$

- $$\varphi_{m,s,k} = \varphi_{s,k} + \varepsilon_{\varphi,s,k} \text{ (measured values)} \tag{36}$$

- $$E[\varepsilon_{Y,s,k}] = 0 \tag{37}$$

- $$E[\varepsilon_{\varphi,s,k}] = 0 \tag{38}$$

- $$E[\varepsilon_{Y,s,k}\,\varepsilon_{Y,t,l}^{T}] = Q_{Y,s,k}\,\delta_{s,t}\,\delta_{k,l}, \tag{39}$$

- $$E[\varepsilon_{\varphi,s,k}\,\varepsilon_{\varphi,t,l}^{T}] = Q_{\varphi,s,k}\,\delta_{s,t}\,\delta_{k,l} \tag{40}$$

- $$E[\varepsilon_{Y,k}\,\varepsilon_{\varphi,l}^{T}] = 0 \quad \forall\, k,l \tag{41}$$

($\varepsilon_{Y,s,k}$ and $\varepsilon_{\varphi,s,k}$ being white measurement noises, normally distributed, with zero mean and covariance matrices $Q_{Y,s,k}$ and $Q_{\varphi,s,k}$). All the unknown parameters of ϑ_C being identified on the basis of (30), an estimation $\hat{C}$ of the matrix C is obtained. Finally, the estimations $\hat{K}_a$ and $\hat{K}_b$ of the matrices K_a and K_b are deduced of equation (23), the existence and uniqueness of the solution being guaranteed by the necessary and sufficient condition of C-identifiability mentioned above.

The signs of the pseudo stoichiometric coefficients in the matrix K are of course imposed by the fact that a given component is consumed in a given reaction (negative coefficient) or is produced in this reaction (positive coefficient). These sign constraints can be verified *a posteriori* at the end of the (unconstrained) identification procedure proposed above. In some particular cases, the sign constraints on the elements of K_a and K_b can be translated in sign constraints on the elements of C (and, consequently, on ϑ_C) thanks to the relation (23). Then the optimisation problem (30) can be solved under these constraints. Finally, there are some cases (especially when the elements of C are nonlinear functions of the elements of K_a and K_b) for which it is not possible anymore to derive the constraints on C from the ones on K_a and K_b. Then the matrix C may be parameterised in function of the elements of K_a and K_b and a nonlinear constrained optimisation problem has to be solved, as shown in Bogaerts (1999).

It has also been proved in Bogaerts (1999) that, provided an approximation of first order, the estimate $\hat{\vartheta}_C$ is unbiased:

$$E[\hat{\vartheta}_C] \approx 0 \tag{42}$$

and that an estimation of the covariance matrix of the parameter estimation errors is given by

$$\hat{E}[\tilde{\vartheta}_C\tilde{\vartheta}_C^T]\approx\left(\sum_{s=1}^{S}\sum_{k=1}^{N_S}\Theta_s(\hat{\vartheta}_C,\hat{\varphi}_{sk})\left(Q_{Ysk}+\vartheta_s^T(\hat{\vartheta}_C)Q_{\varphi,sk}\vartheta_s(\hat{\vartheta}_C)\right)^{-1}\Theta_s^T(\hat{\vartheta}_C,\hat{\varphi}_{sk})\right)^{-1} \tag{43}$$

where

$$\Theta_s(\hat{\vartheta}_C,\hat{\varphi}_{s,k})=\left[\left.\frac{\partial\vartheta_s^{(:,1)T}}{\partial\vartheta_C^T}\right|_{\vartheta_C=\hat{\vartheta}_C}\hat{\varphi}_{s,k}\quad\cdots\quad\left.\frac{\partial\vartheta_s^{(:,N-p)T}}{\partial\vartheta_C^T}\right|_{\vartheta_C=\hat{\vartheta}_C}\hat{\varphi}_{s,k}\right]$$
$$\in\Re^{\dim\vartheta_C x(N-p)} \tag{44}$$

($\vartheta_s^{(:,i)}$ being the ith column of the matrix ϑ_s and $\hat{\varphi}_{s,k}$ being the most likely estimates of the true values $\varphi_{s,k}$, given by

$$\hat{\varphi}_{s,k}=\varphi_{m,s,k}+Q_{\varphi,s,k}\vartheta_s(\hat{\vartheta}_C)$$
$$\left(Q_{Y,s,k}+\vartheta_s^T(\hat{\vartheta}_C)Q_{\varphi,s,k}\vartheta_s(\hat{\vartheta}_C)\right)^{-1}\left(Y_{m,s,k}-\vartheta_s^T(\hat{\vartheta}_C)\varphi_{m,s,k}\right) \tag{45}$$

Finally, note that the nonlinear optimisation problem (30) only guarantees a unique solution if the measurement noises are time-invariant (which means that $Q_{Y,s,k}=Q_Y, Q_{\varphi,s,k}=Q_\varphi, \forall\ s,k$). Nevertheless, it is possible to obtain a unique first initial guess of ϑ_C in a systematic way. It consists either in considering time invariant matrices for the measurement noise covariance or in reducing the problem to a Markov estimate where the covariance matrix $Q_{Y,s,k}$ is diagonal and the covariance matrix $Q_{\varphi,s,k}=O, \forall\ s,k$. Of course this simplification may only serve as unique initial guess as it relies on the assumption that all the measurements contained in $\varphi_{s,k}$ are not corrupted by noise. Although this assumption is most of the time definitely unacceptable, this kind of error is often made in the literature.

4.2.2. Second step: first estimation of the kinetic coefficients

It has been shown that the kinetic model structure (16) can be linearised w.r.t. its parameters thanks to a logarithmic transformation (17). This enables to find a linear least squares estimate of the kinetic coefficients (which necessarily exists, is unique and independent of any initial guess):

$$\hat{\vartheta}_{cin}^{(j)} = \underset{\vartheta_{cin}^{(j)}}{ArgMin} \frac{1}{2} \sum_{s=1}^{S} \sum_{k=1}^{N_s} \left(Y_{m,s,k}^{(j)} - \varphi_{s,k}^{(j)T} \vartheta_{cin}^{(j)} \right)^2 \quad j \in [1, M] \tag{46}$$

where

- $$Y_{s,k}^{(j)} = \ln \hat{\varphi}_j(t_{s,k}) \tag{47}$$

- $$\varphi_{s,k}^{(j)T} = \left[1 \quad \ln \xi_{h...}(t_{s,k}) \quad -\xi_{l...}(t_{s,k}) \right] \tag{48}$$

- $$\vartheta_{cin}^{(j)T} = \left[\ln \alpha_j \quad \gamma_{h...,j} \quad \beta_{l...,j} \right] \tag{49}$$

- and under the constraints
$$\left[\gamma_{h...,j} \quad \beta_{l...,j} \right] \geq 0 \tag{50}$$

Note that the constraints $\gamma_{h...,j} \geq 0$ must be replaced by $\gamma_{h...,j} \geq 1$ if the concentration positiveness must be guaranteed without using saturations with zero lower bound.
In the very usual case where $p = rank\ K = M$, estimates of the reaction rate $\hat{\varphi}_j(t_{s,k})$ can be obtained with the relation

$$\hat{\varphi}(\xi(t_{sk})) = \hat{K}_a^{-1} \left(\left(\frac{d\xi_a(t_{sk})}{dt} \right)^{\wedge} + D(t_{sk})\xi_a(t_{sk}) - F_a(t_{sk}) + Q_a(t_{sk}) \right) \tag{51}$$

where the estimate of the derivative can, for instance, be computed by the analytical derivation of an interpolation model for the vector $\xi_a(t)$. The estimates $\hat{\vartheta}_{cin}^{(j)}$ are based on unreliable assumptions on the measurement errors (errors only on $Y_{s,k}^{(j)}$, with constant standard deviation) and on estimates of the signal derivatives. Therefore, these estimates are just considered as a (unique and systematic) initial guess for the last step of the identification.

4.2.3. Third step: final estimation of the kinetic coefficients (and of some initial concentrations

At this step, the identified pseudo-stoichiometric coefficients (determined in the first step) will not be questioned anymore because they were already deduced from a most likelihood cost function using reliable assumptions. However, it has been shown that the estimate of the kinetic coefficients computed in the second step is based on unreliable assumptions and may only serve as initial guess of a final nonlinear identification that is the aim of this third step. Together with these kinetic coefficients, (part of) the initial concentrations of the simulation model will also be identified, in agreement with the discussion on the initial conditions of a simulation model given in the previous paragraph.

The simulation model {(5),(16)} consists of a nonlinear differential system of the form

$$\frac{d\,x(t)}{dt} = f(x(t), u(t); \vartheta) \tag{52}$$

where

- $$x^T(t) = \xi^T(t) = \begin{bmatrix} \xi_a^T(t) & \xi_b^T(t) \end{bmatrix} \tag{53}$$
  is the state vector containing the concentrations of the components involved in the reaction scheme (1);

- $$u^T(t) = \begin{bmatrix} D(t) & F_1(t) & \dots & F_N(t) \end{bmatrix} \tag{54}$$
  is the input vector containing the dilution rate and the external feed rates;

- $$\vartheta^T = \begin{bmatrix} \alpha_j & \gamma_{h\dots,j} & \beta_{l\dots,j} & \xi_a^T(0) \end{bmatrix} \quad j \in [1, M], h \in R_j^*, l \in P_j^*$$
  (55)
  is the vector of the parameters to be identified (kinetic coefficients and initial concentrations);

- f is the model structure corresponding to relations {(5),(16)}.

Note that the vector ϑ only contains the initial concentrations $\xi_a(0)$, the other ones $\xi_b(0)$ being deduced from relation (27) which reduces, at time $t = 0$, to

$$\xi_b(0) = -C\,\xi_a(0) + z(0) \tag{56}$$

where C and $z(0)$ have been identified in the first step of the identification procedure. On the basis of this property, it is also possible to reduce (especially in the batch case where $D(t)=0\ \forall\ t$) the system of N differential equations (52) in a system of p (rank of matrix K) differential equations, relative to the ξ_a part of the state vector, and $N-p$ algebraic equations deduced from (27), relative to the ξ_b part of the state vector. Details are given in Bogaerts (1999).

Let

$$\xi(t)=g(t,u(t),x(0);\vartheta) \tag{57}$$

be the solution (generally obtained by numerical solving) of the differential system (52) starting from the initial concentrations $\xi(0)$. On the basis of sampled measurements

$$y_{m,s,k}=g(t_{s,k},u(t_{s,k}),x_s(0);\vartheta)+\varepsilon_{y,s,k} \tag{58}$$

($t_{s,k}$ being the kth sample time of the sth experiment) corrupted by white measurement noise $\varepsilon_{y,s,k}$, normally distributed with zero mean and covariance matrix $Q_{s,k}$, the maximum likelihood estimate of ϑ can then be deduced from a nonlinear Markov estimator

$$\hat{\vartheta}=\underset{\vartheta}{ArgMin}\,\frac{1}{2}\sum_{s=1}^{S}\sum_{k=1}^{N_S}\left(y_{m,s,k}-g(t_{s,k},u(t_{s,k}),x_s(0);\vartheta)\right)^T Q_{s,k}^{-1}\left(y_{m,s,k}-g(t_{s,k},u(t_{s,k}),x_s(0);\vartheta)\right) \tag{59}$$

under the constraints

$$\hat{\vartheta}\geq 0 \tag{60}$$

The initial guess of ϑ consists, on the one hand, of the first estimate of the kinetic parameters deduced from the second step of the procedure and, on the other hand, of the measurements of ξ_a at the initial time.

The covariance matrix of the parameter estimation errors can also be estimated in this last step (Bogaerts, 1999):

$$\hat{E}[\tilde{\vartheta}\tilde{\vartheta}^T]\approx\left(\sum_{s=1}^{S}\sum_{k=1}^{N_S}G_\vartheta^T(t_{s,k},u(t_{s,k}),x_s(0);\hat{\vartheta})Q_{s,k}^{-1}G_\vartheta(t_{s,k},u(t_{s,k}),x_s(0);\hat{\vartheta})\right)^{-1} \tag{61}$$

where

$$G_{\vartheta}(t_{s,k}, u(t_{s,k}), x_s(0); \hat{\vartheta}) = \left. \frac{\partial g(t_{s,k}, u(t_{s,k}), x_s(0); \vartheta)}{\partial \vartheta} \right|_{\vartheta=\hat{\vartheta}} \tag{62}$$

This Jacobian is obtained by solving (together with the simulation model (52)) the sensitivity equations

$$\frac{\partial}{\partial t} G_{\vartheta}(t,u(t),x(0);\vartheta) = \frac{\partial f(x,u(t);\vartheta)}{\partial x} G_{\vartheta}(t,u(t),x(0);\vartheta) + \frac{\partial f(x,u(t);\vartheta)}{\partial \vartheta} \tag{63}$$

with the initial condition

$$G_{\vartheta}(0, u(0), x(0); \vartheta) = \frac{\partial x(0)}{\partial \vartheta} = G_{\vartheta 0} \tag{64}$$

where $G_{\vartheta 0}$ is a matrix whose elements are all equalled to zero except the ones giving the partial derivative of the elements of $\xi_a(0) \in x(0)$ w.r.t. the corresponding elements of $\xi_a(0) \in \vartheta$, these partial derivatives being equalled to 1. The Jacobian $G_{\vartheta}(t_{s,k}, u(t_{s,k}), x_s(0); \hat{\vartheta})$ involved in relation (61) is thus obtained by evaluating the numerical solution $G_{\vartheta}(t, u(t), x(0); \vartheta)$ of the system {(52),(63)} for $t = t_{s,k}$ and $\vartheta = \hat{\vartheta}$.

At the end of this third step, all the parameters have been identified: the pseudo-stoichiometric coefficients in the first step (30) and the kinetic coefficients in the third step (59). Hence, the model is completely determined but has of course to be validated (cross validation, study of the correlation matrix of the parametric errors, etc., see in (Bogaerts, 1999)). Note that it is also possible to build confidence intervals for simulation trajectories obtained with the model identified with the above mentioned procedure (Bogaerts, 1999). This allows to quantify the uncertainty in the simulation results coming from the uncertainty on the identified parameters. Finally, the following paragraph provides necessary conditions for the validation of the reaction scheme (1) on the basis of the identified parameters.

4.3. NECESSARY CONDITIONS FOR REACTION SCHEME VALIDATION

The results provided by the three-step identification procedure can be used to determine necessary conditions for the reaction scheme validation. A first one is a "good" level of validation of the linear relations (27) which allow to identify the pseudo-stoichiometric coefficients. Of course this "good" level can be quantified in several ways (e.g., on the basis of linear regression coefficients). If the validation tests fail then it is obvious that the reaction scheme can not be used to reproduce the experiments. However, a good validation is absolutely not a sufficient condition of reaction scheme validation because several reaction schemes may lead to the same linear relations (27).

Another valuable information is provided by the eventual activation of the sign constraints on the activation coefficients $\gamma_{k,j}$ (i.e. $\hat{\gamma}_{k,j}=0$) coupled with the sign of the corresponding pseudo-stoichiometric coefficient $\nu_{k,j}$. Several cases may arise:

- If $\hat{\nu}_{k,j}<0$ and $\hat{\gamma}_{k,j}>0$ (non activated constraint), then the component ξ_k is consumed by the jth reaction and activates this latter. Hence, it consists of a simple reactant.
- However, if $\hat{\nu}_{k,j}<0$ and $\hat{\gamma}_{k,j}=0$ (activated constraint), then the component ξ_k is consumed by the jth reaction but does not activate this latter. This is not very meaningful and could lead to negative concentrations if these latter are not artificially saturated with a zero lower bound. Even if saturations are used, this case is not really acceptable from a physical point of view and highlights probably the narrow limits of the validation field. Consequently, it is preferable to modify such a reaction by taking this reactant away.
- If $\hat{\nu}_{k,j}>0$ and $\hat{\gamma}_{k,j}>0$ (non activated constraint), then the component ξ_k is produced by the jth reaction and activates this latter. Hence, it consists of an autocatalyst for the considered reaction.
- However, if $\hat{\nu}_{k,j}>0$ and $\hat{\gamma}_{k,j}=0$ (activated constraint), then the component ξ_k is produced by the jth reaction but does not activate this latter. This means that this component is a simple product and is not an autocatalyst. If this component corresponds to the biomass, then the considered reaction is out of sense because it should allow the simulation of spontaneous growth of living microorganisms (without any living cell present at the initial time of the experiment).
- If $\hat{\nu}_{k,j}=0$ and $\hat{\gamma}_{k,j}>0$ (non activated constraint), then the component ξ_k is neither consumed nor produced by the jth reaction but activates this latter. Hence, it consists of a catalyst for the considered reaction.
- However, if $\hat{\nu}_{k,j}=0$ and $\hat{\gamma}_{k,j}=0$ (activated constraint), then the component ξ_k is neither consumed nor produced by the jth reaction and does not activate this latter. It does not take place in the considered reaction and may be taken away from it.

In summary, the following propositions are necessary conditions for the reaction scheme validation, which can be verified on the basis of the parameter identification results:

- it is necessary to reach a "good" level of validation of the linear relations (27) which allow to identify the pseudo-stoichiometric coefficients;
- a component which is supposed to be a reactant, a catalyst or an autocatalyst in a given reaction must be characterised by a strictly positive activation coefficient in this reaction.

5. Application on simulated bacteria cultures

In order to test the flexibility of the new general kinetic model structure presented in Section 3, a simulator of batch bacterial cultures is built with the software MATLAB 5.2. The reaction scheme corresponds to the growth reaction

$$(-\nu_S)S \xrightarrow{+} X \tag{65}$$

where

- S is the substrate;
- X is the biomass;
- ν_S is a (negative) pseudo-stoichiometric coefficient;
- $\xrightarrow{+}$ denotes an autocatalytic reaction (X being the autocatalyst).

The mass balances of X and S are then given by

$$\begin{aligned} \frac{d\,X(t)}{d\,t} &= \varphi(S(t), X(t)) \\ \frac{d\,S(t)}{d\,t} &= \nu_S\, \varphi(S(t), X(t)) \end{aligned} \tag{66}$$

where the growth rate

$$\varphi(S(t), X(t)) = \mu(S(t), X(t))\, X(t) \tag{67}$$

is such that the specific growth rate $\mu(S(t), X(t))$ is described by one of the following well known model structures.

- Monod (limitation by S):

$$\mu(S) = \mu_{MAX} \frac{S}{K_m + S} \tag{68}$$

- Tessier (limitation by S):

$$\mu(S) = \mu_{MAX} \left(1 - exp\left\{ -\frac{S}{K_m} \right\} \right) \tag{69}$$

- Ming (limitation by *S*):

$$\mu(S)=\mu_{MAX}\frac{S^2}{K_I+S^2} \tag{70}$$

- Haldane (limitation and inhibition by *S*):

$$\mu(S)=\mu_{MAX}\frac{S}{K_m+S+S^2/K_i} \tag{71}$$

- Contois (limitation by *S* and inhibition by *X*):

$$\mu(S)=\mu_{MAX}\frac{S}{K_C\,X+S} \tag{72}$$

The following numerical values are used:

$$\nu_S=-0.5\,g\,(10^{11}\,X\,)^{-1}\left(=-0.5\,(g\ of\ S)\,(10^{11}\,bacteria\ X\,)^{-1}\right),$$

$$\mu_{MAX}=1{,}4\,h^{-1},$$

$$K_m=12\,g\,l^{-1},$$

$$K_I=36\,g^2\,l^{-2},$$

$$K_i=3\,g\,l^{-1}$$

and $K_C=3\,g\,(10^{11}\,X\,)^{-1}$.

Note that these values of μ_{MAX} and K_m were used by Holmberg (1983) in order to model a culture of bacteria *B. thuringiensis* in a study on the identifiability problems with the Monod law.

The simulator {(66),(67)} (together with one of the laws {(68),...,(72)}) has been used for generating, in each of the five cases of specific growth rate, two experiments with 11 discrete measurement samples of S and X (initial conditions: $C_X(0) = 1{,}4\,10^{11}\,l^{-1}$, $C_S(0) = 9\,g\,l^{-1}$ and $C_X(0) = 1{,}4\,10^{11}\,l^{-1}$, $C_S(0) = 12\,g\,l^{-1}$. For each of the five specific growth rate laws {(68),...,(72)}, the two experiments obtained with the simulator {(66),(67)} (and the corresponding measurements of S and X) are used within the systematic identification methodology (second and third step presented above in Section 4) in order to estimate the kinetic parameters within the general model structure (16) which becomes here:

$$\varphi(S,X) = \alpha\, S^{\gamma_S}\, X^{\gamma_X}\, e^{-\beta_S S}\, e^{-\beta_X X} \tag{73}$$

The aim being to test the flexibility of this structure in order to reproduce the behaviour obtained with 5 different kinetic laws, no noise is added on the "measurements" of X and S provided by the simulator {(66),(67)}. For the same reason, the first step of the identification procedure is not used here and the pseudo-stoichiometric coefficient ν_S is fixed at its "true" value in each case.

Concerning the identification of the kinetic coefficients (α, γ_S, γ_X, β_S and β_X) in relation (73), a linearisation of the structure is performed thanks to a logarithmic transformation:

$$\ln \hat{\varphi}(t_{s,k}) = \alpha + \ln S(t_{s,k})\,\gamma_S + \ln X(t_{s,k})\,\gamma_X - S(t_{s,k})\,\beta_S - X(t_{s,k})\,\beta_X \tag{74}$$

where

- $t_{s,k}$ is the sampling time corresponding to the sample $k \in [1,11]$ of the experiment $s \in [1,2]$;
- $\hat{\varphi}(t_{s,k})$ is an estimation of the reaction rate.

In order to build the estimations $\hat{\varphi}(t_{s,k})$, one has to note that, in the mass balances (66), the reaction rate φ corresponds directly to the derivative of the biomass concentration X w.r.t. time. Hence, an estimation of the reaction rate is given by an estimation of this derivative. Therefore, an Euler approximation is used:

$$\begin{aligned} \hat{\varphi}(t_{s,k}) &= \left(\frac{d\,X(t_{s,k})}{d\,t} \right)^{\wedge} = \frac{X(t_{s,k+1}) - X(t_{s,k})}{t_{s,k+1} - t_{s,k}} \quad \forall k \in [1, N_s - 1] \\ \hat{\varphi}(t_{s,N_S}) &= \left(\frac{d\,X(t_{s,N_S})}{d\,t} \right)^{\wedge} = \frac{X(t_{s,N_S}) - X(t_{s,N_S-1})}{t_{s,N_S} - t_{s,N_S-1}} \end{aligned} \tag{75}$$

Note that this procedure is very rough but is acceptable in this ideal case in which the "measurements" are supposed to be perfect, i.e. not corrupted by noise. The relations {(46),..., (50)} of the second step of the identification procedure become in this case:

$$\hat{\vartheta}_{cin} = \underset{\vartheta_{cin}}{ArgMin} \frac{1}{2} \sum_{s=1}^{2} \sum_{k=1}^{11} \left(Y_{m,s,k} - \varphi_{s,k}^{T} \vartheta_{cin} \right)^2 \tag{76}$$

where

- $$Y_{s,k} = \ln \hat{\varphi}(t_{s,k}) \tag{77}$$

- $$\varphi_{s,k}^{T} = \left[1 \quad \ln S(t_{s,k}) \quad \ln X(t_{s,k}) \quad -S(t_{s,k}) \quad -X(t_{s,k}) \right] \tag{78}$$

- $$\vartheta_{cin}^{T} = \left[\ln \alpha \quad \gamma_S \quad \gamma_X \quad \beta_S \quad \beta_X \right] \tag{79}$$

and under the constraints

$$\gamma_S \geq 0, \gamma_X \geq 0, \beta_S \geq 0, \beta_X \geq 0 \tag{80}$$

This optimisation problem leads to a unique solution, completely independent of any initial guess.

Due to the use of the approximations (75), it is necessary to use the third step of the identification procedure which is only based on the simulation model and on the available measurements. The relations {(52),...,(55)} reduce to:

$$\frac{d\,x(t)}{dt} = f(x(t); \vartheta) \tag{81}$$

where

- $$x^T(t) = \left[S(t) \quad X(t) \right] \tag{82}$$

- $$\vartheta^T = \ln \left[\alpha \quad \gamma_S \quad \gamma_X \quad \beta_S \quad \beta_X \right] \tag{83}$$

- f is the model structure corresponding to the relations $\{(66),(73)\}$.

The solution of this differential system (81), which is obtained numerically, is noted

$$x(t) = g(t, x(0); \vartheta) \tag{84}$$

In agreement with the relations {(59),(60)}, the measurements $y_{m,s,k}$ given by the simulator {(66),(67)} are compared with the numerical solution (84) evaluated at the sampled times $t_{s,k}$:

$$\begin{aligned}\hat{\vartheta} &= \underset{\vartheta}{ArgMin}\, J(\vartheta) \\ &= \underset{\vartheta}{ArgMin} \frac{1}{2} \sum_{s=1}^{2} \sum_{k=1}^{11} \left(y_{m,s,k} - g_s(\varphi_{s,k}, \vartheta)\right)^T \left(y_{m,s,k} - g_s(\varphi_{s,k}, \vartheta)\right)\end{aligned} \tag{85}$$

under the constraints

$$\hat{\vartheta} \geq 0 \tag{86}$$

This nonlinear optimisation problem is initialised with the result (76). The covariance matrix (61) is not computed given the assumption of non noisy measurements. The obtained results are presented in Table 1. One can see that the cost functions (maximum likelihood cost function $J(\vartheta)$ for the final identification (85) of the kinetic coefficients) are significantly decreased when applying the third step on the basis of the results of the second one (hence only using the nonlinear simulation model instead of the Euler approximation which is used in the second step). Results are presented in cross validation (Fig. 1 to 5) starting from the initial conditions $C_X(0) = 1,4\ 10^{11}\ l^{-1}$ and $C_S(0) = 15\ g\ l^{-1}$ (experiment which has not been used for parameter estimation). In these figures, the straight lines correspond to the simulation obtained with the new general kinetic model structure whereas the circles represent "measurements" of the simulator using the Monod-type law. Other simple and cross validation results are presented in Bogaerts (1999). These results are quite convincing and illustrate the flexibility of the general kinetic structure in the sense that it is able to reproduce by its own the behaviour of several different kinetic structures.

Table 1 Estimation of the kinetic coefficients (in each case, the first line corresponds to the second step results and the second line to the third step results of the methodology for parameter estimation)

Specific growth rate	α	γ_S	γ_X	β_S	β_X	$J(\vartheta)$
Monod	0,134	0,900	1,178	0	0,081	36,523
	0,107	0,834	1,182	0,026	0,025	0,076
Tessier	0,192	0,915	1,025	0	0,076	29,458
	0,132	0,851	1,012	0,016	0,005	0,047
Ming	0,069	1,690	0,686	0,032	0,005	21,129
	0,042	1,963	1,196	0,144	0,026	0,041
Haldane	0,038	0,850	1,609	0	0,097	111,35
	0,137	0,752	1,077	0,115	0,012	0,172
Contois	0,256	0,923	0,208	0	0,019	12,489
	0,186	0,840	0,637	0	0,046	0,237

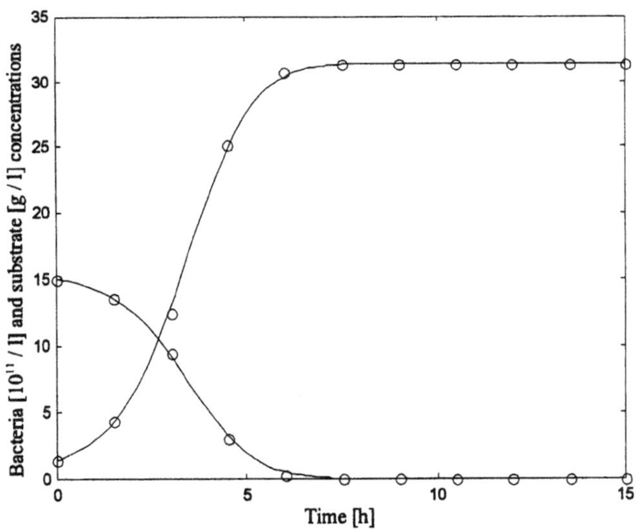

Figure 1. Cross validation of the simulation model (Monod data). The straight lines correspond to the simulation obtained with the new general kinetic model structure whereas the circles represent "measurements" of the simulator using the Monod-type law (Monod law in this case). Cross validation means that the simulation starts from initial conditions different from the measurements used for the estimation.

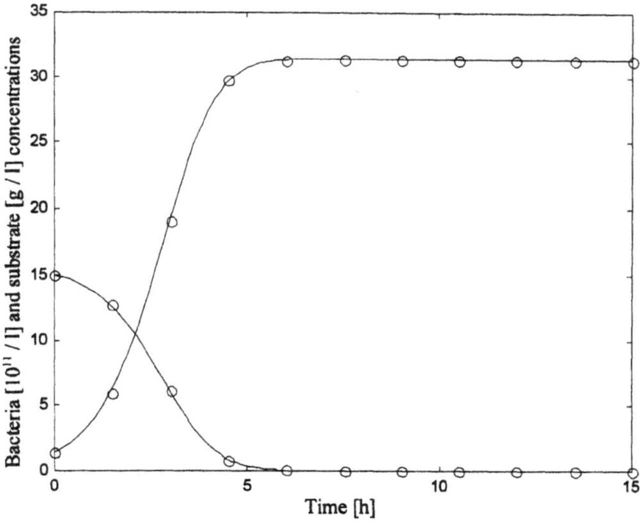

Figure 2. Cross validation of the simulation model (Tessier data)

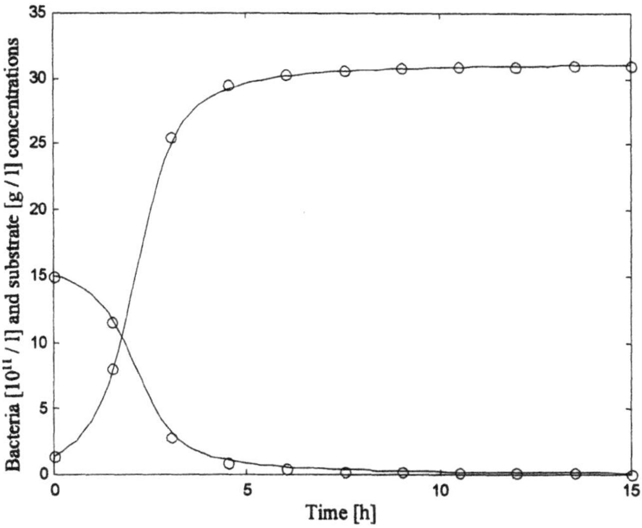

Figure 3. Cross validation of the simulation model (Ming data)

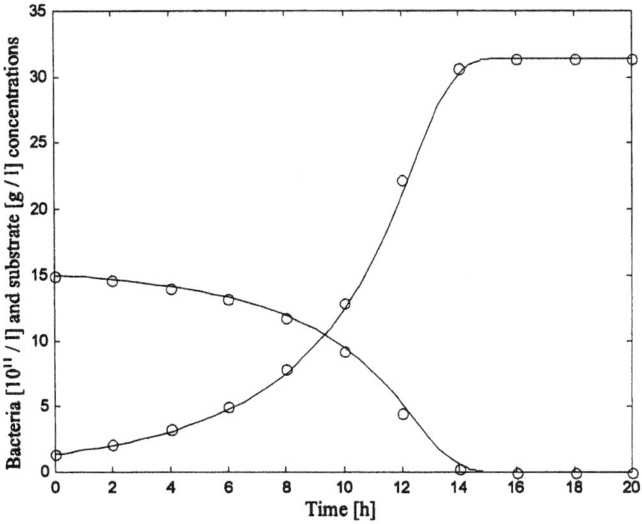

Figure 4. Cross validation of the simulation model (Haldane data)

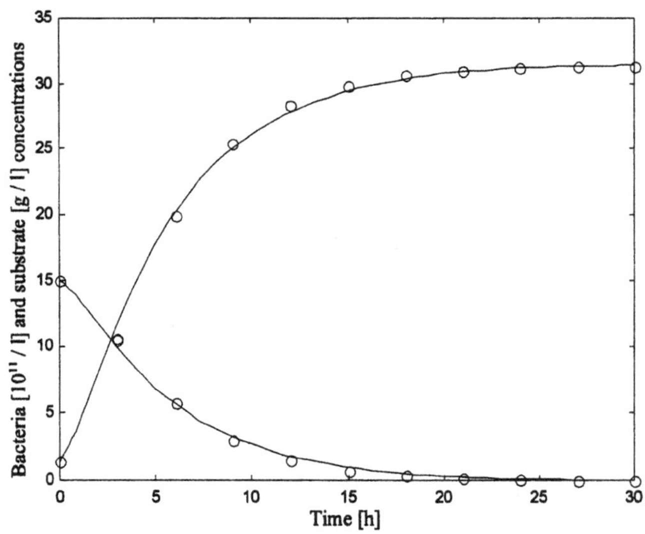

Figure 5. Cross validation of the simulation model (Contois data)

6. Conclusions and perspectives

Mathematical modelling is useful in order to build engineering tools like simulators, software sensors or controllers. For these aims, it is necessary to limit the complexity of the mathematical model structures. Although there exist some general frameworks for mathematical modelling of bioprocesses, they suffer of some significant drawbacks or limitations. In the same way, several drawbacks are present in the use of the usual members of the class of unstructured and unsegregated models, although the level of complexity is well appropriated regarding the modelling aims.

Therefore, a general kinetic model structure, overcoming these problems and exhibiting several interesting properties, is proposed. It has to be used within mass balances for the macroscopic species involved in a reaction scheme that describes the essential phenomena of the culture. Concerning the problem of the parameter identification (pseudo-stoichiometric coefficients and kinetic coefficients), there are also many problems and "lacks" that are encountered in the literature. Hence, a systematic three-step identification procedure has also been proposed. It takes into account the measurement errors (for each signal and at each sampling time, including the initial one) and gives estimation of the covariance for the errors on the identified parameters. Finally, necessary conditions for the reaction scheme validation can be tested on the basis of the identified parameters. The flexibility of the general kinetic model structure and a part of the identification methodology have been illustrated on simulated bacteria cultures.

A detailed illustration of the whole methodology in a real case study (CHO animal cell cultures in spinner flasks) can be found in Bogaerts (1999). Several other applications are currently in progress. In the same context of animal cell cultures, the use of this type of models for building software sensors (for instance for the biomass concentration) can be found in Bogaerts (1999) and in Bogaerts and Hanus (1999, 2000).

Based on the fact that a model has to be thought in terms of the modelling aim, the parameter estimation procedure (namely its third step) has been adapted for the case of models to be used in software sensors (Bogaerts and Vande Wouwer, 2000a, 2000b). A new identification cost function has been proposed, combining a classical maximum likelihood criterion and a scalar function of the state estimation sensitivity matrix. This latter quantifies the ability of the software sensor to reconstruct the state of the process on the basis of the available measurements.

As the macroscopic reaction scheme plays a key role in this mathematical modelling framework (and even in many other ones), the *a priori* determination of this scheme is very important and sometimes very tedious. Therefore, a method for the systematic generation of reaction networks (namely all the reaction networks which are *C*-identifiable) has been developed and will be published in the very near future.

Future research will focus on the problem of the *a posteriori* validation of the *a priori* assumptions on the measurement noises and on the way to handle this information. For instance, if the measurement noise is not really normally distributed some mathematical transformation of the measured signals could lead to new signals which would be normally distributed. Another perspective is to deal not only with the measurement errors but also with some structural errors of the model. Finally, the possibility to extend the use of the kinetic structure and of the parameter estimation method in the field of structured and/or segregated models will also be studied.

References

Barford, J. P., Philips, P. J. and Harbour, C. (1992) Simulation of animal cell metabolism, *Cytotechnology*, *10*, 63-74.

Bastin, G. and Dochain, D. (1990) *On-line estimation and adaptive control of bioreactors*, Elsevier, Amsterdam.

Bastin, G. and Van Impe, J. F. (1995) Nonlinear and adaptive control in biotechnology: a tutorial, *European Journal of Control*, *1*, 37-53.

Bogaerts, Ph. (1999) *Contribution à la modélisation mathématique pour la simulation et l'observation d'états des bioprocédés*, PhD thesis, Université Libre de Bruxelles (Belgium).

Bogaerts Ph. (1999b) A hybrid asymptotic - Kalman observer for bioprocesses, *Bioprocess Eng.*, *20 (No 3)*, 249-255.

Bogaerts, Ph., Castillo, J. and Hanus, R. (1999) A general mathematical modelling technique for bioprocesses in engineering applications, *System Analysis Modelling Simulation*, *35*, 87-113.

Bogaerts, Ph. and Hanus, R. (1999) On-line estimation of biomass concentration in CHO animal cell cultures, *Proceedings of the Third International Symposium on Mathematical Modelling and Simulation in Agricultural and Bio-Industries (M^2SABI'99)*, Uppsala, 97-102.

Bogaerts, Ph. and Hanus, R. (2000) Nonlinear and linearised full horizon state observers - Application to a bioprocess, *IFAC International Symposium on Advanced Control of Chemical Processes (ADCHEM 2000)*, Pisa, *1*, 371-376.

Bogaerts, Ph. and Vande Wouwer, A. (2000a) Parameter identification of a model to be used for state estimation - Application to a bioprocess, *Proceedings of the IEE Control 2000 Conference*, Cambridge, accepted for publication

Bogaerts, Ph. and Vande Wouwer, A. (2000b) , *Modellidentifikation zur zustandsschätzung - Anwendung auf einen bioprozess*, Automatisierungstechnik, *48 (No 5)*, pp. 240-247

Cazzador, L. and Mariani, L. (1993) Growth and production modelling in hybridoma continuous cultures, *Biotechnol. Bioeng.*, *42*, 1322-1330.

Chen, L. and Bastin, G. (1996) Structural identifiability of the yield coefficients in bioprocess models when the reaction rates are unknown, *Math. Biosci.*, 132, 35-67.

Chotteau, V. (1995) A general modelling methodology for animal cell cultures, PhD thesis, Université Catholique de Louvain (Belgium).

Delgado, J., Meruane, J. and Liao, J. C. (1993) Experimental determination of flux control distribution in biochemical systems: in vitro model to analyse transient metabolite concentrations, *Biotechnol. Bioeng.*, **41**, 1121-1128.

Dhurjati, P., Ramkrishna, D., Flickinger, M. C. and Tsao, G. T. (1985) A cybernetic view of microbial growth: modelling of cells as optimal strategists, *Biotechnol. Bioeng.*, 27, 1-9.

Gauthier, J. P., Hammouri, H. and Othman, S. (1992) A simple observer for nonlinear systems - Application to bioreactors, *IEEE Trans. Automat. Contr.*, 37, 875-880.

Glacken, M. W., Adema, E. and Sinskey, A. J. (1988) Mathematical descriptions of hybridoma culture kinetics: I. Initial metabolic rates, *Biotechnol. Bioeng.*, *32*, 491-506.

Goergen, J. L., Marc, A. and Engasser, J. M. (1994) Biochemistry of cells in culture: modelling, in: *Cells and tissue culture: laboratory procedures* (Doyle A. *et al.*, eds.), John Wiley and sons, Chichester, 8B:2.1-8B:2.19.

Holmberg, A. (1983) On the accuracy of estimating the parameters of models containing Michaelis-Menten type non-linearities, in: *Modelling and data analysis in biotechnology and medical engineering* (Vansteenkiste G. C. and Young P. C., eds.), North-Holland, Amsterdam, 202.

Julien, S., Babary, J. P., Lessard, P. and Paul, E. (1998) Theoretical and practical identifiability of an activated sludge process model, *Proceedings of the IMACS Multiconference on Computational Engineering in Systems Applications (CESA '98)*, Nabeul-Hammamet, *1*, 988-992.

Martens, D. E., Sipkema, E. M., de Gooijer, C. D., Beuvery, E. C. and Tramper, J. (1995) A combined cell-cycle and metabolic model for the growth of hybridoma cells in steady-state continuous culture, *Biotechnol. Bioeng.*, *48*, 49-65.

Montague, G. and Morris, J. (1994) Neural-network contributions in biotechnology, *Tibtech*, *12*, 312-323.

Murray-Smith, D. J. (1998) Methods for the external validation of continuous system simulation models: a review, *Mathematical and Computer Modelling of Dynamical Systems*, *4 (No 1)*, 5-31.

Nielsen L. K., Reid S. and Greenfield P. F. (1997) Cell cycle model to describe animal cell size variation and lag between cell number and biomass dynamics, *Biotechnol. Bioeng.*, *56*, 372-379.

Pirt, S. J. (1965) The maintenance energy of bacteria in growing cultures, *Proc. Soc. Lond. Ser. B*, 163, 224-231.

Ramakrishna, R., Ramkrishna, D. and Konopka, A. E. (1996) Cybernetic modelling of growth in mixed, substitutable substrate environments: preferential and simultaneous utilisation, *Biotechnol. Bioeng.*, *52*, 141-151.

Roels, A. (1983) *Energetics and kinetics in biotechnology*, Elsevier Biomedical Press, Amsterdam.

Ryckaert, V. G. and Van Impe, J. F. (1996) Design methodologies for nonlinear observers - Application to biological models, *Proceedings of the IMACS Multiconference on Computational Engineering in Systems Applications (CESA '96)*, Lille, 1, 1052-1057.

Savageau, M. A. (1969a) Biochemical systems analysis. I. Some mathematical properties of the rate law for the component enzymatic reactions, *J. Theor. Biol.*, *25*, 365-369.

Savageau, M. A. (1969b) Biochemical systems analysis. II. The steady-state solutions for an n-pool system using a power-law approximation, *J. Theor. Biol.*, *25*, 370-379.

Some, I., Bogaerts, Ph., Dubois, J., Hanus, R. and Hanocq, M. (1999) Nonlinear identification method for drug stability parameters estimation, *Int. Journal of Pharmaceutics*, 184 (n°2), 165-173.

Syu, M.-J. and Tsao, G. T. (1993) Neural network modelling of batch cell growth pattern, *Biotechnol. Bioeng.*, 42, 376-380.

Tan, Y., Wang, Z.-X. and Marshall, K. C. (1996) Modelling substrate inhibition of microbial growth, *Biotechnol. Bioeng.*, 52, 602-608.

Torres, N. V., Voit, E. O. and González-Alcón, C. (1996) Optimisation of nonlinear biotechnological processes with linear programming: application to citric acid production by *Aspergillus niger*, *Biotechnol. Bioeng.*, *49*, 247-258.

Torres N. V., Voit E. O., González-Alcón C. and Rodríguez F. (1997) An indirect optimisation method for biochemical systems: description of method and application to the maximisation of the rate of ethanol, glycerol, and carbohydrate production in *Saccharomyces cerevisiae*, *Biotechnol. Bioeng.*, *55*, 758-772.

Tsuchiya, H. M., Fredrickson, A. G. and Aris, R. (1966) Dynamics of microbial cell populations, *Adv. Chem. Eng.*, *6*, 125-206.

Tziampazis, E. and Sambanis, A. (1994) Modelling of cell culture processes, *Cytotechnology*, *14*, 191-204.

Van Impe, J. F. and Bastin, G. (1995) Optimal adaptive control of fed-batch fermentation processes, *Contr. Eng. Practice*, 3 (No 7), 939-954.

Walter, E. and Pronzato, L. (1997) *Identification of parametric models from experimental data*, Springer, Paris

Westerhoff, H. V. and Kell, D. B. (1987) Matrix method for determining steps most rate-limiting to metabolic fluxes in biotechnological processes, *Biotechnol. Bioeng.*, *30*, 101-107.

Xie, L. and Wang, D. I. C. (1996a) Material balance studies on animal cell metabolism using a stoichiometrically based reaction network, *Biotechnol. Bioeng.*, *52*, 579-590.

Xie, L. and Wang, D. I. C. (1996b) Energy metabolism and ATP balance in animal cell cultivation using a stoichiometrically based reaction network, *Biotechnol. Bioeng.*, *52*, 591-601.

Zeng, A.-P. and Deckwer, W.-D. (1995) A kinetic model for substrate and energy consumption of microbial growth under substrate-sufficient conditions, *Biotechnol. Prog.*, 11, 71-79.

A MODEL DISCRIMINATION APPROACH FOR DATA ANALYSIS AND EXPERIMENTAL DESIGN

R. TAKORS, D. WEUSTER-BOTZ*, W. WIECHERT, C. WANDREY**
Institute of Biotechnology, Research Centre Juelich, 52425 Juelich, Germany
**Chair of Biochemical Engineering, Munich University of Technology, Boltzmannstr. 15, 85748 Garching, Germany*
***IMR, Department of Simulation, University of Siegen, Paul-Bonatz Str. 9-11, 57068 Siegen, Germany*

Abstract

A general model discrimination approach is presented that enables data based model structure discrimination as well as model discriminating experimental design. Results of closed-loop controlled steady-state fermentations with the methylotrophic yeast *Candida boidinii* are used to clearly discriminate the "right" model out of a group of 10 competing models (53% model probability). Using the identified model the kinetics of batch and fed-batch fermentations with *Candida boidinii* could be described too. The applicability of the model discriminating experimental design approach is shown by simulation results using the kinetics of *Zymomonas mobilis.*

1. Introduction

Microbial fermentation kinetics like growth, substrate consumption and product formation are often described using unstructured kinetic models. These "simple" modelling approaches usually regard cells as "black-box" systems. Thus the kinetic cell behaviour is modelled by taking into account only a relatively few number of physiological state variables like cell-dry-weight, extracellular substrate or product concentrations and by using Monod-type modelling approaches. Due to their simplicity, these models could easily be used to quantitatively describe microbial fermentation courses, to implement model based fermentation control strategies or to simulate the scale-up of fermentation processes.

Unstructured models are usually identified by model parameter fit based on fermentation data that were derived from kinetic experiments. Batch experiments represent the simplest experimental method for the generation of kinetic data. However several authors have pointed out, that large model parameter error bars have to be

M. Hofman and P. Thonart (eds.), Engineering and Manufacturing for Biotechnology, 111–128.

considered (Holmberg, 1981; Nihilitä and Virkkunen, 1977) if these data were used for model fit. Thus, batch experimental design strategies have been investigated to overcome this problem (Yoo et al., 1986).

Beyond it, several authors tested new experimental design approaches for the identification of kinetic models based on fed-batch fermentations. Roels (1983) developed a concept of time varying feeding rates in fedbatch processes to estimate maintenance constants for growth. Focussed on baker's yeast, a special feeding strategy for model parameter estimation was presented by Ejiofor et al. (1994). Using the Fisher information matrix Munack (1989) developed a general methodology for the identification of Monod-type models by fed-batch experiments that was derived from foregoing investigations of Goodwin and Pain (1973). The use of the Fisher information matrix has been shown to be a successful tool for experimental design. Therefore it was tested by Baltes et al. (1994) for fed-batch experimental design, by Munack (1991) to develop optimum sampling strategies and by Schneider and Munack (1995) to estimate bio-process parameters on-line. Furthermore Van Impe et al. (1997) presented an E-optimal experimental design for fed-batch processes and Takors et al. (1997) developed a D-optimal experimental design considering closed-loop substrate controlled steady-state fermentations.

All design strategies have in common, that they are model specific. Only if the appropriate modelling approach is known, an optimal experimental design strategy could be identified. Unfortunately experimentalists are sometimes faced with the problem that principles of microbial kinetics are unknown before kinetic experiments have been carried out. For instance it could be unknown whether product formation is growth coupled or not or whether substrate and/or product inhibition occurs. Thence an alternative experimental design strategy is needed, that takes into account uncertainties of model structure. It should be the task of this approach to design experiments with the aim to clearly discriminate the "right" kinetic model among a group of competing models. This approach is called model discriminating design.

In 1995, Cooney and McDonald presented a test of two different model discrimination approaches, both principally based on a comparison of model responses of a set of four competing kinetic models. As a result, they identified the choice of the discriminating function to be extremely important and favoured the minimum difference between any two model responses for model discrimination.

This paper aims to present a more general approach for model discrimination and model discriminating experimental design. The methodology is based on the calculation of model probabilities derived from the formulation of model system entropy taking into account model parameter inaccuracies as well as model prediction uncertainties. The model discrimination approach is not limited by the number of competing models that could be considered.

The quality of the approach for "simple" model discrimination is shown using experimental data of steady-state *Candida boidinii* fermentations. Experiments with the methylotrophic yeast have been carried out using a closed-loop substrate control for methanol (nutristatic fermentation control) that enables steady-state growth even under substrate inhibiting conditions. It could be shown that the kinetic model that is derived from these steady-state experiments could also be used for the modelling of batch and fed-batch fermentations. Furthermore simulation results using the kinetics of the anaerobic bacterium *Zymomonas mobilis* are presented to show the quality of the approach for model discriminating experimental design.

2. Theoretical concept

For modelling of fermentations a system state vector $\mathbf{x}$ could be defined consisting of variables like cell-dry-weight, substrate and product concentration or liquid and gaseous streams. Using $\mathbf{x}$, a rate vector $\mathbf{f}(\mathbf{x},\boldsymbol{\theta})$ for the description of e.g. μ (spec. growth rate), σ (spec. substrate consumption rate) and π (spec. product formation rate) could be formulated including model parameters like K_S, K_i, μ_{max} etc. in $\boldsymbol{\theta}$

$$\mathbf{f}(\mathbf{x},\boldsymbol{\theta}) = \begin{pmatrix} \mu(\mathbf{x},\boldsymbol{\theta}) \\ \sigma_1(\mathbf{x},\boldsymbol{\theta}) \\ \sigma_2(\mathbf{x},\boldsymbol{\theta}) \\ \vdots \\ \pi_1(\mathbf{x},\boldsymbol{\theta}) \\ \pi_2(\mathbf{x},\boldsymbol{\theta}) \\ \vdots \end{pmatrix} \tag{1}$$

It is a basic characteristic of the model discriminating design approach that model discrimination is achieved sequentially. Based on n ("old") observations $\mathbf{y}_1, \ldots, \mathbf{y}_n$ additional experiments $n+1$, $n+2$, ... are suggested using the system state vector $\mathbf{x}$ as a design vector. Information of the last n fermentations is used for non-linear parameter regression to estimate $\hat{\boldsymbol{\theta}}_n$ and $\mathrm{cov}(\hat{\boldsymbol{\theta}}_n)$.

A set of competing models is defined including all modelling approaches that might be appropriate to describe microbial kinetics of the biological system. For instance, if substrate inhibition may occur, two different growth models (one with and one without substrate inhibition) form a set of competing models for model discriminating design. Model parameters of each model are calculated using the "old" observations.

A detailed description of the model discrimination approach is given by Takors (1997).

2.1. MODEL DISCRIMINATION

Assumed that observations $\mathbf{y}_1, \ldots, \mathbf{y}_n$ of n "old" fermentations are available, it is the task of the model discrimination approach to identify the most appropriate model considering a set of competing models. Thence model probabilities Π_{in} for the i th of m competing models are defined that could be calculated as follows:

$$\Pi_{in} = \frac{\Pi_{in-1} \cdot p_i}{\sum_{i=1}^{m} \Pi_{in-1} \cdot p_i} \tag{2}$$

This Bayes approach was first published by Box and Hill (Box and Hill, 1967) who introduced a model probability density function p_i. The function considers normally distributed measurement errors with constant measurement variance σ^2 and model predictions y_{in}^*

$$p_i = \frac{1}{\sqrt{2\pi\left(\sigma^2 + \sigma_i^{*2}\right)}} \exp\left(-\frac{\left(y_n - y_{in}^*\right)^2}{2\left(\sigma^2 + \sigma_i^{*2}\right)}\right) \tag{3}$$

Additionally a model variance σ_i^{*2} was estimated to include effects of measurement errors for model predictions into model probability calculation.

Often it is useful to consider variable measurement errors instead of constant values. Thus equation (3) could be extended to

$$p_i = \frac{1}{\sqrt{2\pi\left(\sigma_i^2 + \sigma_i^{*2}\right)}} \exp\left(-\frac{\left(y_n - y_{in}^*\right)^2}{2\left(\sigma_i^2 + \sigma_i^{*2}\right)}\right) \tag{4}$$

including the variable measurement variance σ_i^2.

Equation (2) represents a univariate model discrimination approach. However macrokinetic models usually consist of several equations e.g. for growth rate, substrate consumption and product formation. Hence a single model probability could be derived from a multivariate model consisting of k equations as following

$$\overline{\Pi}_{in} = \prod_{l=1}^{k} \Pi_{inl} \tag{5}$$

2.2. MODEL DISCRIMINATING DESIGN

2.2.1. Extended entropy approach

Based on the *a priori* model probability estimation (2), Box and Hill derived a model discriminating design criterion that had to be extended to fulfil biological constraints. In principal, experiments are proposed such that (model) system entropy is reduced which analogously causes an increase of information content. This leads to the maximisation functional

$$D=\sum_{i=1}^{m}\sum_{j=i+1}^{m}\Pi_{in}\Pi_{jn}\left(\int p_i \ln\frac{p_i}{p_j}dy_{n+1}+\int p_j \ln\frac{p_j}{p_i}dy_{n+1}\right) \tag{6}$$

including in pair's considerations of m model probabilities based on $n+1$ probability density functions.

Due to variable measurement errors of equation (4) the integration of equation (6) now leads to equation 7

$$D=\frac{1}{2}\sum_{i=1}^{m}\sum_{j=i+1}^{m}\Pi_{in}\Pi_{jn}\frac{\left(y^*_{in+1}-y^*_{jn+1}\right)^2\left(\sigma_i^2+\sigma_j^2+\sigma_i^{*2}+\sigma_j^{*2}\right)+\left(\left(\sigma_i^2-\sigma_j^2\right)+\left(\sigma_i^{*2}-\sigma_j^{*2}\right)\right)^2}{\left[\left(\sigma_i^2+\sigma_i^{*2}\right)\left(\sigma_j^2+\sigma_j^{*2}\right)\right]}$$

This equation can be transformed to the Box and Hill result assuming $\sigma_i^2=\sigma_j^2=\sigma^2$ (see Box and Hill, 1967).

To extend the originally univariate approach, the following summation functional is used to estimate the experimental design vector $\mathbf{x}^*$

$$\arg\min_{\mathbf{x}}\frac{1}{\sum_{l=1}^{k}D_l^2}=\mathbf{x}^* \tag{8}$$

Hence a multivariate model consisting of k model equations could be used for experimental design.

2.2.2. Model predictive design

It should be pointed out that the extended entropy approach demands the calculation of model variances. Thus model parameter regression must be carried out before the approach could be used. Therefore an alternative design strategy should be developed to overcome this start problem.

Assuming a set of start parameters $\hat{\boldsymbol{\theta}}_i$ model predictions $f_i\left(\mathbf{x}^*,\hat{\boldsymbol{\theta}}_i\right)$ could be calculated using a system state vector $\mathbf{x}^*$. It is the aim of model predictive design to estimate $\mathbf{x}^*$ such that maximum model prediction discrepancies Δf_{ij}^2 of competing models i and j become obvious

$$\Delta f_{ij}^2 = \left(f_i\left(\mathbf{x}^*, \hat{\boldsymbol{\theta}}_i\right) - f_j\left(\mathbf{x}^*, \hat{\boldsymbol{\theta}}_j\right)\right)^2. \tag{9}$$

Thus, if multivariate models with k model equations are considered an optimisation functional R could be defined as

$$\arg\min_{\mathbf{x}} \frac{1}{R} = \arg\min_{\mathbf{x}} \frac{1}{\sum_{l=1}^{k}\sum_{i=1}^{m}\sum_{j=i+1}^{m}\sum_{h=1}^{n+1} \frac{\Delta f_{ijl}^2(\mathbf{x}_h)}{\sigma_{lh}^2}} = \mathbf{x}_{n+1}^* \tag{10}$$

This approach only uses measurement variances σ_{lh}^2 considering n "old" fermentations.

3. Material and methods

3.1. FERMENTATION

Fermentations were carried out at pH 5, temperature 28°C and pressure 1bar using the methylotrophic yeast *Candida boidinii* (ATCC 32195). The yeast was cultivated under aerobic conditions with methanol as only carbon source and formate dehydrogenase (FDH) as internal product. Aerobic stirred reaction vessels with 7.5l (fermenter F1) and 20l (fermenter F2) total volume (Chemap, Switzerland) equipped with standard measuring and control units and an optimised mineral medium for maximum growth (Weuster-Botz and Wandrey, 1994) were used for fermentation. Medium was sterilised by microfiltration (pH-capsule 0.2M, Sartorius, Germany) without antifoam agent, which was autoclaved separately. For titration 4n NaOH was used. To receive on-line information about methanol concentrations a NDIR analyser (Rosemount, USA) was installed in the exhaust gas stream together with a paramagnetic oxygen detector (Rosemount, USA).

For steady-state fermentations *Candida boidinii* was cultivated in 4l reaction volume using a 7.5l fermenter F1. Batch and fed-batch cultivations were carried out in a 20l fermenter F2 using 12l start reaction volume. F1 and F2 were used to start batch and fed-batch fermentations. In F1 *Candida boidinii* was cultivated in steady-state culture with a dilution rate of 0.033 1/h. To start batch or fed-batch fermentations a defined sample was taken out of F1 via a harvesting tube and directly pumped into F2. Thus it was ensured that only "active" cells are used as inoculum for batch or fed-batch experiments. F2 was already filled with 12l cultivation medium. For fed-batch fermentations F2 was equipped with an additional methanol feed only including 16g/l methanol and desalted water.

3.2. ANALYTICAL METHODS

Cell homogenisation for off-line analysis of intracellular FDH is performed with a laboratory vibrator mill (Retsch, Germany) within 15 min using 1.2 g glass beads (diameter 0.5 mm) and 600 l cell sample. After centrifugation the enzyme activity was essayed spectrophotometrically at 340 nm. The 10 mm cuvettes were thermostated at 30°C. The assay mixture was composed of 2.0 ml sodium phosphate buffer (0.1M, pH 7.5), 0.5 ml 0.01M NAD, 0.1 ml sample and 0.5 ml 1.0 M sodium formate. Off-line analysis of methanol was carried out by gas chromatograph (Chrompack, Germany) with a fused silica capillary column. Dry cell mass was determined gravimetrically by use of 0.45 μm filters.

3.3. NUMERICAL AND PROGRAMMING TOOLS

All numerical calculations for experimental design, fermentation analysis and simulation are implemented into the C++ coded program *PARAGLIDE.* Matrix and vector calculations are facilitated by use of ROGUE WAVE libraries (Rogue Wave Software, Inc. Oregon, USA). For graphical applications STARVIEW libraries (STAR DIVISION, Hamburg, Germany) are taken. Numerical optimisations were carried out using the derivation-free Simplex/Nelder-Mead approach.

4. Results and discussion

4.1 MODEL DISCRIMINATION OF STEADY-STATE FERMENTATIONS

The model discrimination approach is tested using experimental results of steady-state fermentations with *Candida boidinii.* The methylotrophic yeast was cultivated with methanol as carbon source under aerobic conditions. Altogether 19 steady-state fermentations were carried out. These experiments were suggested by help of D-optimal design as well as by intuition (Takors et al, 1997). Kinetic results of growth rate μ, substrate consumption rates of methanol σ_{MeOH} and oxygen σ_{O2} and product formation rate of FDH π_{FDH} are presented in figures (1) and (2).

To model microbial kinetics 10 different unstructured approaches are used. They form a set of competing models as introduced in section 2. Derived from former research (Wedy, 1992) a start model (0) was already known:

Based on this model, different modelling approaches are created and put together in a set of competing models. For instance methanol and oxygen effects on growth rate are described by help of Monod-type or Andrews-type kinetic equations. Cell mass/substrate or FDH/cell mass yields are expected to be constant or growth rate dependent. Product formation is assumed to be completely or partially growth coupled. Altogether the number of model parameters varies from 7 to 11. An overview of the competing models is given in table (1).

Table (1): Set of 10 competing models. Abbreviations: mt – Monod-type, at – Andrews-type, wm – maintenance terms considered, no – maintenance terms not considered, gc – growth coupled product formation, pgc – partially growth coupled product formation, yc-constant yield, yv – variable yield, n – number of parameters.

	growth rate μ				substrate consumption σ				product formation π				
	methanol		Oxygen		methanol		oxygen		gc		pgc		
model	mt	at	mt	At	wm	no	wm	no	yc	yv	yc	yv	n
0		•	•		•		•			•			10
1	•		•		•		•			•			9
2		•		•	•		•			•			11
3		•	•		•		•					•	11
4		•	•			•	•			•			9
5		•	•		•			•		•			9
6		•	•			•		•		•			8
7		•	•		•		•		•				9
8		•	•			•		•	•				7
9		•	•		•		•				•		10

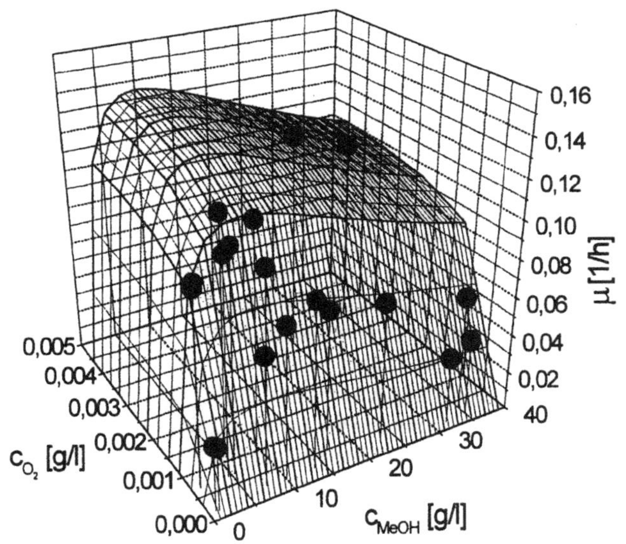

Figure (1): Steady-state experimental results using Candida boidinii – specific growth rate μ as a function of methanol (c_{MeOH}) and dissolved oxygen (c_{O2}) concentration

As indicated in figure (3) model (5) is clearly identified as the most suitable approach obtaining a model probability of 53.7%. This probability is more than double the probability of the second ranked model (0) and more than three times the probability of the third ranked modelling approach (3). Table (1) indicates that model (5) differs from model (0) by the formulation of oxygen/cell mass yield. While model (0) assumes a growth rate dependent oxygen/cell mass yield, model (5) simplifies this relation to a constant yield, which is appropriate to describe the experimental results. As a consequence the number of necessary model parameters is reduced to 9. The third ranked model (3) differs from model (5) by the assumption of a partially growth coupled FDH production instead of a completely growth coupled product formation. Figure (2) indicates that no growth uncoupled product formation could be observed experimentally. This corresponds with metabolic pathway constraints as the enzyme formate dehydrogenase is used in dissimilation for the final oxidation of formate to carbon dioxide.

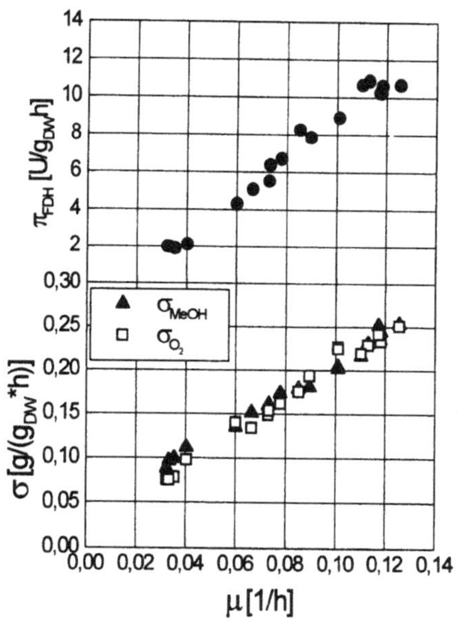

Figure (2): Steady-state experimental results using Candida boidinii – specific FDH formation rate π_{FDH} and specific substrate consumption rates (σ_{MeOH} and σ_{O2}) as functions of growth rate μ.

Summing up, it may be said that the model discrimination approach clearly identified a suitable macrokinetic model in agreement with former research results. A model selection only based on sum-of-squares analysis would not have been successful (see figures above columns in figure (3)). This approach would have lead to the identification of model (2) that assumes a growth inhibition caused by high dissolved oxygen concentrations. The inhibition was not experimentally observed.

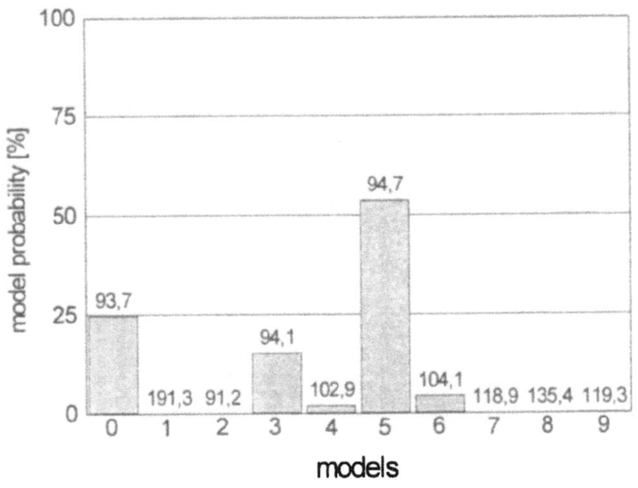

Figure (3): Model probabilities [%] and (weighted) sums of squares [-] for 10 competing models

4.2 BATCH AND FED-BATCH FERMENTATION MODELLING

To test the applicability of the identified model for the description of instationary fermentations, batch and fed-batch experiments were carried out. As pointed out in section 3.1 cell samples were taken out of steady-state conditions (fermenter F1, residence time 30h) to inoculate fermenter F2. As a consequence the inoculum undergoes strong environmental changes from methanol limitation in F1 to optimal growing conditions in F2. These changes result in an undefined lag-time that is needed by the cells to adapt to the new environmental situation. Thus the identified model must be extended to consider the lag-time for cell adaptation.

To prevent large model structure changes the identified model was simply extended with the following approach:

$$\mu = \mu_{\max} \frac{c_{MeOH}}{c_{MeOH} + K_{S,MeOH} + \dfrac{c_{MeOH}^2}{K_{I,MeOH}}} \frac{c_{O2}}{c_{O2} + K_{S,O2}} \left(1 - \exp\left(\frac{t}{t_{lag}}\right)\right) \qquad (18)$$

t_{lag} is defined as the time needed to achieve a 5% decrease of dissolve oxygen concentration in F2 after inoculation. Thus an easy determinable phenomenological parameter is used.

Using this extended model, batch and fed-batch fermentations were simulated based on start conditions that have been determined experimentally immediately after inoculation. Experimental results are presented in figures (4) and (5).

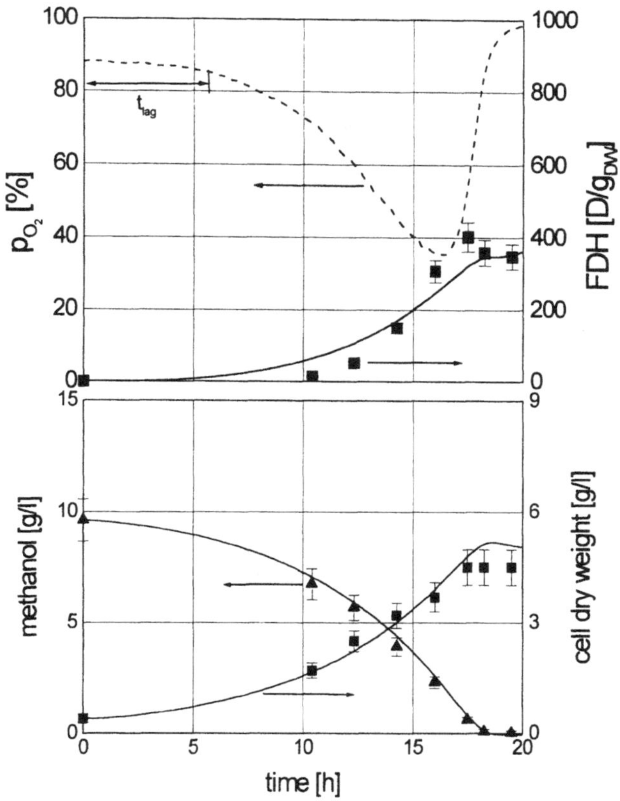

Figure (4): Dissolved oxygen (pO2), FDH, methanol and cell mass courses during batch fermentation with Candida boidinii. Simulation results are indicated with black lines.

As shown simulation results of methanol and cell-dry-weight are in good agreement with experimental results, if a 10% measurement error is assumed. As indicated in figure (5) two feeding phases consisting of 0.05 and 0.1 l/h with 16 g/l methanol were tested. In figures (4) and (5) experimentally determined lag-times (6 and 7h) are presented that were used for simulations. It is shown that especially at the beginning of product formation model predictions do not fit measured values as well as the simulation courses of cell mass and methanol concentration do. This could be a result of missing kinetic information of the preliminary experiments. Figure (2) indicates that no steady-state measurements are available for growth rates lower than 0.033 1/h.

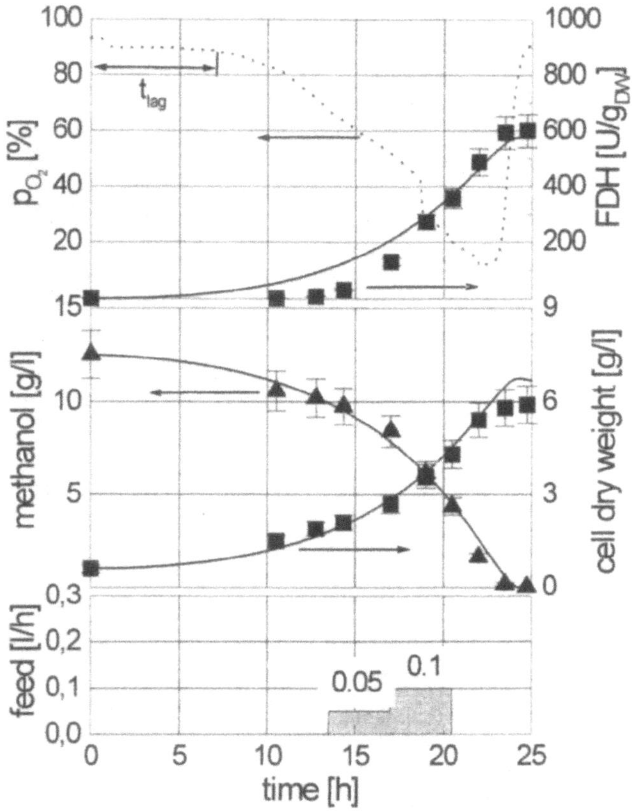

Figure (5): Dissolved oxygen (pO_2), FDH, methanol, cell mass and feeding courses during fed-batch fermentation with Candida boidinii. Simulation results are indicated with black lines.

However a generally good agreement between measured and simulated values could be stated. This allows the conclusion that (at least for *Candida boidinii*) kinetic models derived from steady-state experiments could also be used to describe non steady-state fermentations if the "right" model is identified by model discrimination and (if necessary) model extensions are considered.

4.3 MODEL DISCRIMINATING DESIGN WITH *ZYMOMONAS MOBILIS*

As pointed out in section 2 the model discriminating approach could also be applied for model discriminating design. As an example simulations with the anaerobic bacterium *Zymomonas mobilis* were carried out using the following model (and corresponding parameters) as a basis for microbial kinetic simulations (Weuster-Botz, 1993).

$$\mu = \mu_{max} \frac{c_{Gluc}}{c_{Gluc} + K_{S,Gluc} + \frac{c_{Gluk}^2}{K_{I,Gluc}}} \left(1 - \frac{c_{Gluc}}{P_{EtOH,max}}\right)^N - Y_{x,Gluc,max} \cdot m_{Gluc} \tag{19}$$

$$\sigma_{Gluc} = \frac{\mu}{Y_{x,Gluc}} \tag{20}$$

$$\pi_{EtOH} = \frac{\mu}{Y_{x,EtOH}} \tag{21}$$

$$Y_{x,Gluc} = \frac{\mu \cdot Y_{x,Gluc,max}}{Y_{x,Gluc,max} \cdot m_{Gluc} + \mu} \tag{22}$$

$$Y_{x,EtOH} = \frac{\mu \cdot Y_{x,EtOH,max}}{Y_{x,EtOH,max} \cdot m_{EtOH} + \mu} \tag{23}$$

Glucose is assumed to be the only carbon source and ethanol the fermentation product. High ethanol concentrations cause growth inhibitions as indicated by Levenspiel (1980). Measurements suggested by model discrimination design were simulated using this modelling approach in combination with balancing equations and a randomised "measurement" noise of ± 10%. Furthermore a set of competing models was defined consisting of different growth modelling approaches. Product inhibition was described using the approach of Jerusalimski and Engamberidiev (1969) and glucose effects were assumed to be saturating (Monod-type) or even growth inhibiting (Andrews-type). Table (2) gives an overview about the set of competing models.

Simulations were started with the assumption that no experimental information should be available and all models possess equal model probabilities. Thus at the beginning model predictive design was used, based on model start parameters as indicated in figure (6). After 7 simulated experiments model predictive design was replaced by model discrimination design. In total 12 experiments were simulated. They are presented in figure (7). Additionally an overview of model probability development is given in figure (8).

Table (2): Set of 6 competing (growth) models for Zymomonas mobilis. *Different modelling approaches for glucose and ethanol effects are presented.*

	glucose effect		ethanol effect	
model	Monod	Andrew	Levenspiel	Jerusalimski
1	•		•	
2		•	•	
3	•			•
4		•		•
5	•			
6		•		

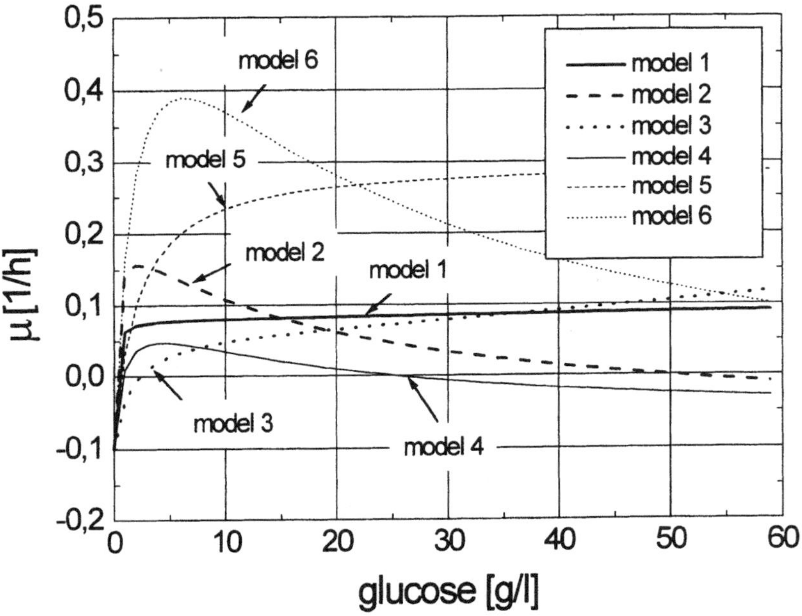

Figure (6): Start configuration of the set of 6 competing Zymomonas mobilis growth models.

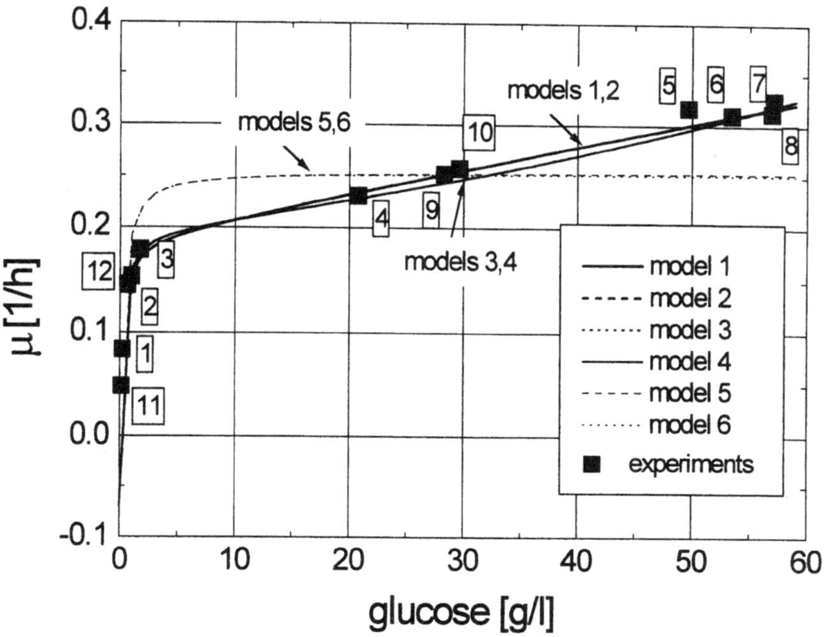

Figure (7): 12 simulated experiments with model discriminating design for 6 competing Zymomonas mobilis models.

After simulation of 7 experiments with model predictive design models (5) and (6) were identified as inappropriate modelling approaches (figure (8)). As a conclusion growth inhibition caused by ethanol must not be neglected for modelling. However no clear model discrimination was achieved with respect to models (1)-(4). This changed after the following three simulated experiments because a glucose inhibiting effect could not be detected ($K_{I,Gluc}$ of the reference model was 913 g/l). Thence model probabilities of models (2) and (4) decreased down to 0%. Simulations (11) and (12) were aimed to discriminate the right product inhibition modelling approach. As most significant model discrepancies are observable at low growth rates, corresponding experimental designs were suggested (figure (7)). Finally the appropriate model could be discriminated with approximately 100% model probability

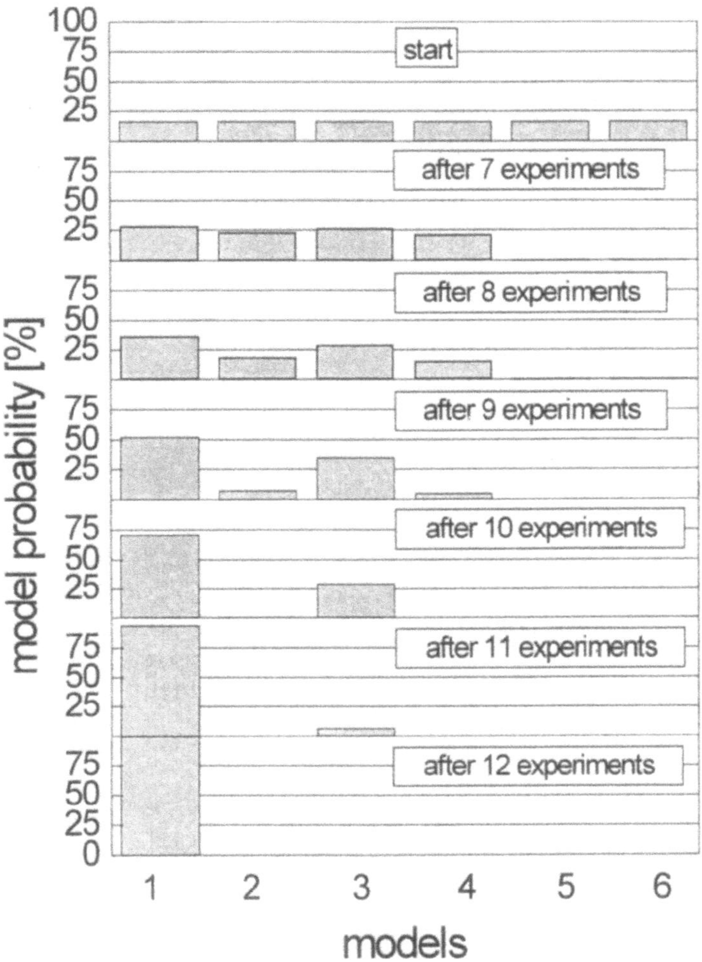

Figure (8): Probabilities of 6 competing Zymomonas mobilis models as a function of simulated experiments

5. Conclusions

A general model discriminating strategy is presented that enables model discrimination based on already available experimental data as well as a model discriminating design. The methodology is generally applicable without limitation concerning model complexity, the number of model parameters or the number of competing models. As a result model probabilities are calculated that represent an expressive value for the suitability of a model.

As model variances are used for calculation, model discrimination based on entropy formulations could only be used if a non-singularity of the Fisher-information matrix is prevented. Thus sufficient experimental information for model parameter regression must be available. As a consequence a simplified model design approach is proposed to overcome this start problem.

Results of experimental data analysis using *Candida* boidinii steady-state fermentations show that the entropy based discrimination approach is well suited to discriminate an appropriate model among a set of competing models. Simulation results with competing *Zymomonas mobilis* models proof that the extended model discriminating design strategy could be used even under poor information start conditions. Furthermore it has been shown that the *Candida boidinii* model which was discriminated by steady-state fermentation results could also be used to describe batch and fed-batch fermentations.

References

Baltes, M.; Schneider, R.; Sturm, C.; Reuss, M. (1994) Optimal Experimental Design for Parameter Estimation in Unstructured Growth Models, *Biotechnol. Prog. 10*, 480-488.

Box, G.E.P. and Hill, W.J. (1967) Discrimination among mechanistic models, *Technometrics 9*, 57-71.

Cooney, M.J. and McDonald, K.A. (1995) Optimal dynamic experiments for bioreactor model discrimination, *Appl. Microbiol. Bioetchnol. 43*, 826-837.

Ejiofor, A.O.; Posten C.H.; Solomon B.O.; Deckwer, W.-D. (1994) A robust fed-batch feeding strategy for optimal parameter estimation for baker's yeast production, *Bioprocess Eng. 11*, 135-144.

Goodwin, G.C. and Payne, R.L. (1973) Designs and characterisation of optimal test signals for linear SISO parameter estimation, *3rd IFAC Symposium on Identification and System Parameter Estimation*, The Hague.

Holmberg, A. (1981) On the practical identifiability of microbial growth models incorporating Michaelis-Menten type non-linearities, *Math. Biosci. 62*, 23-43.

Jerusalimski, N.D. and Engamberidiev, N.B. (1969) *Continuous Cultivation of Microorganisms*, Academic Press, New York.

Levenspiel, O. (1980) The Monod equation: A requisit and generalisation to product inhibition situations, *Biotechnol. Bioeng. 22*, 1671-1687.

Munack, A. (1989) Optimal feeding strategy for the identification of Monod-type models by fed-batch experiments. In: N.M. Fish (ed.) Reprints of the *4th International Congress on Computer Applications in Fermentation Technology*, Ellis Horwood Ltd., Chichester.

Munack, A. (1991) Optimisation of sampling, In: Rehm and Reed (eds.) *Biotechnology Vol. 4*, Verlag Chemie, Weinheim, Germany.

Nihilitä, M. and Virkkunen, J. (1977) Practical identifiability of growth and substrate consumption models, *Biotechnol. Bioeng. 19*, 1831-1850.

Roels, J. A.(1983) *Energetics and kinetics in biotechnology*, Elsevier Biomedical Press, Amsterdam.

Schneider, R. and Munack A. (1995) Improvements in the on-line parameter identification of bioprocesses, Preprints of *the 6th Conference on Computer Applications in Biotchnology*, Garmisch-Partenkirchen, Germany.

Takors, R. (1997) Entwicklung und Einsatz einer Versuchsplanungstechnik zur experimentellen Makrokinetikbestimmung, *Jül-Bericht 3420, Forschungszentrum Jülich GmbH, Jülich, Germany,*, ISSN 0944-2952.

Takors, R.; Wiechert, W.; Weuster-Botz, D. (1997) Experimental Design for the Identification of Macrokinetic Models and Model Discrimination, *Biotechnol. and Bioeng. 56*, 5, 564-576.

Van Impe, J.F.; Versyck, K.J.; Claes, J.E. (1997) Practical identification of unstructured growth kinetics by application of experimental design, *Biotechnol. Prog. 13*, 5, 524-531.

Wedy, M (1992) Reaktionskinetische und regelungstechnische Untersuchungen zur Produktion von Formiatdehydrogenase mit *Candida boidinii*, *Ph.D. thesis*, RWTH, Aachen, Germany.

Weuster-Botz, D. (1993) Continuous ethanol production of *Zymomonas mobilis* in a fluidised bed reactor. Part I Kinetic Studies of immobilisation in macroporous glass beads, *Appl. Microbiol. Biotechnol. 39*, 679-684.

Weuster-Botz, D and Wandrey, C. (1995) Medium optimisation by genetic algorithm for continuous production of formate dehydrogenase, *Process Biochemistry 30*, 563-571.

Yoo, Y.J.; Marino-Galarraga, M.; Hong, J.; Hatch, R.T. (1986) Experimental design for parameter estimation from batch culture, *Biotechnol. Bioeng. 28*, 6, 836-841.

MODEL BASED SEQUENTIAL EXPERIMENTAL DESIGN FOR BIOPROCESS OPTIMISATION - AN OVERVIEW

RALPH BERKHOLZ[1] AND REINHARD GUTHKE[2]
[1] BioControl Jena GmbH, 07745 Jena, Wildenbruchstr. 15, GERMANY
[2] Hans Knöll Institute for Natural Products Research, 07745 Jena, Beutenbergstr. 11, GERMANY

Summary

Model based experimental design for bioprocess optimisation requires transparent, understandable, identifiable models considering the physiological states necessary to obtain high product yields. Knowledge and data based hybrid modelling techniques are suitable to build such models. Two concepts of model based experimental design called direct and indirect experimental design are established. The direct design focuses on experiments being optimal with respect to the process performance but ignoring the relevance of the parameter estimation accuracy. Contrarily the indirect design leads to precise parameter estimates, but may result in unproductive fermentation runs worthless with respect to model validation. Due to the disadvantages of these two design methods the concept of Λ-optimal experimental design was developed. This approach enables experimentalists to suggest experimental set-ups optimal in productivity and parameter estimation accuracy.

1. Introduction

Biotechnological processes are very complex and often poorly understood. Generally the knowledge of the underlying biochemical phenomena is incomplete. Therefore it is impossible to optimise bioprocesses on the basis of theoretical assumptions, only. Experiments have to be performed. Due to the expensive bioprocesses it is necessary to design biotechnological experiments carefully with respect to their aim. For bioprocess optimisation a sequence of experiments has to be designed in order to achieve maximum productivity with minimum experimental effort.

In general the two basic approaches to handle this problem are the empirical experimental design and the model based experimental design. Empirical experimental design means the use of experience, intuition and process related analogies by skilled

M. Hofman and P. Thonart (eds.), Engineering and Manufacturing for Biotechnology, 129–141.

experimentalists to design experimental set-ups. Model based experimental design includes the use of mathematical models to obtain suitable suggestions for efficient experiments. Although it would be very interesting to analyse how skilled experimentalists often design nearly optimal experiments intuitively, in this overview the model based experimental design is considered only.

For model based experimental design the assumption is fundamental that the behaviour of a certain mathematical process model is similar to the process that should be optimised. A model fulfilling this assumption can be used for calculations in order to simulate the real process and to study the influence of different process strategies on the process numerically. So the model acts as a substitute for reality.

Experimental design is always a problem of optimisation concerning the convenient choice of the experimental conditions such as process regime and measurement set-up in order to maximise the efficiency of experiments with respect to the problem to be solved. For model based experimental design the optimisation of the experimental conditions will be carried out by simulations.

It is necessary to formulate a related objective function for the calculation of the efficiency of an experiment with regard to the experimental aim. During the optimisation procedure different experimental conditions will be evaluated according to the resulting value of these objective function. The experimental set-up leading to the maximum (or minimum) value of the objective function represents the optimal experimental design.

Actually there are two basic concepts established to formulate objective functions for the model based experimental design for bioprocess optimisation. The first one called direct experimental design method refers to the process productivity. The second one named indirect experimental design method focuses on the parameter estimation accuracy of the process model used. Both experimental design methods will be discussed later.

This review is organised as follows. Section 2 explains the demands on models concerning their application to experimental design and how to construct these models using knowledge and data based techniques. The two established methods of model based experimental design for bioprocess optimisation (i.e. direct and indirect experimental design method) will be discussed in the sections 3 and 4. A novel experimental design method for bioprocess optimisation called Λ optimal experimental design will be presented in section 5 as a conclusion of the analysis of the advantages and disadvantages of these two concepts. A short experimental example of the application of these design method will be shown in section 6.

1. Bioprocess modelling for experimental design procedures

1.1. BIOPROCESS MODELLING

Usually models for the simulation of dynamic bioprocesses consist of systems of ordinary differential equations for the overall mass balances that include kinetic expressions describing the rate limiting biochemical reactions of the considered process:

$$\dot{\mathbf{x}} = \mathbf{f}(\mathbf{x}, t, \mathbf{u}, \mathbf{p}) \tag{1}$$

where **x** denotes the vector of state variables, **u** the vector of input variables and **p** the vector of kinetic parameters. Since the state variables are often not measurable directly it is necessary to calculate the vector of model output variables **y**:

$$\mathbf{y} = \mathbf{g}(\mathbf{x}(t, \mathbf{u}, \mathbf{p})) \tag{2}$$

Using equation (2) it is possible to compare the predicted model output **y** and the measured process output $\mathbf{y}^M$ in order to evaluate the model performance or to calculate unknown model parameters.

Whereas it is mostly easy to formulate the overall mass balances of a biotechnological system in many cases it is quite difficult to find appropriate mathematical expressions for the biochemical reaction rates. There are two principle approaches to solve that problem, the deterministic and the hybrid modelling. Both concepts are based on the assumption, that only one or few biochemical reaction steps are rate limiting for the whole process. The so called formal kinetic expressions are used for the description of the rates of these reactions in deterministic bioprocess models. Usually these expressions are simple nonlinear algebraic equations such as the well-known Monod equation. Hybrid bioprocess models contain Artificial Neural Networks and/or fuzzy submodels for the calculation of reaction rates.

Generally bioprocess models can be built for different purposes such as process control, reactor design or scale-up. For each of these objectives a model has to satisfy specific requirements. In the following the features of models suitable for experimental design procedures are discussed.

1.2. TRANSPARENCY

The fundamental idea of model based experimental design is the assumption, that the predicted behaviour of the model is similar to that of the real process. That means the model has to predict phenomena of the process that are relevant to the experimental design problem. Generally the admissibility of that assumption can be proved using statistical approaches. However in biotechnological experiments the number and the accuracy of measurement data are mostly insufficient for the application of these statistical hypothesis testing methods. Therefore, the similarity of model behaviour to the real process is assessable empirically only. Skilled experts have to check the biological meaning of the bioprocess model and the underlying hypotheses concerning the rate limiting reaction steps. For that reason the model must be transparent, understandable and explainable.

1.3. RESTRICTED VALIDITY OF BIOPROCESS MODELS

Bioprocess models used for experimental design procedures apply the concept of rate limiting reaction steps. A rate limiting step is determined by the actual and passed physiological states (i.e. biological memory). In general bioprocess models are not valid for all physiological states. Their validity is a restricted or local one. The bioprocess has to be studied and modelled for the relevant physiological state (or for a certain sequence of physiological states). Especially, the model to be applied for experimental design for the optimisation of fermentations has to be valid for the most productive process mode (i.e. the sequence of physiological states).

1.4. IDENTIFIABILITY

Usually unknown kinetic parameters are determined by nonlinear regression methods. It is often impossible to achieve an unambiguous parameter estimation due to the measurement noise, the small number of measurement data and a large number of model parameters. In that case different parameter values lead to nearly the same values of the identification criterion. Then the question arises which parameter values have to be used for the optimisation calculations. Different parameter values may result in different optimal process strategies. Therefore, models used for experimental design must be identifiable.

1.5. KNOWLEDGE AND DATA BASED HYBRID BIOPROCESS MODELLING

Various approaches of bioprocess modelling are established:

- general knowledge from textbooks, e.g. relational knowledge from metabolic pathways of glycolysis or catabolite repression;
- general uncertain knowledge, e.g. the yield coefficient is known to be about 0.5 g/g or smaller for glucose as the sole carbon source in mass balances of microbial biomass growth and glucose consumption;
- process specific knowledge from skilled experts (if available), e.g. dependence of the reaction rates on the process phases;
- process specific knowledge hidden in the measured archived or actual data (if available), i.e. the dependence of reaction rates on environmental conditions (e.g. glucose concentration or pH).

These different kinds of knowledge have to be acquired from literature or experts or have to be discovered from data. They may managed in so called expert systems (e.g. Gensym's G2) or merged into one complex process model. For this merge of different kinds of knowledge those hybrid models are favoured which combine deterministic, fuzzy and data based (e.g. artificial neural network based) modules.

Fuzzy logic (Zadeh, 1965) is a convenient tool to handle uncertainties. Therefore it can be useful to build a hybrid bioprocess model consisting of a system of differential equations for the known mass balances including a fuzzy submodel describing the uncertain kinetic phenomena qualitatively or linguistically in the form of fuzzy rules. The transformation of this qualitative knowledge into quantitative knowledge is carried out by tuning the membership functions of the fuzzy submodel.

This consideration leads us to the question, how to built such kinetic fuzzy submodels. Generally there are two basic approaches to solve that problem, namely manually by an interview of a skilled expert (knowledge acquisition) or automatically by data exploration (data mining as a part of knowledge discovery from data). The success of an expert interview depends on the ability and willingness of the expert to reveal his knowledge. These conditions are not always fulfilled. Therefore great efforts were taken in the last years to develop methods for the extraction of knowledge from stored process data (Guthke and Roßmann, 1991; Guthke, 1992; Guthke and Ludwig, 1994; Guthke *et al*, 1998). This approach - often called data mining - includes two steps, the feature selection combined with fuzzy clustering methods and the fuzzy rule generation. The extracted fuzzy rules may be considered as hypotheses to be evaluated in advance of incorporating them in hybrid bioprocess models.

An example for the fuzzy hybrid bioprocess modelling is given by Babuška *et al.* (1999). It describes the enzymatic penicillin-G conversion by a hybrid model including a fuzzy submodel. The authors showed the application of these approach to the experimental design for the optimisation of the fermentation of the enzyme hyaluronidase in recent own contributions (Berkholz *et al.*, 1999; Berkholz *et al.*, 2000a). In these papers the specific growth rate is described by a fuzzy submodel automatically data-derived using fuzzy-C-means clustering (Bezdek, 1981) and combinatory rule extraction (Guthke, 1992, Guthke and Ludwig, 1994).

Fuzzy hybrid models fulfil the demands on models used for experimental design procedures discussed in the sections above. They are transparent, explainable and evaluable with regard to their validity and relevance for the interesting productive process regions. The incorporated fuzzy submodels can be analysed with respect to their identifiability using sensitivity approaches. Unidentifiable fuzzy submodels have to be reduced stepwise until their output sensitivities become sufficient (Berkholz *et al.*, 2000a).

2. Direct experimental design method

Once an appropriate mathematical description of the considered bioprocess is found it can be applied to model based experimental design. This section refers to the so called direct experimental design method of bioprocess optimisation. This method is named direct method since the objective function is the same for both the optimisation of the experiment conditions and the process optimisation. That means the direct method leads to optimal experimental set-ups with respect to the productivity. Thus, this experimental design method is focussed directly on the primary aim of the experiments. A general expression for the objective function J_P for the direct experimental design is given as follows:

$$J_P(\mathbf{u},\mathbf{p}) = f(\mathbf{y}(\mathbf{u},\mathbf{p},t = t_E)) \tag{3}$$

According to equation (3) the value of J_P is calculated using the predicted model output $\mathbf{y}$ at the free or fixed end t_E of the process and is influenced by the model input $\mathbf{u}$ and the actual parameter estimation **p**. The optimal experimental set-up can be found by searching the optimal model input $\mathbf{u}_{opt}$, that leads to the maximisation of J_P:

$$\mathbf{u}_{opt} = \arg \max_{\mathbf{u}} J_P(\mathbf{y}(\mathbf{u}, \mathbf{p}, t_E)) \tag{4}$$

A detailed literature overview analysing different realisations of equation (3) is given by Schneider (1999).

Examples for the application of this experimental design procedure are discussed by Glassey *et al.* (1994) and Galvanauskas *et al.* (1998). Both contributions describe the experimental design for the optimisation of fermentation of recombinant *E. coli*. The biomass at the end of fermentation is the objective function for both the experimental design and the process performance. Glassey *et al.* use an Artificial Neural Network model whereas Galvanauskas *et al.* describe the bioprocess with a deterministic model.

The main advantage of the direct experimental design method is the ability to propose experiments carried out in the productive region of the considered process. So the model can be validated for the one physiological state or those necessary to reach high product yields.

Disadvantageously the parameter estimation accuracy is not considered by this design method. It is known, that important kinetic parameters are not identifiable using batch experiments (Nihtilä and Virkkunen, 1977; Holmberg, 1982; Holmberg, 1983). Versyck *et al.* (1997) have shown that a fedbatch experiment optimal in productivity may lead to unidentifiable model parameters. So the model itself or the realised experimental conditions might not allow a unique parameter estimation. In this case it is hardly to decide, which parameter set should be used for the optimisation calculations. Different parameter sets will generally lead to different optimisation results.

3. Indirect experimental design method

The indirect experimental design method for bioprocess optimisation focuses on experiments optimal in parameter estimation accuracy. This approach is called indirect method since the model parameters are determined as precise as necessary at first. After estimating the parameters the well adapted model will be used for the optimisation of productivity via simulation without performing further experiments.

The parameter estimation accuracy can be evaluated by calculating functionals of the Fisher information matrix **F**:

$$\mathbf{F}(\mathbf{u}, \mathbf{p}) = \sum_{i=1}^{N} \left[\mathbf{Y_P}(t_i)^T \mathbf{C_M}(t_i)^{-1} \mathbf{Y_P}(t_i) \right] \tag{5}$$

where $\mathbf{Y_P}$ denotes the model output sensitivity matrix, $\mathbf{C_M}$ the measurement error covariance matrix and N the number of measurement points. In the case of a linear regression problem the Fisher information matrix **F** is the inverse of the parameter estimation error covariance matrix $\mathbf{C_P}$, which is defined as

$$\mathbf{C_P} = \begin{pmatrix} \mathrm{var}(p_1) & \mathrm{cov}(p_1, p_2) & \cdots & \mathrm{cov}(p_1, p_m) \\ \mathrm{cov}(p_2, p_1) & \mathrm{var}(p_2) & \cdots & \mathrm{cov}(p_2, p_m) \\ \vdots & \vdots & \ddots & \vdots \\ \mathrm{cov}(p_m, p_1) & \mathrm{cov}(p_m, p_2) & \cdots & \mathrm{var}(p_m) \end{pmatrix} \tag{6}$$

where m denotes the dimension of the parameter vector $\mathbf{p}$ (Ljung, 1987). Therefore in a nonlinear case the Fisher information matrix $\mathbf{F}$ gives an upper bound for the precision of the parameter estimation:

$$\mathbf{C_P} \geq \mathbf{F}^{-1} \tag{7}$$

The model output sensitivity matrix $\mathbf{Y_P}$ is defined as:

$$\mathbf{Y_P}(t_i) = \frac{\partial \mathbf{y}(t_i)}{\partial \mathbf{p}} = \begin{pmatrix} \frac{\partial y_1(t_i)}{\partial p_1} & \cdots & \frac{\partial y_1(t_i)}{\partial p_m} \\ \vdots & \ddots & \vdots \\ \frac{\partial y_n(t_i)}{\partial p_1} & \cdots & \frac{\partial y_n(t_i)}{\partial p_m} \end{pmatrix} \tag{8}$$

where n denotes the dimension of the model output vector $\mathbf{y}$. $\mathbf{Y_P}$ may be obtained from equation (2) by differentiation:

$$\mathbf{Y_P}(t_i) = \frac{\partial \mathbf{g}(t_i)}{\partial \mathbf{p}} + \frac{\partial \mathbf{g}(t_i)}{\partial \mathbf{x}} \frac{\partial \mathbf{x}(t_i)}{\partial \mathbf{p}} = \frac{\partial \mathbf{g}(t_i)}{\partial \mathbf{p}} + \frac{\partial \mathbf{g}(t_i)}{\partial \mathbf{x}} \mathbf{X_P}(t_i) \tag{9}$$

where $\mathbf{X_P}$ regards to the state sensitivity matrix, that may be calculated by differentiation of equation (1):

$$\dot{\mathbf{X}}_\mathbf{P}(t_i) = \frac{\partial \mathbf{f}(t_i)}{\partial \mathbf{p}} + \frac{\partial \mathbf{f}(t_i)}{\partial \mathbf{x}} \frac{\partial \mathbf{x}(t_i)}{\partial \mathbf{p}} = \frac{\partial \mathbf{f}(t_i)}{\partial \mathbf{p}} + \frac{\partial \mathbf{f}(t_i)}{\partial \mathbf{x}} \mathbf{X_P}(t_i) \tag{10}$$

According to equation (6) the parameter estimation accuracy is high if the elements of $\mathbf{C_P}$ have small absolute values. Small variances var(p_i) mean that the estimated parameter values p_i are probably near to their true values. Small absolute values for the parameter covariances cov(p_i, p_j) indicate a low level of linear correlations between the different elements of the parameter set. Due to equation (7) the parameter estimation accuracy is high if the elements of $\mathbf{F}$ have great absolute values.

Experimental set-ups leading to precise parameter estimates are said to be informative. Their information content is high. Thus, the indirect experimental design is focused on the optimisation of the information content. According to equation (5) there are several possibilities to increase the information content of an experimental set-up (Munack, 1995):

- increase in number of measurements N,
- choice of convenient measurement points t_i,
- choice of convenient measurement signals $\mathbf{y}^{\mathbf{M}}$,
- choice of convenient input signals $\mathbf{u}$.

Due to its relevance for process optimisation the choice of convenient input signals $\mathbf{u}$ will be considered only. Then a general expression for the objective function for the indirect experimental design J_S may be formulated as follows:

$$J_S(\mathbf{u},\mathbf{p}) = f(\mathbf{F}(\mathbf{u},\mathbf{p})) \quad (11)$$

The optimal experimental set-up may be found by searching the optimal model input $\mathbf{u}_{opt}$ that leads to the maximisation of J_S:

$$\mathbf{u}_{opt} = \arg\max_{\mathbf{u}} J_S(\mathbf{F}(\mathbf{u},\mathbf{p})) \quad (12)$$

Several expressions for the evaluation of the information content were developed. Some of them are shown in Table 1.

In the last decades a lot of results were published concerning the application of this approach. Some of them will be discussed in the following. Munack (1985) presents the maximisation of information content of experiments carried out in a tower loop reactor by optimising the positions of several pO_2-sensors. Posten and Munack (1990) apply the indirect experimental design to the improved modelling of plant cell suspension cultures. Baltes *et al.* (1994) take into account that bioprocess models are often not valid under transient conditions. Therefore they developed an objective function for the indirect experimental design combining the information content and the degree of stationarity of the process. Takors *et al.* (1997) optimise the parameter estimation accuracy of experiments carried out in a nutristat reactor using D-optimal experimental design. Syddall *et al.* (1998) give an application to improve the parameter estimation of a Penicillin fermentation model.

The main advantage of the indirect experimental design for bioprocess optimisation is that the proposed experimental set-ups lead to unique parameter estimates. Therefore there is no doubt which parameter values should be used during simulation calculations. On the other hand these experiments may be insufficient with respect to productivity. In those cases the experiments will be carried out in process regions out of interest. Thus the resulting experimental data may be worthless with regard to the validation of the state dependent bioprocess model (s. section 2.3).

Table 1. Criteria for the evaluation of the information content that are convenient objective functions for the indirect experimental design for bioprocess optimisation. λ_{min} and λ_{max} are the minimum and maximum eigenvalues of Fisher information matrix

Criterion	Formula	Interpretation
A	$\max\left(\frac{1}{\mathrm{tr}(\mathbf{F}^{-1})}\right)$	minimise the mean variance
D	$\max(\det(\mathbf{F}))$	minimise the volume of the joint confidence ellipsoid
E	$\max(\lambda_{\min}(\mathbf{F}))$	minimise the longest axis of the joint confidence ellipsoid
modified E	$\max\left(\frac{\lambda_{\min}(\mathbf{F})}{\lambda_{\max}(\mathbf{F})}\right)$	minimise the difference between the longest and the shortest axis of the joint confidence ellipsoid

4. Λ-optimal experimental design method

Due to the disadvantages of the both established experimental design methods for bioprocess optimisation discussed above we have proposed a novel experimental design approach called Λ-optimal experimental design (Berkholz *et al.*, 1999). The choice of the Greek letter Λ standing on two feet symbolises the intention to take two objectives into account, namely the process productivity on the one and the parameter estimation accuracy on the other hand. So the Λ-optimal experimental design combines the concepts of direct and indirect experimental design. Therefore a general expression for the objective function for the Λ-optimal experimental design J_Λ may be formulated:

$$J_\Lambda(\mathbf{u},\mathbf{p}) = f(J_P(\mathbf{u},\mathbf{p}), J_S(\mathbf{u},\mathbf{p}),) \tag{13}$$

where J_P and J_S are the objective functions of the direct and the indirect experimental design method respectively. The optimal experimental set-up can be found by searching the optimal model input $\mathbf{u}_{opt}$ that leads to the maximisation of J_Λ:

$$\mathbf{u}_{opt} = \arg\max_{\mathbf{u}} J_\Lambda(\mathbf{u},\mathbf{p}) \tag{14}$$

There are several possibilities to solve this multi-objective optimisation problem. Here a weighted sum of J_P and J_S is applied:

$$J_\Lambda(\mathbf{u},\mathbf{p}) = \varpi \cdot J_P^*(\mathbf{u},\mathbf{p}) + (1-\varpi)\cdot J_S^*(\mathbf{u},\mathbf{p}) \tag{15}$$

The upper index (*) indicates the normalisation of both functionals J_P and J_S due to their different orders of magnitude. It is advisable to normalise J_P and J_S on the interval [0,1]. Doing so the weight factor ϖ is also an element of the interval [0,1] and can easily be selected with respect to the experimental progress. At the beginning of the bioprocess optimisation it is useful to choose a smaller weight. So the Λ-optimal experimental design procedure focuses mainly on the estimation accuracy of the unknown model parameters. Within further experiments the weight ϖ can be increased to set the priority more on the process performance.

The advantage of the Λ-optimal experimental design is the consideration of both the productivity and parameter estimation accuracy. Using this approach it is possible to design experiments allowing a validation of the process model for the interesting productive process region and leading to unique parameter estimates for simulation calculations. The disadvantage of the Λ-optimal experimental design method is the increased computational effort. A software tool using the MATLAB environment is available supporting the design of Λ-optimal experiments for the fermentation optimisation.

5. Experimental Example

In this section a short experimental example for the application of the Λ-optimal experimental design is given considering the design of a single experiment for the optimisation of the hyaluronidase fermentation. Details about the process modelling, the cultivation conditions and the description of the whole sequence of three experiments carried out during the process optimisation can be found in Berkholz *et al.* (2000b).

The objective function for the process performance J_P is expressed by the mass m_P of the product

$$J_P = m_P(t_E) = c_P(t_E) \cdot V(t_E) \tag{16}$$

at the end t_E =12 h of the fermentation process. The objective function for the parameter estimation accuracy J_S bases on the modified E-criterion:

$$J_S = \frac{\lambda_{\min}(\mathbf{F})}{\lambda_{\max}(\mathbf{F})} \tag{17}$$

The only input influencing both J_P and J_S is the substrate dosage rate F_S. In the current process optimisation state F_S is realised quite simple by a pulse-like dosage at the dosage time τ. Applying Λ-optimal design the optimal dosage time τ_{opt} is given by

$$\tau_{opt} = \arg\max_{\tau} J_\Lambda \tag{18}$$

where the objective function J_Λ combining the normalised values of J_P and J_S is:

$$J_{\Lambda} = \varpi \cdot \frac{J_P - \min\limits_{\tau} J_P}{\max\limits_{\tau} J_P - \min\limits_{\tau} J_P} + (1-\varpi) \cdot \frac{J_S - \min\limits_{\tau} J_S}{\max\limits_{\tau} J_S - \min\limits_{\tau} J_S} \tag{19}$$

In figure 1 the criteria J_P^*, J_S^* and J_Λ are shown as functions of the dosage time τ. It is recognisable that both J_P^* and J_S^* have their maximum at different dosage time points τ. The objective function J_Λ is calculated by setting the weight factor ϖ at 0.2. The corresponding optimal dosage time τ_{opt} is 9 h.

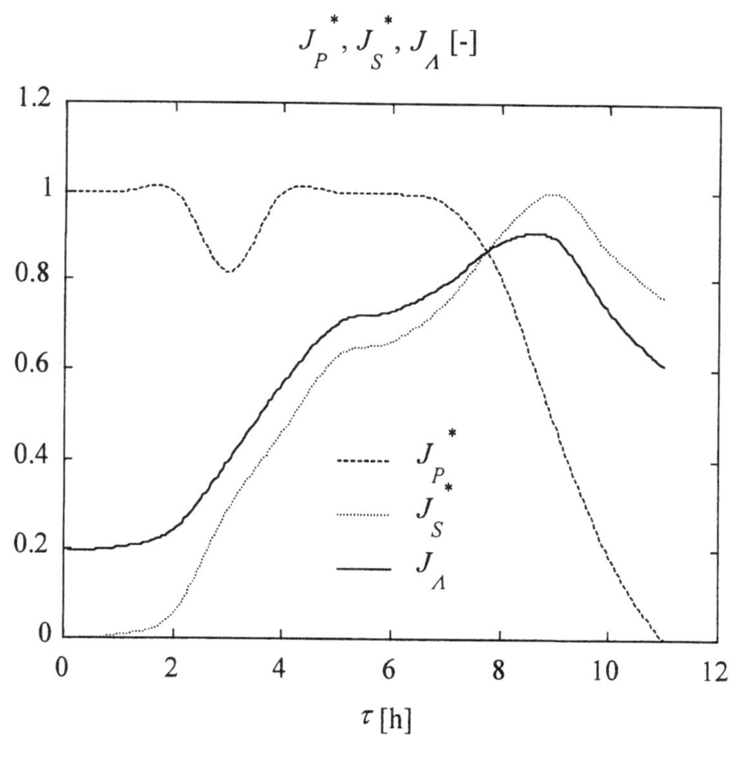

Fig 1. Criteria J_P^*, J_S^* and J_Λ with $\varpi = 0.2$ as functions of the dosage time τ.

In figure 2 the measured data of the designed Λ-optimal experiment and the kinetics of the process model are shown.

It can be seen, that the model fits the experimental data quite well. The process productivity reached during this experiment was about 60 % higher than before. The objectives of the Λ-optimal experimental design namely the experimental validation of the process model in the productive process region and the improving of the parameter estimation have been achieved.

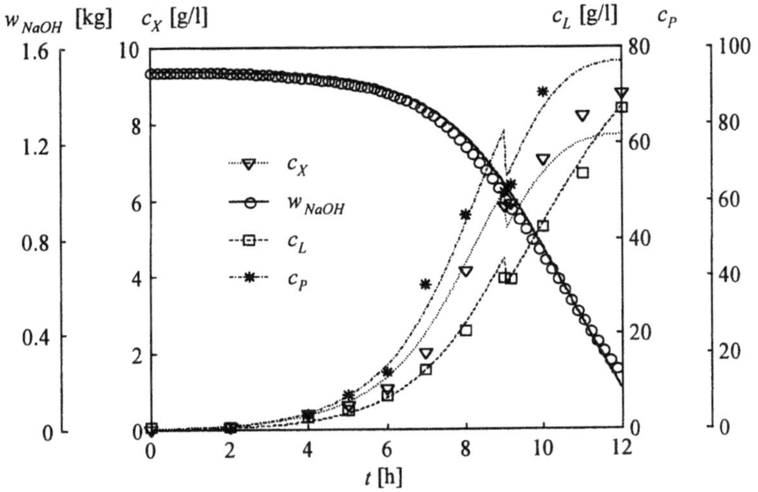

Fig 2. Measured data of the designed Λ-optimal experiment and the kinetics of the process model. c_X: concentration of biomass, w_{NaOH}: weight of alkali in the feeding reservoir, c_L: concentration of lactate, c_P: concentration of hyaluronidase. The substrate pulse was added at the optimal dosage time τ_{opt}= 9 h

References

Babuška, R., H.B. Verbruggen and H.J.L. Can, 1999, Fuzzy Modelling of Enzymatic Penicillin-G Conversion, Engineering Applications of Artificial Intelligence, 12, 59-78.

Baltes, M., R. Schneider, C. Sturm and M. Reuss, 1994, Optimal Experimental Design for Parameter Estimation in Unstructured Growth Models. Biotechnol. Prog. 10, 480-488.

Berkholz, R., R. Guthke and W. Schmidt-Heck, 1999, Sequentielle Versuchsplanung zur Optimierung von Bioprozessen mit Produktbildung im Übergangszustand, DECHEMA/GVC-Tagung „Wechselwirkung zwischen Biologie und Prozeßführung", Erfurt, 10.-11.5.1999.

Berkholz, R., R. Guthke, W. Schmidt-Heck and D. Röhlig, 2000a, Experimental Design for Bioprocess Optimisation: Numerical and Experimental Results, Int. Symp. Biotechnology, Berlin, 3.-8.9.2000.

Berkholz, R., R. Guthke and D. Röhlig, 2000b, Data and Knowledge Based Experimental Design for Bioprocess Optimisation, Enzyme and Microbial Technology, in press.

Bezdek, J., 1981, Pattern Recognition with Fuzzy Objective Function Algorithms, Plenum, New York.

Galvanauskas, R., R. Simutis, N. Volk and A. Lübbert, 1998, Model Based Design of a Biochemical Cultivation Process, Bioprocess Eng., 18, 227-234.

Glassey, J., G.A. Montague, A.C. Ward and B.V. Kara, 1994, Artificial Neural Network Based Experimental Design for Enhancing Fermentation Development, Biotechnol. Bioeng., 44, 397-405.

Guthke, R. and R. Roßmann, 1991, Fermentation Analysis by Clustering. Bioprocess Eng., 6, 157-161.

Guthke, R., 1992, Learning Rules from Fermentation Data, In: Karim, M.N. and G. Stephano-poulos (Eds.), Modelling and Control of Biotechnical Processes, Colorado, USA, 403-405.

Guthke R. and B. Ludwig, 1994, Generation of Rules for Expert Systems by Statistical Methods of Fermentation Data Analysis, Acta Biotechnol., 14, 13-26.

Guthke, R., W. Schmidt-Heck and M. Pfaff, 1998, Knowledge Acquisition and Knowledge Based Control in Bioprocess Engineering, J. Biotechnol., 65, 37-46.

Holmberg, A., 1982, On the Practical Identifiability of Microbial Growth Models Incorporating Michaelis-Menten Type Nonlinearities, Math. Biosci., 60, 23-43.
Holmberg, A., 1983, On Accuracy of Estimating the Parameters of Models Containing Michaelis-Menten Type Nonlinearities, In: Vansteenkiste, G.C. and P.C. Young (Eds.), Modelling and Data Analysis in Biotechnology and Medical Engineering, North-Holland Publishing Company.
Ljung, L., 1987, System Identification: Theory for the User, Prentice-Hall, Englewood Cliffs.
Munack, A., 1985, On the Parameter Identification for Complex Biotechnical Systems, Proc. 1st IFAC Symp. Modelling and Control of Biotechnological Processes, Noordwijkerhout, 159-165.
Munack, A., 1995, Simulation bioverfahrenstechnischer Prozesse, In: Schuler, H. (Eds.), Prozeßsimulation, VCH, Weinheim, New York, Basel, Cambridge.
Nihtilä, M. and J. Virkkunen, 1977, Practical Identifiability of Growth and Substrate Consumption Models, Biotechnol. Bioeng., 19, 1831-1850.
Posten, C. and A. Munack, 1990, Improved Modelling of Plant Cell Suspension Cultures by Optimum Experiment Design, Preprints of the 11th IFAC World Congress „Automatic Control in the Service of Mankind“, Tallinn, Estonia.
Schneider, R., 1999, Untersuchung eines adaptiven prädiktiven Regelungsverfahrens zur Optimierung von bioverfahrenstechnischen Prozessen, Dissertation, VDI-Verlag, Düsseldorf.
Syddall, M.T., G.C. Paul and C.A. Kent, 1998, Improving the Estimation of Parameters of Penicillin Fermentation Models, Preprints of the 7th International Conference on Computer Applications in Biotechnology, 23-28, Osaka, Japan.
Takors, R., W. Wiechert and D. Weuster-Botz, 1997, Experimental Design for the Identification of Macrokinetic Models and Model Discrimination, Biotechn. Bioeng., 56, 564-576.
Versyck, K.J., J.E. Claes and J.F. Van Impe, 1997, Practical Identification of Unstructured Growth Kinetics by Application of Optimal Experimental Design, Biotechnol. Prog., 13, 524-531.
Zadeh, L., A., 1965, Fuzzy Sets, Information and Control, 8, 338-353.

METABOLIC FLUX MODELLING AS A TOOL TO ANALYSE THE BEHAVIOR OF A GENETICALLY MODIFIED STRAIN OF *SACCHAROMYCES CEREVISIAE*

URRIETA-SALTIJERAL J.M., DUSSAP C.G., PONS A., CREULY C. AND GROS J.B.
Laboratoire de Génie Chimique et Biochimique, Université Blaise Pascal. 63177 Aubière cedex-FRANCE - Fax 33 4 73 40 78 29. dussap@gecbio.univ-bpclermont.fr

Abstract

Flux distribution for a wild and a mutant strain of *Saccharomyces cerevisiae* are compared and investigated in terms of metabolic flux calculation and thermodynamic analysis of central metabolism under anaerobic conditions. Starting from a redundant set of measured rates obtained from batch cultures on glucose or fructose as carbon source, an original data reconciliation technique associated with the calculation of metabolic flux is used. Comparative analysis of carbon split in the metabolic network for the mutant yeast strain lacking the glucose6P-dehydrogenase (CD101-1A) and for the reference wild strain (ATCC 7754) allows to conclude that the pentose phosphate is in priority devoted to its anabolic function rather than to the production of NADPH cofactors. This last function seems to be as well assumed by the specific NADP acetaldehyde dehydrogenase enzyme ; this explains the significantly higher production of acetate by the mutant strain.

1. Introduction

Cellular functions of a micro-organism are closely related to the environmental conditions (temperature, pH, nature and concentrations of substrates) and to the genetic particularities of the strain. Also, the chemical energy conversion systems are intimately linked to the macroscopic behaviour of the strain in a bioreactor. Providing that variations are not lethal, changes in the environment of the cell as well as its genetic material are followed by modifications, leading to observe changes of cellular functions, such as biomass composition and new metabolites synthesis. A useful tool to investigate these metabolic possibilities is the quantification of intracellular flux in the metabolic pathways. This methodology has been confirmed for investigating the

M. Hofman and P. Thonart (eds.), Engineering and Manufacturing for Biotechnology, 143–156.

catabolic, anabolic and the energy conversion pathways of a micro-organism giving an integrated view of the metabolism on the basis of intracellular metabolites conversion constraints (Vallino and Stephanopoulos, 1990 ; Nielsen and Villadsen, 1994 ; Pons *et al*, 1996 ; Lee and Papoutsakis, 1999).

During the last decade, metabolic engineering has been widely developed to propose a rational analysis of metabolic pathways and to help in deep understanding of cellular metabolism. Metabolic engineering associates genetic engineering and microbial physiology in a mathematical model with the aim to predict the rise of the yield of desired product as well as the decrease of an unwanted metabolite (Goel *et al*, 1999). Quantification of metabolic flux stands as an important aspect of these studies in providing a direct and accurate picture of the transformation of a particular substrate by a defined micro-organism (Vallino and Stephanopoulos, 1993). It is based on the proposal of a metabolic network including the main functions of the cell and taking into account the eventual genetic modifications. The step of validation of the stoichiometries stipulated in the network stands as a major point, in order to be used in the area of metabolic engineering.

However, until now, only few studies based on metabolic flux analysis are agreed through the use of experimental data (Vanrolleghem *et al*, 1996 ; Pramanik and Keasling, 1997). In majority, existing models are structured on the basis of *a priori* knowledge of the metabolism of the micro-organism under study with biochemical reactions and energetic parameters mainly established from literature data (Çalik *et al*, 1999). To validate the stipulated network, experimental strategy must at least include a study of the environmental culture conditions on the micro-organism behaviour.

An extended study including these aspects was performed with the yeast *Saccharomyces cerevisiae* , a reference strain and a genetically modified one being under study with various environmental conditions.

Saccharomyces cerevisiae is a versatile organism that is used for production of a whole range of different products. The physiology and the cell function strongly depend on the environment of the cells and on the genetic characteristics of the strains, leading to a great diversity of possible behaviours. An important part of research on *S. cerevisiae* physiology has been focused on its aerobic growth whereas anaerobic metabolism is generally considered to be simpler, leading to a *c.a.* equimolar production of carbon dioxide and ethanol. However in many yeast fermentation processes, especially in food industry, production of other by-products in much smaller amounts may become important, calling for a thorough understanding of yeast metabolism in anaerobic conditions. The tool used in the approach for evaluating the cell metabolism is the intracellular flux analysis based on the mass balance technique.

The present study concerns the following aspects :

- obtaining complete experimental sets concerning the carbon compounds excreted during anaerobic growth of two *Saccharomyces cerevisiae* strains (one ATCC 7754 strain and a genetically modified strain) cultivated on glucose or fructose in batch fermentor ;
- determination of specific growth rates and products yields ;
- selection of a consistent metabolic network from a flux computation technique using a data reconciliation method and an associated statistical analysis ;

- comparison of intracellular metabolic fluxes obtained in the 4 cases (2 strains, 2 substrates) in terms of metabolic bottlenecks and of distance from thermodynamic equilibrium.

2. Materials and methods

2.1. MICROORGANISMS AND GROWTH CONDITIONS

An ordinary baker yeast (ATCC 7754) and a glucose-6P dehydrogenase null mutant strain (CD101-1A: MAT alpha, his3, leu2, ura3, ade2, trp1, met19::URA3) were under study. The mutant strain was constructed and furnished to us by the Centre de Génétique Moléculaire d'Orsay, France (Thomas *et al*, 1991). Anaerobic batch cultures of the yeasts were done at 30°C in an automatically controlled fermentor Biostat ED (B.BRAUN, Germany) with a volume of 4 litres and a stirring speed of 500 rpm. The pH was regulated at 5 with 2N-NaOH addition. The mineral culture medium was prepared according to Kristiansen (1994): $(NH_4)_2SO_4$ (12 g l^{-1}), KH_2PO_4 (3.2 g l^{-1}), sodium glutamate (2 g l^{-1}), KCl (120 mg l^{-1}), $CaCl_2$ (60 mg l^{-1}), $MgCl_2$ (520 mg l^{-1}), $FeSO_4$ (35 mg l^{-1}), $MnSO_4$ (3.8 mg l^{-1}), $CuSO_4$ (0.5 mg l^{-1}) and supplemented with vitamins and trace metals. A mixture of adenine and amino acids (histidine, leucine, tryptophane and methionine) was added to the cultures of the mutant strain at initial concentration enabling final biomass concentration of *c.a.* 20 g DW l^{-1} (Oura, 1983). It is well established that addition of sterols and unsaturated fatty acids to the medium is required for optimal growth of *S. cerevisiae* under strictly anaerobic conditions (Andreasen and Stier, 1953, 1954). The carbon sources used were glucose or fructose with initial concentration in the reactor of 100 g l^{-1}.

2.2. ANALYSIS OF METABOLITES

Glucose, fructose, ethanol, glycerol, succinic acid, lactic acid, acetic acid and pyruvic acid were determined with High Performance Liquid Chromatography fitted with two ionic exclusion columns (PHENOMENEX Rezex ROA-300x7.8mm). Cell mass was measured by dry weight (24 h drying in an oven at 110°C). Carbon dioxide evolution was determined by integration of the gas flow rate measured by a mass flow-meter (Tylan, Mettler-Toledo). All the measurements were expressed as volumetric concentrations compensated for sampling, and dilution effects.

2.3. FLUX ESTIMATION AND STATISTICAL ANALYSIS

Metabolic fluxes were calculated from a mass balance technique. This method takes into account the stoichiometric reactions obtained by an analysis of the internal behaviour of the micro-organism. In the mass flux balance-based analysis, a pseudo-steady-state (PSS) approximation for the metabolic intermediates is assumed (Vallino and Stephanopoulos, 1990, Nielsen and Villadsen, 1994). The bioreaction network is used to determine the rates of production and consumption of each metabolite in the network ($R_{metabolite}$) as a function of all the unknown fluxes (J_i for the ith reaction). The accumulation rate of a metabolite in a metabolic network is given by the summation of all reactions producing that metabolite minus the reactions consuming it :

$$R_j(t) = \sum_m \alpha_{jm} J_m(t) - \sum_k \alpha_{jk} J_k(t) \tag{1}$$

where α_{jm} are the stoichiometric coefficients, $J_m(t)$ is the flux through reaction m and $R_j(t)$ is the accumulation rate of metabolite j.

The set of equations formed from such balances for each metabolite in the network is represented in matrix notation by :

$$\mathbf{AJ}(t) = \mathbf{R}(t) \tag{2}$$

A is the (c x r) matrix of stoichiometric coefficients (c metabolites reactions) of all the reactions involved in the metabolism. Each column in the matrix represents a metabolic reaction.

The c metabolic constituents are divided into two categories :

- the n exchangeable compounds, which are exchanged with the growth medium ; from a macroscopic point of view, these compounds are the products and the substrates involved during the cellular growth phase ; their accumulation rates are linked to growth yields ;
- the m non-exchangeable compounds, which are only involved inside the cell ; for these the accumulation rate is assumed to be equal to zero. This is the pseudo-steady state hypothesis that expresses that the rate of accumulation of intracellular compounds is negligible compared to the rates of production / consumption of exchangeable compounds (Vallino and Stephanopoulos, 1990).

Since metabolic networks generally contain more than 100 reactions, cyclic and/or parallel pathways may have been introduced (Fell, 1990). Such cycles have to be detected, then suppressed before the step of flux calculation. Cycle detection can be performed through a mathematical procedure based on the graph methodology (Veverka and Madron, 1997). Cycle elimination is obtained by supplying d new rows containing pertinent information founded on the metabolic capacities of the micro-organism under study.

Assuming that during a sufficiently long period of growth the yields remain constant (some of them being measured), and that non-exchangeable compounds are not accumulated inside the cell, the previous equation (2) is split in two parts : the first part $\mathbf{A}_1$ corresponds to the known accumulation rates $\mathbf{R}_1$ (non-exchangeable compounds, measured yields and the d rows relevant to cycle elimination), the second part $\mathbf{A}_2$ being the remaining rows corresponding to the unknown rates $\mathbf{R}_2$ (Pons *et al*, 1996).

The total number of supplied information must at least be equal to the number of columns *i.e.* of unknown fluxes **J**. In case of equality, the provided matrix $\mathbf{A}_1$ is of full rank, and fluxes are calculated by inversion of the matrix. In case of redundant information (more independent rows than columns) a data reconciliation technique is worked using the conservation laws of non-exchangeable compounds as constraints. To solve this problem of optimisation under constraints, we use a Lagrange method in minimising a quadratic criterion Φ calculated between the n' measured values of the yields contained in $\mathbf{R}_1$ and the computed values $\hat{\mathbf{R}}_1 = \mathbf{A}_1\,\mathbf{J}$ (Dussap *et al*, 1997) :

$$\Phi = {}^t(\mathbf{R}_1 - \mathbf{A}_1\,\mathbf{J})\,\mathbf{w}\,(\mathbf{R}_1 - \mathbf{A}_1\,\mathbf{J}) \qquad (3)$$

where $\mathbf{w}$ is the diagonal weighted matrix of the n' experimental values in $\mathbf{R}_1$ which accounts for the accuracy of the determination. In the case under study, variances σ^2 of the measured yields are known and elements of $\mathbf{w}$ are given by:

$$w_i = \frac{1}{\sigma_i^2} \qquad i = 1, n' \qquad (4)$$

The Lagrange function to be considered is given by :

$$L = \Phi - {}^t\Lambda\,\mathbf{A'}_1\,\mathbf{J} \qquad (5)$$

where $\mathbf{A'}_1$ is a matrix (m+d , r) and Λ is a vector (m+d , 1). $\mathbf{A'}_1$ contains all the rows corresponding to the m non-exchangeable compounds plus the d relationships between some of the fluxes necessary to eliminate to metabolic cycles included in the network (Dussap *et al*, 1997).

The r flux of $\mathbf{J}$ and (m+d) values of Λ vector are calculated by solving the system :

$$\frac{\partial L}{\partial \mathbf{J}} = -2\,{}^t\mathbf{A}_1\,\mathbf{w}\,(\mathbf{R}_1 - \mathbf{A}_1\,\mathbf{J}) - {}^t\mathbf{A'}_1\,\Lambda = 0 \qquad \text{(r relations)}$$

$$\mathbf{A'}_1\,\mathbf{J} = 0 \qquad \text{(m+d relations)}$$

The solution is given by :

$$\mathbf{J} = \mathbf{V}\left[\mathbf{G} - {}^t\mathbf{A'}_1\,\mathbf{H}\,\mathbf{A'}_1\,\mathbf{V}\,\mathbf{G}\right] \qquad (6)$$

where **V**, **H** and **G** matrices are defined as :

$$\mathbf{V}\,(r, r) = ({}^t\mathbf{A}_1\,\mathbf{w}\,\mathbf{A}_1)^{-1} \qquad (7)$$

$$\mathbf{G}\,(r, 1) = \mathbf{A}_1\,\mathbf{w}\,\mathbf{R}_1 \qquad (8)$$

$$H(m+d,n') = (\mathbf{A'}_1 \mathbf{V}\ {}^t\mathbf{A'}_1)^{-1} \qquad (9)$$

The re-calculated conversion rates values are given by $\hat{\mathbf{R}}_1 = \mathbf{A}_1\ \mathbf{J}$, where the (m+d) drawn values are found by construction. The standard deviations of the calculated flux and rates are estimated by diagonals elements of matrices of covariance :

$$\text{Co var}(\mathbf{J}) = \frac{\Phi}{m+d+n'-r}\mathbf{V} \qquad (10)$$

$$\text{Co var}(\hat{\mathbf{R}}) = \mathbf{A}\ \text{Covar}(\mathbf{J})\ {}'\mathbf{A} \qquad (11)$$

This set of equations gives the best estimates of all the yields and all of the fluxes. Statistical analysis also provides the standard deviations of the estimation (Equations 10, 11).

3. Results and discussion

3.1. GROWTH YIELDS DETERMINATION

The two strains produced in addition to biomass, carbon dioxide and ethanol, significant quantities of glycerol, lactate, acetate, pyruvate and succinate in various proportions depending on the strain and on the carbon substrate. No production of acetoin, acetaldehyde or other compounds specific of anaerobic metabolism have been detected.

The measured liquid phase concentrations have been plotted against total carbon substrate (glucose or fructose) consumption. The results obtained indicate that the data could be treated by linear regression (constant yields) over a period of 25 h corresponding to substrate consumption of 150 g l^{-1}. Specific growth rates have been calculated by semi-log plotting of biomass concentration versus culture time. The reproducibility of the experiments has also been checked by comparison of two sets of data obtained in similar conditions. The average values of carbon compounds yields and the maximum specific growth rates are reported in table 1. Importantly, the calculation also includes estimations of the standard deviations (Himmelblau, 1968). These results correspond to a global carbon and available electron recovery from 95 to 100 %.
Significant differences between the cultures were observed :

- higher growth rates of the ATCC 7754 strain ;
- higher growth rates on glucose than on fructose substrate for the ATCC 7754 strain;
- higher production of glycerol, acetate and succinate by the mutant strain ; a 10 fold value is observed for acetate ;
- less synthesis of biomass by the mutant strain ;
- small relative differences for the two main products carbon dioxide and ethanol.

Table 1 : Experimental values of conversion yields and maximum specific growth rate of S. cerevisiae *ATCC 7765 and* S. cerevisiae *CD101-1A cultivated on glucose or fructose in anaerobic conditions (standard deviations in brackets).*

	Glucose		Fructose	
Yield (g g^{-1})	**ATCC**	**CD101-1A**	**ATCC**	**CD101-1A**
Biomass x 100	9.0 (0.2)	5.6 (0.2)	7.5 (0.2)	5.1 (0.2)
Ethanol x 10	4.0 *(0.1)*	3.9 *(0.2)*	4.2 *(0.1)*	4.0 *(0.1)*
Carbon dioxide x 10	4.11 *(0.09)*	3.8 *(0.1)*	3.91 *(0.08)*	4.1 *(0.2)*
Glycerol x 100	6.4 *(0.2)*	9.4 *(0.4)*	6.5 *(0.2)*	10.8 *(0.3)*
Acetate x 1000	1.5 *(0.3)*	12.0 *(1)*	1.7 *(0.3)*	12.8 *(0.7)*
Lactate x 1000	4.0 *(0.3)*	1.1 *(0.3)*	3.7 *(0.3)*	2.3 *(0.2)*
Succinate x 1000	2.9 *(0.4)*	4.7 *(0.4)*	2.0 *(0.4)*	3.4 *(0.8)*
Pyruvate x 1000	2.9 *(0.4)*	2.9 *(0.7)*	2.6 *(0.8)*	2.2 *(0.6)*
μ_{max} (h^{-1})	0.28 *(0.01)*	0.075 *(0.007)*	0.20 *(0.02)*	0.091 *(0.006)*

3.2. SELECTION OF A RELIABLE METABOLIC NETWORK

Assuming the detailed composition of yeast cell from Oura (1983), the following global composition of biomass was established :

$$C\,H_{1.62}\,O_{0.52}\,N_{0.15}\,S_{0.0012}\,P_{0.0095} \tag{12}$$

According to the known synthesis pathways of amino acids, lipids, nucleic acids and carbohydrates, a metabolic network of 117 reactions was built where compartmentation was not considered. This network includes several possible anaplerotic pathways *e.g.*

pyruvate carboxylase, malic enzyme, PEP carboxykinase, glyoxylic shunt, and parallel sequences of reactions such as phosphofructokinase / fructose1,6-biphosphatase and NAD dependent isocitrate dehydrogenase / NADP dependent isocitrate dehydrogenase. The presence of so-called substrate cycles (Fell, 1990) has therefore been revealed, imposing to choose among several sets of operative enzymes in order to manage metabolic flux calculation. Importantly, the yields reconciliation using the metabolic network as internal constraint (including stoichiometry conservation) was performed with the values of the yields weighted by their experimental variance. This means that the final quadratic criterion can be interpreted as a variance ratio (variance of the lack of adequacy of the model divided by the variance of experimental error) enabling to take a rational decision for selecting the correct model. In the present study, the dimensionless criteria are always lower than $F_{0.975}$ (3 , 20) = 3.90, which is the value of the Fisher, test distribution (3 degrees of freedom for the reconciliation method and 20 degrees of freedom for the determination of each yield).

The best results, consistent on a statistical basis, are obtained if pyruvate carboxylase, malic enzyme, ICDH-NADP, F1,6 biphosphatase on fructose and phosphofructokinase on glucose are chosen as operative enzymes for all conditions considering in addition that for the mutant strain the G6Pdehydrogenase is not working. The results of the identification are given in the table 2 for the main key reactions, knowing that the experimental results and the selected network lead to thermodynamically consistent values for all the rates *e.g.* no negative fluxes in irreversible reactions.

3.3. DISCUSSION

From the previous calculations, several points can be outlined :

- 2-oxoglutarate dehydrogenase specific rate is almost zero for the ATCC 7754 strain if glucose or fructose are used whereas the mutant strain exhibits a significant positive value on fructose;
- Pyruvate kinase specific rate is twofold higher for the ATCC strain than for the mutant strain, not depending on the substrate;
- Glucose-6-phosphate-dehydrogenase (G6PDH) specific rate of ATCC 7754 strain is not affected by the nature of substrate ; the very low value corresponds to a split of carbon in the oxidative branch of the pentose phosphate pathway of 2 – 3 % of the total carbon flux. This is in agreement with the results of Lagunas and Gancedo (1973);
- A central role is played by the enzyme acetaldehyde dehydrogenase NADP dependent in the synthesis of the so-called reduced cofactors by the mutant strain ; its specific rate is significantly different from zero with the mutant strain compared to the reference strain for which it can be assumed that the enzyme is not operative.

Table 2. Metabolic flux of key reactions (mmol g^{-1} h^{-1}) in the wild and mutant strains on glucose or fructose. Standard deviations are reported in brackets (%).

	Glucose		Fructose	
Enzyme	ATCC 7754	CD101-1A	ATCC 7754	CD101-1A
Glucose -6-Phosphate Dehydrogenase	0.4 (8 %)	0 (0 %)	0.4 (30 %)	0 (0 %)
Phosphoglucose Isomerase	16.1 (0.2 %)	6.3 (0.5 %)	-1.0 (13 %)	-0.2 (7 %)
Phosphofructokinase	16.2 (0.1 %)	6.3 (0.5 %)	0 (0 %)	0 (0 %)
F1-6 biphosphatase	0 (0 %)	0 (0 %)	0.8 (15 %)	0.2 (7 %)
Pyruvate kinase	29.9 (0.2 %)	11.5 (1 %)	26.9 (1 %)	16.3 (0.3 %)
2-Oxoglutarate Dehydrogenase	0.02 (28 %)	0.05 (20 %)	0.004 (600 %)	0.23 (4 %)
Pyruvate Carboxylase	1.7 (4.6 %)	0.5 (17 %)	1.2 (45 %)	0.61 (7 %)
Malic Enzyme	1.1 (6 %)	0.3 (21 %)	0.7 (66 %)	0.2 (18 %)
Isocitrate Dehydrogenase (NADP)	0.31 (3 %)	0.12 (13 %)	0.24 (18 %)	0.32 (3 %)
Acetaldehyde Dehydrogenase (NADP)	0.02 (77 %)	0.27 (19 %)	0.03 (108 %)	0.23 (7 %)
Overall maintenance $ATP + H_2O \rightarrow ADP + Pi$	18.2 (1 %)	8.1 (7 %)	16.3 (14 %)	11.4 (2 %)

3.4. THERMODYNAMIC ANALYSIS

As the major differences in the topology of metabolism for the considered situations are located at the upper part of glycolysis (phosphofructokinase, fructose 1-6 biphosphatase, G6PDH), a thermodynamic analysis has been performed for the following reactions :

G6P	→	F6P	(reaction 1)
F6P + ATP	→	F1-6 biP + ADP	(reaction 2)

F1-6 biP + H_2O	→	F6P + Pi	(reaction 3)
ATP + H_2O	→	ADP + Pi	(reaction 4)

The rate laws express reaction velocities as function of concentrations. In an approach based on the principles of non equilibrium thermodynamics (Heinrich and Schuster, 1996 ; Dussap, 1988), velocities are expressed in terms of thermodynamic forces. Basically the reactions are supposed to be driven by reaction affinities in a multilinear way just as heat flow by temperature gradients and mass diffusion by concentration gradients.

The reaction affinity A_i is defined as the negative change in Gibbs free energy accompanying the reaction i. Using the Gibbs energies of formation of the compounds (Ould-Moulaye *et al*, 1999), the following values of the reaction affinities (kJ mole^{-1}) were obtained (T= 30°C, intracellular pH = 7.5, Mg^{2+}= 1 mM, ionic strength 0.2 M, Pi = 1mM, ATP/ADP= 4) :

$$A_1 = -2.7 + 2.52 \ln \frac{G6P}{F6P}$$

$$A_2 = 22.4 + 2.52 \ln \frac{F6P}{F1,6bP}$$

$$A_3 = 15.3 - 2.52 \ln \frac{F6P}{F1,6bP}$$

$$A_4 = 37.7$$

It must be outlined that such calculations need:

- to select Gibbs energies values of formation for the compounds from a unique reference state;
- to account for the detailed composition of the solution including all ionic species resulting from acid dissociation and complexation (Ould-Moulaye *et al*, 1999).

The thermokinetics approach is particularly appropriate for reactions the detailed kinetics of which are incompletely known in physiological conditions (intracellular conditions). In this case, the linear approximation in the vicinity of thermodynamic equilibrium (-5 kJ.mol^{-1} < A < 5 kJ.mol^{-1}) leads to:

$$J_i = \sum_j L_{ij} A_j$$

The first reaction catalysed by phosphoglucose isomerase obviously corresponds to this case :

$$J_1 = L_{11} A_1$$

From the results of metabolic flux calculation (Table 2) and considering that for the highest specific value of reaction rate (ATCC strain on glucose) the G6P/F6P ratio is 1.5 times the equilibrium value, one obtains:

$L_{11} = 15\ mol^2\ J^{-1}\ g^{-1}\ h^{-1}$

$$\left(\frac{G6P}{F6P}\right)_{ATCCglucose} = 4.4 \qquad \left(\frac{G6P}{F6P}\right)_{CD101-1Aglucose} = 3.4$$

$$\left(\frac{G6P}{F6P}\right)_{ATCCfructose} = 2.8 \qquad \left(\frac{G6P}{F6P}\right)_{CD101-1Afructose} = 2.9$$

The previous results show that the G6P/F6P ratio is slightly modified for the different culture conditions although the reactions rates are very different. Even if the concentrations ratios are calculated from an arbitrary value $\left[\frac{G6P}{F6P}\right]_{ATCCglucose} = 1.5\left[\frac{G6P}{F6P}\right]_{equilibrium}$, the main result is that such an approach leads to consider that a small change in intracellular concentrations may considerably modify the rates. Conversely, it may be concluded that the intracellular concentrations are finely tuned and are characteristic of a given physiological state.

The two other reactions catalysed by Phosphofructokinase and F1,6 biphosphatase do not proceed in the vicinity of thermodynamic equilibrium, knowing that F6P and F1,6bP concentrations are of the same order of magnitude which cannot compensate for the offset values of A_2 and A_3 (22.4 and 15.3 kJ mol^{-1} respectively). Moreover, this two enzymes system results in a futile cycle or substrate cycle (Fell, 1990) which globally hydrolyses ATP (reaction 4).

As previously indicated, the metabolic flux calculation (Table 2) leads to compute an overall resultant flux and a global maintenance flux which accounts for all futile cycles involved in the metabolism and other nondescript processes in the metabolic network such as membrane transport.

Starting from the hypothesis that for a purely futile cycle behaviour of this two enzymes system, the affinities of the two reactions might be equal, we calculated that the overall flux (J_2 on glucose, $-J_3$ on fructose) could be correlated as previously in a linear expression of the difference of the affinities such as :

$$J_2 = L_{23}(A_2 - A_3)$$

From the above, the purely futile cycle behaviour corresponds to $A_2 = A_3$ so that :

$$\left(\frac{F6P}{F1,6bP}\right)_{equilibrium} = 0.24$$

Considering as previously that for the reference situation (ATCC strain cultivated on glucose) that the concentration ratio F6P/F1,6bP is 1.5 times the equilibrium ratio, one obtains :

$$L_{23} = 8.3\,\mathrm{mol^2 J^{-1} g^{-1} h^{-1}}$$

$$\left(\frac{\mathrm{F6P}}{\mathrm{F1,6bP}}\right)_{\mathrm{ATCCglucose}} = 0.36 \qquad \left(\frac{\mathrm{F6P}}{\mathrm{F1,6bP}}\right)_{\mathrm{CD101-1Aglucose}} = 0.28$$

$$\left(\frac{\mathrm{F6P}}{\mathrm{F1,6bP}}\right)_{\mathrm{ATCCfructose}} = 0.24 \qquad \left(\frac{\mathrm{F6P}}{\mathrm{F1,6bP}}\right)_{\mathrm{CD101-1Afructose}} = 0.24$$

Starting from a reference value of 1 mmol g^{-1} for G6P intracellular concentration, the intracellular concentration of F6P and F1,6bP are calculated in table 3.

Table 3 : Calculated intracellular concentrations of F6P and F1,6bP in the various cases (G6P basis 1 mmol g^{-1})

	Equilibrium	ATCCglucose	CD101glucose	ATCCfructose	CD101fructose
G6P (mmol g^{-1})	1	1	1	1	1
F6P (mmol g^{-1})	0.34	0.23	0.29	0.36	0.34
F1,6bP (mmol g^{-1})	1.43	0.63	1.05	1.49	1.44

The results show that the F1,6bP concentrations are always greater on fructose than on glucose and that the intracellular concentrations vary in narrow ranges for the different situations investigated.

4. Conclusion and perspectives

The study of a strain lacking the glucose-6P dehydrogenase enzyme was interesting to a thorough understanding of energetic metabolism of the yeast *Saccharomyces cerevisiae*. Such a genetic mutation is not lethal which suggests that one or more points are disposable to produce NADPH,H^+ species, generally obtained at the level of the pentose phosphate pathway. The statistical analysis of a complete set of experimental results for anaerobic growth of the reference (ATCC 77554) and the mutant (CD101-1A) strains enable to draw a map of specific intracellular rates, including the estimation of the reliability of the predictions. This analysis conducts to the selection of 3 enzymes as possible points of NADPH,H^+ production. The selected network allows a satisfactory qualitative interpretation of the possible development of the mutant strain which seems to undergo NADPH,H^+ production through acetate synthesis with the operation of the

NADP dependent acetaldehyde dehydrogenase. The flux computation confirms that, in yeast, the pentose phosphate pathway is mainly devoted to anabolic functions. These results allow to conclude that pentose phosphate pathway as well as NADP dependent acetaldehyde dehydrogenase stand as limiting rates providing control among all other reactions in the network.

The analysis of the upper part of glycolysis in terms of linear thermodynamics of irreversible processes has enabled to calculate the intracellular concentrations on the basis of an assumed reference situation of exponential growth of the ATCC strain on glucose. Obviously the computed values need to be compared to experimental measurements of intracellular metabolites concentrations. However, the obtained calculated results indicate that slight variations of intracellular concentrations correspond to completely different distributions of intracellular flux. Such a result is consistent with the general assumption that the intracellular concentrations of key metabolites are highly regulated variables. This leads to conclude that such a phenomenological approach, via thermodynamics of irreversible processes, is an interesting tool of the investigation of cellular metabolism.

References

Andreasen A.A. and Stier T.J.B. (1953). Anaerobic nutrition of *Saccharomyces cerevisiae* I. Ergosterol requirement for growth in a defined medium. J. Cell. Comp. Physiology, 41, 23-26.

Andreasen A.A. and Stier T.J.B. (1954). Anaerobic nutrition of *Saccharomyces cerevisiae* II. Unsaturated fatty acid requirement for growth in a defined medium. J. Cell. Comp. Physiology, 43, 271-281.

Çalik P., Çalik G., Takaç S., Özdamar T.H. (1999). Metabolic flux analysis for serine alkaline protease fermentation by *Bacillus licheniformis* in a defined medium : Effects of the oxygen transfer rate. Biotech. Bioeng., 64, 151-167.

Dussap C.G. (1988). Etude thermodynamique et cinétique de la production de polysaccharides microbiens par fermentation en limitation par le transfert d'oxygène. Thèse de Docteur ès Sciences Physiques de l' Université Blaise Pascal - Clermont-Ferrand, France.

Dussap C.G., Pons A., Péquignot C., Gros J.B. (1997). Application d'une méthode de réconciliation de données à des cultures menées en fermenteur en discontinu. Récents Progrés en Génie des Procédés. Lavoisier Technique et Documentation, vol. 11(57), 7-12.

Fell D.A. (1990). Substrate cycles : theoretical aspects of their role in metabolism. Comments Theoretical Biology, 1, 341-357.

Goel A., Lee J., Domach M.M., Ataai M.M. (1999). Metabolic fluxes, pools, and enzyme measurements suggest a tighter coupling of energetics and biosynthetic reactions associated with reduced pyruvate kinase flux. Biotech. Bioeng., 64, 129-134.

Heinrich R. and Schuster S. (1996). The regulation of cellular systems. Chapman & Hall, New York.

Himmelblau D.M., (1968). Process analysis by statistical methods. John Wiley & Sons, New York.

Kristiansen B. (1994). Integrated design of a fermentation plant, VCH - Weinheim.

Lagunas R. and Gancedo C. (1973). Reduced pyridine-nucleotides balance in glucose-growing *Saccharomyces cerevisiae*. Eur. J. Biochem., 37, 90-94.

Lee S.Y. and Papoutsakis E.T. (1999). Metabolic Engineering. Marcel Dekker Inc, New York.

Nielsen J. and Villadsen J. (1994). Bioreaction Engineering Principles. Plenum Press, New York.

Ould-Moulaye C.B., Dussap C.G., Gros J.B. (1999). Estimation of Gibbs energy changes of central metabolism reactions. Biotechnol. Techniques, 13, 187- 193.

Oura E. (1983). Formation of biomass from carbohydrates. In Biotechnology, H.J. Rehm and G. Reed (eds.), Verlag Chemie, vol 3, pp. 3-16.

Pons A., Dussap C.G., Péquignot C., Gros J.B. (1996). Metabolic flux distribution in *Corynebacterium melassecola* ATCC 17965 for various carbon sources. Biotech. Bioeng., 51, 177-189.

Pramanik J. and Keasling J. D. (1997). Stoichiometric model of *Escherichia coli* metabolism : incorporation of growth-rate dependent biomass composition and mechanistic energy requirements. Biotech. Bioeng., 56, 398-421.

Thomas D., Cherest H., Surdin-Kerjan Y. (1991). Identification of the structural gene for glucose-6-phosphate dehydrogenase in yeast. Inactivation leads to a nutritional requirement for organic sulphur. The EMBO Journal, 10, 547-553.

Vallino J.J. and Stephanopoulos G. (1990). Flux distribution in cellular bioreaction networks : application to lysine fermentations. in Frontiers in Bioprocessing, S.K. Sikdar (ed.), CRC Press, pp. 205-219.

Vallino, J.J. and Stephanopoulos G. (1993). Metabolic flux distributions in *Corynebacterium glutamicum* during growth and lysine overproduction. Biotech. Bioeng., 41, 633-646.

Vanrolleghem P.A., de Jong-Gubbels P., van Gulik M.W., Pronk J.T., van Dijken P.J., Heijnen S. (1996). Validation of a metabolic network for *Saccharomyces cerevisiae* using mixed substrate studies. Biotechnol. Prog., 12, 434-448.

Veverka V.V. and Madron F. (1997). Material and energy balancing in the process industries. Elsevier, Amsterdam.

METABOLIC INVESTIGATION OF AN ANAEROBIC CELLULOLYTIC BACTERIUM : *FIBROBACTER SUCCINOGENES*

C. CREULY, A. PONS, AND C.G. DUSSAP
Laboratoire de Génie Chimique et Biochimique. Université Blaise Pascal. Clermont-ferrand II. 24 avenue des Landais. F63177 Aubière cedex.

Abstract

Fibrobacter succinogenes, an anaerobic bacteria is cultivated in a batch reactor. A detailed stoichiometric model of metabolism is developed. It includes all reactions of catabolic pathways that have been proved to exist and anabolic pathways. The measured conversion yields are correlated in terms of metabolic flux distribution using a mathematical technique of yields calculation associated with a methodology of data reconciliation. This approach validates the metabolic network built on *Fibrobacter succinogenes*.

1. Introduction

Microbial cellulases and hemicellulases are widely used in different industrial activities, such as in textile, detergent, brewery or wood-processing, and also in the treatment of domestic wastes and in biological treatment of fibrous feeds in the non-ruminant livestock industry. However, these enzymes are not very efficient for the degradation of highly lignified plant cell walls because cellulose and hemicelluloses are cross-linked to lignin which is very difficult to degrade, and protects cellulose and hemicelluloses against enzymatic hydrolysis (Selinger et al., 1996).

Ruminant animals possesses rumen bacteria which developed a symbiotic relationship to digest lignocellulosic substrates (Hungate,1950). *Fibrobacter succinogenes* is a major fibrolytic bacterium found in the rumens of cattle and sheep.

With the aim of developing a biotechnological process for the degradation of lignocellulosic residues, we propose to take advantage of the high potentiality of *Fibrobacter succinogenes*. Its enzymatic equipment explains these specific performances ; it includes a very efficient cellulolytic system, ferulic acid and acetylxylane esterases, arabinofuranosidases, xylanases and glucuronidases (Chesson and Forsberg, 1997). This strictly anaerobic bacterium uses cellulose, glucose and

M. Hofman and P. Thonart (eds.), Engineering and Manufacturing for Biotechnology, 157–167.

cellobiose as carbon and energy sources, and produces succinate, acetate and few formate (Stewart and Flint, 1989).

The development of a high-performance bioreactor is based on a general concept of metabolic engineering : the idea is to estimate metabolic fluxes in *F. succinogenes* to direct bacterial metabolism towards the production of biomass and enzymes of interest. The aim of this paper is to show how, using the data concerning the biochemistry of *F. succinogenes* and the measured conversion yields, the special metabolic pathways specific of those anaerobic bacteria could be validated. The supposed metabolic network involves 96 stoichiometric reactions. Validation of the model is obtained by comparison between the theoretical yields of carbon elements calculated by a data reconciliation technique and the experimental yields measured during anaerobic cultures of *F. succinogenes* in a bench scale bioreactor.

2. Material and method

2.1. STRAIN AND CULTIVATION

Fibrobacter succinogenes S85 (ATCC 19169) was originally isolated from the bovine rumen (Bryant and Doetsch, 1954) and has been maintained as pure culture in laboratory ever since. It was grown anaerobically under 100% CO_2 in a medium containing (per litre): 450 mg KH_2PO_4, 450 mg K_2HPO_4, 900 mg NaCl, 1.8 g $(NH_4)_2SO_4$, 90 mg $MgSO_4$, 90 mg $CaCl_2$, 3 mg $MnSO_4,6H_2O$, 0.3 mg $CoCl_2,6H_2O$, 8 mg $FeSO_4,7H_2O$, 0.25 mg Biotin, 0.005 mg para-aminobenzoic acid, 500 mg Cysteine, 4 g Na_2CO_3, 8 g Glucose and a volatile fatty acid mixture (Gaudet et al, 1992). 1.2 litre of the above medium was sterilised (120°C, 20 min) in the reactor (2 litres total volume).

2.2. EXPERIMENTAL DESIGN

After redox potential reduction at -350 mV and temperature equilibration at 37°C, the thermostated, stirred (set at 40 rpm) fermentor (1.5 l) was inoculated with 200 ml of an overnight culture. pH and redox were measured on line. Growth was monitored by the increase of optical density (600nm). Growth was correlated with the decrease of the pH since the initial value (6.9) until value 5.5. At this value, pH was readjusted at neutral value by Na_2CO_3 addition. In the same order, a substrate feed was performed when glucose concentration became limiting (fed-batch technique)

2.3. METABOLITES ASSAYS

Cell mass was determined by dry weight at 100°C after a centrifugation step. Extracellular metabolites (glucose, succinate, acetate, volatile fatty acids) were quantified in cell free samples by high pressure-liquid chromatography (1100 series Hewlett Packard, 1047A refractometer analyser) fitted with two ionic exclusion columns (Phenomenex rezex organic acid - 300*7.8 mm) maintained at 80 °C and isocratic eluted with solvent 5 mM H_2SO_4.

2.4. FLUX ESTIMATION AND STATISTICAL ANALYSIS

Metabolic fluxes were calculated from a mass balance technique. This method takes into account the stoichiometric reactions obtained by an analysis of the internal behaviour of the microorganism. The metabolic reactions network is depicted in terms of matrices and vectors algebra for representing the metabolites balances. The bioreaction network (Vallino and Stephanopoulos, 1990) is used to determine the rates of production and consumption of each metabolite in the network ($\mathbf{R}_{\text{metabolite}}$) as a function of all the unknown fluxes ($\mathbf{J}_i$ for the ith reaction). The accumulation rate of a metabolite in a metabolic network is given by the summation of all reactions producing that metabolite minus the reactions consuming it.

$$R_j(t) = \sum_m \alpha_{jm} J_m(t) - \sum_k \alpha_{jk} J_k(t) \qquad (1)$$

where a_{Jm} are the stoichiometric coefficients, $J_m(t)$ is the flux through reaction m and the $R_j(t)$ is the accumulation rate of metabolite j.

The set of equations formed from such balances for each metabolite in the network is represented in matrix notation by :

$$\mathbf{A}\ \mathbf{J}(t) = \mathbf{R}(t) \qquad (2)$$

A is the (c x r) matrix of stoichiometric coefficients (c metabolites and r metabolic reactions) of all the reactions involved in the metabolism. Each column in the matrix represents a metabolic reaction.

The c metabolic constituents are divided into two categories :

- the n exchangeable compounds, which are exchanged with the growth medium, from a macroscopic point of view, these compounds are the products and the substrates involved during the cellular growth phase ; their accumulation rates are linked to growth yields ;
- the m non-exchangeable compounds, which are only involved inside the cell ; for these the accumulation rate is assumed to be equal to zero.

This is a pseudo-steady state hypothesis that expresses that the rate of accumulation of intracellular compounds is negligible compared to the rates of production / consumption of exchangeable compounds (Vallino and Stephanopoulos, 1990). Assuming that during a sufficiently long period of growth the yields remain constant (some of them being measured), and that non-exchangeable compounds are not accumulated inside the cell, the previous Equation (2) is split in two parts: the first part corresponds to the known accumulation rates (non-exchangeable compounds and measured yields), the second part being the remaining rows corresponding to the unknown yields (Pons et al, 1996).

The number of supplied information must at least be greater than the number of columns *i.e.* of unknown fluxes J. In case of equality, provided matrix **A** is of full rank, fluxes are calculated by inversion of the matrix.

In case of redundant information (more independent rows than columns) a data reconciliation technique is worked using the conservation laws of non-exchangeable compounds as constraints. To solve this problem of optimisation under constraints, we use a Lagrange method in minimising a quadratic criterion Φ calculated between the n' measured values of the yields contained in $\mathbf{R_1}$ and the computed values $\hat{\mathbf{R}}_1 = \mathbf{A_1 J}$ (Dussap et al., 1997).

$$\Phi = {}^t(\mathbf{R_1} - \mathbf{A_1 J})\, \mathbf{w}\, (\mathbf{R_1} - \mathbf{A_1 J}) \tag{3}$$

where **w** is the diagonal weighted matrix of the n' experimental values in $\mathbf{R_1}$ which account for the accuracy of the determination. In the case under study, variances σ^2 of the measured yields are known and elements of **w** are given by:

$$w_i = \frac{1}{\sigma_i^2} \qquad i = 1, n' \tag{4}$$

The Lagrange function to be considered is given by :

$$L = \Phi - {}^t\Lambda\, \mathbf{A'_1}\, \mathbf{J} \tag{5}$$

where $\mathbf{A'_1}$ is a matrix (m+d , r) and Λ is a vector (m+d , 1). $\mathbf{A'_1}$ contains all the rows corresponding to the m non-exchangeable compounds plus the d relationships between some of the fluxes necessary to eliminate to metabolic cycles included in the network (Dussap et al., 1997). This makes (m+d) rows for $\mathbf{A'_1}$.

r flux of **J** and (m+d) values of Λ vector are calculated by solving the system:

$$\frac{\partial L}{\partial \mathbf{J}} = -2\, {}^t\mathbf{A_1}\, \mathbf{w}\, (\mathbf{R_1} - \mathbf{A_1 J}) - {}^t\mathbf{A'_1}\, \Lambda = 0 \qquad \text{(r relations)}$$

$$\mathbf{A'_1 J} = 0 \qquad \text{(m+d relations)}$$

The solution is given by:

$$\mathbf{J} = \mathbf{V}\left[\mathbf{G} - {}^t\mathbf{A'_1}\, \mathbf{H}\, \mathbf{A'_1}\, \mathbf{V}\, \mathbf{G}\right] \tag{6}$$

where **V**, **H** and **G** matrices are defined as:

$$\mathbf{V}(r, r) = ({}^t\mathbf{A_1}\, \mathbf{w}\, \mathbf{A_1})^{-1} \tag{7}$$

$$\mathbf{G}(r, 1) = \mathbf{A_1}\, \mathbf{w}\, \mathbf{R_1} \tag{8}$$

$$\mathbf{H}\,(\mathrm{m+d}\,,\mathrm{n'}) = (\,\mathbf{A'}_1\,\mathbf{V}\,'\mathbf{A'}_1\,)^{-1} \tag{9}$$

The re-calculated conversion rates values are given by $\hat{\mathbf{R}}_1 = \mathbf{A}_1\ \mathbf{J}$ where, the (m+d) drawn values are found by construction. The standard deviations of the calculated flux and rates are estimated by diagonal elements of matrices of covariance:

$$\mathrm{Covar}\,(\mathbf{J}) = \frac{\Phi}{m + d + n' - r}\ \mathbf{V} \tag{10}$$

$$\mathrm{Covar}\,(\hat{\mathbf{R}}) = \mathrm{A}\ \mathrm{Covar}\,(\mathrm{J})\ '\,\mathbf{A} \tag{11}$$

This set of Equations (10,11) leads to compute the best estimates of all the yields and all of the fluxes. Statistical analysis also provides the standard deviations of the estimation.

3. Results and discussion

3.1. METABOLIC NETWORK

The purpose of this work is to present a novel approach on this bacterium by developing a detailed stoichiometric model of anaerobic metabolism that includes a more complete database of known reactions involved in the central catabolism of glucose and previously established anabolic reactions based on the general knowledge of bacterial metabolism (Gottschalk, 1986) (Figure 1).

The aim is to check if the present reported biochemical pathways could be consistently integrated in the overall growth metabolism without involving thermodynamically impossible reactions. The major features of catabolic pathways, which have been described, are as follows:

- Glucose is transported across the cytoplasmic membrane through independent constitutive transporters that are sodium dependent. In the cytoplasm, glucose is phosphorylated by a GTP-dependent glucokinase (Glass and Sherwood, 1994). Cultures that were provided with glucose produced cellobiose, and cellobiose gave rise to cellotriose. Gaudet et al (1992) showed that *Fibrobacter succinogenes* continuously synthesised and degraded glycogen during all phases of growth, but its dependency on glycogen catabolism was not defined.
- As it possesses fructose 1,6-biphosphate aldolase and glyceraldehyde 3-phosphate dehydrogenase, *Fibrobacter succinogenes* is assumed to ferment hexoses by the Embden-Meyerhof-Parnas pathway (Joyner and Baldwin, 1966; Miller, 1978) until pyruvate. Phosphoenolpyruvate is carboxylated to oxaloacetate by a GDP- specific PEP carboxykinase.
- Oxaloacetate is converted to malate by a pyridine nucleotide-dependent malate dehydrogenase (Table 1; J20).

- Fumarase activity (Table 1; J14) was not demonstrated but it is probably present to produce fumarate from malate. Fumarate is reduced by a flavin-dependent, membrane bound fumarate reductase to produce the major fermentation product, succinate. The reduction of fumarate with reduced flavins is likely to involve cytochrome b (Miller, 1978). Flavin nucleotides mediate electron transport between pyruvate and fumarate (Table 1; three coupled reactions J9, J79, J77).

In addition, *Fibrobacter succinogenes* possesses the essential enzymes of the non-oxidative branch of the pentose phosphate pathway (Matte et al, 1992), though known to be unable to metabolise pentoses. Enzymatic researches on the oxidative branch give negative results. This specificity imposes to set glutamate dehydrogenase (Table 1; J22) on NAD/NADH dependence and isocitrate dehydrogenase NADP / NADPH specific (Table 1; J82). Matheron et al (1999) have isolated a specific enzyme that produce alanine from pyruvate and NH_3 (Table 1; J80).

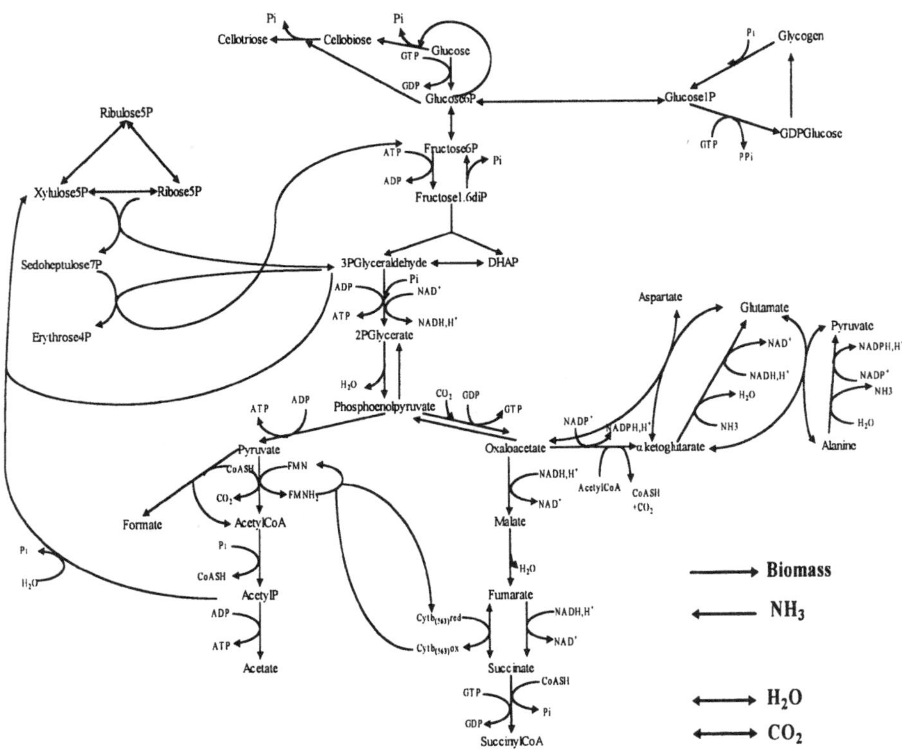

Figure 1: Metabolic network built for Fibrobacter succinogenes

Table 1: Catabolic reactions considered in the network

Flux	Stoichiometric reactions
J1	glucose + GTP → glucose-6P + GDP
J2	glucose-6-P → fructose-6-P
J3	fructose-6-P + ATP → fructose-1,6-diP + ADP
J4	fructose-1,6-diP → dihydroxyacetone-P + glyceraldehyde-3-P
J5	glyceraldehyde-3-P → dihydroxyacetone-P
J6	glyceraldehyde-3-P + NAD + ADP + Pi → 3-P-glycerate + ATP + $NADH_2$
J7	3-P-glycerate → P-enolpyruvate + H_2O
J8	P-enolpyruvate + ADP → pyruvate + ATP
J9	pyruvate + CoASH + FMN → acetyl-CoA + $FMNH_2$ + CO_2
J10	pyruvate + CoASH → acetyl-CoA + formiate
J11	acetyl-CoA + Pi → acetyl-1P + CoASH
J12	acetyl-1P + ADP → acetate + ATP
J13	fumarate + $NADH_2$ → succinate + NAD
J14	fumarate + H_2O → malate
J15	malate + NAD → oxaloacetate + $NADH_2$
J16	ribulose-5P → ribose-5P
J17	ribulose-5P → xylulose-5P
J18	xylulose-5P + ribose-5P → glyceraldehyde-3-P + sedoheptulose-7-P
J19	glyceraldehyde-3-P + sedoheptulose-7-P → fructose-6-P + erythrose-4-P
J20	P-enolpyruvate + CO_2 + GDP → oxaloacetate + GTP
J76	xylulose-5P + Pi → glyceraldehyde-3-P + acetyl-1P + H_2O
J77	$Cytb_{red}$ + fumarate → $Cytb_{ox}$ + succinate
J78	xylulose-5P → ribose-5P
J79	$Cytb_{red}$ + $FMNH_2$ → $Cytb_{ox}$ + FMN
J80	pyruvate + $NADPH_2$ + NH_3 → alanine + NADP + H_2O
J81	CoASH + GTP + succinate → succinylCoA + GDP + Pi
J82	oxaloacetate + acetyl-CoA + NADP + H_2o → α-ketoglutarate + CoASH + $NADPH_2$ + CO_2
J83	glucose + glucose-6P → cellobiose + Pi
J84	glucose-6P + cellobiose → cellotriose + Pi
J85	cellobiose + H_2O → 2 glucose
J21-J75	**Anabolic network reactions**

3.2. FLUX CALCULATION

The biochemical composition and the macrocomponents biochemical composition of *Fibrobacter succinogenes* are unknown. The data representing an average composition of the bacteria *Escherichia coli* established by Neidhardt (1987) have been used:

$$C\,H_{1.577}\,O_{0.388}\,N_{0.257}\,S_{0.006}\,P_{0.022} \tag{12}$$

The matrix built with the above metabolic network is a 108 x 96 matrix, with 108 compounds, 96 reactions, 13 exchangeable and 95 non-exchangeable compounds. After all steps of computing analysis, the rank of the square matrix is 96. We have considered eight exchangeable compounds to make data reconciliation: glucose, acetate, succinate, formate, cellobiose, cellotriose, NH_3 and biomass.
Yield products are calculated and used as experimental data in the program of metabolic flux calculation associated with the data reconciliation method. In Table 2, the key fluxes of central metabolism are reported. The fluxes are thermodynamically consistent, *i.e.* no negative flows in irreversible reactions.

Table 2: Examples of fluxes in central metabolism

Fluxes	Enzymes	mol /100 mol substrate	standard deviation %
J1	hexokinase	90.4	2.5
J4	F 1-6 diP aldolase	75.0	6.3
J7	Enolase	140	6.3
J8	Pyruvate kinase	55.8	10
J20	PEP carboxykinase	81.8	5.5
J10	Pyruvate formate lyase	8.2	52
J12	Acetate kinase	28.9	10
J13	Succinate dehydrogenase	34.6	8
Jbio		6.2	14

3.3. VALIDATION

Validation of this model is obtained by comparison between the theoretical yields of carbon elements calculated by computed flux program and the experimental yields measured during anaerobic culture of *Fibrobacter succinogenes* (Table 3). This table shows a good correlation between yields values of the main products of metabolism (succinate, acetate and formate). These results validate the metabolic network proposed in Figure 1. An other interesting information given by the flux data is that carbon dioxide is consumed and not produced by this bacteria with a consumption ratio of 0.09 g CO_2 / g glucose.

Table 3: Comparison of experimental and theoretical yields obtained by metabolic fluxes computing.

Mass yields	Experimental value g product /g substrate	Theoretical value g product / g substrate	Standard deviation %
Biomass / Glucose	0.25	0.16	14
Succinate / Glucose	0.46	0.48	5.8
Acetate / Glucose	0.11	0.10	9.4
Formate / Glucose	0.02	0.02	52
Cellobiose / Glucose	0 (estimated)	0.07	104
Cellotriose / Glucose	0 (estimated)	0.16	49

The major difference is observed on the estimation of biomass between the experimental and computed yields. The best value may be the calculated value because it is correlated to the actual consumption of NH_3 that validates the previous biomass elemental composition. The mass balance calculation imposes directly a calculated yield of 5.58 g biomass / g NH_3. The deviation with biomass measurement could be assigned to the experimental indetermination of glycogen and cellodextrins. In fact, when extracellular sugar concentration is high, part of the sugar is stored as glycogen (Gaudet et al, 1992). This glycogen storage can represent as much as 70% of the total dry mass of the bacteria and it seems to be included in our experimental value of biomass determined by dry matter.

Glucose is also released as cellodextrins *via* cellobiose into the external medium (Wells et al, 1995). These authors evidenced cellodextrin synthesis from cellobiose as substrate but cellodextrins were also synthesised when glucose was the sole carbon source (Matheron et al, 1996). This entails that the first step of cellodextrins synthesis is catalysed by the cellobiose phosphorylase that is able to condense one glucose and one glucose1-phosphate into cellobiose. Larger cellodextrins may then be synthesised by cellobiase activity. This polysaccharide would be precipitated with cell mass during the step of centrifugation and subsequently majored experimental value of biomass.

4. Conclusions and perspectives

This fed-batch process enables to produce more than 0.15 g biomass per g glucose. This value would be improved by a more complete understanding of *Fibrobacter succinogenes* metabolism. The first step was the development of the metabolic network. There is a satisfactory agreement between theoretical and experimental mass balance data. Though the oxidative part of pentose pathway has not been included in the network, the calculated distribution of metabolic flux satisfies thermodynamic consistency for all reactions. The pentose requirements of metabolism are supplied through glyceraldehyde 3-P and acetyl 1-P as metabolic intermediates (Table 1 - J76) which is a specific reaction of these bacteria.

This utilisation of a metabolism model to simulate the growth of this microorganism presents a new approach for strict anaerobic bacteria and allows to predict some data that are difficult to measure in anaerobic conditions, such as the carbon dioxide uptake yield. In the same way, we show the difficulty to obtain an experimental value of the biomass concentration; the model calculates this value.

The metabolic model should be a useful tool to provide information about how the overall flux distribution will be affected by various growth conditions and specially with various substrates. Glucose and cellobiose, final products of cellulose degradation, are taken up and metabolised by the cells into succinate, acetate and formate (Gaudet et al., 1992). In vivo ^{13}C-NMR spectroscopy has been used successfully to investigate metabolism of various microorganisms (Matheron et al., 1996). This technique allows identification of ^{13}C-enriched molecules and, more precisely, localisation of the labelling in the molecule. Thus, the use of a ^{13}C specifically labelled substrate allows a detailed description of the metabolic pathway that it enters. The quantitative determination of metabolic fluxes by NMR experiments has shown the reversibility of different metabolic pathways: reversibility of glycolysis, reversibility of the succinate synthesis pathway and futile cycle of glycogen (Matheron et al., 1999). But all these measurements have been carried out with resting cells in order to maintain suitable in vivo NMR conditions. This study, performed from an overall determination of yields and a biochemically structured model, enables to extend the previous results for resting cells to a growth situation obtained in controlled bioreactor.

In a near future, this approach could be improved by focusing attention and developing theoretical tools in parallel with experimental bioreactor in two directions:

- further investigations about biochemically structured metabolism : ^{1}H and ^{13}C NMR experiments have to be carried out with the aim of collecting new information about the metabolism of the rumen cellulolytic bacterial strains, particularly the carbohydrate metabolism which seems to present interesting specificity ;
- development of quantitative aspects of polymerised and lignified substrates digestion, such as digestion of cellobiose, crystalline cellulose and straw feed.

For more complex substrates, we will have to include in the model the simultaneous but differential consumption of sugars. For example, the relative contribution of glucose and cellobiose to metabolite production, glycogen storage, and cellodextrins synthesis is not known but it would have to be predicted.

This will allow direct monitoring of the metabolism towards the production of biomass and esterases in conditions of bioreactors performances. The metabolic control aspects will certainly help for the development and control of a biotechnological process that is efficient in degrading lignocellulosic wastes.

Acknowledgement

The Centre National de la Recherche Scientifique, France, supported this work.

References

Bryant M.P. and Doetsch R.N. (1954) A study of actively cellulolytic rod-shaped bacteria of the bovine rumen, *J.Dairy Sci.*, 37, 1176-1183.

Chesson A. and Forsberg C.W. (1997) Polysaccharide degradation by the rumen microorganisms, Hobson, P.N., Stewart, C.S., *the Rumen Microbial Ecosystem*, 2nd ed. Blackie Academic. London, 329-381.

Dussap C.G., Pons A., Pequignot C. and Gros J.B. (1997) Application d'une méthode de réconciliation de données à des cultures menées en fermenteur discontinu, *Récents Progrès en Génie des Procédés*, 11, 7-12.

Gaudet G., Forano E., Dauphin G. and Delort A.M. (1992) Futile cycle of glycogen in *Fibrobacter succinogenes* as shown by in situ ^{1}H-NMR and ^{13}C-NMR investigation, *Eur.j.Biochem.*, 207, 155-162.

Glass T.L. and Sherwood J.S. (1994) Phosphorylation of glucose by a guanosine-5'-triphosphate (GTP)-dependent glucokinase in *Fibrobacter succinogenes* S85, *Arch. Microbiol.*,162,180-186.

Gottschalk G. (1986) *Bacterial metabolism.* 2nd ed, Springer-Verlag, New-York.

Hungate R.E. (1950) The anaerobic mesophilic cellulolytic bacteria, *Bacteriol.Rev.*, 14, 1-49.

Joyner Jr A.E. and Baldwin R.L. (1966) Enzymatic studies of pure cultures of rumen microorganisms, *J.Bacteriol.*, 92, 1321-1330.

Matte A., Forsberg C.W. and Verrinder A.M. (1992) Enzymes associated with metabolism of xylose and other pentoses by *Prevotella (Bacteroides) ruminicola* strains, *Selenomenas ruminantium* D and *Fibrobacter succinogenes* S85, *Can.J.Microbiol.*, 38, 370-376.

Matheron C., Delort A.M., Gaudet G. and Forano E. (1996) Simultaneous but differential metabolism of glucose and cellobiose in *Fibrobacter succinogenes* cells, studied by in vivo 13C-NMR, *Can.J.Microbiol.*,42, 1091-1099.

Matheron C., Delort A.M. and Forano E. (1999) Interactions between carbon and nitrogen metabolism in *Fibrobacter succinogenes* S85 : a ^{1}H and ^{13}C nuclear magnetic resonance and enzymatic studies, *Appl.Environ. Microbiol.*, 65, 1941-1948.

Miller T.L. (1978) The pathway of formation of acetate and succinate from pyruvate by *Bacteroides succinogenes*, *Arch.Microbiol.*, 117, 145-152.

Neidhardt F.C. (1987) Chemical composition of *Escherichia coli*, pp-3-6. In : F.C. Neidhardt (ed.), *Escherichia coli* and *Salmonella typhimurium*, vol.1. American Society for Microbiology, Washington, DC.

Pons A., Dussap C.G., Pequignot C. and Gros J.B. (1996) Metabolic flux distribution in *Corynebacterium melassecola* ATCC 17965 for various carbon sources, *Biotechnol.Bioeng.*,51, 177-189.

Selinger L.B., Forsberg C.W., Cheng K.J. (1996) The rumen : a unique source of enzymes for enhancing livestock production, *Anaerobe*, 2, 263-284.

Stewart C.S. and Flint H.J. (1989) *Bacteroides (Fibrobacter) succinogenes.* A cellulolytic anaerobic bacterium from the gastrointestinal tract, *Appl.Microbiol.Biotechnol.*, 30, 433-439.

Vallino J. J. and Stehanopoulos G. (1990) Flux distribution in cellular bioreaction networks : application to lysine fermentations. pp-205-219. In : S.K.Sikdar, M.Bier, and P.Todd.(eds.), *Frontiers in bioprocessing.*CRC Press, Boaca Raton, FL.

Wells J.E., Russel J.B., Shi Y. and Weimer P. J. (1995) Cellodextrin efflux by the cellulolytic ruminal bacterium *Fibrobacter succinogenes* and its potential role in the growth of non adherent bacteria, *Appl.Environ.Microbiol.*, 61, 1757-1762.

PART III
INTEGRATED PROCESSES

CROSSFLOW ULTRAFILTRATION OF *BACILLUS LICHENIFORMIS* FERMENTATION MEDIUM TO SEPARATE PROTEASE ENZYMES

SERPIL TAKAÇ, SEMA ELMAS*, PINAR ÇALIK[+], TUNÇER H. ÖZDAMAR
Ankara University Biotechnology Research Center, Industrial Biotechnology Department, Tandoğan 06100 Ankara, Turkey
**Cumhuriyet University, Department of Chemical Engineering, 58140 Sivas, Turkey*
+Middle East Technical University, Department of Chemical Engineering 06531 Ankara, Turkey

Abstract

Separation conditions for the serine alkali protease (SAP) enzyme from the neutral protease and amylase enzymes of Bacillus licheniformis cells were investigated in a crossflow ultrafiltration system by using 30 000 Da MWCO polysulphone membrane. The effects of initial enzyme concentration and recirculation velocity on the permeate flux, on the total resistance, and on the recovery yield of SAP in the permeate were investigated. High permeate flux was obtained at high recirculation velocity but at low initial enzyme concentration, where the SAP enzyme activity was best recovered at low velocity and enzyme concentration conditions.

1. Introduction

Crossflow ultrafiltration provides an attractive means of separation of enzymes and other proteins from fermentation medium since it prevents sensible molecules from being denaturated by heat or chemicals. Membrane based processes are more easily operated and scaled up in comparison to other bioseparation processes such as chromatography and electrophoresis. However, the low flux due to concentration polarisation and fouling problems are the main disadvantages of the ultrafiltration. The separation of proteins in dilute solution by ultrafiltration is generally effective only for macromolecules of considerable size difference; and the recovery of macromolecules from cell suspensions is limited by concentration polarisation, resistances associated with protein rejection, adsorption or pore plugging. However, by manipulating interactions between the protein and the membrane, ultrafiltration performance can be

M. Hofman and P. Thonart (eds.), Engineering and Manufacturing for Biotechnology, 171–179.

enhanced. Process parameters such as protein concentration, transmembrane pressure (TMP), pH, ionic strength, temperature, shear rate, membrane material and structure have significant effects on ultrafiltration (Flaschel *et al.*, 1983).

The literature covers information on the separation of different protein macromolecules that reports on the fouling mechanism (Pradanos *et al.*, 1996; Boyd and Zydney 1998), adsorption mechanism (Gekas *et al.*, 1993; Pradanos and Hernandez, 1996), transport phenomena (Henriksen and Hassager 1993; Saksena and Zydney 1997) and process parameters (Pradanos *et al.*, 1992 and 1994; Pradanos and Hernandez 1995). Serine alkaline proteases (SAP; E.C.3.4.21.14) produced by the genus *Bacillus* are the most important group of industrial enzymes (Kalisz 1988) and the separation yield of extracellular SAP from biomass, other proteins, salts and impurities is important. Recovery of protease enzymes from fermentation medium usually includes stepwise procedure where ultrafiltration is an intermediate step. Gnosspelius (1978) reported 96% recovery of the protease enzyme from *Myxococcus virences* in the ultrafiltration step of the purification sequence. Manachini *et al.*, (1988) investigated the separation of *Bacillus termoruber* alkaline protease from fermentation medium and reported 85% yield in the ultrafiltration step of the separation sequence. Sheehan *et al.*, (1990) developed a two-stage pilot scale membrane filtration process to recover bacterial cells followed by ultrafiltration of an extracellular protease from fermentation broth and investigated the effect of transmembrane pressure and recirculation rate on the ultrafiltration performance. Bohdziewich (1994) studied the purification of Proteopol BP-S, a commercial preparation of proteolytic enzymes, with polyacrylonitrile membranes and reported the optimum values for TMP, flow velocity and temperature for the ultrafiltration. Although the information on the ultrafiltration of protease enzymes from fermentation medium is limited, a number of studies on the ultrafiltration conditions of several proteins provide important insights into how to select the optimal conditions for effective protein separation. Rodgers and Sparks (1991) investigated the effect of negative transmembrane pressure pulsing on solute rejection for an albumin and gamma-globulin mixture in ultrafiltration. Reis *et al.*, (1997) studied various limitations of membrane systems for protein purification such as buffer composition, fluid dynamics; and reported on the optimisation of high performance tangential flow filtration of proteins. Burns and Zydney (1999) investigated the effect of solution pH on the transport of globular proteins with different surface-charge characteristics and molecular weight through ultrafiltration membranes.

In our previous study, we reported the separation from fermentation medium of extracellular SAP enzyme produced by *Bacillus licheniformis* using a crossflow ultrafiltration system. 30 000 and 10 000 Da MWCO polysulphone membranes were used to separate SAP from high molecular weight enzymes and low molecular weight organic acids and amino acids, respectively (Takaç *et al.*, 2000). The effects of transmembrane pressure, recirculation velocity and enzyme concentration on the permeate flux, on the activities of enzymes, and on the recovery of SAP were investigated. In the present work, we performed further studies to develop the ultrafiltration process conditions to separate SAP from accompanying high molecular weight enzymes i.e., neutral protease and amylase enzymes in the *B.licheniformis* fermentation medium by using 30 000 Da MWCO polysulphone membrane. We investigated the effect of enzyme concentration and recirculation velocity on the permeate flux, on the total resistance, and on the recovery yield of SAP.

2. Materials and methods

2.1. EXPERIMENTAL RUNS

B. licheniformis (DSM 1969) cells were grown and inoculated as described elsewhere (Çalık *et al.,* 1998). The shake flask culture was transferred into the fermentation medium that contained (g dm^{-3}): either glucose, 6 or citric acid, 9; $(NH_4)_2HPO_4$, 4.714 and KH_2PO_4 ,2. The initial pH of the medium was adjusted to 7.6 with 0.04 mol dm^{-3} NaH_2PO_4-Na_2HPO_4 buffer. The laboratory-scale 3.5 dm^3 batch bioreactors (Chemap CF 3000, Switzerland) were operated at 37°C temperature, 750 min^{-1} agitation rate and 1 vvm aeration rate conditions for 40 h in the cultivation for the production of protease enzymes. After harvesting cells by centrifugation at 8000xg at +4°C (RC28S; Sorvall, Wilmington, DE), the fermentation medium was subjected to crossflow ultrafiltration in a flat modular configuration ultrafiltration device (Sartocon Mini SM 17521, Sartorius, Germany) by using 30 000 Da MWCO asymmetric polysulphone membrane. The experimental set-up is given elsewhere (Takaç *et al.,* 2000). The solutions of different initial enzyme concentrations were tangentially driven over the membrane surface and recirculated with several rates. The permeate flux was collected in the permeate tank while the retentate was recycled back to the feed-retentate tank. The variation in the permeate flux with time was followed and the results were analysed with the cake resistance model. Ultrafiltration runs were terminated when a constant permeate flux was obtained. At the end of each experiment, the activity of SAP enzyme was measured off-line by taking samples from the permeate and feed-retentate tanks (Elmas 1997). Following each run, the membrane was rinsed with distilled water, with Sartocon Cleaning Agent 17639 (1.5%) and with formaldehyde solution (2-3%) in sequence. Prior to ultrafiltration runs, the water flux was measured in order to calculate the hydraulic resistance for determining the cleaning efficiency.

2.2. ANALYSES

SAP activity was measured in borate buffer (pH=10) with casein (0.5%) at 37°C. 2 cm^3 of casein solution (containing 0.037 g dm^{-3} EDTA) was incubated with 1 cm^3 of enzyme solution for 20 min. Precipitated protein was removed by centrifugation for 15 min at 9000xg and then the absorbance of the supernatant was determined at λ=275nm (UV-Shimadzu 160A, Tokyo, Japan). One unit of enzyme activity was defined as the activity that liberates 4 nmol tyrosine per min per cm^{-3}. The protein content of the samples that was assumed to be equal to the total enzyme concentration was determined by the method of Lowry (Lowry *et al.,* 1951) with BSA as the standard.

2.3. CAKE RESISTANCE MODEL

Crossflow ultrafiltration is a pressure-driven process and gel polarisation, cake resistance, and osmotic pressure models characterise the process. According to the cake resistance model the solute rejected at the membrane surface results in the accumulation of molecules on the membrane, which leads to formation of a cake layer. The permeate flux (J) that is defined as the volumetric flow rate per unit area of the membrane may be expressed by Eq. (1):

$$J = \frac{\Delta P - \Delta \Pi}{R_{tot}} \tag{1}$$

where R_{tot} is total resistance, which is sum of the resistance of the membrane and the resistance due to cake layer; ΔP is the transmembrane pressure; and ΔΠ is the osmotic pressure difference. Since the osmotic pressure for macrosolutes are usually low, Eq.(1) is reduced to Eq.(2).

$$J = \frac{\Delta P}{R_{tot}} \tag{2}$$

In these equations ΔP is defined as:

$$\Delta P = \frac{P_i + P_o}{2} - P_p \tag{3}$$

where P_i is the measured pressure at the inlet of the membrane, P_o is the measured pressure at the retentate outlet, and P_p is the permeate pressure that is usually neglected due to its low value.

In this study, we calculated the total resistance R_{tot} by using the constant permeate flux obtained after a period of ultrafiltration time in Eq.(2). Since the hydraulic resistance of the membrane was constant in all experiments as a result of effective cleaning, the magnitudes of R_{tot} values showed us the effects of process parameters on the ultrafiltration performance.

3. Results and Discussion

3.1. EFFECT OF INITIAL ENZYME CONCENTRATION

The effect of initial total enzyme concentration on the permeate flux was investigated for C_E=0.082, 0.153 and 0.166 g dm^{-3} values at 10±2.5 kPa TMP and 0.5 m s^{-1} recirculation velocity. Different gradual decreases in the fluxes were seen depending on the enzyme concentration of the solution (Fig.1). Higher molecules than the MWCO of the membrane plug the pores leading to a decrease in membrane pore volumes, which cause an initial decrease in the permeate flux. The following cake layer formation, which results in concentration polarisation, continues the decrease in the permeate flux. Since the cake layer is formed earlier and thicker in concentrated solutions, lower ultimate permeate fluxes were obtained in comparison with diluted solutions. The increase in the total resistance R_{tot} with enzyme concentration is given in Table 1. We separated more concentrated enzyme solutions (from 0.178 to 0.347 g dm^{-3} at v=0.32 m s^{-1} and TMP=10 kPa) in our previous study and observed the similar trend in flux declines (Takaç *et al.*, 2000).

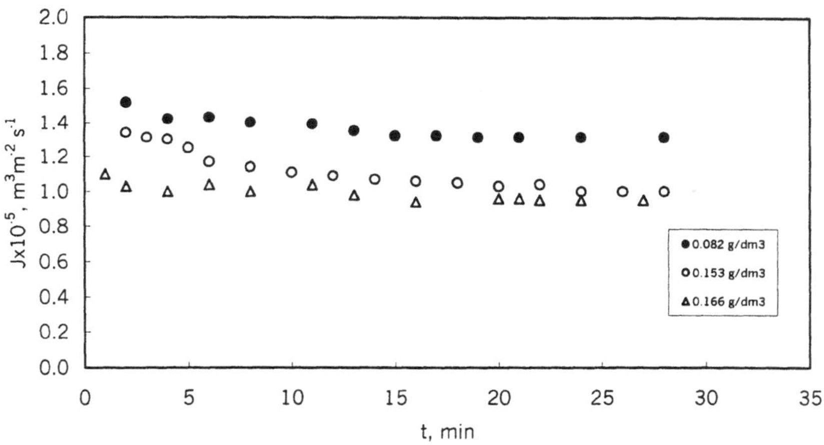

Fig.1. Effect of initial enzyme concentration on the permeate flux (TMP=10 kPa, v=0.5 m s^{-1})

Table 1. Effect of initial enzyme concentration on the total resistance and on the recovery yield of SAP in the permeate (TMP=10 kPa, v= 0.5 m s^{-1})

C_E, g dm^{-3}	R_tx10^{-9}, Pa m s^{-1}	Recovery yield of SAP in the permeate, %
0.082	0.76	82.5
0.153	1.00	54.3
0.166	1.05	55.8

3.2. EFFECTS OF RECIRCULATION VELOCITY AND TRANSMEMBRANE PRESSURE

Hydrodynamic conditions such as recirculation velocity and TMP are among the major parameters that affect the ultrafiltration performance. The influence of recirculation velocity on the permeate flux was investigated at v=0.07, 0.37 and 0.52 m s^{-1} values for 10±2.5 kPa TMP and 0.31 g dm^{-3} initial total enzyme concentration. The increase in the recirculation velocity increased the permeate flux (Fig.2). The velocity directly affected the shear rate and increased the rate of removal of cake from the membrane surface. The initial decrease in the flux at high recirculation velocities was less than observed at low values since the cake layer formation delayed at high rates. The decrease in the total resistance R_{tot} with recirculation velocity is given in Table 2. The results obtained in this study are in accordance with those obtained in our previous study for 0.1, 0.2, 0.3 and 0.5 m s^{-1} velocities at 0.105 g dm^{-3} enzyme concentration and for 0.38 and 0.50 m s^{-1} velocities at 0.28 g dm^{-3} enzyme concentration (Takaç *et al., 2000*).

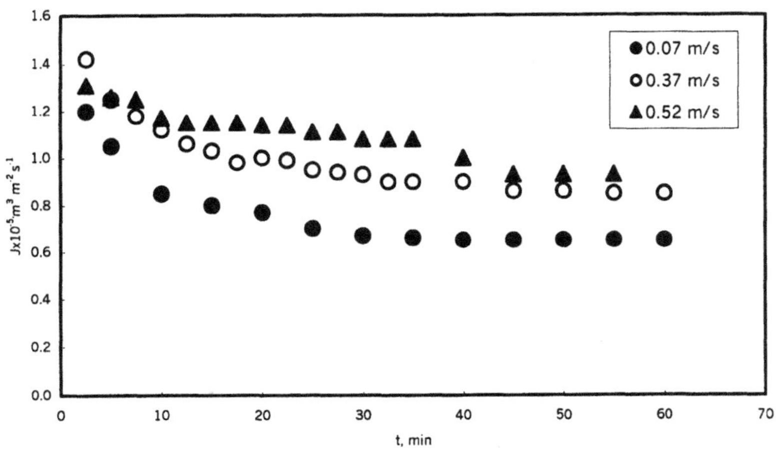

Fig. 2. Effect of recirculation velocity on the permeate flux (TMP=10 kPa, C_E=0.31 g dm^{-3})

Table 2. Effect of recirculation velocity on the total resistance and on the recovery yield of SAP in the permeate (TMP=10 kPa, C_E=0.25 g dm^{-3})

v, m s^{-1}	$R_t x10^{-9}$, Pa m s^{-1}	Recovery yield of SAP in the permeate, %
0.07	1.56	82.4
0.37	1.17	68.5
0.50	0.76	71.6

Transmembrane pressure serves as the driving force for ultrafiltration and increasing TMP results in increased flux at low pressures. However, above a limiting pressure the increase in flux decreases due to concentration polarisation (Lojkine *et al.*, 1992). In our previous study, we investigated the effect of transmembrane pressure on the permeate flux for 10±2.5 and 20±2.5 kPa initial pressures at 0.38 and 0.50 m s^{-1} recirculation velocities and 0.280 g/dm^3 initial total enzyme concentration (Takaç *et al.*, 2000). In each case, the permeate flux reached a constant value following an initial decrease and 20kPa TMP resulted in higher permeate fluxes. The cake layer resistance also decreased with increasing TMP. Throughout this study, however, initial TMP was kept constant at a low value i.e., at 10 kPa to reduce the rejection of molecules on the membrane surface; therefore, the effect of pressure on the ultrafiltration performance did not observe. Since not only increasing TPM but the values of recirculation velocity and enzyme concentration also affect the critical TMP value a detailed study is required to reduce the cake formation and to increase the permeate flux.

3.3. THE RECOVERED ACTIVITY OF SAP ENZYME AFTER SEPARATION

Ultrafiltration of enzymes differs from that of proteins in terms of loosing part or all of their catalytic activity in course of time during separation; and the conditions that give the highest permeate flux are not always optimum for the recovered enzyme activity after separation. We determined the activity of SAP by taking samples from the permeate tank in the end of experiments to compare it with the initial activity in the feed/retentate tank. The effects of enzyme concentration and recirculation velocity on the recovery yield of SAP in the permeate are given in Tables 1 and 2, respectively. Better separation of SAP from neutral protease and amylase enzymes with high recovery yield was achieved at low enzyme concentration and low recirculation velocity; where the increase in concentration decreased the recovery yield more drastically than the velocity caused. At the conditions used in the present study, SAP was best separated from high molecular weight enzymes with the recovery of 82% of its initial activity (Tables 1 and 2).

4. Conclusions

The present paper and first part of our previous paper (Takaç *et al.*, 2000) describe the influences of crossflow ultrafiltration conditions for the separation of SAP from neutral protease and amylase enzymes of *B. lichenifirmus* fermentation medium. Although high permeate flux ascertains high performance of an ultrafiltration process, the catalytic stability is the crucial factor in enzyme separations. In our both studies, a decrease in the permeate flux with time in different quantities depending on operation conditions was observed. Faster filtration rates were obtained with lower enzyme concentrations due to lower cake growth rates. The increase in recirculation velocity, however, increased the flux since high velocity limits the cake growth or polarisation layer formation. Accordingly, total resistance to flow in the ultrafiltration module calculated in this study increased with enzyme concentration and decreased with recirculation velocity. Due to possible shear inactivation, adsorption losses at the membrane surface or changes in the ionic environment during the ultrafiltration, the activity of the SAP enzyme decreased after each separation. The recovered SAP activity in the permeate decreased with recirculation velocity. Our results in the present study also show that initial total enzyme concentration more affected the recovered enzyme activity than recirculation velocity caused. This phenomenon may be explained by the complex structural interactions between enzyme molecules, which are more influenced by concentration.

Acknowledgements

The financial supports of The Scientific and Technical Research Center of Turkey (TÜBİTAK) Contract No MISAG-61 and SPO (Turkey) Contract No 96K120400 are gratefully acknowledged. Cumhuriyet University is also acknowledged for giving S.Elmas a leave grant for her MSc studies in Ankara University.

References

Bohdziewich J. (1994) Ultrafiltration of technical proteolytic enzymes, *Process Biochem.*, 29, 109-118.

Boyd,R.F. and Zydney,A.L. (1998) Analysis of protein fouling during ultrafiltration using a two-layer membrane model, *Biotechnol. Bioeng.*, 59(4), 451-460.

Burns, D.B. and Zydney, A.L. (1999) Effect of solution pH on protein transport through ultrafiltration membrane, *Biotechnol.Bioeng.*, 64(1), 27-37.

Çalık, P., Çalık, G. and Özdamar, T.H. (1998) Oxygen transfer effects in serine alkaline protease fermentation by *Bacillus licheniformis*: Use of citric acid as the carbon source, *Enzyme Microb. Technol.* 23, 451-461

Elmas, S.(1997) Separation of protease enzymes from the reaction medium by ultrafiltration, M.S.Thesis, Ankara University, Ankara.

Flaschel, E., Wandrey, C. and Kula, M-R. (1983) Ultrafiltration for the separation of biocatalysts. *Advances in Biochemical Engineering/ Biotechnology*, 26, 73-82.

Gekas,V., Aimar,P., Lafaille,J-P. & Sanchez,V. (1993) A simulation study of the adsorption concentration polarisation interplay in protein ultrafiltration, *Chem.Eng.Sci.*, 48(15), 2753-2765.

Gnosspelius, G. (1978) Purification and properties of an extracellular protease from *Myxococcus virences, J.Bacteriol.*, 133(1), 17-25.

Henriksen, P. and Hassager, O. (1993) Simulation of transport phenomena in ultrafiltration, *Chem.Eng.Sci.*, 48(17), 2983-2999.

Kalisz, H.M. (1988) Microbial Proteinases, *Advances in Biochemical Engineering/ Biotechnology*, 36, 1-65.

Lojkine,M.H., Field,R.W. and Howel,J.A. (1992) Crossflow microfiltration of cell suspensions: A review of models with emphasis on particle size effects, *TransIChemE C*, 70, 149-164.

Lowry, O.H., Rosebrough, N.T., Farr, A.L., Randall, R.J. (1951) Protein measurement with folin phenol reagent. *J.Biol. Chem.*, 193, 265-275.

Manachini, P.L., Fortina, M.G. and Parini, C.(1988) Thermostable alkaline protease produced by *Bacillus thermoruber* - a new species of *Bacillus, Appl.Microbiol.Biotechnol.*, 28, 409-413.

Pradanos, P., Arribas, J.I. and Hernandez, A. (1992) Hydraulic permeability, mass transfer and retention of PEGs in cross-flow ultrafiltration through a symmetric microporous membrane. *Sep.Sci.Tech.*, 27 (15), 2121-2142.

Pradanos, P., Arribas, J.I. and Hernandez, A. (1994) Retention of proteins in cross-flow UF through asymmetric inorganic membranes, *AIChEJ*, 40(11), 1901-1910.

Pradanos, P. and Hernandez, A. (1995). Cross-flow ultrafiltration of proteins through asymmetric polysulphonic membranes: I Retention curves and pore size distributions. *Biotechnol. Bioeng.*, 47, 617-625.

Pradanos, P. and Hernandez, A. (1996) Pore size distributions of polysulphonic UF membranes and protein adsorption. *Sep.Sci.Tech.*, 31(17), 2419-2441.

Pradanos, P., Hernandez, A., Calvo, J.I. and Tejerina, F. (1996) Mechanism of protein fouling in cross-flow UF through an asymmetric inorganic membrane, *J. Memb. Sci.*, 114, 115-126.
Reis, R., Gadam, S., Frautschy, L.N., Orlando, S., Goodrich, E.M., Saksena, S., Kuriyel, R., Simpson, C.M., Pearl, S. and Zydney, A.L. (1997) High performance tangential flow filtration, *Biotechnol.Bioeng.*, 56(1), 71-82.
Roders, V.G. J. and Sparks, R.E. (1991) Reduction of membrane fouling in the ultrafiltration of binary protein mixture, *AIChEJ.*, 37(10), 1517-1528.
Saksena, S. and Zydney, A.L. (1997), Influence of protein-protein interactions on bulk mass transport during ultrafiltration, *J.Membrane Science*, 125, 93-108.
Sheehan, J.J., Hamilton, B.K. and Levy, P.F. (1990) Pilot-scale membrane filtration process for the recovery of an extracellular bacterial protease. In: *Downstream Processing and Bioseparations* (Hamel, J-F.P., Hunter, J.B.& Sikdar, S.K, Eds.). *ACS Symposium Series,* Washington DC, 131-155.
Takaç, S., Elmas, S., Çalık, P., Özdamar, T.H. (2000), Separation of protease enzymes of *Bacillus licheniformis* from the fermentation medium by crossflow ultrafiltration, *J.Chem.Tech. Biotechnol.*, 75, 491-499.

Nomenclature

BSA	bovine serum albumin
C_E	enzyme concentration, g dm^{-3}
Da	dalton (1.67x10^{-24} g)
J	permeate flux, m^3 m^{-2} s^{-1}
MWCO	molecule weight cut-off, Da
P_i	inlet pressure, kPa
P_o	outlet pressure, kPa
P_p	permeate pressure
R_{tot}	total resistance, Pa s m^{-1}
SAP	serine alkaline protease
TMP	transmembrane pressure, kPa
t	ultrafiltration time, min
v	recirculation velocity, m s^{-1}
vvm	volume volume^{-1} min^{-1}
ΔP	TMP, kPa
$\Delta \Pi$	osmotic pressure difference

PART IV
MONITORING AND CONTROL

EVALUATING K_LA DURING FERMENTATION USING MANY METHODS SIMULTANEOUSLY

K. POULIOT[1], J. THIBAULT[2], A. GARNIER[1], G. ACUNA LEIVA[3]
[1]Department of Chemical Engineering, Laval University, Sainte-Foy (Quebec), Canada G1K 7P4; [2]Department of Chemical Engineering, University of Ottawa, Ottawa (Ontario) Canada K1N 6N5; [3]Departamento de Ingenieria Informática, Universidad de Santiago de Chile, Avda. Ecuador 3659, Casilla 10233, Santiago, Chile

Abstract

The oxygen mass transfer coefficient often serves to compare the efficiency of bioreactors and their mixing devices as well as being an important scale-up factor. In submerged fermentation, four methods are available to estimate the overall oxygen mass transfer coefficient (K_La): the dynamic method, the stationary method based on a previous determination of the oxygen uptake rate ($Q_{O2}X$), the gaseous oxygen balance and the carbon dioxide balance. Each method provides a distinct estimation of the value of K_La. Data reconciliation was used to obtain a more probable value of K_La during the production of *Saccharomyces cerevisiae*, performed in 22.5-litre fed-batch bioreactor. The estimate of K_La is obtained by minimising an objective function that includes measurement terms and oxygen conservation models, each being weighted according to their level of confidence. Weighting factors of measurement terms were taken as their respective inverse variance whereas weighting factors of oxygen conservation models were obtained using Monte Carlo simulations. Results show that more coherent and precise estimations of K_La are obtained.

1. Introduction

The supply of oxygen is a critical factor in all aerobic fermentations. An insufficient oxygen transfer leads to a decrease of microbial growth and product formation. In order to assess if particular equipment would be able to supply oxygen at a non-limiting rate, it is essential to have a good estimate of the oxygen mass transfer coefficient (K_La). In

M. Hofman and P. Thonart (eds.), Engineering and Manufacturing for Biotechnology, 183–201.

submerged fermentation, the oxygen mass transfer coefficient serves to compare the efficiency of bioreactors and their mixing devices. It is also one of the most important scale-up factors [1,2].

Many methods for the determination of K_La in submerged fermentation have been proposed. The majority of investigations have however been performed with water and other model fluids, in an attempt to mimic as closely as possible conditions encountered in fermentation systems. These investigations are very useful because conditions are well defined and can be rigorously controlled, and provide fairly good estimates of the oxygen mass transfer that can be used in design calculation. The determination of oxygen absorption from air into the fermentation broth should however be assessed under actual operating conditions of fermenters since the rate of oxygen absorption into a culture medium can be greatly affected by the presence of microorganisms, substrate, substances excreted by microorganisms and antifoam [3]. K_La values in fermenters often differ substantially from values predicted for oxygen absorption into water or simple aqueous solutions even when differences in liquid physical properties such as viscosity and diffusivity are taken into account [4]. Methods for K_La determination during the course of fermentation are normally classified as dynamic or steady state methods. The technique of dynamic measurements normally consists of following the dissolved oxygen concentration during a step change in the inlet gas concentration. Only a fast response dissolved oxygen probe is required to obtain the necessary data. The steady-state methods are based either on a global oxygen balance or a global carbon dioxide balance in the gas phase of the bioreactor. A fast response dissolved oxygen probe, and oxygen and carbon dioxide sensors are required. An additional steady-state method based on a prior estimation of the oxygen uptake rate ($Q_{O2}X$) is also used to estimate K_La. These methods are described with more details in the Materials and Methods section.

In principle, the value of K_La obtained should be independent of the method employed and K_La values estimated with the dynamic and steady state methods should indeed be identical. In fact, this ideal situation is rarely met and each method provides a distinct estimation of K_La. This may lead to some design problems in scaling-up fermenters if the scale-up method that is used is to maintain a constant oxygen mass transfer coefficient. Moreover, the accuracy of K_La values obtained by those methods is based on the reliability of the measured data. Brown [5] indicates that even after a measuring device has been developed, there still remains problem of the reliability of the signal value. The calibration of the sensor and stability of the reading over a period of time are a part of the sensitivity problems that have to be taken into account. A large number of measured variables contain some degree of error. Therefore, data reconciliation techniques could be used with advantage to come up with the most probable value of K_La. In data reconciliation, both the reliability of data measurements and the accuracy of each estimation method are taken into consideration. Data reconciliation essentially consists of writing and minimising an objective function that considers the level of confidence on the various measurements and the oxygen conservation models.

In the present investigation, K_La has been evaluated by four different methods for the culture of *Saccharomyces cerevisiae* in a fedbatch bioreactor and data reconciliation has been used to determine a better estimate of the oxygen mass transfer coefficient. The objective function is composed of the weighted sum of 12 measurement terms and 6 terms for oxygen conservation models. The weight associated to each measurement term is the inverse of the measurement variance and the weights of conservation models have been estimated using a Monte Carlo method. The paper is divided as follow: after a description of the experimental system and a review of the methods for measuring K_La, the data reconciliation technique is presented and the main results are presented and discussed.

2. Materials and methods

2.1. ORGANISM AND MEDIUM

The strain used in this study was *Saccharomyces cerevisiae*. The culture medium composition was: 0.5 g peptone, 3 g yeast extract, 1.25 g KH_2PO_4, 1.25 g K_2HPO_4, 0.45 g $MgSO_4{\bullet}7H_2O$, 5.33 g $(NH_4)_2SO_4$, and 1.1 g glucose per litre of water. Two bags of 8 g of Fleischmann's quick-rise yeast were used as inoculum and were added to an Erlenmeyer flask containing 750 ml of the medium given above. The cells were incubated on an orbital shaker at 25°C for 1.5 hours before being added to the fermenter.

Glucose was used as the carbon and energy source. The growth behaviour of *Saccharomyces cerevisiae* is strongly influenced by glucose concentration. To avoid the Crabtree effect, glucose was fed in order to maintain a low concentration within the bioreactor. The Crabtree effect occurs at large glucose concentrations in an aerobic environment. The glucose is then predominantly fermented instead of being oxidised and ethanol and carbon dioxide are produced. Since one of the methods used to estimate K_La value is based on the carbon dioxide production rate, this phenomenon would induce a systematic error in the K_La evaluation. Glucose oxidation can only be predominant in continuous cultures or fedbatch fermentations [6]. Thus, fed-batch fermentation has been performed under specific conditions. An optimisation procedure has been used to determine the glucose feeding rate as function of time. The objective was to maintain the respiratory quotient (RQ) around unity [7,8]. RQ corresponds to the ratio of the carbon dioxide evolution rate to the oxygen uptake rate. Therefore, a peristaltic pump (Masterflex Model 7521-50 with no. 13 C-Flex tubing) continuously fed a solution of 200 g/L glucose according to the rate determined by an optimisation routine. The cultivation medium, the Erlenmeyer flask containing the medium for the inoculum and the glucose solution were sterilised separately at 121°C and 200 kPa for 25 minutes. Antifoam agent (Dow Corning, Emulsion C for food grade) has been used whenever necessary.

2.2. EXPERIMENTAL SYSTEM

Fermentations were carried out in a baffled stirred tank reactor constructed in our laboratory. The fermenter is made of two concentric stainless steel columns. The annular section, in which water continuously circulates, is used as a heat exchanger to maintain a constant temperature inside the fermenter. The central column has an internal diameter of 228 mm and a height of 550 mm. The bioreactor has a total volume of 22.5 L.

The impeller was driven by a mechanical system composed of a motor (90 VDC, 1800 RPM, ½ HP, Frame 56C, Model 8293, Pacific-Scientific) and a ten-to-one speed reducer assembly (Model 201657, Doerr Electric). A speed controller (Multi-Drive Model KBMD-240D, KB Electronics, Inc.) connected to the motor allows a variable speed in both clockwise and counter-clockwise rotation. Three Rushton turbines were mounted on the central shaft. Each turbine has 6 blades mounted on the periphery of a 50-mm diameter disk. Four baffles were placed inside the mixing vessel to favour turbulence and to prevent the formation of vortices.

Dissolved oxygen was measured with an amperometric oxygen electrode (Ingold, Model P/N 40179-02). The mass flow rate of compressed air or nitrogen, fed at the base of the column, was controlled with a mass flow meter (Matheson, Model 8272-0414). The gas sparger was a perforated plate that contains one hundred uniformly distributed holes, 1 mm in diameter. The concentration in the off gases was measured on-line. The O_2 was detected by paramagnetism (Maihak, Multor 610) while CO_2 was detected by infrared (Maihak, Multor 610). A silica gel column was used to dehumidify the exhaust air of the fermenter. Periodic measurements of the inlet air composition have also been performed. The data acquisition was done by a multiplexer connected to a personal computer.

In this investigation, the operating conditions have been kept constant throughout the fermentation. The stirring speed was 400 RPM and the airflow rate was 10 L/min. The volume of the medium in the fermenter varies from 15 L to 19.1 L since a glucose solution was continuously added.

2.3. REVIEW OF THE METHODS FOR MEASURING K_LA DURING THE COURSE OF FERMENTATION

Two experimental on-line methods have been proposed to determine the value of K_La during the course of fermentation: the dynamic method and the overall gas balance. These two methods make use of the fate of the dissolved oxygen within the fermenter that is given by the following equation:

$$\frac{dC_L}{dt} = K_La\left(C_L^* - C_L\right) - Q_{O2}X \tag{1}$$

This equation states that the rate of change of the dissolved oxygen in the fermenter is equal to the rate of oxygen mass transfer from the gas to the liquid phase minus the rate of oxygen utilisation by the microorganisms.

2.3.1. Dynamic method

In the dynamic method, first reported by Taguchi and Humphrey [9], the oxygen uptake rate (OUR or $Q_{O2}X$) and K_La are determined in turn using the following procedure. The gas supply and the agitation are stopped momentarily to cut the oxygen supply to the liquid phase so that the rate of decrease of dissolved oxygen is caused entirely by the OUR. The decrease in dissolved oxygen is usually linear and the slope of the plot of C_L as a function of time provides a direct estimate of the oxygen uptake rate (Figure 1). The underlying hypothesis is that the rate of oxygen utilisation is unaffected by the absence of air bubbling and agitation, and lower dissolved oxygen concentration.

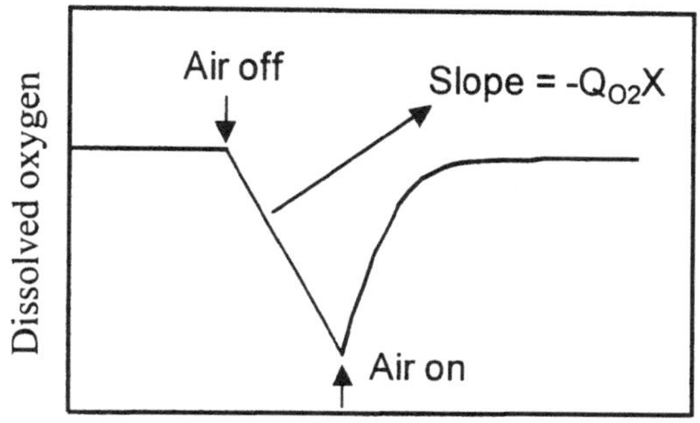

Fig. 1 Determination of total oxygen consumption rate using the dynamic method.

Before the dissolved oxygen concentration reaches its critical lower limit, aeration and agitation are resumed and the dissolved oxygen concentration normally returns to its initial level. K_La can be estimated using Equation (1), reformulated in terms of the dissolved oxygen concentration.

$$C_L = C_L^* - \frac{1}{K_L a}\left(\frac{dC_L}{dt} + Q_{O2}X\right) \tag{2}$$

Equation (2) is the equation of a straight line and K_La can be readily obtained from the slope of this line. Estimations of $Q_{O2}X$ and K_La using this method are somewhat dependent on the section of the curves that are used for their evaluation. In addition, the dynamics of the dissolved oxygen probe is not taken into account and could induce an important bias for higher values of K_La. The probe dynamics can be included by solving simultaneously the two differential equations (Equation 1 and a first order differential equation for the probe dynamics) to form an analytical solution or by solving the two dynamic differential equations using finite difference methods. It is important to point out that the estimate of K_La using the dynamic method does not depend on the estimation of $Q_{O2}X$. Indeed, it can easily be shown that the dynamic response, after aeration and agitation are resumed, is independent of $Q_{O2}X$ apart from having an influence on the final level of dissolved oxygen.

In our experiments, a small variant of this method has been used. Aeration was cut momentarily whereas the agitation was reduced to 30 RPM in order to maintain cells in suspension. Indeed, if the agitation and the aeration are both stopped, cells decant and oxygen consumption is higher at the bottom of the fermenter. The probe cannot measure this consumption since it is located near the middle of the fermenter. With this low agitation rate, both K_La and surface aeration are negligible [10] so that the first term of the right hand side of Equation 1 can be safely put to zero and $Q_{O2}X$ evaluated precisely.

2.3.2. Steady-state methods

Three steady-state methods are available to determine K_La. The gaseous oxygen mass balance method, the gaseous carbon dioxide mass balance technique and a method that needs a prior estimation of $Q_{O2}X$. Under pseudo-steady-state conditions, Equation (1) allows to readily calculate K_La using the following equation:

$$K_L a = \frac{Q_{O2} X}{C_L^* - C_L^0} \tag{3}$$

A previous estimation of $Q_{O2}X$ is however needed and can be obtained from the dynamic method. This steady-state method will be referred as the stationary method.

Under steady-state conditions, the oxygen deficit of the gas stream across the fermenter must be equal to the oxygen uptake rate. Equation 4 allows determining K_La by the gaseous oxygen mass balance technique:

$$K_L a = \frac{\dfrac{1}{V_L}\left(\dfrac{P_1}{R\,T_1} Q_{1,G}\, y_{1,O2} - \dfrac{P_2}{R\,T_2} Q_{2,G}\, y_{2,O2}\right)}{(C_L^* - C_L^0)} \tag{4}$$

Finally, the gaseous carbon dioxide production rate can also be used as described by the following equation:

$$K_L a = \frac{\dfrac{1}{RQ}\dfrac{1}{V_L}\left(\dfrac{P_2}{R\,T_2} Q_{2,G}\, y_{2,CO2} - \dfrac{P_1}{R\,T_1} Q_{1,G}\, y_{1,CO2}\right)}{(C_L^* - C_L^0)} \tag{5}$$

To use this method, an estimation of the respiratory quotient (RQ) must be available a priori. A good estimate of RQ is available for a large number of fermentations or can also be estimated from past fermentations.

Since it is now more common for fermentation systems to be equipped with an O_2/CO_2 monitor or mass spectrometer [11], up to four methods are available to determine K_La, thereby leading to four different estimates. A simple average of the four values could be taken to give a unique and more precise value of K_La. However, as some methods are more accurate than others at different stages of fermentation, averaging K_La is not the best method to achieve a more precise value. Instead, a data reconciliation technique could be used with advantage to resolve this problem. This technique, briefly described in the next section, considers the precision of each measurement and each estimation method to provide the best estimate of the K_La value.

2.4. DATA RECONCILIATION METHOD

The availability of accurate process data is no doubt the prerequisite condition to successful cost accounting, process control, statistical quality analysis, and performance evaluation. Unfortunately, measurements of variables in most processes are generally subject to significant random and non-random errors. It is therefore important that these measurement errors be corrected before using the measured variables for final process analysis and control. Data reconciliation, as an on-line optimisation method, is sometimes used in process industries to reduce these errors in order that the adjusted or reconciled values of the process measurements around a given system obey conservation laws as well as other physical and chemical constraints. Steady-state material and/or energy balances are commonly used as the constraints. It is important to point out that, to perform process data reconciliation, the measured data must be redundant; that is, there exist more measured data than are necessary to satisfy system balances.

A large amount of research work has been devoted to data reconciliation problems [12-18]. The measured data are generally contaminated with different types of system noises and, at the same time, system models are seldom a perfect representation of the underlying behaviour of the process so that data reconciliation has to be performed by taking into account both the measurement errors and the process modelling errors. To achieve this dual objective, general criteria for data reconciliation can be defined to take simultaneously into account all measured variables and all conservation models, each one being affected by a weight that corresponds to the level of confidence that one has in each measurement and each conservation model. This dual objective has been used successfully for mineral processing plants [14,18].

In the case of the fermentation system described earlier, a series of measured or estimated variables is required for the determination of the overall oxygen mass transfer coefficient (K_La) by at least one method. These variables are: the pressure (P), the temperature (T), the time constant of the dissolved oxygen probe (τ_P), the saturated dissolved oxygen concentration ($\tilde{C}_L^*$), the dissolved oxygen concentration (C_L), the liquid volume (V_L), the respiratory quotient (RQ), the gas flow rate (Q_G), the oxygen mole fraction in the inlet gas ($y_{1,O2}$), the oxygen mole fraction in the outlet gas ($y_{2,O2}$), the carbon dioxide mole fraction in the inlet gas ($y_{1,CO2}$), and the carbon dioxide mole fraction in the outlet gas ($y_{2,CO2}$). These represent a total of 12 variables that are either measured or estimated, and to each of them is associated a certain level of accuracy. In addition, six oxygen mass conservation equation models that make use of these variables are available. It is therefore possible to write an objective cost function that is comprised of the weighted sum of all 12 measurements or estimations, and all oxygen conservation models. To better understand the make up of the complete objective function, it has been divided into partial objective functions. The first partial objective function (J_1) considers all measurements or estimations that are involved in the determination of K_La by one method or another:

$$J_1 = \alpha_1 \left(\hat{P} - P\right)^2 + \alpha_2 \left(\hat{T} - T\right)^2 + \alpha_3 \left(\hat{\tau}_P - \tau_P\right)^2 + \alpha_4 \left(\hat{\tilde{C}}_L^* - \tilde{C}_L^*\right)^2 + \alpha_5 \left(\hat{C}_L^0 - C_L^0\right)^2 + \alpha_6 \left(\hat{V}_L - V_L\right)^2 + \alpha_7 \left(\hat{RQ} - RQ\right)^2 + \alpha_8 \left(\hat{Q}_G - Q_G\right)^2 + \alpha_9 \left(\hat{y}_{1,O2} - y_{1,O2}\right)^2 + \alpha_{10} \left(\hat{y}_{2,O2} - y_{2,O2}\right)^2 + \alpha_{11} \left(\hat{y}_{1,CO2} - y_{1,CO2}\right)^2 + \alpha_{12} \left(\hat{y}_{2,CO2} - y_{2,CO2}\right)^2 \quad (6)$$

The hat indicates the estimated values. This partial objective function implies that the estimated values should be kept as close as possible to their respective measured values.

The next partial objective function (J_2) considers the estimation of $Q_{O2}X$ from the slope of the first portion of the curve associated to the dynamic method, that is where agitation has ceased, thereby rendering the supply of dissolved oxygen negligible.

$$J_2 = \alpha_{13} \left(\frac{dC_L}{dt} + \hat{Q_{O2}X} \right)^2 \quad (7)$$

The partial objective function (J_3) of the second portion of the curve associated with the dynamic method and which provides an estimate of K_La is given by the following equation:

$$J_3 = \alpha_{14} \left(C_L - \left(\hat{C}_L^* - \frac{1}{\hat{K_La}} \left[\frac{dC_L}{dt} + \hat{Q_{O2}X} \right] \right) \right)^2 \quad (8)$$

Partial objective functions J_2 and J_3 can simply be solved by evaluating respectively the slope of the rate of change in dissolved oxygen when aeration and agitation are stopped, and the slope of Equation (2) when aeration and agitation are resumed. In this investigation, Equation (1) was solved by finite differences simultaneously with the dynamics of the dissolved oxygen probe for both portions of the dynamic curve. For the decreasing portion of the curve, K_La was set to zero. The dynamics of the dissolved oxygen probe was simulated using the following first order equation:

$$\hat{\tau}_P \frac{d\hat{C}_P}{dt} = \hat{C}_L - \hat{C}_P \quad (9)$$

The partial objective functions J_2 and J_3 were therefore set equal, for the two portions of the dynamic response curve, to the average difference between the predicted and experimental probe dissolved oxygen concentrations. The predicted dissolved oxygen concentration, as measured by the probe, was simulated by finite differences using the most current estimates of $\hat{C}_L^*, \hat{C}_L^0, \widehat{K_La}, \widehat{Q_{O2}X}$, *and* $\hat{\tau}_P$.

The partial objective function (J_4) associated to the estimation of K_La using Equation (3) is given by:

$$J_4 = \alpha_{15}\left(\widehat{K_La} - \frac{\widehat{Q_{O2}X}}{\hat{C}_L^* - \hat{C}_L^0}\right)^2 \tag{10}$$

The next two partial objective functions (J_5 and J_6) are very similar to the previous partial objective function. The difference lies in the way the estimation of the oxygen uptake rate is performed. The two partial objective functions use the information provided by O_2 and CO_2 gas balances:

$$J_5 = \alpha_{16}\left(\widehat{K_La} - \frac{1}{\hat{C}_L^* - \hat{C}_L^0}\frac{1}{\hat{V}_L}\frac{\hat{P}}{R\hat{T}}\hat{Q}_G\left(\hat{y}_{1,O2} - \hat{y}_{2,O2}\right)\right)^2 \tag{11}$$

$$J_6 = \alpha_{17}\left(\widehat{K_La} - \frac{1}{\hat{C}_L^* - \hat{C}_L^0}\frac{1}{\hat{V}_L}\frac{1}{\widehat{RQ}}\frac{\hat{P}}{R\hat{T}}\hat{Q}_G\left(\hat{y}_{2,CO2} - \hat{y}_{1,CO2}\right)\right)^2 \tag{12}$$

It is assumed that the gas flow rate, the pressure and the temperature were identical for the inlet and outlet gas streams. This is a reasonable assumption since the respiratory coefficient is close to unity, and the outlet stream is dehumidified before being analysed.

Since the saturated dissolved oxygen concentration depends on the average gaseous oxygen concentration within the fermenter, on the total pressure and the chemical components of the fermentation broth, an additional term (J_7) was added to the overall objective function to include its estimate:

$$J_7 = \alpha_{18}\left(\hat{C}_L^* - \hat{\tilde{C}}_L^*(\hat{T},S)\frac{\hat{y}_{1,O2} + \hat{y}_{2,O2}}{2\,\tilde{y}_{O2}}\frac{\hat{P}}{\tilde{P}}\right)^2 \tag{13}$$

The value of the saturation dissolved oxygen concentration ($\hat{\tilde{C}}_L^*$) was evaluated at the temperature of the fermenter broth and corrected for its initial chemical composition [19].

All the partial objective functions can now be joined together to form the overall data reconciliation objective function expressed by the following equation:

$$J = J_1 + \beta\left(J_2 + J_3 + J_4 + J_5 + J_6 + J_7\right) \tag{14}$$

The parameter β is an additional weighing factor that is used to put more or less emphasis on the oxygen conservation models with respect to the measurements.

A net advantage of using data reconciliation methods is that the user is forced to examine the precision of all measured variables and conservation models, and the relative precision of all terms of the objective function are considered in estimating a unique value of K_La.

2.4.1. Weighting factors

To solve the objective function and obtain the most probable value of K_La, weighting factors (α_i, $i \in (1, 18)$) must be specified according to the respective level of confidence on each measurement and oxygen conservation model. Assuming that measurement errors are normally distributed, the level of confidence of a process variable is related to its variance. The weighting factor for each measurement term was therefore set equal to its inverse variance. The estimation of the variance of each measurement considers the accuracy of the sensing device, calibration errors, stability of the signal over a period of time, and measurement errors. Table 1 gives the estimated precision ($\pm 3\sigma$) of each process variable that is used in oxygen conservation models to determine K_La. Table 1 also gives the range of variation of each process variable during a typical fermentation performed in this investigation. The saturated dissolved oxygen concentration ($\tilde{C}_L^*$) takes into account the concentration of nutrients, which affect the oxygen solubility [19]. The time constant of the dissolved oxygen probe (τ_P) was evaluated with separate experiments in water under identical conditions of agitation and aeration.

The estimation of K_La values using oxygen conservation models largely depends on the precision of measurements used in each conservation model. Weighting factors of conservation models must therefore be determined considering the individual precision of each measured variable. Since the precision of each conservation model varies during the course of fermentation, it is necessary to determine weighting factors for each individual experimental test performed at regular intervals. For instance, in the last portion of the exponential growth phase, provided that all other operating conditions remain identical, estimation of K_La using the oxygen and carbon dioxide mass balance is more accurate because the differences between the input and the output concentrations are larger. To obtain the respective weighting factors for each individual experiment, a Monte Carlo simulation method was used. Monte Carlo simulation is a mathematical technique for numerically solving differential or algebraic equations. It is used extensively in science to solve many problems for which no other solutions exist. It consists of using random numbers to generate a large number of possible scenarios and the results of the many scenarios are analysed statistically to obtain information on estimated average value and variability of possible results. The answers are always approximate, but with sufficient number of scenarios, tend to converge to the theoretical answers. In the present investigation, 1000 simulations were performed to evaluate partial objective functions J_2 to J_7. In each simulation, the 12 process variables were generated in the range of $\pm 3\sigma$ of each variable measured at each fermentation time using random gaussian numbers. Then, the variance of each partial objective function was evaluated in order to provide the relative accuracy of each oxygen conservation term in the overall objective function. The inverse of the calculated variance of each partial objective function was used as the weighting factor.

Table 1 Range of variation of each process variable during a typical fermentation and estimated precision ($\pm 3\sigma$) of each variable.

Parameter	*Range of variation*	*Precision ($\pm$ 3σ)*
P	101 300 Pa	± 3000
T	303.2 K	± 1
τ_P	4.0 s	± 0.5
$\tilde{C}_L^*$	7.4 ppm	± 0.2
C_L	0 - 7.4 ppm	± 0.2
V_L	15.0 – 19.1 L	± 0.1
RQ	1.2	± 0.2
Q_G	10.0 L/min	± 0.1
$y_{1,O2}$	0.209 mol/m^3	± 0.001
$y_{2,O2}$	0.183 - 0.209 mol/m^3	± 0.002
$y_{1,CO2}$	0.0000 mol/m^3	± 0.0005
$y_{2,CO2}$	0.0000 – 0.0315 mol/m^3	± 0.001

3. Results and discussion

Data reconciliation methods are frequently used on-line to provide more coherent and accurate values of process variables that are subsequently used in control strategies and yield calculations. In the particular application of K_La estimation, an instantaneous value of K_La can only be obtained from the gas stream analysis and a better and more coherent value can be determined only after the completion of the dynamic method, when an estimate of $Q_{O2}X$ is available and the value of K_La can be estimated using the four different methods.

To obtain an instantaneous value of K_La, both the oxygen and carbon dioxide mass balance methods can be used. However, at the beginning of the fermentation, the differences of oxygen and carbon dioxide concentrations between the inlet and the outlet gas streams are generally low. The accurate estimation of K_La by these two methods strongly relies on the precision of the O_2 and CO_2 sensors, and of the mass flow meter. K_La estimation using the oxygen mass balance method can be directly calculated from experimental measurements whereas the estimation with carbon dioxide mass balance method requires the a priori knowledge of the respiratory quotient (RQ).

In this investigation, the purpose of using a fed-batch system to produce *S. cerevisiae* was to maintain the respiratory quotient as close as possible to unity during the course of the fermentation. A RQ of unity corresponds to the purely oxidative metabolism under which only carbon dioxide is formed, without ethanol formation. During aerobic growth of *S. cerevisiae*, ethanol is formed when the glucose concentration exceeds a certain critical level. In practice, under stringent control of glucose concentration, the RQ slightly exceeds unity due to the anabolic process and it is approximately equal to 1.04 [20,21]. Because the value of RQ must be known a priori in order to use the carbon dioxide mass balance method to estimate K_La, its value was estimated from prior experiments. The nominal value, used in the data reconciliation objective function, was 1.2 with a precision of ±0.2 (Table 1). The RQ value remained in this range for the majority of the fermentation and it was therefore assumed constant. Toward the end of fermentation, RQ values had the tendency to increase as high as 1.6. The influence of RQ will be discussed later.

The overall oxygen mass transfer coefficient has been estimated by the four methods at a time interval of approximately 1.5 h for a total fermentation time of 24 h. For each set of data, Monte Carlo simulations were performed to determine the respective weighting factors for each partial objective function. The overall objective function was minimised using a quasi-Newton optimisation routine in order to obtain estimated values of the 12 measurements of Table 1 as well as estimates of $Q_{O2}X$ and K_La, for a total of 14 parameters. Initial estimates of $Q_{O2}X$ and K_La were obtained graphically using the original method proposed by Taguchi and Humphrey [9]. In the present investigation, equations associated with the dynamic method were solved by finite differences in order to include the dynamics of the dissolved oxygen probe. The sequence of data associated with the dynamic method was split into two segments, and two partial objective functions (J_2 and J_3) were defined to minimise the sum of squares of the differences between the experimental and predicted responses of the dissolved oxygen probe. Figure 2 compares the predicted response, using the converged values of K_La and $Q_{O2}X$, with the experimental data for the test performed at a fermentation time of 13.5 h. It is clear that the converged values lead to a good fit of the experimental data.

The estimation of K_La with the three other steady-state methods depends on the converged values of the 12 measurements and the converged value of $Q_{O2}X$. Figure 3 illustrates the variation of the converged oxygen consumption rate as a function of time. It shows that the maximum metabolic activity was reached at around 16 h. Given all the weighting factors, the data reconciliation routine determines the best estimate of K_La to minimise the overall objective function. Figure 4 compares K_La values obtained by data reconciliation and values obtained by taking a simple arithmetic average of the four initial K_La values. The initial values correspond to the values estimated by each method before applying data reconciliation. Results indicate that the average K_La value increases with time whereas the converged K_La remains fairly constant. It is normally accepted that for a non viscous fermentation, such as the production of *S. cerevisiae*, K_La remains relatively constant during the course of fermentation if the air flow rate, the agitation speed and all the other experimental conditions are kept constant. Small K_La changes could occur since it is slightly affected by cell concentration, metabolite production, and addition of antifoaming agent [3,22]. These changes could lead to slight K_La variation but never to the extent depicted by the variation of the arithmetic average. Converged K_La values obtained by data reconciliation also varied but swayed around 0.053 s^{-1} with a standard deviation of ± 15%. It is postulated that data reconciliation provides more precise K_La values since it takes into account the possible errors of the various measurements and the relative accuracy of each oxygen conservation model. For a fermentation time superior to 20 h, high estimated values of K_La obtained by the simple arithmetic average are partly due to the respiratory quotient estimation. After a fermentation time of 20 h, the average respiratory quotient is around 1.60 whereas it was assumed constant during the entire fermentation with a value of 1.20. This implies an overestimation of K_La values by the carbon dioxide method and, as a result, the arithmetic average value of K_La is significantly biased. Data reconciliation has allowed taking into account this error.

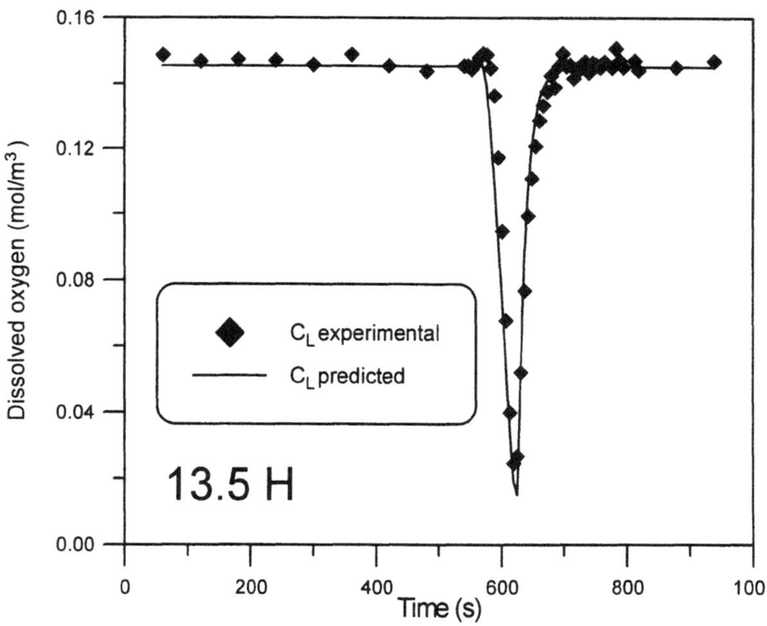

Fig. 2 Dissolved oxygen concentration variation as a function of time when K_La is estimated by the dynamic method.

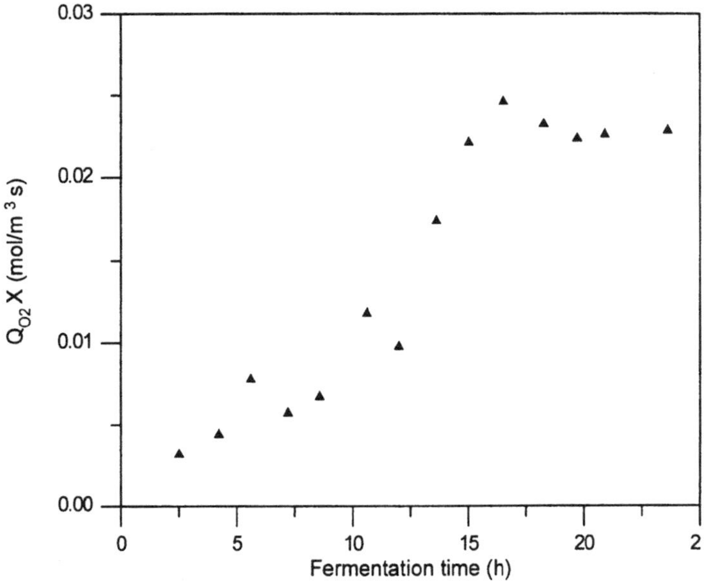

Fig. 3 Variation of the converged oxygen consumption rate as a function of time.

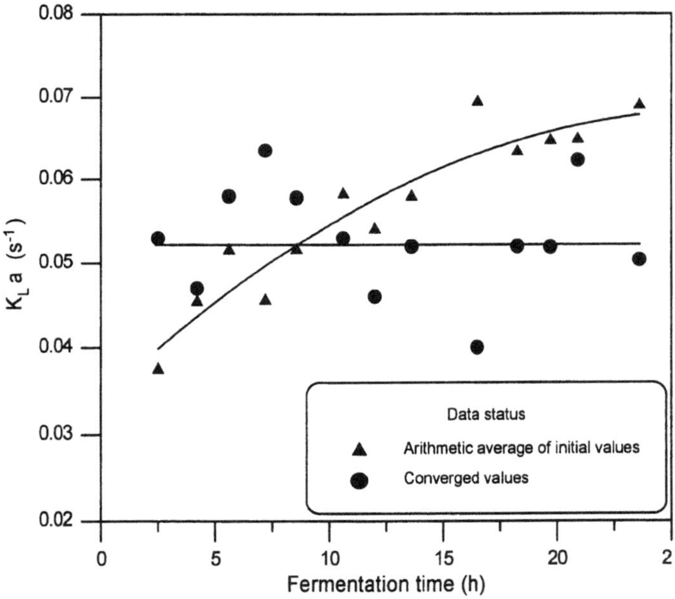

Fig. 4 Converged K_La values obtained by data reconciliation and values obtained by a simple average of the four initial values.

Figure 5 presents the converged K_La values and those obtained by each of the four methods at five different fermentation times. Values presented for the dynamic method are not reconciled. They were obtained using the original graphical method proposed by Taguchi and Humphrey [9]. The estimation of these K_La values does not take into account the measurement errors and the dynamics of the dissolved oxygen probe. As a result, variations of estimated K_La values are significantly larger than the reconciled values. The estimations of K_La with the other three methods were calculated using the converged values of the 12 parameters and of the value of $Q_{O2}X$. The stationary method is the one that provides values that have the smallest variability and that are closer to final reconciled values. This good prediction is due to the excellent accuracy in evaluating the dissolved oxygen concentration and the oxygen utilisation rate that are used in the stationary oxygen conservation model (Eq. 3). K_La values estimated by the two gas mass balance methods tend to increase during the course of fermentation. The increase is more pronounced for K_La values estimated with the difference in CO_2 concentration. This overestimation is due to the RQ estimation. After 20 h of fermentation, the average respiratory quotient is in the vicinity of 1.60 instead of the assumed constant value of 1.20. The data reconciliation technique modifies slightly the value of RQ to reduce the overestimation of K_La. Ways to improve the estimation of K_La by the CO_2 gas balance method are: (1) to reduce the weighting factor of the estimated RQ term in the objective function, toward the end of fermentation, to allow the optimisation routine to make greater changes in RQ, and (2) to estimate on-line the current value of RQ some time before a K_La estimation is performed instead of assuming a constant value during the entire fermentation.

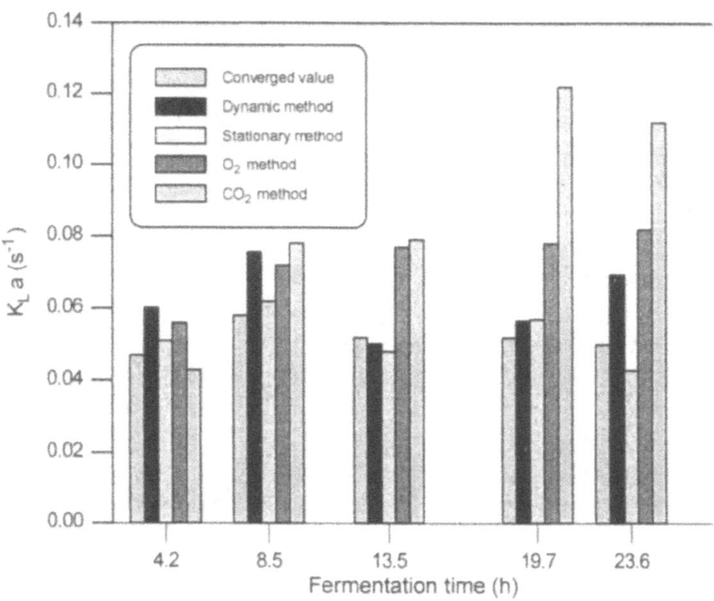

Fig. 5 Converged K_La values and values obtained by each of the methods at five different fermentation times.

Since data reconciliation forces the user to examine the precision of all measured variables and conservation models, an additional weighting factor β (Eq.14) was incorporated in the model. This weighting factor serves to put more or less emphasis on the oxygen conservation models with respect to the measurements. Results presented previously were obtained for a weighing factor β equal to 1. Table 2 presents K_La values for a variation of β between 0.001 and 1000 at a fermentation time of 8.5 h. When more emphasis is put on models, K_La obtained by each method tends to be closer to the converged value. This however implies that the optimisation algorithm can make important changes in the values of the 12 parameters, and if the value of β is excessively high, unrealistic values of variables are obtained. On the other hand, when significantly less emphasis is put on the models, the objective function boils down in minimising the sum of squares between the estimated and measured values. A weighting factor smaller than 0.01 or greater than 1000 has no further impact on the outcome of the estimation of K_La. The selection of the value of β is partly subjective but the results of Table 2 show that the range of variation of K_La with β is not excessively large so that no serious damage would occur if a value of β slightly different than the optimal would be used. Users can choose to put more or less importance on the variables or the models but, in general, it is recommended to use a weighting factor β equal to the unity since Monte Carlo simulations determine the level of confidence for the oxygen conservation models and take into account the precision of each parameter.

Few additional considerations for the evaluation of K_La using the technique presented in this paper need to be addressed. First, it was assumed that the fermentation broth had a uniform concentration of dissolved oxygen and cells, as well as a value of K_La independent of position within the fermenter. In the present investigation, these are reasonable assumptions because of the relatively high intensity of mixing and the low viscosity of the fermentation broth. For a highly viscous fermentation broth, where large variations of the concentration of dissolved oxygen and K_La normally occur, all four estimation methods would provide erroneous values of the overall oxygen mass transfer coefficient because they all rely on the measurement of the dissolved oxygen concentration by the oxygen probe. For instance, if the probe is in a region of lower dissolved oxygen concentration than the bulk average, K_La estimated by the two gas mass balance methods would be underestimated. Secondly, if the activity of the biochemical reaction is high enough to significantly reduce the level of dissolved oxygen in the fermentation broth, then the estimation with the dynamic method would be impossible and only the two methods based on the gas mass balances would be applicable. Under these circumstances, the value of K_La would be estimated fairly accurately because of the O_2 and CO_2 concentration differences would be higher. Thirdly, if the biochemical activity is slow as in the case of animal and vegetal cell cultures, the stationary method and the methods based on O_2 and CO_2 mass balances would greatly suffer in precision. In that case, the dynamic method would provide a more accurate estimate of K_La. However, the data reconciliation technique presented in this paper would automatically, through weighting factors found in a Monte Carlo simulation, put significantly more weight on the dynamic method terms (partial objective functions J_2 and J_3).

Table 2 Variations of K_La in function of the weighing factor β for a fermentation time of 8.5 h.

β	K_La (s^{-1})			
	Converged	*Stationary*	*O_2 Balance*	*CO_2 Balance*
0.001	0.075	0.052	0.060	0.065
0.01	0.075	0.052	0.060	0.065
0.1	0.059	0.054	0.076	0.055
1	0.058	0.059	0.071	0.062
10	0.059	0.060	0.057	0.062
1000	0.059	0.060	0.056	0.062

4. Conclusion

In this investigation, data reconciliation was used to estimate a more probable value of K_La during the course of fermentation. Data reconciliation is based on statistical adjustment of redundant process data to obey laws of mass and energy conservation principles. In submerged fermentation, there are four possible methods available to estimate the overall oxygen mass transfer coefficient (K_La) in the course of fermentation: the dynamic method, the stationary method, the gaseous oxygen balance and carbon dioxide balance. Each method provides a distinct value of K_La. Data reconciliation technique was therefore used to obtain the most probable K_La value that takes into account both measurement and process modelling errors. An objective function, composed of the weighted sum of 12 measurement terms and 6 terms for oxygen conservation models, was minimised using a quasi-Newton optimisation routine. Weighting factors of each measurement term can be easily determined whereas Monte Carlo simulation was used to determine the relative weighting factors for each oxygen conservation model.

The converged estimated K_La value was 0.053 s^{-1} with a standard deviation of 15 %. Results clearly showed that a simple arithmetic average of the distinct values, obtained with the four methods, is not adequate to estimate the proper value of K_La. A net advantage of using data reconciliation is that the user is forced to examine the precision of all measured variables and oxygen conservation models.

Acknowledgements

The Natural Science and Engineering Research Council of Canada, FONDECYT 1980667, and FONDECYT 7980028 have supported this work.

References

1. Jarai, M.: Factors Affecting the Scale-up of Aerated Fermentation Processes. Int. Chem. Eng. 19 (1979) 710.
2. Moo-Young, M.; Blanch, H.W.: Design of Biochemical Reactors – Mass transfer Criteria for Simple and Complex Systems. In: Advances in Biochemical Engineering Biotechnology Mineral Engineering. 8 (7) (1995) 753.
3. Yagi, H.; Yoshida, F.: Oxygen Absorption in Fermenters – Effects of Surfactants, Antifoaming Agents, and Sterilised Cells. J. Ferment. Technol. 52 (1974) 905.
4. Gauthier, L.; Thibault, J.; LeDuy A.: Measuring K_La with Randomly Pulsed Dynamic Method. Biotechnol. Bioeng., 37 (1991) 889.
5. Brown, D.E.: Bioprocess Measurements and Control. Chem. Ind., 16 Sep. (1991) 678.
6. Yamane, T.; Shimizu S.: Adv. Bioch. Eng./Biotechnol. 30 (1984) 147.
7. Copella, S.J.; Dhurjati P.: A Detailed Analysis of *Saccharomyces cerevisiae* Growth Kinetics in Batch, Fed-batch, and Hollow-fibre Bioreactors. Chem. Eng. J. 41 (1989)B27.
8. Wang, H.Y.; Cooney, C.L.; Wang D.I.C.: Computer-Aided Baker's Yeast Fermentations. Biotechnol. Bioeng. 19 (1977) 69.
9. Taguchi, H.; Humphrey, A.E.: Dynamic Measurement of the Volumetric Oxygen Transfer Coefficient in Fermentation Systems. J. Ferment. 44 (1966) 881.
10. Gagnon, H.; Lounes, M.; Thibault, J.: Power consumption and mass transfer in agitated gas-liquid columns: a comparative study. Can. J. Chem. Eng. 76 (1998) 379-389.
11. Heinzle, E.: Present and Potential Applications of Mass Spectrometry for Bioprocess Research and Control. J. Biotechnol. 25 (1992) 81.
12. Crowe, C.M.: Observability and Redundancy of Process Data for Steady State Reconciliation. Chem. Eng. Sci. 44 (1989) 2909.
13. Hodouin, D.; Everell, M.D.: A Hierarchical Procedure for Adjustment and Material Balancing of Mineral Process Data. Int. J. Mineral Processing. 7 (1980) 91.
14. Hodouin, D.; Bazin, C.; Makni, S.: On-Line Reconciliation of Mineral Processing Data. Proc. of the AIME/SME Symposium - Emerging Computer Techniques for the Mineral Industry," Reno, Nevada, Feb. 1993.
15. Hodouin, D.; Bazin, C.; Makni, S.: Dynamic Material Balance Algorithm: Application to Industrial Flotation Circuits. SME / AIME Annual Meeting, Phoenix, Arizona, March 1996.
16. Liebman, M.J.; Edgar, T.F.; Lasdon, L.S.: Efficient Data Reconciliation and Estimation for Dynamic Processes Using Nonlinear Programming Techniques Computer Chem. Eng. 16 (1992) 963.
17. Mah, R.S.H.: Chemical Process Structures and Information Flows. Batterworths, Boston 1990.
18. Makni, S.; Hodouin, D.; Bazin, C.: A Recursive Node Imbalance Method Incorporating a Model of Flowrate Dynamics for On-Line Material Balance of Complex Flowsheets. Mineral Eng. 8(7) (1995) 753.
19. Schumpe, A.; Adler, I.; Deckwer, W.D.: Solubility of Oxygen in Electrolyte Solutions. Biotechnol. Bioeng. 20 (1978) 145.
20. Shuler, M.L.; Kargi, F.: Bioprocess Engineering – Basic Principles. Prentice Hall PTR, Englewood Cliffs, New Jersey 1992.
21. Nielsen, J.; Villadsen, J.: Bioreaction Engineering Principles. Plenum Press, New York, 1994.
22. Lavery, M.; Nienow, A.W.: Oxygen Transfer in Animal Cells Culture Medium. Biotechnol. Bioeng. 30 (1986) 368.

Nomenclature

C_L dissolved oxygen concentration (mol/m^3)

C_L^0 pseudo-steady-state dissolved oxygen concentration recorded at the initiation of the dynamic method (mol/m^3)

C_L^* dissolved oxygen concentration in equilibrium with mean gaseous oxygen concentration (mol/m^3)

C_P	dissolved oxygen concentration recorded by the probe (mol/m^3)
K_La	overall oxygen mass transfer coefficient (s^{-1})
J	objective function
P	pressure (Pa)
Q_G	gas flow rate (m^3/s)
$Q_{O2}X$	oxygen uptake rate (mol/m^3 s)
R	gas constant (8.306 Pa m^3/(mol K))
S	substrate (mol/m^3)
RQ	respiratory quotient
T	temperature (K)
V_L	liquid volume in the fermenter (m^3)
y	gaseous mole fraction

GREEK LETTERS

α	weighting factor associated to each term in the objective function
β	relative weighting factor between measurements and conservation models
τ_P	time constant of the dissolved oxygen probe (s)

SUBSCRIPTS

1	inlet stream
2	outlet stream
CO_2	carbon dioxide
O_2	oxygen

SYMBOLS

^	estimated values
~	corresponds to ideal or theoretical conditions

RESPIRATION QUOTIENT: ESTIMATION DURING BATCH CULTIVATION IN BICARBONATE BUFFERED MEDIA

RONALD NEELEMAN
Wageningen University, Systems and Control Group
Bomenweg 4, 6703 HD, Wageningen, The Netherlands.

Abstract

The Respiration Quotient (*RQ*) is a key metabolic parameter for cell cultures and is usually determined from gas analysis only. In bicarbonate buffered media the carbon dioxide balance is affected by accumulation and hence the *RQ* can not directly be calculated from gas measurements. A Kalman Filter as software sensor that estimates the CER can cope with these buffering capacities and thus is used for determining the *RQ*. The model used by the Kalman Filter lumps all carbonate in the liquid to one term in order to eliminate the role of a priori knowledge of cell and medium kinetics without affecting the performance.

1. Introduction

Animal cells, yeast cells and aerobic microbial cells oxidise organic compounds into water, carbon dioxide and other organic compounds to gain energy for their maintenance and growth. As such these organisms consume oxygen and produce carbon dioxide with rates called the oxygen uptake rate (*OUR*) and the carbon dioxide evolution rate (*CER*). These rates are direct indicators of metabolic activity. Their ratio called the respiration quotient ($RQ = CER\ /\ OUR$) varies with the nature of the substrates and products of the organism. Bonarius et al. (1995) and Royce (1992) argue that for cell cultures the *RQ* can be considered as a key metabolic parameter making it possible to detect on what medium substrate the organism grows.

Furthermore, stoichiometric coefficients, e.g. yields, are usually determined using elemental balancing (conservation of chemical elements). This set of balance equations can mostly not be solved since the number of unknown coefficients exceeds the number of balance equations. Additional information is required and with the right set of extra measurements the equations become solvable. Measuring the *RQ* introduces such information to solve the balance equations and calculate the stoichiometric coefficients.

M. Hofman and P. Thonart (eds.), Engineering and Manufacturing for Biotechnology, 203–216.

By means of *RQ* measurements it has been possible to close mass balances and to determine metabolic flux distributions for yeast (Vallino, 1993). *RQ* data has also been used for on-line bioreactor control, for example to minimise glucose effects (Wang, 1979) or to optimise substrate consumption in yeast (Wu, 1993). Its accurate evaluation is hence of great importance.

The *RQ*-value can be calculated using the oxygen and carbon dioxide concentrations in the gas stream into and from the reactor headspace (Bonarius et al., 1995). However, the effect of pH control action on carbon dioxide evolved from the medium troubles the measurement of the *CER*. Royce (1992) describes this phenomenon together with a solution to compensate for these so-called pH-effects. Besides these pH-effects, buffering capacity (e.g. bicarbonate) of the medium also troubles on-line determination of the CER. When the reactor is not in steady state (e.g. batch cultivation), direct measurement of carbon dioxide concentrations in the inlet and outlet of the reactor headspace does not satisfy. Dissociation and accumulation of carbon dioxide in the medium and headspace disturbs the steady state balance and results in an incorrect *CER*- and thus *RQ*-calculation.

To manage such problems software-sensors based on standard measurements are combined with mathematical observers to derive the internal states of the system. Stephanopoulos and San (1984) gave an extensive discussion on such observer-based software-sensors for the reconstruction of the process states. For example the achieved amount of biomass and substrate concentrations are derived from oxygen and carbon dioxide measurements. Other important work in this area was produced by Bastin and Dochain (1990) and concerned observer-based software-sensors for adaptive control. All the work in this area was focussed on biomass and substrate estimation and requires more or less detailed a priori knowledge of the system. However, for monitoring and control purposes it is not always necessary to know the states, in a lot of applications only an indicator is needed on what substrate is being used or is necessary to be added. This chapter describes and discusses a Kalman Filtering algorithm as software-sensor for estimating the *RQ*-value for batch cell cultivation from carbon dioxide measurements as presented by Neeleman (1999). The sensor will take account for buffering capacities of the media for carbon dioxide.

2. Gas concentrations in batch-wise cell cultures

For the on-line calculation of the *RQ*, the *OUR* and *CER* have to be known (measured or estimated) on-line. Two different methods are applied in the determination of these rates. First the *OUR* will be deduced directly from gas analyses and liquid measurements. Since there are no satisfactory sensors for dissolved carbon dioxide and the accumulation of carbon dioxide can not be discarded an estimator is developed for the on-line estimation of the *CER*. Finally the ratio of these rates will give the *RQ*-value.

2.1. OXYGEN UPTAKE RATE (OUR)

In their general form the mass balances for oxygen in the gas and liquid phases include accumulation terms for oxygen. However, oxygen is sparingly soluble in aqueous solutions and usually kept constant by a controller. Therefore accumulation of the dissolved oxygen concentration will be very small and makes the *OUR* practically equal to the transfer rate of oxygen over the gas-liquid interface. As a consequence the *OUR* can be deduced directly from gas and liquid analyses only:

$$OUR = k_L^{O_2} a \left(\frac{P - p_W}{H^{O_2}} X_{O_2} - [O_2]_L \right) \quad (1)$$

For this expression it is assumed that the gas and the liquid phase are sufficiently mixed and that the gas flow rate is high so that $X_{O_2}^{in} \approx X_{O_2}^{out}$. X_{O_2} represents the fraction of oxygen in the headspace of the bioreactor, $[O_2]_L$ the oxygen concentration in the liquid phase (mol.l^{-1}), P the headspace pressure (Pa) and p_W the partial vapour pressure of water, which has to be taken into account because the oxygen fraction is expressed on a dry basis. H^{O_2} is Henry's constant for oxygen (Pa.l.mol^{-1}). The mass transfer coefficient for oxygen transfer, $k_L^{O_2} a$, is expressed in h^{-1} and can experimentally be determined using the dynamic method (van 't Riet, 1991).

2.2. CARBON DIOXIDE EQUILIBRIUM IN THE GAS PHASE

The transfer of carbon dioxide across the gas-liquid interface is a physical process (as for oxygen), which is liquid-film limited:

$$CTR = -\, k_L^{CO_2} a \left(\frac{P - p_W}{H^{CO_2}} X_{CO_2} - [CO_2]_L \right) \quad (2)$$

where X_{CO_2} is the fraction of carbon dioxide in the headspace, $[CO_2]_L$ is the carbon dioxide concentration in the liquid phase (mol.l^{-1}), and H^{CO_2} the Henry's constant for carbon dioxide (Pa.l.mol^{-1}). In this expression it is assumed that both the gas-phase and the liquid-phase are well mixed. The mass transfer coefficient for carbon dioxide, $k_L^{CO_2} a$, is difficult to determine by the dynamic method. Transfer kinetics could be influenced by pH control and pCO_2 electrodes suffer from poor response times. However, the ratio of $k_L a$ values for CO_2 and O_2 is proportional to the ratio of their liquid phase diffusivities, and thus of the square root of their mole mass:

$$\frac{k_L^{O_2} a}{k_L^{CO_2} a} \cong \frac{D_{O_2}}{D_{CO_2}} \cong \sqrt{\frac{M_{O_2}}{M_{CO_2}}} = 0.89 \quad (3)$$

Since the $k_L^{O_2}a$ can be determined experimentally as mentioned in the previous paragraph and the mole masses are known, the $k_L^{CO_2}a$ can be calculated. The total mass balance for the CO_2 concentration in the headspace is given by

$$\frac{dX_{CO_2}}{dt} = -\frac{F_g RTX_{CO_2}}{V_H(P - p_w)} - k_L^{CO_2}a\left(\frac{P - p_w}{H^{CO_2}}X_{CO_2} - [CO_2]_L\right)\frac{V_L RT}{V_H(P - p_w)} \tag{4}$$

with F_g the gas flow (mol.h^{-1}), assuming absence of carbon dioxide in the incoming gas. Here R is the gas constant (kJ.mol^{-1}.K^{-1}), T the temperature (K) and V_H the headspace volume (l).

2.3. CARBON DIOXIDE EQUILIBRIUM IN THE LIQUID PHASE

In contrast to the assumptions for the *OUR* there is a discrepancy between the carbon dioxide transfer rate (*CTR*) across the gas-liquid interface, available from gas analyses, and the carbon dioxide evolution rate (*CER*) of the biomass in the bioreactor. The *CER* cannot be measured directly as carbon dioxide has a much higher solubility than oxygen that is enhanced by its hydrolysis to bicarbonate. Changes in the concentrations of dissolved carbon dioxide and bicarbonate results in differences between the *CTR* and *CER* (Royce, 1992), which increase with the bicarbonate buffering capacity of the medium. By measuring carbon dioxide in the off-gas of the bioreactor, a substantial amount of carbon dioxide coming from the buffer system will inevitably be measured, and not only the carbon dioxide produced by the cells (Bonarius et al., 1995). Figure 1 shows the diffusion of carbon dioxide through the cell membrane and the subsequently following reactions with k_1 and k_2 as the rate constants for the indicated reactions (h^{-1}) and K_C is the carbonic acid dissociation constant (mol.l^{-1}). Further dissociation, of bicarbonate into carbonate is negligible for the *pH* range used during standard cell cultures (pH < 7.8). The concentration of carbonic acid is always very small in comparison to that of dissolved carbon dioxide, $[CO_2]_L$ .The mass balance for carbon dioxide and bicarbonate is:

$$\left(\frac{d[CO_2]}{dt} + \frac{d\left[HCO_3^-\right]}{dt}\right) = CER - CTR \tag{5}$$

with *[CO2]* the concentration of carbon dioxide and *[HCO3-]* the concentration bicarbonate both expressed in mol.l^{-1}. As there is no sink for bicarbonate ions other than by dehydration, the time scale of changes in cell culture is long enough to ensure that the reactions involved in the dehydration of bicarbonate are close to equilibrium. The rate constants associated with carbonic acid dissociation are so large that this reaction can be considered to be at equilibrium.

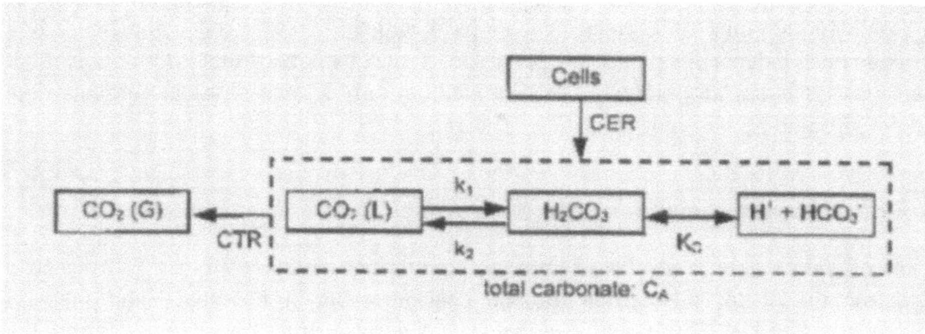

Figure 1. *Diagram of the CO_2 balance in a cell culture. Showing the rate constants k_1 and k_2 and the dissociation constant K_C. All components in the dashed area are lumped to C_A, which is called the 'total carbonate concentration'.*

During batch cultivations the *pH* is controlled closely to the desired set-point. Therefore it is not necessary to model both the carbon dioxide and the bicarbonate concentration separately, because carbon dioxide and bicarbonate are in equilibrium; the dissociation constant for hydrolysis of bicarbonate (mol.l^{-1}) is given by:

$$K_A = \frac{\left[HCO_3^-\right] \times \left[H^+\right]}{\left[CO_2\right]} \tag{6}$$

Now, a lumped variable C_A, called 'total carbonate concentration', can be introduced:

$$C_A = \left[HCO_3^-\right] + \left[CO_2\right] \tag{7}$$

and thus the carbon dioxide concentration in the liquid is given by:

$$\left[CO_2\right] = \frac{C_A}{1 + \frac{K_A}{\left[H^+\right]}} \tag{8}$$

The dashed area in Figure 1 represents the lumped concentration. Combining Equations 2, 5, 7 and 8 gives the mass balance for the 'total carbonate concentration':

$$\frac{dC_A}{dt} = CER + k_L^{CO_2} a \left(\frac{P - p_w}{H^{CO_2}} X_{CO_2} - \frac{C_A}{1 + \frac{K_A}{[H^+]}} \right) - \frac{F_s}{V_L} C_A \tag{9}$$

The last term in this equation, with sample flow rate F_S, is inserted to take the loss of C_A due to sampling in account. Equations 4 and 9 represent the complete carbon dioxide mass balance for liquid and gas phase. It must be noticed that information about cell kinetics is not necessary and for the medium only known physical constants like H^{CO2}, K_A and R, are used.

3. Software sensor design

The software sensor is not based upon an equilibrium model ($CTR = CER$) but on a dynamic model for the carbon dioxide concentrations in headspace and medium (Equations 4 and 9). This dynamic model is used to reconstruct the actual *CER* from the measured carbon dioxide in the headspace.

3.1. DYNAMIC MODEL

In systems theory there are two concepts, which must be clearly distinguished from each other. The plant is the actual physical system that needs to be observed. The model is the mathematical description of the physical system, which is used for the filter design stage. Within the control-engineering field it is common to rewrite a model to its state-space representation. In such a presentation all relations are linear or linearised and written in matrix-notation. The input, state, output and noise of the model are given as vectors. The input vector usually contains the manipulatable variables of the model, the state vector consists of those variables that develop through time dependant of each other and of the input, and the output usually contains all the measured variables. The total system can be written as an *ABCD*-system:

$$\begin{aligned} \frac{dx}{dt} &= Ax + Bu \\ y &= Cx + Du \end{aligned} \tag{10}$$

where u is the input vector with m inputs, x the state vector with n states and y the output vector with k outputs. The matrices A and B contain information about the way the states develop through time and with the matrices C and D the output can be calculated. A is a $n{\cdot}n$, B a $n{\cdot}m$, C a $k{\cdot}n$ and D a $k{\cdot}m$ matrix.

In this situation the model, consisting of Equations 4 and 9, are extended with an extra differential equation used for estimating the *CER*. The prediction of the *CER* is a zero mean random walk process, i.e. its derivative is zero:

$$\frac{dCER}{dt} = 0 \tag{11}$$

For use in a discrete Kalman Filter algorithm the model is discretised and rewritten to a state-space representation by defining the input, state and output vectors. The state vector consists of the *CER*, the lumped concentration of both dissolved carbon dioxide and bicarbonate (C_A) and the molar fraction of carbon dioxide in the headspace (X_{CO_2}). There is no input vector and the output vector is constructed from the measured variables, which is only the molar fraction of carbon dioxide in the off-gas.

This model is linear in the state variables for the equations so the *ABCD*-matrices can be developed straightforward. A consequence of computer application for process monitoring is data sampling. Only process values at discrete time moments are available. Therefore, instead of continuous time models, the process behaviour is given by the discrete time equivalents. And since there is no input, there are no *B* and *D* matrices. Now the state vector, *A* matrix and *C* vector are:

$$
\begin{gathered}
x = \begin{bmatrix} C_A & X_{CO_2} & CER \end{bmatrix}^T \\
A = \begin{bmatrix}
1 - \dfrac{k_L^{CO_2} a \Delta t}{1 + \dfrac{K_A}{[H^+]}} - \dfrac{F_s \Delta t}{V_L} & \dfrac{k_L^{CO_2} a (P - p_w) \Delta t}{H^{CO_2}} & \Delta t \\
\dfrac{k_L^{CO_2} a V_L RT \Delta t}{(P - p_w) V_H (1 + \dfrac{K_A}{[H^+]})} & 1 - \dfrac{F_g RT \Delta t}{(P - p_w) V_H} - \dfrac{k_L^{CO_2} a V_L RT \Delta t}{H^{CO_2} V_H} & 0 \\
0 & 0 & 1
\end{bmatrix} \\
C = \begin{bmatrix} 0 & 1 & 0 \end{bmatrix}
\end{gathered} \tag{12}
$$

3.2. THE KALMAN FILTER ALGORITHM

Both the plant and the model use the same known input. The state of the plant is, of course, unknown and its output can be measured. The state and output of the model is based on the predictions of the model, as can be seen in Table 1, where w_k and v_k represent process noise and measurement noise vectors, respectively, they are assumed to be random with mean values zero and variances *Q* and *R*.

The Kalman Filtering algorithm is depicted in Figure 2. In this figure a distinction is made between the cultivation process which gives the data and the algorithm that processes the data. The algorithm has two steps, respectively the time update and the measurement update. In the time update a one sample ahead prediction is made for the state and output variables (respectively $\overline{x}_{k+1}, \overline{y}_{k+1}$) and the prediction variance of the states ($\overline{P}_{k+1}$). The actual values of the input variables (u_k) and the available estimated results of the Kalman filter ($\hat{x}_k, \hat{P}_k$) at sample moment *k* are used for this prediction.

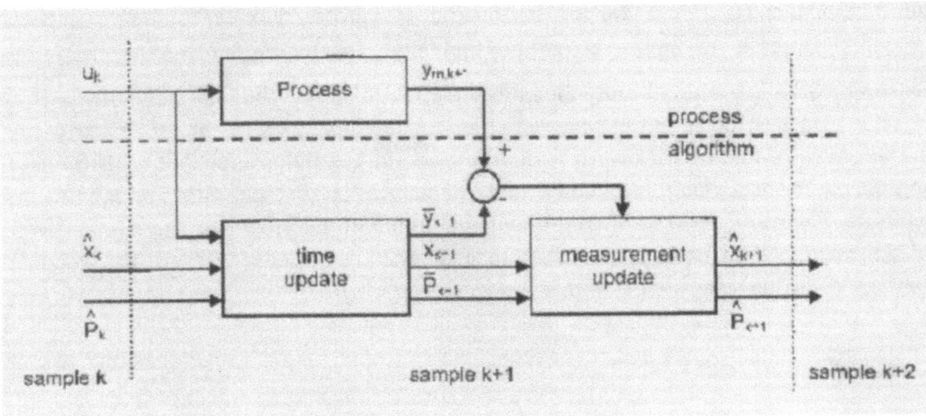

Figure 2. Schematic representation of the Kalman filter calculations. It shows the different steps to be taken at each time instant k. The dashed horizontal line makes distinction between the algorithm and process, the dotted vertical lines symbolise the succeeding sample moments.

The next step is the measurement update, which takes place as soon as new data becomes available and where the prediction of the output variable ($\bar{y}_{k+1}$) is corrected. The correction is proportional to the error between measured process output ($y_{m,k+1}$) and predicted output ($\bar{y}_{k+1}$). The magnitude of the correction gain varies for succeeding samples and aims to minimise the error covariance ($\hat{P}_k$) for the state and output variables. Now, the measurement update gives the best estimate ($\hat{x}_{k+1}$) of the states and its variance ($\hat{P}_{k+1}$). The values for the state estimate ($\hat{x}_{k+1}$) are used as the software sensor output. The entire set of these equations comprises what is called the discrete time Kalman Filter (Chen, 1993 and Lewis, 1986) and is summarised in Table 1.

A few remarkable advantageous features of the Kalman filtering algorithm can easily be observed. First, each calculation step only requires the last estimate and a new set of measurement data. The essential advantage of such simple "step-by-step" structure of the computational scheme is that there is no need to store all old results and measurement data for each up-dating state estimate, and this saves computer memory and processor time, especially in real-time (on-line) applications. Second, all recursive formulas of the algorithm are straightforward and linear, consisting of only matrix multiplication and addition, and a single matrix inversion in the calculation of the Kalman gain.

Table 1: Discrete time Kalman Filter algorithm

System and measurement model	
$x_{k+1} = A_k x_k + B_k u_k + w_k$	$w_k \sim (0, Q_k)$
$y_k = C_k x_k + v_k$	$v_k \sim (0, R_k)$
Initialisation	
$P_0 = P_{x_0},\ \hat{x}_0 = \bar{x}_0$	
Time update (one sample-ahead prediction)	
$\bar{x}_{k+1} = A_k \hat{x}_k + B_k u_k$	
$\bar{y}_{k+1} = C_k \bar{x}_{k+1}$	
$\bar{P}_{k+1} = A_k \hat{P}_k A_k^T + Q_k$	
Measurement update (after including the measurement)	
$\hat{P}_{k+1} = \bar{P}_{k+1} - \bar{P}_{k+1} C_{k+1}^T R_{k+1} C_{k+1} \bar{P}_{k+1}$	
$K_k = \hat{P}_{k+1} C_{k+1}^T R_{k+1}^{-1}$	
$\hat{x}_{k+1} = \bar{x}_{k+1} + K_k (y_{m,k+1} - \bar{y}_{k+1})$	

4. Application of the software sensor

The Kalman Filtering algorithm is used as software sensor for the CER. First the performance was validated by an experiment, then the application and use was proven by a series of experiments and finally two experiments show the robustness of the software sensor in coping with disturbances.

4.1. VALIDATION OF THE SOFTWARE SENSOR

Before applying the software-sensor, its performance needs to be validated. The bottleneck in the *RQ*-estimation procedure is that the software sensor must be able to detect the *CER* correctly. To validate the sensor at this point medium, to which a known amount of bicarbonate was added, was pumped from a storage vessel into the reactor. So the carbon production by cells was imitated by pumping an exactly known amount of bicarbonate into the reactor. If the software sensor operates well it should be able to detect this amount of bicarbonate correctly. Moreover, as the software sensor gives also the value of the total carbonate content (C_A), the performance can also be checked by analysis of the total carbonate concentration in the solution.

During the experiment there was a minimum of headspace in the storage vessel, which limits the loss of bicarbonate from the liquid phase. As a result the total carbonate concentration in the feed to the reactor remained constant. Feeding of the bicarbonate solution to the fixed volume reactor by a continuous pumping system imitated the *CER* by cells. From the measurements of the total carbonate concentration (C_A) in the feed and reactor together with the feed rate, the actual *CER* was calculated. During the experiment the feed flow rate was manipulated in order to obtain several levels for the *CER* and C_A.

Figure 3 shows the C_A and *CER* obtained by the software sensor compared with the measured C_A and the "imitated" *CER* (obtained by pumping the bicarbonate medium in the reactor). At 600 minutes the reactor contained an amount of bicarbonate and the feed flow and software sensor were started. The sensor results of the first five minutes are affected by the start-up of the sensor. The total carbonate concentration, C_A, in the reactor decreased during the first 500 minutes due to its high start concentration. At 3200 minutes together with the feed flow rate the *CER* was increased and at 4700 minutes it was decreased again. During the whole experiment there is a good match between the C_A and *CER* values obtained from the software sensor and the laboratory measurements. The *CER* signal from the software sensor in the second part of the experiment reveals some variations, which are a consequence of variations in the measured carbon dioxide concentrations in the off-gas. The good match between the obtained results confirms the use of the software sensor for C_A and *CER* measurement.

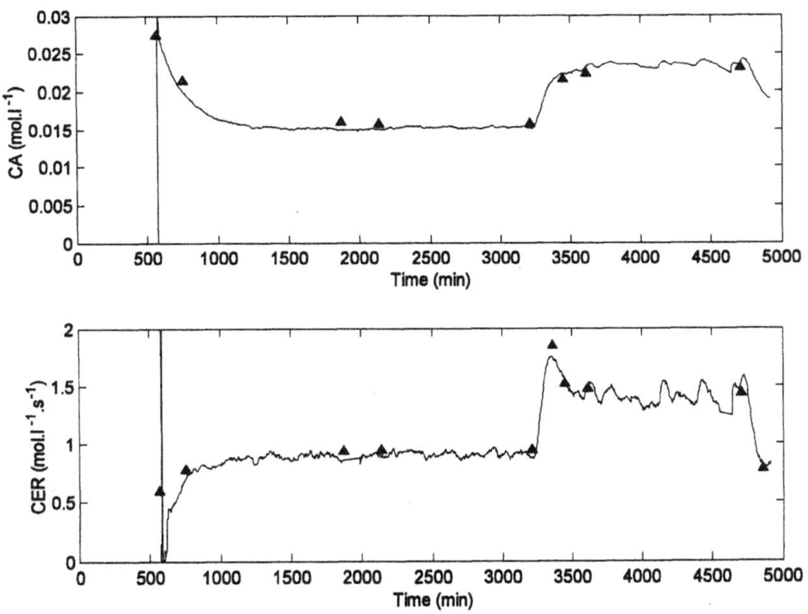

Figure 3. Validation of the software sensor, imitation of the CER of cells by feeding a given amount of bicarbonate to the reactor. A: Measured (▲) and estimated (solid line) total carbonate concentrations, C_A. B: Actual (▲) and estimated (solid line) CER. Data from Neeleman (1999).

4.2. APPLICATION TO CELL CULTIVATION

The on-line application of the *CER* software sensor was evaluated for 18 batch cultivations of insect cells and 2 mammalian cell cultivations. Figure 4A shows standard results for the measured carbon dioxide in the off-gas and the *CER*-values from the software sensor during one of the insect cell cultivations. The *CER*-values are in the range that can be expected for the used cells. Due to a recalibration of the pH sensor at 5100 minutes the pH controller reacted suddenly, which in turn affected the measured carbon dioxide and the estimated *CER*. In this experiment, together with the estimation of the *CER*, the C_A is estimated. Figure 4B shows the estimated C_A and the laboratory measurements, which correspond well. The mean error between the estimated and measured 'total carbonate concentration' is 0.014 $mmol.l^{-1}$ with a standard deviation of 0.066 $mmol.l^{-1}$. With the estimated *CER* and calculated *OUR* the *RQ* could be calculated.

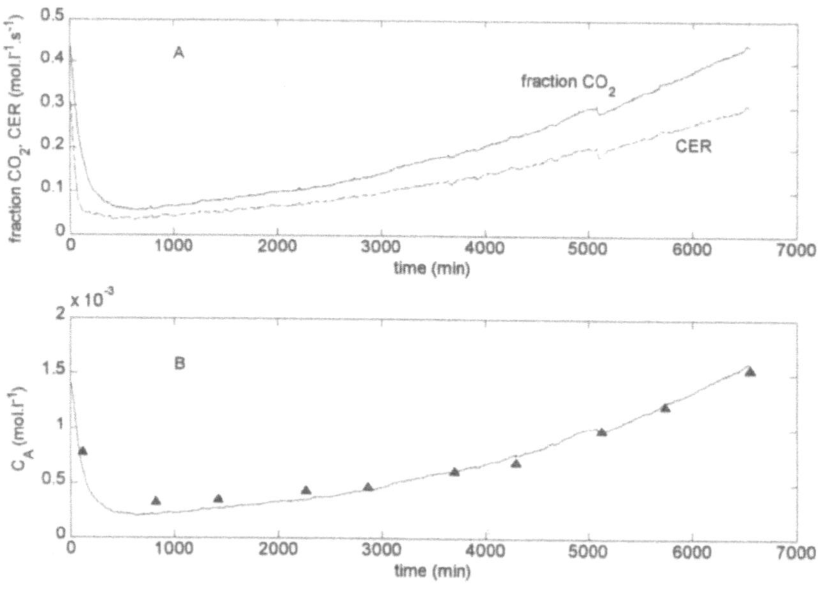

Figure 4. Batch wise insect cell cultivation. A: Measured molar fraction of carbon dioxide in the off-gas (solid line) and the estimated CER (dashed line). B: Measured total carbonate C_A (▲) and the estimated total carbonate (solid line).Data from Neeleman (1999)

Figure 5 shows the estimated *RQ* for a part of a batch together with the measured substrate concentrations: glutamine and glucose. The peaks in the *RQ* signal fall together with the sampling moments when some C_A containing medium is replaced by the same amount of C_A free medium. Figure 5 illustrates that when glucose and glutamine (till 4500 minutes) are in sufficiently high concentration available the organism grows well with a *RQ* around 1.0. After 5000 minutes the glucose is depleted and now glutamine is the main substrate for the organism. The *RQ*-value that falls from 1.0 to 0.8 reflects the change in the physiological state of the organism due to the lack of glucose. Thus, with the ability to estimate the *RQ* on-line the physiological state of the cells can be monitored on-line and be used as an indicator for adding extra substrates. This finding confirms the statement of Royce (1992) and Bonarius et al. (1995) that the *RQ* can be used as parameter to detect the metabolic activity of the cell.

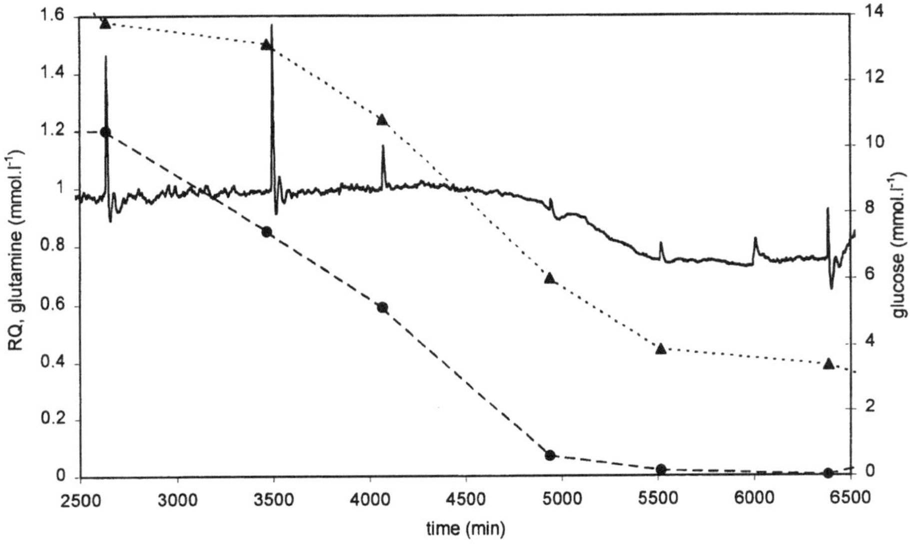

Figure 5. The estimated RQ *(solid line) and the concentrations of glutamine (▲, left axis) and glucose (•, right axis) during batch wise cultivation of insect cells. Data from Neeleman (1999)*

4.3. ROBUSTNESS OF THE SOFTWARE SENSOR

Examples of *RQ* estimates for other cultures are given in figure 6. Figure 6A shows the estimated *RQ* during a batch where at 2700 minutes the communication between the computer and the PLC failed for approximately 1 hour. The Kalman filter waited 1 hour before a new measurement update could be calculated. After this disturbance the software sensor quickly recovered the same *RQ* value. Figure 6B shows the estimated *RQ* during another batch where at 1800 and 2600 minutes huge samples were taken and replaced with fresh medium. Samples were taken by pressurising the headspace to withdraw the liquid. This way of sample taking gives a major impact on the headspace pressure and CO_2 fraction. Furthermore, the fresh medium instantly changes the 'total carbonate concentration' in the medium. Again it can be seen that the software sensor quickly recovers from these disturbances.

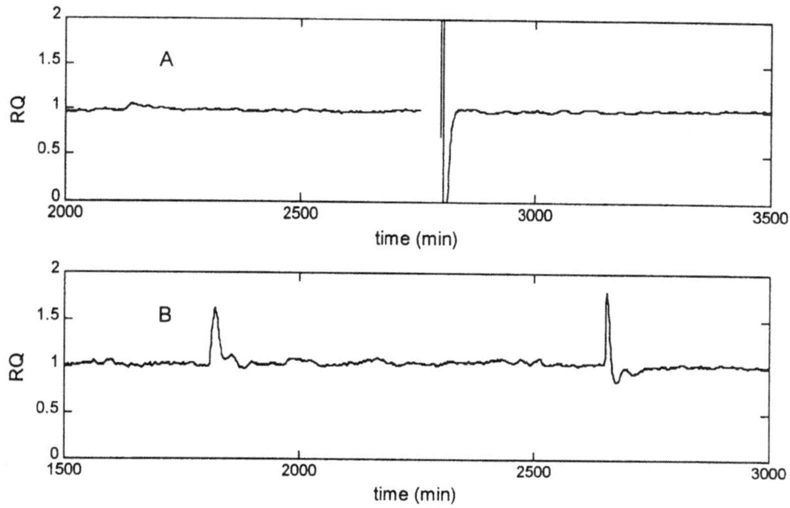

Figure 6. Two examples of estimated RQ values during batches where the Kalman filter had to cope with disturbances. A: The estimated RQ during a batch where at 2700 minutes the communication failed for 1 hour. B: The estimated RQ during a batch where at 1800 and 2600 minutes huge samples were replaced with the same amount of fresh medium.

5. Concluding remarks

For batch cell cultivation in media which buffer carbon dioxide, the carbon dioxide evolution rate (*CER*) and respiration quotient (*RQ*) cannot be obtained from an equilibrium model only. As an alternative a software sensor is used. Main features of the sensor are:

- The *CER* detection is based on a dynamic model that covers the transient phases for bicarbonate buffering media and batch wise operations. In the model all carbon dioxide and bicarbonate in the liquid phase are lumped to a single component "total carbonate" and as a result no detailed knowledge of the reaction kinetics is necessary; only physical constants are required.
- The sensor uses a discrete time Kalman Filter algorithm.
- In a validation experiment where cell carbon dioxide production was imitated, the software sensor proved to be successful for estimation of the *CER* from on-line carbon dioxide measurements in the off-gas.
- Results obtained from experiments with insect cells showed that the *RQ*-value was close to 1.0, a value that is common for cells growing on glucose and glutamine as main substrates. As the physiological state of the organism changes due to substrate limitations the *RQ*-value reflects this change.

- Results of experiments where disturbances occurred showed that the Kalman Filter could cope efficiently with communication failures and sudden medium or headspace changes.

Important difference with the work of Stephanopoulos and San (1984) and Bastin and Dochain (1990) is that a minimum of a priori system knowledge is needed for the estimation of the *RQ* value. In contrast to a limited amount of information that can be obtained from laboratory samples the software sensor produces a continuous stream of *RQ* and *CER* values to monitor the state of the organism. Therefore it is a powerful tool to be used for various cell cultivations.

References

Bastin, G., and Dochain, D. (1990) On-line Estimation and Adaptive Control of Bioreactors. Elsevier Science Publishing Co. Amsterdam.

Bonarius, H.P.J., de Gooijer, C.D., Tramper, J. and Schmid, G. (1995) Determination of the Respiration Quotient in Mammalian Cell Culture in Bicarbonate Buffered Media. Biotechnology and Bioengineering, Vol. 45, Pp. 524-535.

Chen, G. (1994) Approximate Kalman Filtering. World Scientific. Singapore.

Lewis, F. (1986) Optimal estimation. Wiley Interscience, USA.

Neeleman, R, End, E.J. van den and Boxtel, A.J.B. van (1999). Estimation of Respiration Quotient in Bicarbonate Buffered Media. Ninth European Congress on Biotechnology, Brussels, Belgium.

van 't Riet, K. and Tramper, H. (1991) Basic Bioreactor Design. Marcel Dekker, Inc., New York, USA.

Royce, P.N. (1992) Effects of Changes in the pH and Carbon Dioxide Evolution Rate on the Measured Respiratory Quotient of Fermentations. Biotechnology and Bioengineering, Vol. 40, Pp. 1129-1138.

Stephanopoulos, G. and San, K. (1984) On-Line Bioreactor Identification. I. Theory. Biotechnology and Bioengineering, Vol. 26, Pp. 1176-1188.

Vallino, J.J., Stephanopoulos, G. (1993) Metabolic flux distributions in *Corynebacterium glutamicum* during growth and lysine overproduction. Biotechnology and Bioengineering, Vol. 41, Pp. 633-646.

Wang, H.Y., Cooney, C.L. and Wang, D.I.C. (1979) Computer Control of bakers' yeast production. Biotechnology and Bioengineering, Vol. 21, Pp. 975-995.

Wu, W.-T., Wang, P.-M. (1993) On-line optimal control for ethanol production. Journal of Biotechnology, Vol. 29, Pp. 257-266.

FERMENTATION PHASE DETECTION USING FUZZY CLUSTERING TECHNIQUES AND NEURAL NETWORKS FOR IMPROVED CONTROL

TAKOI K. HAMRITA AND SHU WANG
Dept. of Biological and Agricultural Engineering
University of Georgia takoi@bae.uga.edu, Fax (706) 542-8806

1. Introduction

Phase detection is important for supervision and control of fermentation processes. A new approach for phase detection is developed and applied to simulations of a Gluconic acid fermentation. The method uses fuzzy clustering to identify culture phases off-line. Using clustering results, neural networks are trained and used for on-line phase detection. The fermentation industry is becoming increasingly important as many fermentation products are being commercialised and used in the pharmaceutical, food, and chemical industries. A well-controlled fermentation process can reduce production costs, increase yield and maintain quality of metabolic products (Shimizu, 1996). During fermentation the microbial species continuously undergo physiological changes. Many researchers define these as different physiological states or phases of the microbial population (Konstantinov et al., 1989; Karim et al., 1997). A new approach, physiological state control, which accounts for fermentation physiological changes, has been proposed recently (Konstantinov et al., 1989). This approach decomposes the fermentation process into several phases. In every phase the cell population expresses stable characteristics and an invariant control strategy is applied to each phase. One of the most important aspects of this new approach is the ability to determine on-line the current phase of the fermentation (Konstantinov et al., 1989). Moreover, since phase transitions occur gradually and in a smooth fashion, it is necessary to develop techniques which identify the phases and the smooth transitions between them in real-time.

Fuzzy clustering has been shown to identify variable structure behaviour of dynamic processes through a study concerning modelling of sleep dynamics (Kosanovic et al., 1994). The author also pointed out that this method could be used to model physiological processes, where several dynamics act together to produce an overall process (Kosanovic et al., 1995). The only constraint is that the system be quasi-stationary and that the variables used in clustering reflect the quasi-stationary dynamics. As pointed out by Konstantinov et al. (1989), fermentations do exhibit quasi-stationary

M. Hofman and P. Thonart (eds.), Engineering and Manufacturing for Biotechnology, 217–226.

behaviour and certain process variables, especially physiological ones, are indicative of this behaviour.

Hence, in this chapter, a new technique for fermentation phase detection is developed using fuzzy clustering. Clustering results are used to train neural networks to recognise different process phases in real-time. Since the Elman-Jordan network structure is well suited for temporal data, it is used in this study. The resulting real-time phase detection technique based on fuzzy clustering and neural networks is an easier, more effective alternative to using heuristic expert knowledge and fuzzy logic and can be the basis for much improved supervisory, diagnostic and control systems for fermentation. Excellent results have been obtained using simulated data.

2. Fermentation phase detection

2.1. OFF-LINE PHASE DETECTION USING FUZZY CLUSTERING

2.1.1. Variable selection for phase detection

The selection of variables for phase detection should be implemented only after accumulation of a sufficient base of knowledge of physiology and the behaviour characteristics of the population. These variables should have clear relationship with the phase state of the fermentation and have a low-level noise corruption (Konstantinov et al., 1991). Specific Metabolic Rates, especially specific growth rate, are popular variables for phase detection (Konstantinov et al., 1989). But in many cases, the calculation of the desired variables for phase detection might be difficult, inaccurate or impossible. "Then these variables can be replaced with other variables which describe, in a known manner, the phenomena of interest." (Konstantinov et al., 1992).

2.1.2. Fuzzy clustering for off-line phase detection of penicillin-G fed-batch fermentation

To test effectiveness of the fuzzy clustering algorithm in identifying process phases off-line, we will first test this technique using simulations of a process for which phases are known a priori. The Penicillin-G fedbatch fermentation is chosen since a multiphase mathematical model is available for it in the literature (Menezes et al., 1994). There are two phases in this fermentation, growth phase and production phase (Menezes et al., 1994). In the growth phase the process produces large quantities of biomass, predominantly utilising the substrate in the initial media. As the initial substrate becomes exhausted, feed additions being made to the bioreactor are increased. During the production phase the substrate additions are maintained at a rate which keeps the substrate concentration at a low level. This results in a low growth rate (Ignova et al., 1996). In this phase most Penicillin is produced (Menezes et al., 1994).
Figure 1 shows a simulation of the model in (Menezes et al., 1994). Under the initial conditions used, the phase transition time of this fermentation is known to occur around the 70th hour.

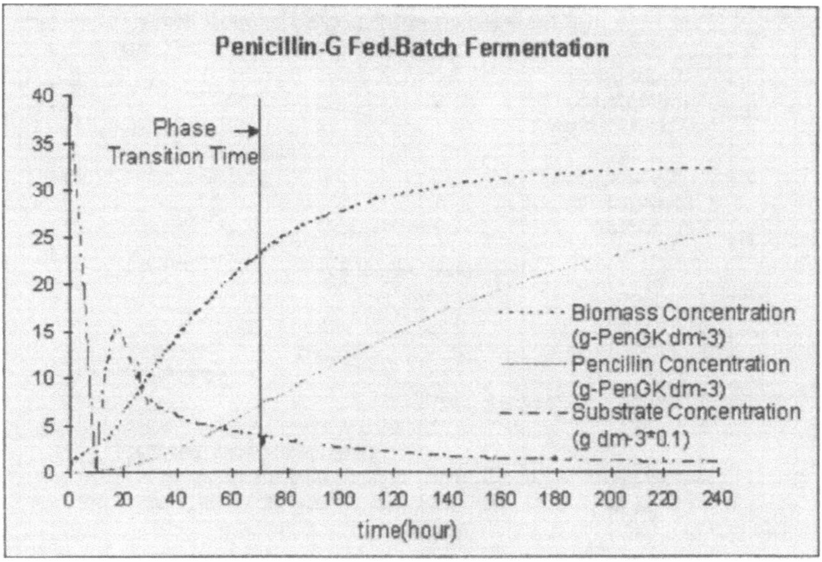

Figure 1. Penicillin-G fed-batch fermentation

Fuzzy clustering is applied to this process to identify the already known phase transition time. The fuzzy clustering algorithm proposed by Kaufman (1990) is used. The NCSS 6.0 software package1 is used to implement the clustering algorithm. Seven variables are used for phase detection and they are: specific biomass growth rate, specific substrate consumption rate, biomass concentration, live hyphae concentration, dead hyphae concentration, production concentration and time, see figure 2(a). After converting fuzzy clustering results into crisp results, the identified phase transition time is at 70.31 hours. Compared with the known phase transition time at the 70th hour, it is apparent that fuzzy clustering was effective in phase detection, see figure 2(b). Since these results are encouraging, we will proceed by applying fuzzy clustering to a more complex fermentation, that of Gluconic acid batch fermentation by the microorganism *Pseudomonas ovalis*.

1 Number Crunch Statistical System. Kaysville, Utah 84037

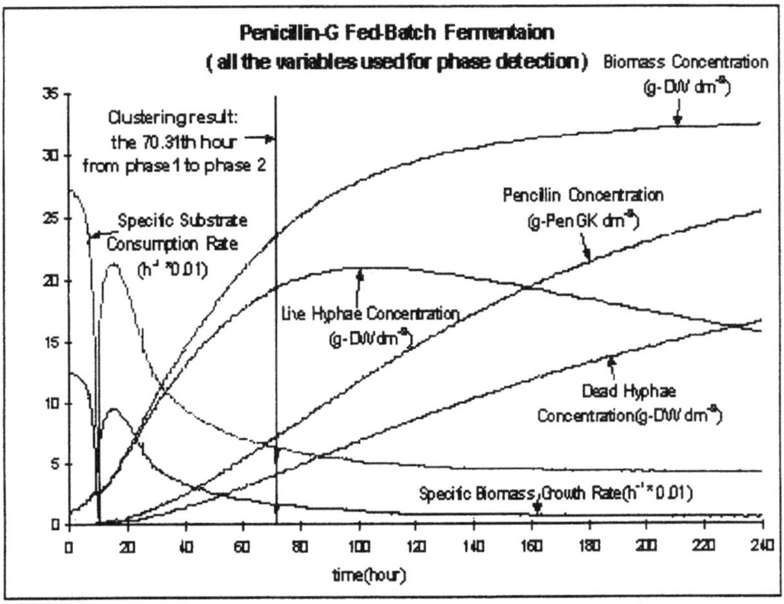

(a)

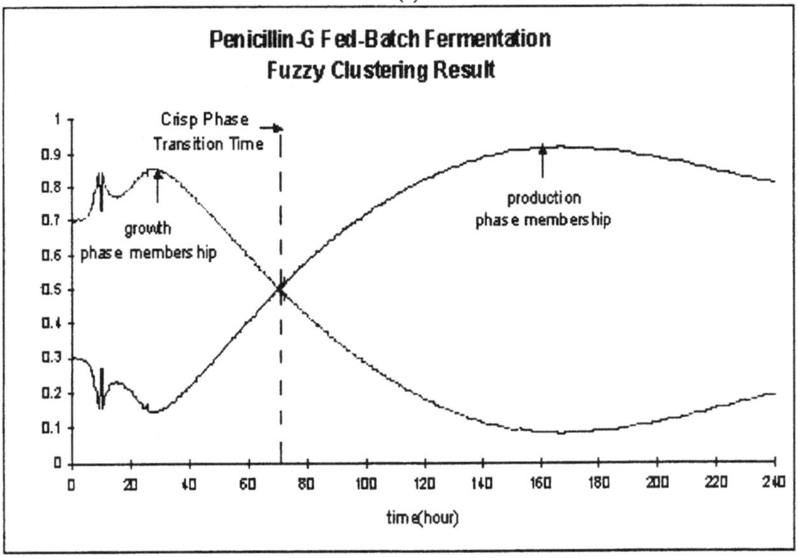

(b)

Figure 2
(a) Variables used as inputs to the fuzzy clustering algorithm
(b) Fuzzy clustering for the Penicillin-G fed-batch fermentation

2.1.3. Fuzzy clustering for off-line phase detection of gluconic acid batch fermentation

The nonlinear model provided by Foss at al. (1995) is used in this study to simulate the Gluconic acid batch fermentation. This Gluconic acid fermentation consists of three phases. At the beginning of the batch, the production of Gluconolactone is small due to the small concentration of cells. Hence the production of Gluconic acid is small due to the low concentration of Gluconolactone. This phase is characterised by relatively high concentrations of both dissolved oxygen (DO) and glucose. In the intermediate phase of the fermentation, the production of cells and Gluconolactone proceeds at high rate. Some Gluconic acid is produced. This phase is characterised by a relatively low concentration of DO and a decreasing concentration of glucose. During the final phase of the fermentation, the production of cell and Gluconolactone is reduced due to shortage of glucose. The only significant reaction is production of Gluconic acid from Gluconolactone. This phase is characterised by low glucose concentration and high DO concentration. Hence, there is strong evidence that phases for this process can be characterised by the concentrations of DO and glucose (Foss et al., 1995).

Fuzzy clustering is applied to identify the three phases of this fermentation process. Eight process simulations were carried out. Glucose and DO concentrations were used as inputs to the clustering algorithm. The initial states of cell and glucose concentrations were chosen from the intervals [0.4, 0.5] and [40, 50]. Table 1 shows the initial conditions under which these simulations were conducted as well as the applied noise levels. Fuzzy clustering results for the eight process simulations are shown in table 2 where fuzzy phase results were converted to crisp phase results. Phase transition times identified by clustering are different for all eight simulations. This is expected since each simulation was carried out using different initial conditions and noise levels.

Table 1. Initial Conditions and Noise Levels Used in Gluconic Acid Simulation

Simulation Process Number	Cell Concentration (UOD ml^{-1})	Glucose Concentration (g $litre^{-1}$)	Noise Level
1	0.4	40	2%
2	0.4	50	6%
3	0.5	40	3%
4	0.5	50	5%
5	0.42	48	2%
6	0.43	47	3%
7	0.45	43	1%
8	0.45	45	5%

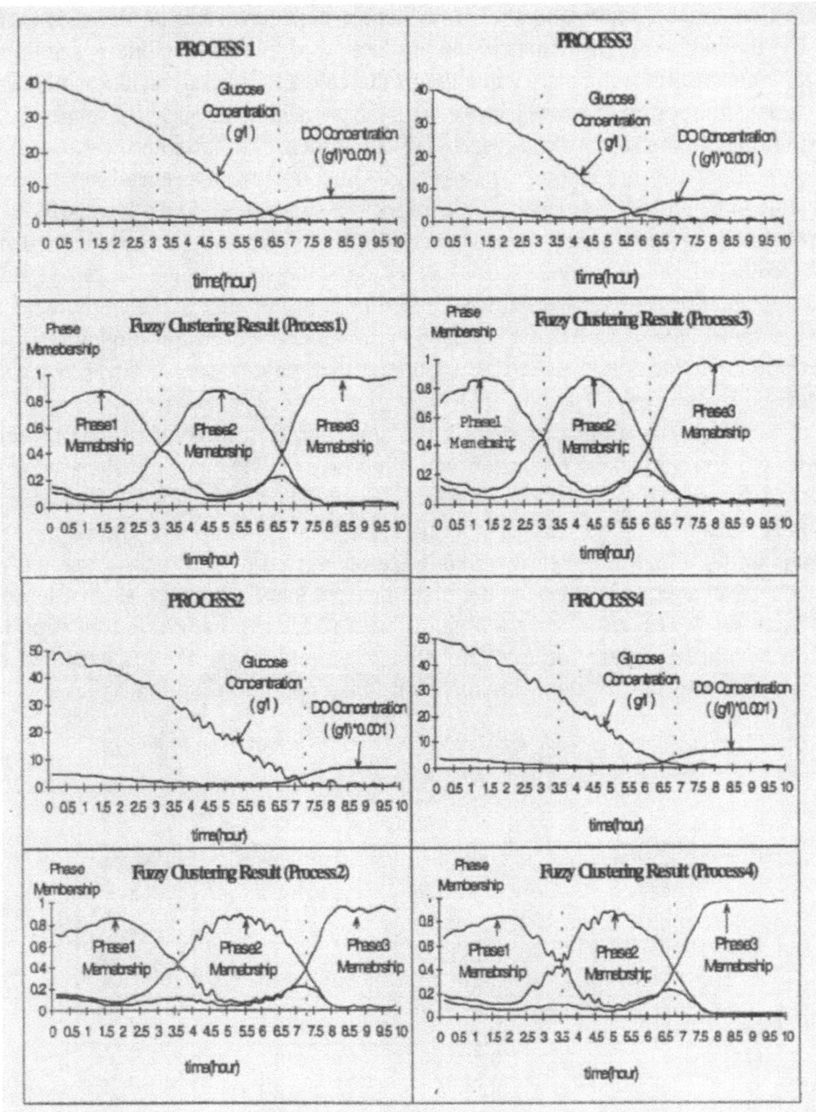

Figure 3. Fuzzy Clustering Results Phase1: High DO and Glucose Concentration Phase2: Low DO concentration and Glucose concentration is decreasingPhase3: High DO concentration and low Glucose concentration

Table 2. Phase Transition Time Identified by Fuzzy Clustering

Simulation Process Number	Phase Transition Time (Phase 1 to Phase 2)	Phase Transition Time (Phase 2 to Phase 3)
1	3.20	6.65
2	3.60	7.30
3	3.05	6.15
4	3.50	6.80
5	3.40	7.05
6	3.25	7.00
7	3.25	6.75
8	3.30	6.70

Figure 3 shows the fuzzy clustering results. The relationships between Glucose and DO concentrations for each of the identified phases are consistent with those identified by expert knowledge (Foss et al., 1995). These relationships are summarised at the bottom of Figure 3. It is important to note that clustering not only identified the phases of the fermentation but it also identified the smooth transition regions between them.

2.2. NEURAL NETWORKS FOR ON-LINE FUZZY PHASE DETECTIOI

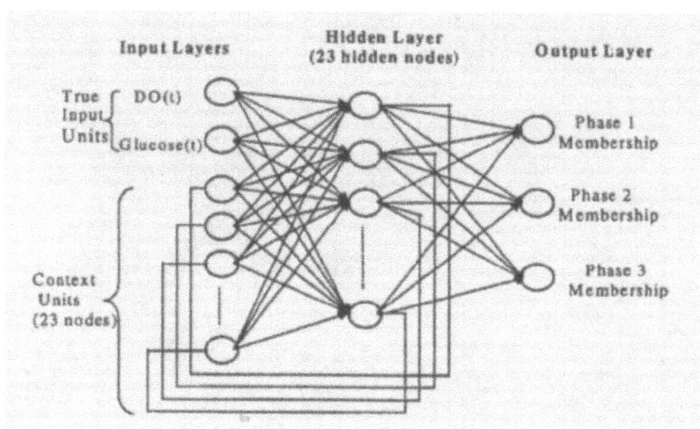

Figure 4. Elman-Jordan Network Structure for On-Line Fuzzy Phase Detection

Fuzzy clustering results were used to train NNs for on-line phase detection of the Gluconic acid fermentation. Since the Elman-Jordan NN is well suited for classifying time-series data, it is used in this experiment for phase detection. The input layer consists two true inputs nodes, a time sample of Glucose and DO concentrations. The remaining nodes of the input layer are fed back from the hidden layer. The NN consists of three outputs with each output representing the degree of membership of any time sample to one of the three phases. The optimum number of hidden nodes for this network was experimenting determined to be nine. The NN is implemented us Neuroshell 2.0[2]. Figure 4 shows the architecture of this network.

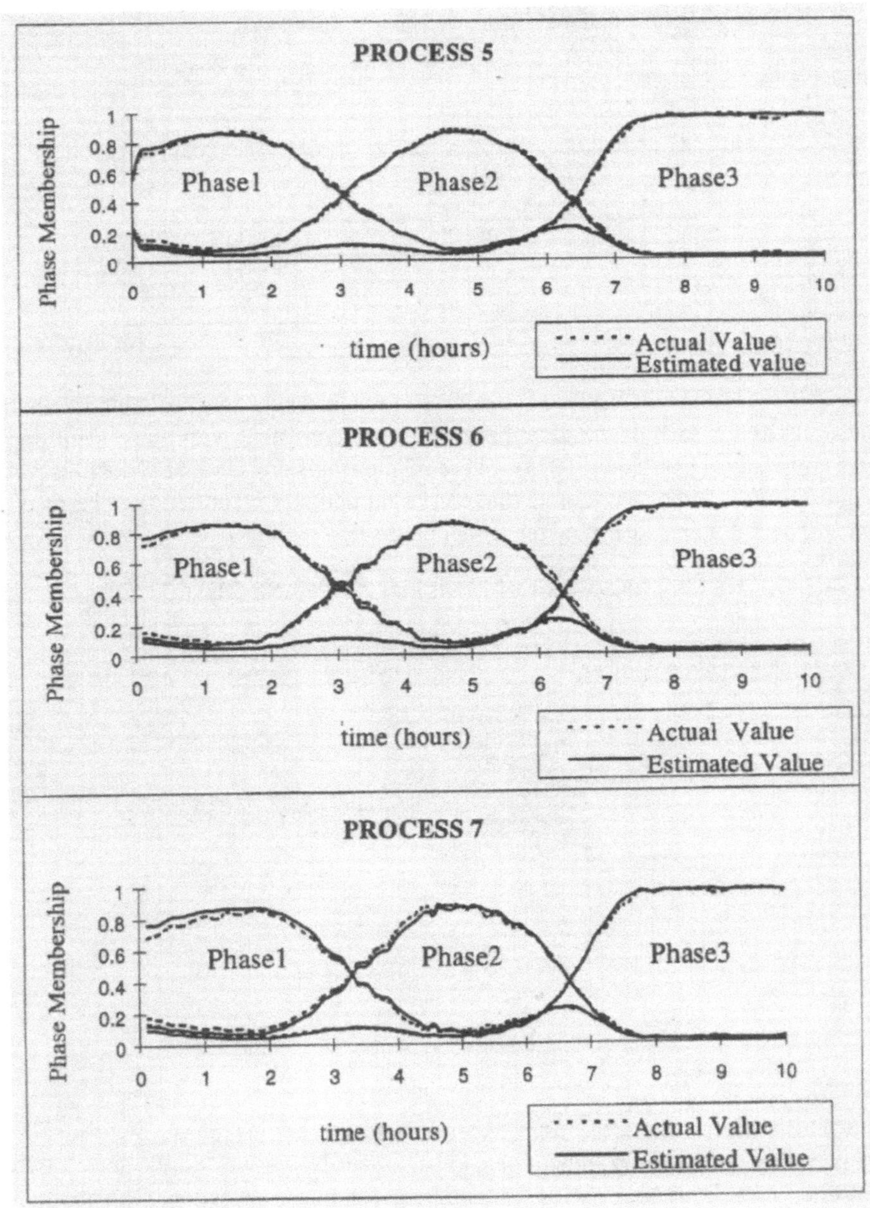

Figure 5. Neural Network Phase Detection Results

---

2 Ward System Group, Inc. Frederick, MD 21702

Process simulations one, two, and three in table 1 are used by the NN as training set. Process simulation four is used by the NN as test set to check for convergence of the network. The sequential training approach, which was shown in (Karim et al., 1992) to give best results, is used here to train the NN. The NN is first trained with data from one simulation process out of the training set. After processing 61 data patterns, the network temporarily stops training, reads the test set and computes an average error for it. Test set average error graph is obtained with the test set average errors plotted against the numbers of training patterns. By watching the test set average error graph, training is stopped when the test set average error is no longer decreasing and over-fitting happens. Then the NN weights are saved and used as initial weights for the next data set from another process simulation out of the training set. This training process continues iteratively until the NN can identify the phases of process simulation four effectively. Process simulations five, six, seven and eight in table 1 are used to evaluate effectiveness of the NN in detecting the phases of process simulations it has not seen before. Figure 5 shows phase detection results of the NN for process simulations five, six, and seven compared to those obtained using fuzzy clustering. The R^2 values for each of the membership functions for all three process simulations are over 0.99. With each new time sample of glucose and DO concentrations, the NN predicts precisely to what degree the fermentation is in phase one, two, or three. Since the three testing process simulations were performed with different initial conditions and noise levels than the four training ones, phase transition times were also different for these simulations. The NN successfully identified these new transition times.

3. Conclusion

A new technique for off-line and on-line fermentation phase detection was developed and successfully applied to simulations of a Gluconic acid fermentation. This technique which is based on fuzzy clustering and NNs is a better alternative to simply using expert knowledge for phase detection. In addition to detecting the phases of a fermentation, the technique has the advantage of identifying the smooth transitions between the phases. This information could be used as a basis for developing controllers with smooth control during phase transitions. The technique was tested with simulations at various initial conditions and noise levels and proved to be insensitive to noise.

References

1. Foss B.A., T.A. Johansen and A.V. Sorensen (1995). Nonlinear predictive control using local models applied to a batch fermentation process. *Control Engineering Practice* 3(3), 389-396.
2. Ignova M., G.C. Paul, J. Glassey, A.C. Ward, G.A. Montague, C.R. Thomas and M.N. Karim (1996). Toward intelligent process supervision: industrial penicillin fermentation case study. *Computers Chem. Eng.* 20, Suppl.. S545-S550.
3. Karim M.N. and S.L. Rivera (1992). Artificial intelligence neural networks in bioprocess state estimation. *Adv. Biochem. Eng. Biotechnol.* 46, 1-22.
4. Karim M.N., T. Yoshida, S.L. Rivera, V.M. Saucedo, B. Eikens and G-S OH (1997). Global and local neural network models in biotechnology: application to different cultivation processes. *Journal of Fermentation and Bioengineering*. 33(1), 1-11.

5. Kaufman L. (1990). *Finding groups in data: an introduction to cluster analysis,* John Wiley, New York.
6. Kosanovic B.R., L.F. Chaparro, M-G. Sun, R.J. Sclabassi (1994). Physical system modelling using temporal fuzzy sets. *Proc. of the Inter. Joint Conf. of NAFIPS/IFIS/NASA '94,* 429-433.
7. Kosanovic B., L.F. Chaparro and R.J. Sclabassi. (1995). Hidden process modelling. *Proceedings ICASSP-95* (5), 2935-2938.
8. Konstantinov K.B. and T. Yoshida (1991a). A knowledge-based pattern recognition approach for real-time diagnosis and control of fermentation processes as variable structure plants. *IEEE Trans. On Systems, Man and Cybernetics* 21(4), 908-914.
9. Konstantinov K.B., R. Aarts and T. Yoshida (1992). On the selection of variables representing the physiological state control of cell cultures. *IFAC Modelling and Control of Biotechnical Processes, Colorado, USA*, 263-266.
10. Konstantinov K.B. and T. Yoshida. 1989. Physiological state control of fermentation processes. *Biotechnology and Bioengineering* 33, 1145-1156.
11. Menezes J.C. and S.S. Alves (1994). Mathematical modelling of industrial pilot-plant Penicillin-G fed-batch fermentations. *J. Chem. tech. Biotechnol.* 61, 123-138.
12. Shimizu K. 1996. A tutorial review on bioprocess systems engineering. *Computers Chem. Engng.* 20(6/7), 915-941.

SIMULATION, DESIGN AND MODEL BASED PREDICTIVE CONTROL OF PHOTOBIOREACTORS

J.-F. CORNET*, C.G. DUSSAP* AND J.-J. LECLERCQ**
** Université Blaise Pascal, Laboratoire de Génie Chimique et Biochimique (LGCB), 24 avenue des Landais, F63177 Aubière Cedex.*
*** ADERSA, 10 rue de la Croix Martre, F91873 Massy.*

Abstract

A simple generalised two-flux approach is presented for modelling radiant light transfer in photobioreactors. A predictive method to obtain optical properties for the medium, based on the Lorenz-Mie theory is discussed. In the same way, a biochemically structured approach is proposed to predict stoichiometries of reactions, including energetic aspects. The formulation for coupling available light and kinetic rates is then proposed by defining a working illuminated volume. The obtained model, compared to experimental results in many different conditions, is proved to be a good tool for simulation, design and predictive control of photobioreactors.

1. Introduction

Engineering of photobioreactors (PBRs) becomes a field of increasing importance, for production of valuable products from micro-algae, for CO_2 exhaustion or as a part of Closed Ecological Life Support Systems for food production and O_2 regeneration. In order to develop processes in axenic conditions, and highly operative in quality and productivity, artificial PBRs illuminated with lamps are preferred to solar open or semi-closed PBRs. Such artificial reactors can then be fully controlled if predictive models exist for biomass quality and productivity.

For the mathematical modelling of photobioreactors it is necessary to understand and formulate the coupling between the metabolism of micro-organisms and the physical phenomenon of light transfer inside the culture medium. PBRs are governed by radiant light energy availability, which is highly heterogeneous within the culture volume (Aiba, 1982; Cornet *et al.*, 1992; Cassano *et al.*, 1995; Acien-Fernandez *et al.*, 1997). This spatial heterogeneity causes varying local reaction rates, which makes it necessary to derive local equations and calculate the mean volumetric growth rate by

M. Hofman and P. Thonart (eds.), Engineering and Manufacturing for Biotechnology, 227–238.

integration over the working illuminated volume in the reactor (Cornet *et al.*, 1992; 1995 and 1998).

The aim of this paper is to show how, with considerable theoretical efforts, it is possible to obtain general knowledge models with a reduced number of empirical coefficients. These models can then be used in a predictive manner as powerful tools for simulation, design or control of photoreactors. The paper gives an overview of theoretical and experimental results obtained in the last decade in the field of PBR modelling.

2. Modelling photobioreactors

Modelling PBRs appears as a difficult task because of the heterogeneity of the radiation field inside the reactor. First, due to the absorption and scattering of light by micro-organisms, the radiant light energy available $I_{\lambda\Sigma}$ (W/m^3) is unequally distributed inside the vessel. Second, at a given point, the specific intensities I_λ (W/m^3) depend on the phase function for scattering and have different values over a 4π solid angle. This leads to a complex formulation of the problem in order to calculate the local radiant light energy available from the radiative transfer theory.

It is therefore necessary to formulate the coupling between light energy transfer in PBRs and local biomass growth rates and stoichiometries, leading to define zones in which metabolic activity occurs, and to volumetrically average local kinetic rates. This paper presents several experimental and theoretical developments obtained for two micro-organisms, *Spirulina platensis* and *Rhodospirillum rubrum*, cultivated in different PBRs.

2.1. RADIATIVE TRANSFER FORMULATION

Solving the general tridimensional form of the equation of radiative transfer in the given geometry of a reactor is a very difficult task. This requires Monte Carlo or finite element methods (Spadoni *et al.*, 1978; Aiba, 1982; Siegel and Howell, 1992; Cornet *et al.*, 1994) which are highly time consuming, limiting such an approach to accurate simulation of PBRs. Nevertheless, as the authors previously showed, many practical cases can be formulated with monodimensional approximation for radiative transfer (Cornet *et al.*, 1992; 1995 and 1998). For example, a generalised two-flux method, derived from the assumptions of Schuster (1905), can be used with a sufficient accuracy if the radiative coefficients are properly determined (Brewster and Tien, 1982; Wang and Tien, 1983; Koenigsdorff *et al.*, 1991, Cornet *et al.*, 1995; Brucato and Rizzuti, 1997). The main advantages for using a simplified monodimensional approximation are first that analytical solutions to the radiative transfer problem exist (Cornet *et al.*, 1995); and second that only mean values in intensities are used, corresponding to the physical quantities required in modelling the process.

Let introduce the total radiant light energy available at a point over the solid angle ω by $I_{\lambda\Sigma} = \iint_{4\pi} I_\lambda d\omega$, and the mean quantities over the considered visible spectrum $\Delta\lambda$ by

$I, I_\Sigma = \int_{\Delta\lambda} I_\lambda, I_{\lambda\Sigma} d\lambda$ (W/m^2). In this case, the profile of radiant energy available for the micro-organisms in a cylindrical reactor of radius R, radially illuminated with a mean incident flux F_0 is given by:

$$\frac{I_\Sigma}{F_0} = \frac{R}{r} \frac{2\cosh(\delta r)}{\cosh(\delta R) + \alpha \sinh(\delta R)} \tag{1}$$

where:

$$\alpha = \sqrt{\frac{A}{A + 2bS}} \tag{2}$$

$$\delta = \sqrt{A(A + 2bS)} \tag{3}$$

We have introduced the volumetric absorption and scattering Schuster coefficients A and S, easily related to actual coefficients a and s by:

$$\begin{aligned} A &= 2a \\ S &= 2s \end{aligned} \tag{4}$$

It is clear that these coefficients are mean coefficients in wavelength on the considered spectrum $\Delta\lambda$ for a given micro-organism, obtained by:

$$a, s = \frac{1}{\Delta\lambda} \int_{\Delta\lambda} a_\lambda, s_\lambda \, d\lambda \ (\mathrm{m}^{-1}) \tag{5}$$

Moreover, the coefficient b, appearing in equations (2-3) is the back-scattered fraction of light, obtained from the phase function of the medium and given by the Lorenz-Mie theory (Brewster and Tien, 1982; Wang and Tien, 1983; Koenigsdorff *et al.*, 1991); that is:

$$b = \int_{\frac{\pi}{2}}^{\pi} \int_0^{2\pi} p(\theta, \varphi, \theta', \varphi') \sin\theta' d\theta' d\varphi' \tag{6}$$

where quotes indicates the scattered direction. This integral corresponds only to an incident beam parallel to the r-axis, and in fact, it is necessary to take into account all the incident directions leading, for an equivalent sphere, to the double integral (Koenigsdorff *et al.*, 1991):

$$b = 2\pi \int_0^{\frac{\pi}{2}} \int_{\frac{\pi}{2}}^{\pi} p(\theta,\theta') \sin\theta' d\theta' \sin\theta d\theta \qquad (7)$$

Finally, in equation (1), the mean incident flux F_0, describing the boundary condition of the radiative transfer problem is clearly a key parameter. It can be obtained either by chemical actinometry (Cornet *et al.*, 1997), or from integral measurements with a spherical sensor (Cornet *et al.*, 1995).
Obviously, equation (1), which has been obtained for different geometries (rectangular, cylindrical, spherical, annular region...- see Cornet *et al.*, 1995 -) is fully predictive if the coefficients a, s and b, called the optical properties of the medium, can be obtained theoretically. For a micro-organism, the absorption coefficient a is an intrinsic property which depends on the pigment content, and can be calculated by convolution for each wavelength from data banks, once this content is known. The Lorenz-Mie theory then provides an excellent basis to compute the wavelength dependent scattering coefficients s_λ, and the phase function $p(\theta,\varphi,\theta',\varphi')$.

2.2. COMPUTING THE OPTICAL PROPERTIES

From the basic electromagnetic characteristics of the micro-organism, i.e. the refractive index of the medium n_m and the complex refractive index of the particle $m = n + \kappa\, i$, the Lorenz-Mie theory enables to calculate, with tedious computation (Van de Hulst, 1981; Bohren and Huffman, 1983), the optical properties necessary to formulate the radiative transfer model. The wavelength dependent properties are obtained from the definition of the size parameter x, given by:

$$x = \frac{\pi D_{eq}\, n_m}{\lambda_0} \qquad (8)$$

where D_{eq} is the equivalent diameter (equivalent sphere) of the micro-organism, and λ_0 is the considered wavelength in the vacuum at which the computation is performed.

The details of these computations are not given in this paper, but as example, results were obtained for the calculation of the scattering efficiencies Q_{SCA} for two micro-organisms, *Rs. rubrum* and *S. platensis* (Figure 1). The actual computation requires to know the size distribution *f(x)* for the corresponding micro-organism, enabling the assessment of the mean scattering efficiency from:

$$\overline{Q}_{SCA} = \int_0^{\infty} Q_{SCA}(x)\, f(x)\, dx \qquad (9)$$

The scattering efficiency is then easily related to the scattering volumetric coefficient by:

$$s_\lambda = \frac{\pi D_{eq}^2}{4} N_p \overline{Q}_{SCA} \tag{10}$$

where N_p is the number of micro-organisms per unit volume.

The calculation was performed for two extreme cases, a non-absorbed wavelength (perfect dielectric) and a wavelength at a maximum of absorption (from the highest value of κ). It clearly appears that, in the range of interest for the size parameter in the visible spectrum, most of the micro-organisms can be considered as a perfect dielectric with a maximum deviation less than 10% (Figure 1). This enables to use a simplified engineering equation for the calculation of the scattering efficiency Q_{SCA}, available when the ratio of refractive indexes $\overline{n} = n/n_m$ tends to 1 (Van de Hulst, 1981, Cornet *et al.*, 1996):

$$Q_{SCA} = \left[2 - \frac{4}{\rho}\sin\psi + \frac{4}{\rho^2}(1-\cos\psi)\right] \tag{11}$$

where $\psi = 2x(\overline{n} - 1)$, and which is probably one of the most famous and useful relation in this field.

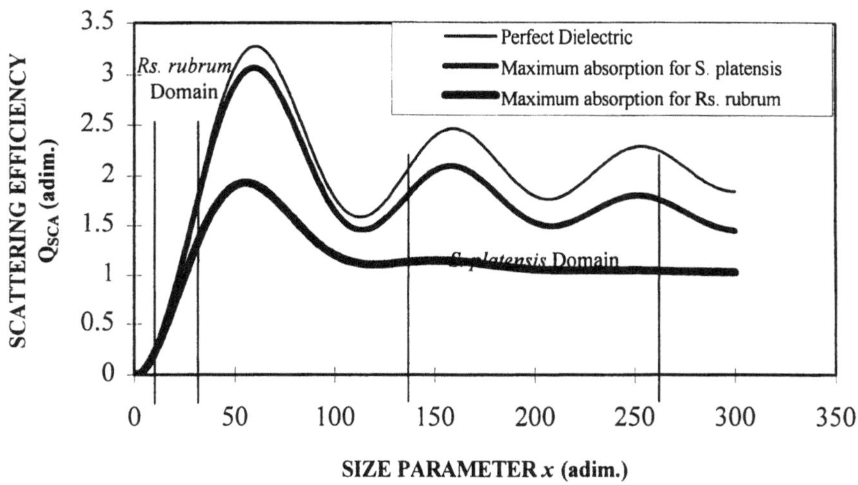

Figure 1: Scattering efficiency Q_{SCA} *versus size parameter* x. *Effects of absorption on coefficients computation in the respective domains for Spirulina platensis and Rhodospirillum rubrum.*

These results can be used to determine the wavelength dependent absorption and scattering mass coefficients defined by $Ea_\lambda = a_\lambda / C_X$ and $Es_\lambda = s_\lambda / C_X$ (m^2/kg) where C_X is the biomass concentration. These coefficients should be used in equation (4) instead of the volumetric coefficients a_λ and s_λ because in batch cultivations, the biomass concentration is time dependent. An example of this determination was performed for *Rs. rubrum*, on which Ea_λ is obtained experimentally and Es_λ is theoretically computed from equation (11) (Cornet *et al.*, 1996) (Figure 2). Typically, these results enable to perform the integration of equation (5), and to determine mean coefficients in wavelength.

Finally, from the Lorenz-Mie theory (Bohren and Huffman, 1983), we have computed phase functions for the same micro-organisms and in the same extreme conditions (Figure 3). Clearly, as it is well known, the phase function for scattering of micro-organisms is strongly peaked in the forward direction ($\theta = 0$). From these data, relation (7) gives respectively for *Rs. rubrum* and *S. platensis*, $b = 6.23\ 10^{-2}$ ($D_{eq} = 21$ μm) and $3.73\ 10^{-2}$ ($D_{eq} = 2.75$ μm).

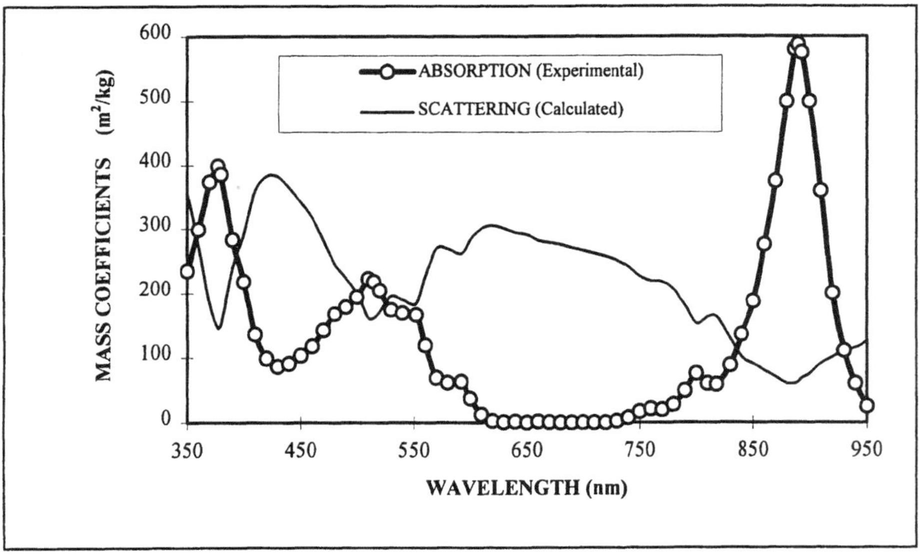

Figure 2: Calculation of absorption and scattering coefficients for Rhodospirillum rubrum by the Lorenz-Mie theory.

2.3. COUPLING RADIATIVE TRANSFER WITH RATES AND STOICHIOMETRY

Once a correct formulation of the radiative transfer has been done, it is necessary to properly define the coupling between the local total energy available I_Σ inside the reactor and the local rates. It is not so trivial because I_Σ influences also the stoichiometry of the produced biomass. As an example, we give here theoretical results experimentally validated and obtained from the biochemically structured approach, using the phenomenological thermodynamics of irreversible processes (Dussap, 1988, Cornet *et al.*, 1998). For a considered available light energy, this approach leads to the following structured stoichiometric equation for *S. platensis*:

$$\begin{aligned} & CO_2 + 0.7\,H_2O + 0.154\,HNO_3 + 0.007\,H_2SO_4 \\ & + 0.005\,H_3PO_4 + 12.41\,h\upsilon \\ & \rightarrow CH_{1.583}O_{0.514}N_{0.154}S_{0.007}P_{0.005} + 1.3455\,O_2 \end{aligned} \qquad (12)$$

in which the number of quanta is obtained from the thermodynamically calculated value of the *in vivo* P/2e$^-$ ratio (Cornet *et al.*, 1998). Because this kind of equation can be theoretically established for each value of I_Σ, it is fully predictive for calculating the stoichiometric energetic conversion yield η from the number of photons involved in the reaction. For example, we obtained from equation (12) $\eta = 8.6\ 10^{-9}$ kg/J for *S. platensis* and with a similar equation, $\eta = 2.5\ 10^{-8}$ kg/J for *Rs. rubrum*.

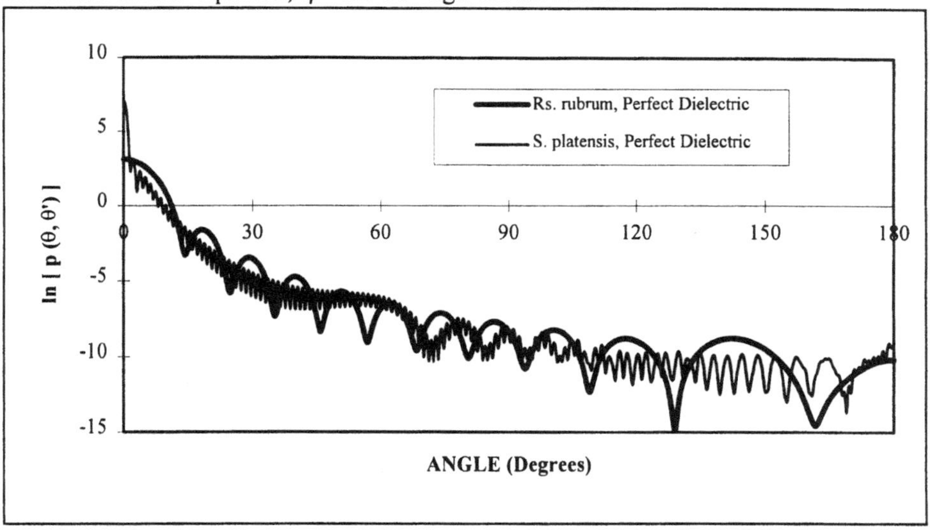

Figure 3: Phase functions (displayed as their logarithm) versus the angle of scattering for Rhodospirillum rubrum (D_{eq} = 2.75 μm) and Spirulina platensis (D_{eq} = 21 μm) and calculated from the Lorenz-Mie theory.

Actually, photosensitised reactions does not operate at the maximum conversion yield, because, the primary quantum yield ρ strongly decreases when I_Σ increases. One can postulates that ρ is given by a relation of the form:

$$\rho = \rho_{max} \frac{K}{I_\Sigma + K} \qquad (13)$$

in which ρ_{max} is the well-known maximum quantum yield at the thermodynamics optimum. The coefficient K is a characteristic of the pigment content of the photosynthetic antenna and remains difficult to obtain theoretically. Nevertheless, this coefficient, which is the sole to obtain experimentally for the proposed model, can be easily determined (Cornet *et al.* 1998). Then, the local biomass volumetric reaction rate in the illuminated volume of the reactor is given by:

$$r_X = \phi \, a I_\Sigma = \rho \eta a I_\Sigma \tag{14}$$

Because, the kinetic coefficients previously discussed are only valid in the illuminated zone of the reactor, it is obvious that the mean averaged observed volumetric rate in the reactor will be given by (Cornet *et al.*, 1998 and 2000):

$$< r_x >= f_I \, (\beta + \gamma) \frac{1}{V_{UT}} \iiint_{V_{UT}} r_X \, dV \tag{15}$$

The working illuminated volume V_{UT} and the illuminated fraction γ are easily obtained from equation (1) (Cornet *et al.*, 1995). The dark operative volume fraction β has been introduced for photoheterotrophic micro-organisms, as *Rs. rubrum*, because in this case, a metabolic activity can occur for short residence time of cells in darkness, from a reverse electron transfer (Cornet *et al.*, 2000). It is also easily determined from the appearance of a constant growth rate on batch cultures (Cornet *et al.*, 2000). Finally, the illuminated surface fraction f_I enables to describe cases in which only part of the photoreactor is illuminated, as it is often the case on industrial reactors.

3. Results and discussions

The formulation of the radiative transfer problem on a physical basis, together with a correct understanding of the coupling between light transfer, stoichiometry and rates at the level of the cell, then integrated at the scale of the whole process, leads to the proposed knowledge model. This approach presents at least the two following advantages:

- it is not specific of a given micro-organism or of a photoreactor geometry, so it is quite general;
- it is fully robust and predictive (only one coefficient of the model remains to be experimentally determined).

Consequently, this model can be successfully used to simulate experimental results in PBRs where the monodimensional approximation is justified.

3.1. SIMULATION AND DESIGN

During the last decade, the above model has been applied on different geometries (rectangular, cylindrical, annular region), mixing types (rushton turbine, air-lift,...) and volumes (1 to 100 litres) of artificial PBRs operating in batch and continuous mode with *S. platensis* and *Rs. rubrum*, and with incident fluxes F_0 varying by a factor 100 (4 to 400 W/m^2). The standard deviation observed never exceeded 10% (Cornet, 1997; Cornet *et al.*, 1992, 1994; 1995, 1998 and 2000). It succeeded also as a basic tool for scaling up by a factor 10 an airlift PBR with a constant productivity.

Figure 4 shows an example of comparison between model and experimental results on a 10 L cylindrical photobioreactor radially illuminated and during a step in incident light flux F_0. Clearly, the agreement between the biomass concentration and the predictive values of the model is excellent, both on steady and transient states.

It must be emphasised that more refined numerical tools exist regarding the radiative transfer formulation if a high accuracy is required for local simulation and design. They use Monte Carlo or finite elements methods with wavelength dependent coefficients (Spadoni *et al.*, 1978; Aiba, 1982, Cornet *et al.*, 1994).

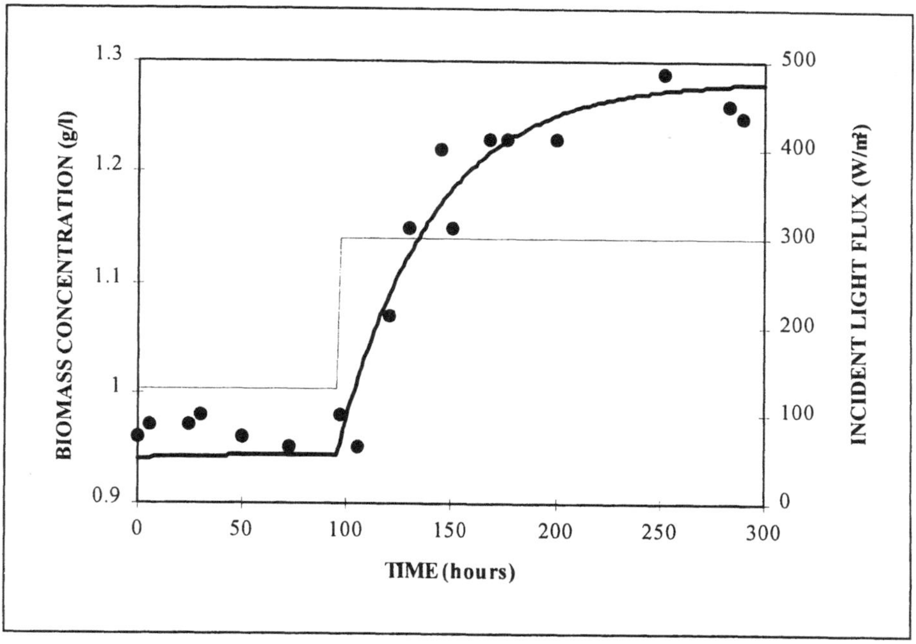

Figure 4: Comparison between experimental results in total biomass concentration (closed circles) and the proposed model (thick line) for a cylindrical 10 L photobioreactor cultivating Spirulina platensis *in continuous mode. The dilution rate was 0.025 h^{-1} and the experiment displays a step in incident light flux (narrow line) onto the reactor from 130 to 300 W/m^2. The illuminated surface fraction for this reactor was 0.6.*

3.2. MODEL BASED PREDICTIVE CONTROL

Model based control of processes requires predictive models with short calculation times. For this reason, the proposed model (leading to analytical solution for radiative transfer) appears as a good candidate to be used as a model based predictive control of PBRs. Very good results have then been obtained (Leclercq, 1998; Cornet *et al.*, 1999) with the two previous micro-organisms involved in the MELiSSA project of ESA (Figure 5).

4. Conclusions and perspectives

A model for simulation, design and model based predictive control of PBRs was presented and discussed. It relies on a monodimensional formulation of the radiative transfer problem using a generalised two-flux method, providing analytical solutions for available light energy profiles inside the reactor in a given geometry. The optical properties of the medium were shown to be theoretically obtained from the Lorenz-Mie theory, giving the model's coefficients by a predictive mean. This approach was used to define the coupling between the available light energy and both the local kinetics and stoichiometry. This led to the concept of working illuminated volume inside the reactor, allowing to calculate the mean volumetric growth rates in biomass.

Such a physically and biochemically consistent model appears fully predictive because calculations can be performed from the experimental knowledge of only one coefficient relative to the primary quantum yield of a given micro-organism. Simulations of many experimental results proved the model very robust. Thus, good results were also obtained in model based predictive control of PBRs.

In a near future, this approach could be improved by focusing attention and developing theoretical tools in three main directions:

- further investigations about biochemically structured metabolism are necessary for different metabolic conditions (photoautotrophy with one and two photosystems, photoheterotrophy). This requires advanced formulations in the domains of metabolic network analysis for photosynthetic micro-organisms together with energetics and irreversible thermodynamics analysis of photosynthesis;
- the formulation of the coupling between kinetic rates and radiative transfer in cases where transient states exist with short time dark efficient zones in the PBR (as it is the case for photoheterotrophic metabolisms) needs also to be further investigated. Using populations balances formalism then could be a good alternative for this;
- optical radiative properties remain difficult to assess for many photosynthetic micro-organisms, and there is a great necessity in developing data banks of *in vivo* pigment absorption properties, together with analytical and numerical tools for computation of Lorenz-Mie series with actual shapes of micro-organisms (cylinders, spheroids,...).

The elucidation of these different points is a prerequisite step for the demonstration that, besides their industrial interests, photosynthetic micro-organisms for which, at a given complexity, knowledge models can proceed with a lesser number of degree of freedom than for heterotrophic micro-organisms, are ideal case study for advanced modelling in the field of biochemical reactors.

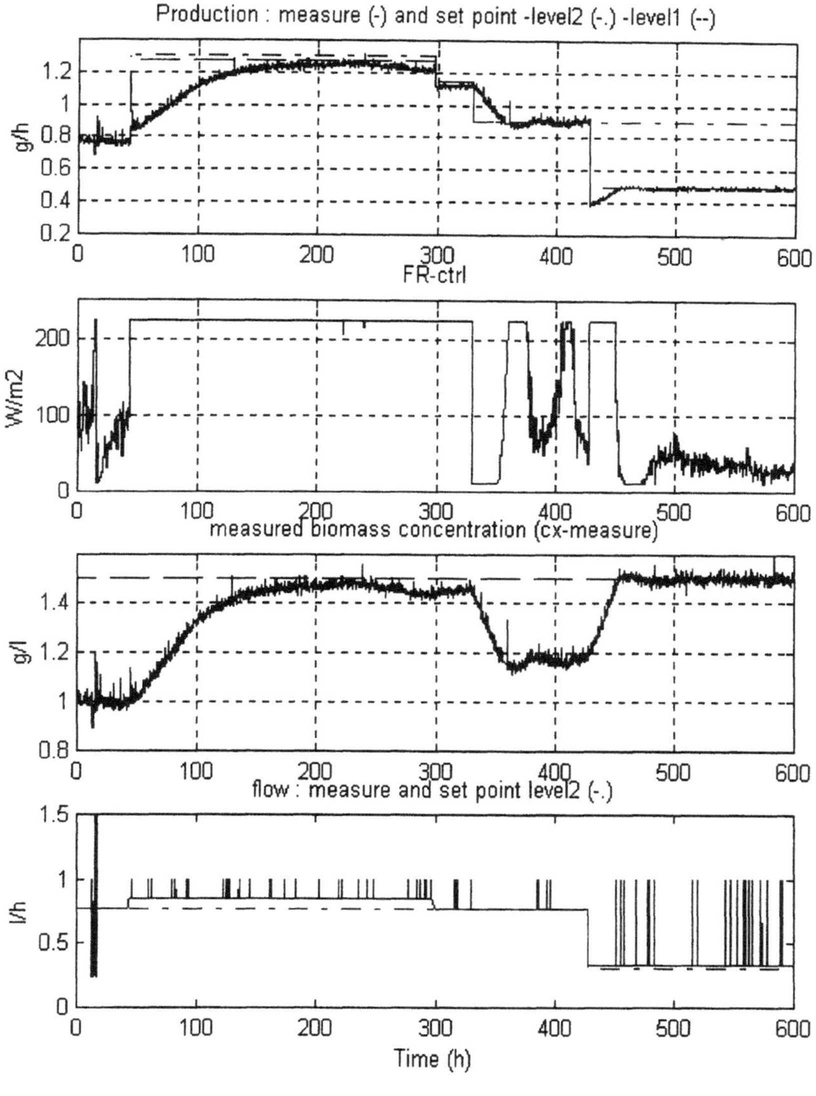

Figure 5: Example of predictive functional control on 80 L airlift photobioreactor cultivating Spirulina platensis. *Good results are obtained by varying the productivity set point with incident light flux as manipulated variable (these results were obtained at the University Autonomous of Barcelona, department of Chemical Engineering, MELiSSA laboratory ; from Mengual* et al., *2000)*

Acknowledgement

This work is supported by the European Space Agency (ESA) through the MELiSSA Project.

References

Acien Fernandez F. G., Garcia Camacho F., Sanchez Perez J. A., Fernandez Sevilla J.M., Molina Grima E. (1997). A Model for light distribution and average solar irradiance inside outdoor tubular photobioreactors for the microalgal mass culture. *Biotech. Bioeng.* 55, 701-714.

Aiba S. (1982). Growth kinetics of photosynthetic microorganisms. *Adv. Biochem. Eng.* 23, 85-156.

Bohren C.F., Huffman D.R. (1983). *Absorption and scattering of light by small particles*, John Wiley & Sons Inc.

Brewster M.Q., Tien C.L. (1982). Radiative transfer in packed fluidised beds: dependent versus independent scattering. *J. Heat Transfer* 104, 573-579.

Brucatto A., Rizzuti L. (1997). Simplified modelling of radiant fields in heterogeneous photoreactors. *Ind. Eng. Chem. Res.* 36, 4748-4755.

Cassano A. E., Martin C. A., Brandi R.J., Alfano O. M. (1995). Photoreactor analysis and design: Fundamentals and Applications. *Ind. Eng. Chem. Res.* 34, 2155-2201.

Cornet J.-F., Dussap C.G., Dubertret G. (1992). A structured model for simulation of cultures of the cyanobacterium *Spirulina platensis* in photobioreactors. *Biotech. Bioeng.* 40, 817-825.

Cornet J.-F., Dussap C.G., Gros J.-B. (1994). Conversion of radiant light energy in photobioreactors. *A.I.Ch.E. Journal.* 40, 1055-1066.

Cornet J.-F., Dussap C.G., Gros J.-B., Binois C., Lasseur C. (1995). A simplified monodimensional approach for modelling coupling between radiant light transfer and growth kinetics in photobioreactors. *Chem. Eng. Science.* 50, 1489-1500.

Cornet J.-F. (1996). Model parameters for growth of Rhodospirillum rubrum under light limitation in photobioreactors. *Technical Note 23.4.* ESA contract PRF 141315, M.O.U. ECT/FG/CB/95.205.

Cornet J.-F. (1997). *Analysis of a photobioreactor performances for sizing a consumer compartment.* ESA report, Biorat Project. Consulting agreement 2011/97.

Cornet J.-F., Marty A., Gros J.-B. (1997). A revised technique for the determination of mean incident light fluxes on photobioreactors. *Biotechnol. Prog.* 13, 408-415.

Cornet J.-F., Dussap C.G., Gros J.-B. (1998). Kinetics and energetics of photosynthetic microorganisms in photobioreactors. *Adv. Biochem. Eng. Biotech.* 59, 153-224.

Cornet J.-F, Dussap C. G., Leclercq J.-J. (1999). Simulation and model based predictive control of photobioreactors. In: *9th European Congress in Biotechnology - ECB9*, July 1999, Brussels.

Cornet J.-F., Albiol J. (2000). Modelling photoheterotrophic growth kinetics of *Rhodospirillum rubrum* in rectangular photobioreactors. *Biotech. Prog.* 16, 199-207.

Dussap C.G. (1988). *These de Doctorat es Sciences Physiques*, Univ. Blaise Pascal. Serie E, n° 409.

Koenigsdorff R., Miller F., Ziegler R. (1991). Calculation of scattering fractions for use in radiative flux models. *Int. J. Heat Mass Transfer.* 34, 2673-2676.

Leclercq J.J. (1998). *Technical Note 38.2.* ESA contract PO 171686, M.O.U. ECT/FG/MMM/97.012.

Mengual X., Albiol J., Godia F. (2000). General Purpose Station 98. ESA Contract, MELiSSA Project ; *Technical Note 43.7.*

Schuster A. (1905). Radiation through a foggy atmosphere. *Astrophys. J.* 21, 1-22.

Siegel R., Howell J.R. (1992). *Thermal radiation heat transfer.* Hemisphere Publishing Corp. 3rd ed.

Spadoni G., Bandini E., Santarelli F. (1978). Scattering effects in photosensitised reactions. *Chem. Eng. Science.* 33, 517-524.

Van de Hulst H.C. (1981). *Light scattering by small particles.* Dover publications Inc. 2nd ed.

Wang K.Y., Tien C.L. (1983). Radiative heat transfer through opacified fibres and powders. *J. Quant. Spectrosc. Radiat. Transfer.* 30, 213-223.

PART V
REACTOR ENGINEERING

BIOREACTORS FOR SPACE : BIOTECHNOLOGY OF THE NEXT CENTURY

ISABELLE WALTHER*, BART VAN DER SCHOOT, MARC BOILLAT AND AUGUSTO COGOLI*
**Space Biology Group ETH-Technopark Technoparkstr. 1 CH-8005 Zurich*
Phone: ++41 1 445 12 80 Fax: ++41 1 445 12 71
E-mail walther@spacebiol.ethz.ch

Summary

Space biology is a young and rapidly developing discipline comprising basic research and biotechnology. In the next decades biotechnology in space will play a prominent role in the International Space Station (ISS). Therefore, there is an increasing demand for sophisticated instrumentation to satisfy the requirements of the future projects in space biology. Bioreactors will be needed to supply fresh living material (cells and tissues) either to study still obscure basic biological mechanisms or to develop profitable bioprocesses which will take advantage of the peculiar microgravity conditions. Instruments especially developed for space may be the starting point of new technology uses and lead to interesting spin-offs for Earth-bound research.

1. Introduction

Space biology is a relatively young science that has evolved from scientists' need to better understand the effects of a space environment on living systems. At the beginning of space flight, the main concern was the health of the astronauts; therefore almost all experiments were oriented toward physiology and medicine. Though these two domains are still investigated, the interest in basic research and biotechnology in space has risen drastically in the last years.

On Earth, biotechnology has already been used for centuries to produce or to modify food products such as cheese, wine or beer. But it is in the last 30 years that it has really flourished, thanks the use not only of microorganisms, but also of plants and mammalian cells. With an expanding role in health, environmental protection and agriculture, biotechnology is expected to have a significant impact on our lives in the next decades. One of the key elements for the achievement of biotechnological investigations is the reproducibility of the bioprocess. The usual way is to keep the

M. Hofman and P. Thonart (eds.), Engineering and Manufacturing for Biotechnology, 241–251.

environmental factors under control by automation and regulation of the cultivation process. For this purpose, the cultivation is performed in a bioreactor that allows the control of the physical parameters of the culture such as temperature, mixing, aeration and pH. The use of a continuous cultivation mode allows avoiding a constant change of the nutritive environment of the cells as in a batch culture.

In space, the use of bioreactors has been very limited to date, but with the upcoming of the international space station, a special attention has evolved for the development of life support systems allowing the recycling of expendable materials (i.e. water, air) and the treatment of waste by-products [1,2,3] as well as for the cultivation of microorganisms, mammalian cells and tissues for food production or medical utilisation.

2. Space bioreactors : instrument

The bioreactors normally used on Earth cannot be used in space for several reasons. First, most of the materials utilised for the fabrication of a bioreactor are not accepted in space for safety reason. No plastic easily flammable, no large piece of glass, nor any strong acid or base is allowed. Second, the design of the space apparatus is restricted by limitations of size, weight, and power. Third, the absence of sedimentation in space obliges, for example, to fill completely the cultivation chamber of the reactor (zero headspace) to avoid the presence of a gas bubble in the system. In fact, the bubble will not go to the top of the chamber as at 1*g* but will float somewhere in the middle of the bioreactor creating interference. Finally, as convection movements are also reduced to near zero, nutrient, oxygen and waste products should be efficiently transported by means of medium exchange, perfusion or slow mixing. For these reasons, new types of bioreactors specifically adapted to space investigations had to be developed. Several types of cultivation systems have been designed (Table 1) or are currently under development. The sophistication grade of the devices presented below varies greatly. Some of the instruments consist of simple cultivation chambers with automatic perfusion system for the exchange of medium. Some are much more elaborated and allow automatic regulation of cultivation parameters (pH, oxygen), sampling or fixation of the cells during flight.

Table 1: Synopsis of the characteristics of the different space cultivation systems

Instrument	Flight	Characteristics
Woodlawn Wanderer 9	Skylab, 1973	Perfusion chamber for adherent cells, fixation possible in flight, only thermal regulation

Space Tissue Loss (STL-A) Cell Culture Module (CCM)	9 Shuttle flights (1992-1996)	Hollow-fibre cartridges for adherent and non-adherent cells perfused with oxygenated medium, spent media sampling, aliquot or whole culture fixation, thermal regulation CCM is an up-graded STL-A with pH regulation and a 4°C compartment
Space Tissue Loss (STL-B)	3 Shuttle flights (1994-1996)	as STL-A but with video and microscope
Rotating Wall Vessels (RWV)	Shuttle 1991,1995 MIR Station, 1996	Zero headspace bioreactor with rotating vessel wall and gas diffusion membrane, for adherent and non-adherent cells, very low shearing force, sampling port, both fed-batch and perfused systems. In the perfused RWV, dissolved oxygen, pH and temperature are regulated.
Dynamic Cell Culture System (DCCS)	Biokosmos 9, 1989 Shuttle, 1992	Cultivation chamber with medium exchange (osmotic pump), no regulation
Swiss Bioreactor (SBR I)	2 Shuttle flights 1994, 1996	Zero headspace bioreactor for yeast cells, flexible continuous medium exchange (piezoelectric pump), sampling port, flow rate and pressure sensor, pH regulation, on-line data transfer

2.1. LARGE SPACE BIOREACTORS

The first bioreactor-like instrument, which has flown in space, was the so-called *Woodlawn Wanderer 9* apparatus [4]. It consisted of a fully automated perfusion chamber with devices for light microscopy and a motion picture camera. It was installed aboard the US space station Skylab in 1973. The Space Tissue Loss (STL) system, based on the hollow-fibre technology, provides flexible feeding capabilities, thermal regulation and chemical fixation of the cells. It fits in a mid-deck locker of the Space Shuttle and was used on several flights between 1992 and 1996. To our knowledge only the results of the first flight of the STL-A apparatus with mammalian cells in 1992 have been presented [5]. A set of bioreactors, all based on the principle of the rotating wall vessel (RWV), was developed by NASA at the Johnson Space Centre. Both fed-batch and perfusion systems are available, and oxygenation is achieved by diffusion through a silicone membrane. The first flight of a RWV bioreactor took place with mammalian cells in 1991. This instrument, which provides a very low shear force environment, is now widely used for microgravity simulations on Earth.

Two other devices have been developed recently in Switzerland under ESA (European Space Agency) contract, but have not been flown yet. In the first apparatus, the cell culture is held in a chamber between two plates, each of which carries dividers that interlock to partially divide the chamber. The plates are flat with cylindrical rims. Rotating one of the plates causing the spirals to approach and separate from each other agitates the culture. A portion of at least one plate is porous allowing the oxygen to diffuse into the culture (Patent FR 2724280, 1996). The second apparatus is a bioreactor with a culture volume of 200 ml; it contains an agitator equipped with permeable tubing to provide microgravity-compatible gas exchange together with medium supply and removal. The bioreactor is further equipped with temperature control (Peltier), pH control, dissolved oxygen and pressure sensors, sampling ports and optical access ports. Its total mass is 30 kg including pumps and electronic box (ESA communication).

2.2. MINIATURE SPACE BIOREACTORS

The instruments described above have the disadvantage of being of a rather large size and certain have no control capacities. Because of their volume they could not be used by. European investigators In fact the two experiment containers used up to now by the European Space Agency (ESA) are of very limited size (65 cc, respectively 385 cc, with dimensions of 80x40x20 mm, respectively 87x63x63 mm).

To palliate this lack, we develop several small culture systems. The first culture chamber was still uncontrolled, but fresh medium was continuously supplied by an osmotic pump [6]. The second instrument was a miniaturised bioreactor named (SBRI) running in a continuous mode, regulated to procure controlled environmental growth conditions and allowing the delivery, on-line, of biological parameters [7,8]. The construction of this bioreactor in such a reduced volume was only possible thanks the use of silicon microtechnology. A short description of these instruments is given in the next chapters.

2.2.1. The DCCS

The Dynamic Cell Culture System (DCCS, Fig. 1) has been developed by the Group of Space Biology at the ETH Zurich in collaboration with Contraves AG. This system is designed for the cultivation of mammalian and plant cells and fits into one Biorack Type I container (80x40x20 mm). The DCCS consists of a culture chamber of 200 µl supplied continuously with fresh medium by an osmotic pump at a flow rate of 1 µl h^{-1} [6]. The DCCS was first tested in a 14-day flight onboard the Soviet biosatellite Biokosmos 9 in 1989. In this mission, the growth and development of plant protoplasts were studied [9]. The system worked satisfactorily. The DCCS was tested a second time on the IML-1 mission in 1992. In this investigation, the growth and the effect of microgravity on hamster kidney cells was studied [10]. After 8 days of cultivation, the cell growth was compared between the perfusion and the batch chambers. The cells grew better and they produced more tissue plasminogen activator in the perfusion that in the batch chamber. No effect of the microgravity on the cells per se was observable.

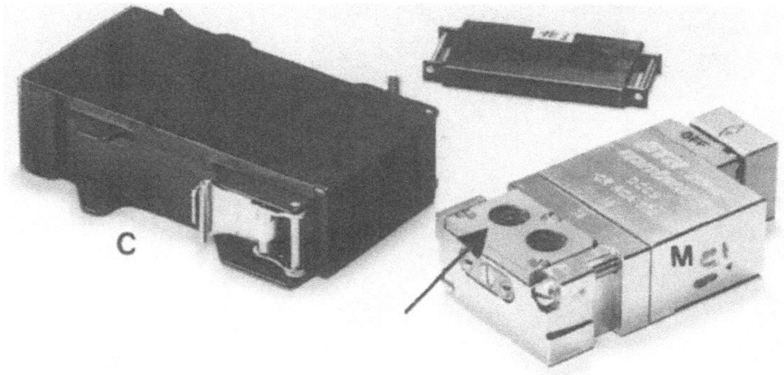

Figure 1. Dynamic cell cultivation system DCCS. On the left is the container Type I (c). On the right the DCCS with the two circular windows (arrow) over the cultivation chambers. The osmotic pump was located inside of the metallic bloc (M).

2.1.2. The Swiss space bioreactor: SBR I

The promising results observed with the DCCS concerning the perfusion chamber drove us to develop a more sophisticated device: a bioreactor for continuous cultivation in space. This controlled miniaturised instrument (Fig. 2) was built in collaboration with the company Mecanex S.A. and the Institute of Microtechnology of the University of Neuchâtel (Switzerland). The SBR I was developed for the cultivation of yeast cells. It fits in one Biorack Type II container (87x63x63 mm) and has a weight of 610 g, medium included [7]. The 3-ml culture chamber is supplied continuously with fresh medium (100 ml) by means of a piezoelectric silicon micropump; a flow and a pressure sensor insure a constant delivery of the fresh medium. The flow rate can be modified among 0.2-1.2 ml h^{-1} during cultivation. The pH, temperature and Redox potential values are measured by a microsensor. All the measured data (pH, T°, Redox, flow rate, pressure) are available on-line. The pH is regulated electrochemically to avoid the use of a strong base solution. The culture is agitated by means of a magnetic stirrer if required. Samples were withdrawn with a syringe through a rubber sampling port. This instrument flew aboard the Shuttle missions STS-65 and STS-76, in 1994 and 1996 respectively, to investigate the effect of stirring on the cultivation of yeast cells in space [8]. In fact, the absence of convection in microgravity leads to the formation of gradients in the culture, which might affect the growth of the cells. The gradients cannot form when the culture is stirred. The performances of the SBRI in space flights are presented in the next chapter.

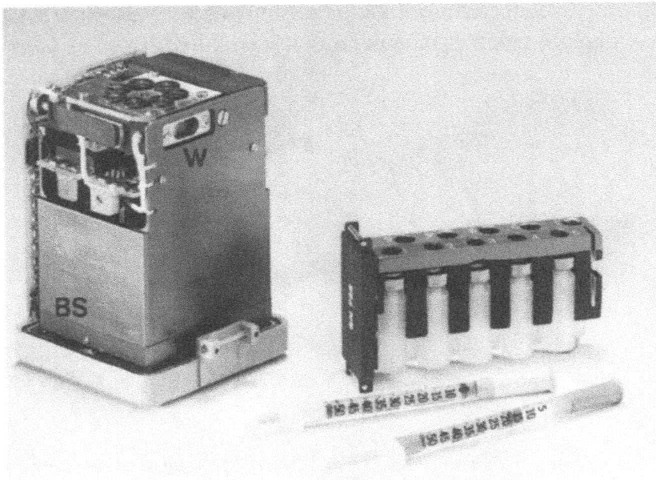

Figure 2. SBR I: Miniaturised bioreactor with, on its left, two syringes for the sampling and the sample bottles. The cultivation chamber is on the top of the bioreactor. The inspection window is located on the right upper side (W). The sensors, which are integrated into the chamber wall at the opposite of the window, are not visible on this picture. The basis structure (BS) contains the fresh and the used medium. Total height: 8.5 cm.

3. Space bioreactor SBRI : performances in flight

In view of the complexity of the instrument and the limited available space in the experiment container, it is obvious that the use of miniature, microfabricated components offers a distinct advantage. In the following sections, the individual elements will be described in more detail and special attention will be given to the way they are assembled in the bioreactor.

3.1. LIQUID HANDLING

The cell culture has to be continuously supplied with fresh nutrient medium at an adjustable rate. For this, a piezoelectric silicon membrane pump is used. The design of the micropump is based on the original work described in [11]. It consists of a sandwiched glass-silicon-glass structure with the in- and outlet in the bottom glass plate. Fluidic connections are made by clamping the pump onto a base plate containing channels, using O-rings to ensure a leak free mounting. The base plate is machined from Vespel® polyimide, a material that has been used for many of the mechanical parts in the bioreactor. Vespel® has an excellent mechanical strength and dimensional stability and can easily be machined. In the first flight of the bioreactor, we noticed that the delivered flow of the pump was highly dependent on the output pressure that has to be supplied.

To overcome this pressure dependence, the improved version of the bioreactor for the second flight was outfitted with a flow sensor. This sensor is used in a closed loop control system that adjusts the driving voltage of the pump to obtain the required output. The flow sensor, based on a modified commercial piezoresistive low-pressure sensor, uses a differential pressure measurement across a flow restriction (Fig. 3).

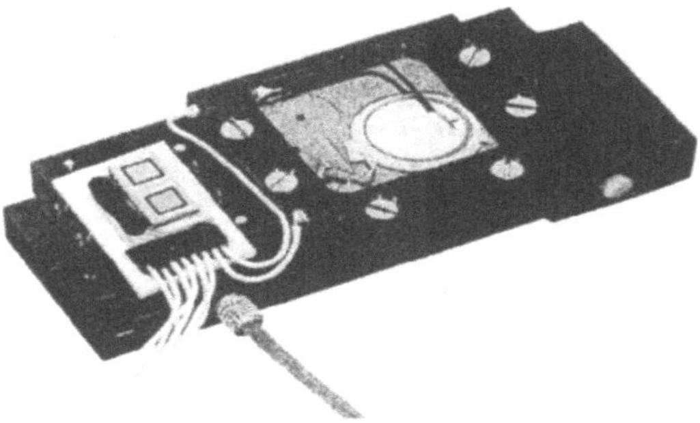

Figure 3. Micropump and flow sensor. The piezoelectric device is clearly visible as a disk on the right of the pump. The left part of the structure (open) contains the flow sensor with the electrical connector and the tube bringing the medium to the culture chamber.

As a final step in the processing of the sensor wafers, a flow restrictor is etched between two adjacent pressure sensors on the backside. The silicon wafer is subsequently bonded on a glass wafer with through holes for the fluidic connections. The sensor is then diced from the wafer and mounted stress-free on a ceramic substrate, again with through holes for the fluidic connections, using silicone rubber joints. Electrical contacts are made to a metalisation pattern that is screen-printed on the substrate. The mounting of the sensor is thus fully compatible with conventional hybrid electronic assembly techniques that ensure a high reliability at a modest cost. The complete sensor is connected in a fashion similar to the pump, by clamping the ceramic substrate onto a base in which fluidic channels are machined.

Its accuracy is over 2% and it has a full-scale sensitivity of 5 ml h^{-1}. The flow was regular and stable at the five different rates.

3.2. CHEMICAL MEASUREMENT AND CONTROL

To ensure optimal growth conditions, the pH of the yeast culture has to be tightly controlled. During normal growth, the organisms produce carbon dioxide and other acidic products which lower the pH of the nutrient medium. In general it is sufficient to neutralise these acids by adding an alkaline solution, and it is not necessary to be able to compensate pH in the opposite direction. This, however, would require an additional dosing system as well as special measures to safely contain concentrated alkali. Therefore, an alternative, electrolytic method (Fig. 4) has been developed which details have been presented in [12].

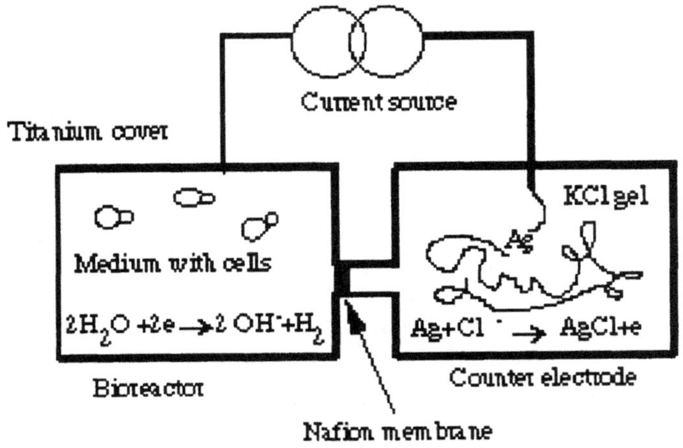

Figure 4. Principle of the electrochemical pH regulation. K^+ ions are formed in the counter electrode, they pass through the Nafion membrane to the reactor chamber and combined with the OH^- ions to form KOH.

The various elements of the pH-control system are assembled so that they are becoming part of the chamber. The sensor chip is mounted on a carrier that is inserted in a hole in the wall of the chamber so that the sensor is virtually part of the wall. This allows direct contact with the culture without creation of dead angles. The sensor is operated in conjunction with a silver / silver-chloride reference electrode connected with the solution through a porous junction.

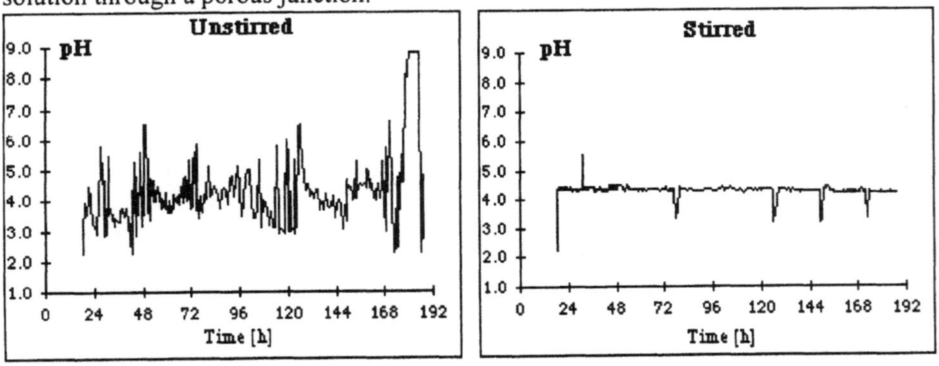

Figure 5. pH values for both ground bioreactors. In the stirred one, the sampling times are very well visible (drops). About 1 ml of fresh medium at pH 2.5 is delivered to the chamber (pH 4.5) to replace the sampling volume, a pH drop is resulting especially because no compensation of pH is effected during sample withdrawal (no electrical connection of the bioreactor).

3.3. SYSTEM CONTROL

The Biorack facility provides power for the experiment as well as the possibility to retrieve analog data signals that are sent to ground control and also stored for later use. The experiment itself has to be autonomous for all measurement and control functions. The system is governed by a microcontroller. Regulation of the pump flow is achieved by an analog PI control that adjusts the driving voltage of the piezoceramic actuator that is pulsed at a fixed frequency. The control set point is provided through a digital-to-analog converter (DAC) driven by the microcontroller. A second DAC provides multiplexed analog data to Biorack that are stored and also already available during the flight for the scientists on the ground. The electronic circuits also comprise amplifiers for the various sensors and a controlled current source for pH regulation. The pH control system is galvanically separated from the sensor circuitry to ensure that the current flows only through the electrodes meant for this purpose. The electronic circuits are realised using commercially available components, mostly in surface mount technology. Consequently, the electronics take up a rather large part of the available instrument volume. However, this approach allows for maximum flexibility at minimum cost in the design of the bioreactor. This is especially important in view of the limited time available to develop the flight instrument.

3.4. BIOLOGICAL ANALYSES

The main results of the biological analyses have been published previously [12,14]. In résumé, they showed that the general metabolism glucose consumption and alcohol production) and morphology of the cells were similar in space as on ground, but that the cultivation conditions (stirred/unstirred) had an important effect on the cells' behaviour on the ground as well as in space where the convection movements are quasi absent.
At the morphological level, no difference was observed between the 0*g* and 1*g* cells, but interestingly it was observed that the specific bud scar positioning was more altered under unstirred and microgravity conditions. In fact, a statistically significant higher percentage (10%) of cells grown under unstirred conditions show a random distribution of scars compared to the samples withdrawn from the stirred culture where scars are localised at both poles in the majority of the cells. The space samples had also a higher percentage (10%) of cells with random distribution of scars as the ground ones. This result might be correlated with the fact that cytoskeleton, which plays an important role in the process of bud emergence, is affected in microgravity [13].

All the biological analyses showed that not only the gravity changes play an important role in space, but also that the cultivation conditions are interacting significantly in the way cells react to this new environment. A comparison of cultivation (stirred versus unstirred at 0g) showed that the stirred environment in space is more favourable for the yeast cells than the unstirred one.

4. Conclusions and perspectives

Up to now bioreactors were used in space for basic research and to carry out a preliminary screening of biological systems candidates for bioprocessing in microgravity. While interesting observations were made, no biotechnological breakthrough was achieved so far. Nevertheless it can be said that independently of their design, their origin or their mode of function, bioreactors will be increasingly used in space in the future. Several types of bioreactors will be required: small-sized reactors (3-50 ml) of the type describe here for basic research and pilot bioprocesses, medium-sized reactors (1 l) for established bioprocesses, large bioreactors (100 l) as a mandatory component for the production of food and of the recycling of consumable material (closed ecological life-support systems) to support long-duration human life in space [14]. In co-operation with the medical community, they will be also used to prepare better models of different types of tissue, for the investigations of cell growth, and probably for the engineering of larger 3-dimensional cell constructs.

Moreover the development of bioreactors for space laboratories, the trend to miniaturisation and automation will be a technological challenge that has and will probably also favour spin-offs for Earth. One example is the use of the RWV that provides low-shear and a 3-D environment. On the basis of the amount of published reports, it appears that the RWV, which was initially developed for space, is used now in Earth-bound rather than in space studies. At 1*g*, the RWV system simulates microgravity allowing the cultivation of shear-stress sensitive cells with increased intracellular interactions. It appears suitable for the production of tissue-like aggregates of cells that resemble liver cells, colon cells or chondrocytes [15,16,17]. Another example is the use of the micro- and silicon-technology and the pH-regulation of the SBRI. The application of silicon microtechnology for the development of instrumentation for space laboratories, the International Space Station in particular, will be of primary importance in the next decades. Long-term experiments will need stable and reliable chemical and biological sensors for the measurement of parameters such as glucose consumption, concentration of dissolved O_2 and CO_2, metabolite production (*i.e.* alcohol, lactate) as well as environmental control systems (pH, temperature, waste management).

Currently under development is a new version of the Swiss bioreactor, with a larger working volume (7 ml), as well as the additional investigation chambers allowing an *in situ* heat-shock to investigate the stress response of the yeast cells in microgravity. This experiment is planned to fly in the new facility BioPack in STS–107 in 2001.

Eventually, the microtechnology will also be part of our newest project recently selected by ESA within the application and commercialisation program of the International Space Station. This project, with the collaboration of several universities (Hanover, Udine, Münster, Hamburg) and the participation as industrial partner of Sulzer Medica (Switzerland), will lead to the development of procedures of *in vitro* organogenesis of pancreatic islets, thyroid tissue, liver, vessels and especially cartilage, the study of the mechanism of organogenesis in low-g and the set-up of procedures for the production of implants for medical applications. Flights in space are foreseen in 2001 and 2003.

Acknowledgements

The developments of the DCCS and of the SBR I performed in our laboratory have been supported by grants of the Prodex Program of ESA, of the ETH Zurich, of Contraves AG (Zurich), and of Bio-Strath AG (Zurich).

References

1. Skoog, A.I. and Broullet, A.O. (1981) Trends in space life support. *Acta Astronautica* 8, 1135-1146.
2. Wolf, L. (1996) Bioregeneration in space, *Adv. in Space Biol. and Med.* 5, 341-356.
3. Muller, C.N. and Porter, R.J. (1996) Evaluation of a bioreactor system used to treat human waste streams on space station, Abstract of the general meeting of the American society for microbiology, 400.
4. Montgomery, P.O.B., Cook, J.E., Reynolds, R.C., Paul, J.S., Hayflick, L., Stock, D., Schulz, W.W., Kimsey, S.,Thirolf, R.G., Rogers, T., and Campbell, D. (1978) The response of single human cells to zero gravity, *In Vitro* 14, 165-173.
5. Kulesh, D.D., Anderson, L.H., Wilson, B., Otis, E.J., Elgin, D.M., Barker, M.J., Mehm, W., and Kearney, G.P. (1994) Space shuttle flight (STS-45) of L8 myoblast cells results in the isolation of a nonfusing cell line variant, *J. Cell. Biochem.* 55, 520-544.
6. Gmünder, F.K., Nordau, C.-G., Tschopp, A., Huber, B., and Cogoli, A. (1988) Dynamic cell culture system a new cell cultivation instrument for biological experiments in space, *J. Biotechnol.* 7, 217-228.
7. Walther, I., van der Schoot, B., Jeanneret, S., Arquint, P., de Rooij, N.F., Gass, V., Bechler, B., Lorenzi, G. and Cogoli, A. (1994) Development of a miniature bioreactor for continuous culture in a space laboratory, *J. Biotechnol.* 38, 21-32.
8. Walther, I., Bechler, B., Müller, O., Hunzinger, E., and Cogoli, A. (1996) Cultivation of *Saccharomyces cerevisiae* in a bioreactor in microgravity, *J. Biotechnol.* 47, 113-127.
9. Rasmussen, O., Gmünder, F.G., Tairbekov, M., Kordyum, E.L., Lozovaya, V.V., Baggerund, C., and Iversen, T.-H. (1990). In: Proceedings of the fourth European symposium on life sciences research in space, Trieste, ESA SP-307, 527.
10. Lorenzi, G., Gmünder, F., and Cogoli, A. (1993) Cultivation of hamster kidney cells in a Dynamic Cell Culture system in Space, *Microgr. sci. technol.* 6, 34-38.
11. Van Lintel, H.T.G., van der Pol, F.C.M., and Bouwstra S. (1998) A piezoelectric micropump based on micromachining of silicon, *Sensors and Actuators* 15, 153-167.
12. Walther, I., van der Schoot, B., Boillat, M., Müller, O., and Cogoli, A. (1999) Microtechnology in Space Bioreactors. *Chimia* 53, 75-80.
13. Cogoli-Greuter, M., Pippia, P., Sciola, L., and Cogoli, A. (1994) Lymphocytes on sounding rocket flights. *J. Gravit. Physiol.* 1, 90-91.
14. Bréchignac, F. and Schiller, P. (1992) Pilot CELSS based on maltose-excreting Chlorella: concept and overview on the technical developments, *Adv. in Space Research* 12, 33-36.
15. Jessup, J.M, Goodwin, T.J., and Spaulding, G. (1993) Prospects for use of microgravity-based bioreactors to study three-dimensional host-tumour interactions in human neoplasm, *J. Cell. Biochem.* 55, 290-300.
16. Duke, P.J., Daane, E.L., and Montufar-Solis, D. (1993) Studies of chondrogenensis in rotating systems, *J. Cell. Biochem.* 55, 274-282.
17. Lewis, M.L., Moriarity, D.M., and Campbell, P.S. (1993) Use of microgravity bioreactors for development of an *in vitro* rat salivary gland cell culture model, *J. Cell. Biochem.* 51, 265-273.

PART VI
IMMOBILISATION AND PERMEABILISATION

STATE OF THE ART DEVELOPMENTS IN IMMOBILISED YEAST TECHNOLOGY FOR BREWING

C.A. MASSCHELEIN[1] AND J. VANDENBUSSCHE[2]
[1] Katholieke Universiteit Leuven, Center for malting and Brewing Science, Kardinaal Mercierlaan 92, B-3001 Heverlee, Belgium
[2] Meura Delta and Meura s.a., Chaussée d'Antoing 55, B-7500 Tournai, Belgium

Although the large volume tank has been widely accepted, there has been a great deal of debate within the brewery world as to the desirability of moving towards fermenters of even larger capacity. To take full advantage of capital costs it is clear that tank size should be as large as possible. However, tank volume is limited by process restrictions including multiproduct schedules and a necessary balance between the fermenter size and the brew length. Another major disadvantage of greater volume processing is an increasing degree of inflexibility with respect to production planning and efficient plant utilisation during sales off peak periods. Thus in terms of process economics small scale high rate fermentation systems, able to be stepped up to meet peak output when necessary could well be the key of future profitability [21].

Immobilised brewing yeast cell technology provides the opportunity to improve volumetric productivity compared with traditional free cell, batch fermentation. In recent years, immobilisation of microbiological cells by entrapment within natural polymers, by adsorption to solid supports or by ionic attraction has become a rapidly expanding research area. Various workers have described immobilised systems for application in the brewing process. The removal of flavour matching constraints makes immobilised cell technology and ideal choice for continuous industrial ethanol fermentation. However, for such strategies to be successful for beer production, there is an increasing need to find a solution that satisfies both productivity and the quality of the final product. In this regard, commercial viability of immobilised cell systems depends on optimising the interrelated factors of cell physiology, mass transfer, immobilisation procedures and reactor design to ensure high specific rate of fermentation independently of yeast growth and to consistently produce beer with the desired sensory and analytical profile.

1. Process requirements for high turnover rates in brewery fermentations

In contrast to homogeneous isotropic fermentation systems where volumetric productivity vary directly to yeast concentration and temperature, fermentation rates in

M. Hofman and P. Thonart (eds.), Engineering and Manufacturing for Biotechnology, 255–276.

heterogeneous batch systems are mainly controlled by the amount of yeast in suspension as a result of the gas lift action taking place in the fermenter. Thus, to maximise volumetric productivity, fermentation parameters and fermenter design must be optimised with respect to agitation and yeast concentration. The power generated by the gas-lift action increasing logarithmically with the liquid depth, faster process times will be obtained in tall thin cylindro-conical vessels. This would be an ideal situation provided that environmental conditions created by deep fermentation are not adversely affecting beer flavour.

Unfortunately, high degrees of agitation lead to beers, which lack body and have a poor aroma. Complete correction of flavour profiles in very deep fermenters has been found difficult [19]. Another approach to achieve high cell densities is cell immobilisation. The immobilisation of viable cells can be defined as any technique that limits their free migration.

For large industrial applications the following factors must be considered when choosing a cell immobilisation system:

- The support material must be readily available and affordable.
- The system should be efficient; easy to operate and give good yields.
- The cells should have a prolonged viability in the support, which should not be toxic to the cells.
- The support material should allow for high cell loading (weight cells/weight support)
- The kinetic behaviour of the loaded support should be understood, and not hinder the fermentation. This includes diffusional limits, local pH, and inhibitor accumulation.
- Any modifications of metabolic process associated with the carrier should be realised and accounted for.

Immobilised yeast cells represent a highly flexible catalyst system, due to the many enzyme activities, which they contain, and the various physiological and metabolic conditions in which they can be used. (figure 1)

However, to take full advantage of these technological opportunities we need to build up a sufficient understanding of the process behaviour and response to micro-environmental effects within the immobilising matrix to permit feedback control.

The volumetric rate of reaction for a fermenter is given by:

$$\frac{\text{rate}}{\text{unit volume}} = \frac{\text{rate}}{\text{mass of cells}} * \frac{\text{mass of cells}}{\text{unit volume}}$$

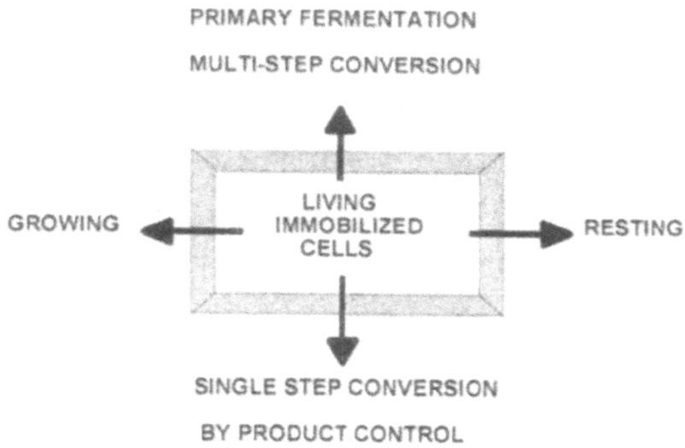

Figure 1 : Technological Potential of Immobilised Yeast Systems for Application in the Brewery Industry.

Thus, to maximise volumetric productivity, fermentation parameters and reactor design must be optimised with respect to the rate of fermentation per unit yeast and biomass concentration.

Immobilised cell technology provides the opportunity to improve volumetric productivity by achieving high concentrations of catalytic biomass in a controllable form. This would be an ideal situation provided that environmental conditions within the immobilising matrix are not adversely affecting cellular activity. Unfortunately, maximisation of biomass concentration gives rise to transport associated limitations, which have been shown to impact negatively on yeast growth and catalyst efficiency. Thus, to allow increased throughput in immobilised cell reactors yeast concentrations must be enhanced significantly over that attainable in free cell systems.

To overcome diffusion hindrance, matrix design has to go toward highly porous and rather small particles. Discussion with respect to mass transfer limitations concerns the entire reactor. Thus matrix and reactor design rank equally in defining catalytic efficiency [18;20].

2. Matrix design for application in the brewing process

Mechanical integrity, biomass hold-up and hydrodynamic characteristics of the supporting material in the reactor are of major importance in determining the commercial feasibility of large scale applications. Cell mobility can be restricted by passive and active techniques. Passive techniques based on the selfaggegrating capabilities of very flocculent yeast strains have been applied to cell cycle reactors for the continuous beer production [22]. However, mass transport and mass transfer problems related to the rheological properties of high-density free cell suspension are the major drawbacks of such systems.

Another suitable approach for application in brewing is the immobilisation in or on preformed supports, either porous or no porous. From the microorganisms point of view it is also the most gentle fixation method as no changes in the cultivation conditions are necessary to produce the immobilised biocatalyst. A direct advantage of a porous matrix is the possibility for regeneration and reactivation of cellular activity by repeated growth within the matrix. Most preformed porous supports represent a combined form of cell immobilisation involving adsorption, cell growth, self-aggregation of cell populations and finally entrapment of the aggregate within the porous network of the carriers.

Moreover, a cellular secretion of adhesion polymers that stabilise the attachment often follows initial attraction. While this situation is satisfactory for cell retention it raises the question to what extent cells immobilised in the pore space are replaced by newly yeast grown cells. This aspect is important because age distribution of the free and immobilised cell population could have a bearing on their respective catalytic activity. Precisely, how immobilising mechanisms apply to various matrix structures and to what extent glycolytic activity is affected by such conditions is unclear and difficult to anticipate. Evidently, more fundamental information is needed concerning the interrelated factors, which affect yeast physiology in immobilised systems.

For detailed discussions of the fundamentals of immobilised yeast cells for continuous beer fermentation a comprehensive review is presented by Pilkington *et al.* [34]. Cells immobilised on non-porous surfaces in direct contact with the bulk liquid substrates reduce mass transfer problems associated with more intensive immobilisation techniques. However, the maximum flow rate through such reactor may create shear-enforced detachment from the support.

Active techniques all involve as a first stage the production in a reactor of a culture of freely suspended cells, which are subsequently immobilised in a separate reactor. Entrapment in polymeric gels such as alginate [7;18;28;42] and kappa-carrageenan [25;26;29] has been vigorously studied. Simplicity of operation and high biomass hold-up are the major advantages of this method. Less desirable characteristics of alginate support material are loss of mechanical integrity by either dissolution or breakdown due to abrasion and internal gas accumulation.

Considering process intensification as the primary requirement of cell immobilisation it may be concluded that entrapment in kappa-carrageenan gel beads as well as the adsorption of cells to solid surfaces such as porous glass, ceramic, silicon carbide or the attachment of cells to modified surfaces such as DEAE cellulose are suitable means of immobilising yeast for large scale industrial application.

3. Reactor design for application in the brewing process

Reactor design is a crucial parameter in the implementation of immobilised systems. Layout and operation should meet the following requirements:

- large mass transfer rate for wort nutrients at low energy input.
- the reactor must be capable of being constructed in large units.
- a simple and robust design, which is characterised by low construction costs, easy
- to keep sterile, low maintenance costs, and high on-stream availability.

Reactor types commonly used for continuously production of beer include packed bed, fluidised bed, gas-lift draft tube and loop reactors.

The packed bed reactor is the simplest stationary configuration but less efficient than mixed reactors especially at high CO_2 evolution rates where CO_2 venting is a prerequisite for optimal functioning. However, the possibility of maintaining a plug flow may have the advantage of keeping non-limiting substrate concentrations or subinhibitory levels of toxic products. The successful application of plug-flow continuous systems to the reduction of vicinal diketones substantiates this view.

The liquid fluidised bed reactor is usually a tall thin reactor in which fermentation medium is circulated upward through the immobilised cell bed. Abrasion caused by particle-to-particle and particle to the inside wall of the reactor contact may lead to immobilised cell aggregate breakdown. This reactor type is best used with support particles that are significantly more dense than the fermentation liquor, because a less dense particle would be carried upward in this configuration and result in less efficient mixing and lower mass transfer rates.

The gas-lift draft tube reactor consists of a cylindrical vertical vessel with a slenderness ratio of 1:5-1:10, and an inner concentric draft tube. The draft tube is fixed inside the reactor in such a position that there is a flow connection at the bottom and the top of the reactor. Thus, a directed flow through and round about the draft tube is facilitated. The internal liquid circulation is induced by means of a gas sparger in the type of bioreactor used for primary fermentation. For optimal mixing and high mass transfer rates support particles should have a wet density near that of the liquid phase. An appropriate choice for this type of application is the use of hydrogel carriers such as kappa-carrageenan.

In the loop bioreactor designed for continuous primary fermentation of beer yeast cells are immobilised on porous rod matrices containing numerous internal channels. The fermentation liquor flows from the bottom of the fermenter through both the internal channels and around the matrices for contact with the immobilised yeast to the top of reactor with external recycle. This loop reactor design has the advantage of combining mechanical strength, low shear, excellent mixing minimising diffusional limitations and easy scale-up.

Considering efficient material transport and CO_2 venting at high CO_2 evolution rates. It may be concluded that fluidised bed and gas/liquid circulating bioreactors are best suited for large-scale commercial operation.

4. Reactor configuration for continuous immobilised yeast fermentation systems

Among the various immobilised systems which have been described for continuous primary fermentation two classes of process can be usefully distinguished:

- a single stage process in which essentially beer is mixed with immobilised yeast cells and simultaneously wort is introduced at one point while excess beer containing the free cells leaked from the immobilised matrix is allowed to escape from another point.

- (ii) a two or multistage process into which wort is mixed with immobilised yeast in the first reactor and the residence time adjusted such that wort sugars are partly utilised in the first stage and the reaction completed in the second or following stages.

It is important to recognise a basic difference between these two reactor configurations. A single stage process result in fermenting the incoming wort in a medium which is essentially beer and thus under conditions where growth is very limited.

It is well established that glycolytic flux in a batch fermentation is associated with growth in a way that specific productivity reach a maximum during the growth phase and progressively decline as the yeast enters the stationary phase [24]. Therefore, specific sugar consumption rates in single stage immobilised systems operating continuously under steady state conditions at low sugar and nutrient concentrations will be limited by substrate availability emphasising low volumetric productivity and high residence times.

A process which employs a series of more than one reactor converts wort into beer by series of steps involving compositional changes as well as growth and non growth conditions as in conventional batch fermentations.

Considering cellular activity and its effect on reactor productivity, bulk phase free cells leaked from the matrix and immobilised cells have to be distinguished for both stages. As a result the following four situations have to be taken into consideration:

- « first stage » actively growing « free cells » with high specific productivity
- « first stage » actively growing « immobilised cells » with lowered specific productivity due to restricted material transport
- « second stage » slow growing to resting « free cells » with reduced specific productivity depending on substrate availability
- « second stage » immobilised cells having presumably the lowest specific productivity.

Depending on the structure of the matrix and the resulting equilibrium between free and immobilised cells the following two strategies may be considered for optimal performance of a two-stage system for beer fermentation:

- a two-stage immobilised configuration using flow rates to match final attenuation. In this situation the high cell densities may compensate the low specific productivity of the second-stage reactor. Further study is required to determine to what extent this advantage is affected by the increased cost of carrier material.
- a two-stage configuration where the immobilised cells provide a continuous and consistent inoculum to a second-stage free cell system. Synchronisation of both stages is easily achieved by adjusting the working volume of the free cell reactor. A major advantage of the immobilised/free cell configuration is that capital cost is significantly reduced for the same beer throughput rate compared to a two-stage immobilised concept.

5. Flavour development and control in immobilised yeast systems

Following the well-known work of Jones and Pierce [14] of the mid 1960's when sequential removal of amino acids from wort was realised, numerous studies have been undertaken to understand the complex interaction of amino acid permeases in order to control the formation of flavour active compounds in brewery fermentations. Amino acid metabolism is obviously critical to beer quality being closely related to the production of vicinal diketones, hydrogen sulphide and higher alcohols [22]. Packed bed fermentations using alginate immobilised lager yeast have been shown to be associated with severe limitations at the levels of amino acid uptake [6], fermentation performance [35], oxygen transfer [17], synthesis of membrane lipids [17] and formation of higher alcohols and esters [35].

These problems were overcome by using the Kirin two-stage free/immobilised cell system [28] or by taking advantage of the superior mass transfer capabilities of fluidised bed [4;6] and gas-lift draft tube fermenters or multichannel loop reactor systems. Thus, in terms of beer quality and fermentation efficiency, optimal mixing will have to be a primary concern in future work on immobilised system design.

A factor common to all high rate processes for primary fermentation is to achieve rapid and efficient removal of vicinal diketones and precursors. The formation of vicinal diketones has been well described, it results from the oxidative decarboxylation of excess α-acetohydroxy acids leaked from the isoleucine-valine biosynthesis pathway.

The industrial exploitation of genetically modified yeast strains to overcome diacetyl problems is a possible way in which the brewing industry could move today [24;36]. It is, however, most likely that application of recombinant DNA technology to brewer's yeast will be delayed by regulatory requirements and above all by the exclusive rule imposed by tradition and consumer consideration.

In the short-term, greater promise for rapid and complete removal of vicinal diketones may lie with immobilised cell technology. Conversion of α-acetohydroxy acids to vicinal diketones being the rate limiting step, effective control of diacetyl and 2,3 pentanedione levels in the finished beer may be expected by increasing the rate of chemical decarboxylation. This may be achieved by heating the beer after yeast separation. It has been shown that 7 minutes at 90°C are required for complete conversion of α-acetolactate [31]. Interestingly, under strictly anaerobic conditions, the reaction proceeds directly to acetoin [12].

In this connection, the two alternative approaches which have recently be presented at the EBC congress held in Maastricht in 1997 are the rapid conversion of acetolactate into acetoin using either encapsulated α-acetolactate decarboxylase [9] or aluminiumsilicate zeolite as a catalyst [3]. In terms of process economics it is clear that, if such treatments prove to be effective, lower cost alternatives to the two-stage heat/cell immobilisation processes could become a reality.

Pajunen *et al.* [33] have described a commercially used maturation system employing immobilised yeast on granular-derivatised cellulose (DEAE cellulose) material spezyme ® produced by Cultor (Finland). Sintered glass and silicon carbide carriers have also been successfully used for continuous maturation with immobilised yeast [4].

It has been reported that diacetylreductase has a much higher affinity for diacetyl than alcoholdeshydrogenase which is preferentially reducing acetoin [37]. It is important to note, however, that coenzyme regeneration is a prerequisite for optimal functioning of these reaction sequences. Therefore, immobilised cells must contain intact coenzyme regeneration systems so that high activity can be maintained in the process stream. The long-term stability of the diacetyl reducing capacity of immobilised resting cells highlights the many operational advantages of using mild immobilisation methods. An almost similar situation exists in the fast flowing immobilised yeast systems used for the production of low-and alcohol-free beers. Indeed, fast and effective regeneration of NADH and NADPH has been found critical for the conversation of the flavour-potent wort aldehydes to the low flavour-intensive alcohols [8;38].

6. Technological potential of options for immobilised yeast application in the brewing industry

The immobilisation of yeast cells for successful application in brewing implicates the retention of whole catalytic cells within a bioreactor. In order to be a viable alternative to traditional free cell fermentation and maturation systems, immobilised cells must have considerably long working lifetimes, characteristically measured in weeks or months. Mass transfer limitations of substrate into and products out of the immobilised cells and associated matrix are of critical interest [23]. Anticipated criteria for the commercial feasibility of using immobilised cell systems are presented in table 1.

This section describes recent advances in immobilised cell technology and the current commercial options practically applicable to the brewing process for continuous fermentation and/or maturation as well as for the production of malt beverages with defined organoleptic/analytical spectra.

For detailed discussions of immobilised yeast and applications in the brewing industry a comprehensive review has recently be presented by Mensour *et al.* [27].

Table 1 : Process requirements for application of immobilised cell technology to beer production.

Requirement	Comment
Low capital cost	High productivity Mechanically simple
Low operating cost	Continuous operation Simple operation Low energy input
Operational control and flexibility	Controlled oxygenation Controlled yeast growth Rapid start-up after shut down Control of contamination
Quality control and flexibility	Desired flavour profile Consistent product Wide choice of yeast Complete attenuation

6.1. IMMOBILISED PRIMARY FERMENTATION

The reactors commonly used for continuous primary fermentation include packed bed, gas lift draft tube and loop reactors.

6.1.1. Packed bed reactor systems

KIRIN BREWERY Co., Ltd (Japan) developed a two stage free/immobilised process witch employs a continuous stirred tank reactor (CSTR) for the first stage and a packed bed reactor for the second stage (figure 2) [13].

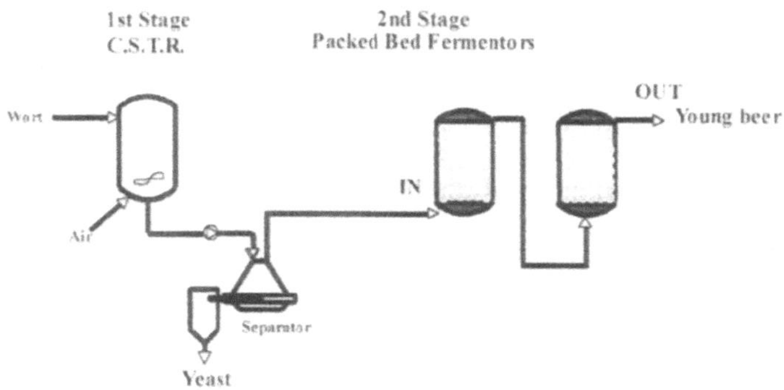

Figure 2 : Scheme of Kirin's free/immobilised cell bioreactor system (adapted from Inoue [15])

Their objective in initiating this technology was to convert wort into beer by a series of steps characterised by cyclic variations in yeast growth and thus to develop a process which resembles the conventional free cell batch fermentation with respect to the metabolic regulatory mechanisms involved in flavour formation.

The free cell chemostat is operated at 13°C. with continuous air sparging (0.017 v.v.m.) in order to achieve a final extract of 8%. The partly fermented beer is centrifuged (less than 10^6 cells/ml) and fed to the PBR consisting of a cylindro-conical fermenter with cooling jackets. Bioceramic porous beads with a central pore size of 10-20 µm, a bulk density of 0.35 g/cm^3 and a surface area of 3.13 m^{2}g^{-1}. The void volume of the reactor is 40% (vol/vol). The flow rate was controlled to maintain the residual extract between 1.8 and 2.5%. The 20 L pilot plant was scaled up to 5 HL in 1989 and finally to 100 HL in 1991. Major problems encountered with the larger system were operational difficulties related to temperature control and fluid channelling and finally higher capital and operating costs than expected.

Nevertheless, it was decided to proceed with the construction of the restaurant brewery "Boga Boga" located on Saipan Island in the north Mariana Islands. The continuous plant with a maximum daily production volume of 5 HL has been operational since 1992. While the Kirin system may allow some increase in productivity over the conventional batch process, the added complexity resulting from the chemostat and the need for mechanical centrifuge as well as cooling rods in the packed bed fermenter will probably offset this advantage.

The research team of the HARTWALL Brewing C° from Finland in conjunction with the VTT Biotechnology and Food Research group was assessing the potential of applying porous glass beads (Siran®) within a two stage packed bed reactor to accomplish the primary fermentation [11;16]. Further study is necessary to determine overall process stability before any evaluation of the potential of this technology can be made.

6.1.2. Gas lift draft tube reactor systems

A novel continuous beer fermentation system based on the superior mixing and surface exposure for mass transfer found in gas lift draft tube reactors and the use of Kappa-carrageenan immobilisation carrier has been developed by LABATT Breweries of Canada in collaboration with the Department of Chemical and Biochemical Engineering at the University of Western Ontario (figure 3) [25;26;29].

A 50-L pilot plant was designed and installed for use with the carrageenan beads. To this effect a continuous bead production process based on static mixers was engineered [29]. Beads within a 0.2 to 1.4 mm size range are produced at a maximum throughput of 10 L per hour per static mixer (Canadian Patent 2,133,789,1994).

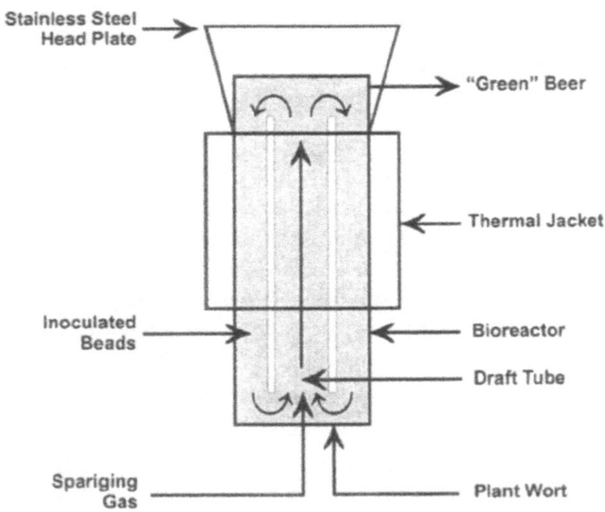

Figure 3 : 50 L gas lift draft tube bioreactor at Labatt (Mensour et al. *[25]).*

A mixture of air and carbon dioxide is utilised as sparging gas. The proportion of air to CO_2 determines the level of yeast growth within the bioreactor as well as the flavour profile of the finished product. The yeast concentration increases from 10^9 to 10^{10} cells per ml of gel and remains constant for several months.

The level of free cells released from the beads is 10^7 yeast cells per ml of liquid medium. The residence time for continuous operation at full attenuation is 20-40 hours against a batch fermentation time of 5-6 days. In terms of process evaluation criteria, the gas lift draft tube immobilised yeast bioreactor yields a very high productivity, low energy system. However, a detailed economic analysis is required before the merits of the system can be fully assessed for large-scale industrial applications.

6.1.3. Loop reactor systems

MEURA DELTA, a Belgian R&D engineering company, in association with the research team of CERIA, has developed a two-stage immobilised/free cell loop reactor system for the continuous production of beer [1;2;15;39]. The immobilising carrier is made of sintered silicon carbide particles into a highly porous cylindrical module with 37 internal channels. This 900 mm matrix with outer diameter of 26 mm and 2 mm internal diameter channels has a pore volume of 180 ml.

Advantages for using this particular design are that :

- silicon carbide is an inert material, neutral in taste and food approved with high mechanical strength and chemical resistance.
- it is suitable for CIP cleaning, re-usable and steam sterilisable.
- it is easy to scale-up by modular assembly.
- it has an ideal preformed shape for optimal mass transfer.
- the open pored structure with a pore size distribution between 40 and 60 μm allows rapid colonisation and efficient gas venting at high CO_2 evaluation rates.

The first stage consists of a cylindrical vertical vessel in which the silicon carbide matrices are fixed in such a position that the flow of wort is directed within a loop from the bottom of the fermenter through both the internal channels and round about the matrices to the top of the reactor. The circulation is induced by means of a pump and recirculation rate adjusted in order to secure complete mixing and optimal mass transfer.

This particular design has the advantages of both fluidised bed (efficient mixing and gas venting) and a packed bed (mechanically simple) but not the disadvantages of the fluidised bed (high shear with abrasion of carrier material and scale-up problems) nor of the packed bed (plugging, gas flooding and lack of mixing with concomitant diffusion limitations).

An additional advantage of this flow pattern is that wort may be used without any pre-treatment allowing steady-state production over a period of six months. Moreover, cleaning in place is considerably simplified by the possibility of backward and forward flushing. The second stage of the immobilised/free cell reactor system consists of a cylindro-conical vessel. A loop reactor design is used to ensure complete mixing. This is achieved by pumping the fermenting wort taken from the bottom of the fermenter at the top and periphery of the cone.

Recirculation rate and inlet velocity are adjusted to allow a uniform distribution of cells and substrates inside the reactor. This mechanically simple design permits to maintain sterile operating conditions. The value of a two-stage configuration combining immobilising and free cell reactor systems is that the cells produced in the first stage under constant controlled conditions may be maintained in the growth-limiting environment of the second stage for prolonged periods of time.

The net growth in the second stage will be low but washout does not occur because the first stage continually supplies cells. The experimental set-up is presented in figure 4.

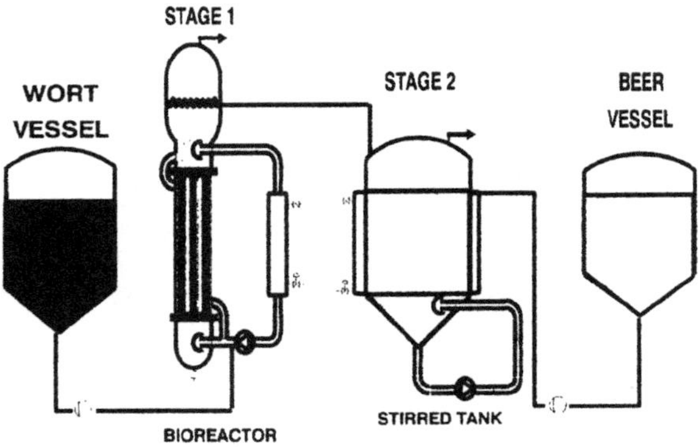

Figure 4 : Meura Delta immobilised/free cell loop reactor system (Krikilion et al. *[15]).*

The system was found to be stable and the beer produced had a composition and flavour profile similar to that produced using traditional methods of manufacture. Considering a two-stage immobilised/free cell configuration apparent attenuation values are respectively 35 and 75% using 16°Plato wort and rates of 0.085 l/h.

Thus, 7 Hl high gravity beer may be produced on a year basis for a working volume of 5.2 l so that the volumetric productivity of the system may be rated at 135 Hl/Hl-year or 180 Hl after adjusting the original gravity at 12°Plato. The productivity of a plant batch fermentation would be 48 Hl/Hl-year assuming one week to complete attenuation and thus 3.75 times lower as compared to the immobilised/free cell reactor system.

Reduced process times and elimination of lag times and high peak load levels on electricity, steam and other services are clearly advantageous for a fully continuous process. The great advantage intrinsic to immobilised techniques is the availability of high concentrations of catalytic biomass in a controllable form resulting in potentially faster process times. However, immobilised systems have also some important drawbacks, particularly diffusional limitations that have been shown to impact negatively on yeast performance [17]. It can, indeed, be seen from the results in table 2 that using specific sugar consumption rates of 0.034 g/h 10^9 cells more than 60% of the conversion must be ascribed to the free cells although immobilised cell concentration are about 5 times higher than the free cell concentration (187.106 against 40.106 cells/ml).

Table 2 : Fermentation performance of the Meura-Delta immobilised yeast fermenter.

Parameter:	Stage I	Stage II
Reactor volume (l)	1.6	3.6
Wort flow rate (l/h)	0.085	0.085
Residence time (h)	19	42
Sugar concentration (g/l)	110	65
Sugar feed rate (g/l)	9.35	5.52
Residual sugar concentration (g/l)	65	5
Residual sugar feed rate (g/h)	5.52	0.43
Sugar consumption rate (g/h)	3.83	5.09
Free cell concentration (10^6/ml)	40	50
Immobilised cell concentration (10^6/ml)	187	0
Total free cells (10^9)	64	180
Total immobilised cells (10^9)	300	0
Sugar consumption rate of free cells (g/h)	2.18	5.09
Sugar consumption rate of immobilised cells(g/h)	1.64	0

Consequently, specific sugar consumption rates for the immobilised cells are only 0.0055 against 0.034 for the newly grown free cells. This reduced specific activity is largely due to substrate transport associated limitations which are known to increase with increased cell loading such that maximisation of biomass concentration will be more than offset by a concomitant loss in specific activity. This rate limiting effect is expected to be even more pronounced at low substrate concentrations prevailing in the second stage.

These views find support in the results obtained with a two-stage immobilised system showing that about 85% of the wort sugars fed to the second stage are converted by the free cells. Taking into account both capital cost and catalyst efficiency implications the immobilised/free cell fermenter configuration appears to be the best compromise for industrial application. It is also the most appropriate way to mimic the sequential growth and stationary phase conditions involved in a traditional batch fermentation which are known to be important for flavour development.

6.2. FAST FLOWING IMMOBILISED YEAST SYSTEMS FOR THE PRODUCTION OF LOW AND ALCOHOL-FREE BEER

Suppression of alcohol formation by arrested batch fermentation is widely accepted as a basic principle for the production of alcohol-free beer. Arrested batch fermentation appears attractive in terms of low capital costs and operating simplicity. It fails when rated on product quality and flavour consistency. Moreover, beers thus produced are often characterised by an undesirable wort taste and aroma.

The reason for this is that in the early stages of traditional batch fermentation the physiological state of the yeast is varying with the time so that the full potential to reduce the wort aldehydes is never achieved. Optimal steady state conditions can only be maintained by the application of fully continuous processes, preferably combined with the concept of cell immobilisation in order to allow operation beyond the nominal washout flow rate.

Thus in terms of process economics, product quality and operating flexibility small scale high rate immobilised cell systems offer an interesting challenge to traditional batch processing for the production of alcohol-free beer with the desired sensory and analytical profile. The reactors most commonly used for the continuous production of alcohol-free beer with immobilised cells include packed bed, fluidised bed and gas lift loop reactors.

6.2.1 Packed bed reactors

Controlled ethanol production for a low and non-alcohol beer has been successfully achieved by partial fermentation through DEAE cellulose immobilised yeast columns [41].

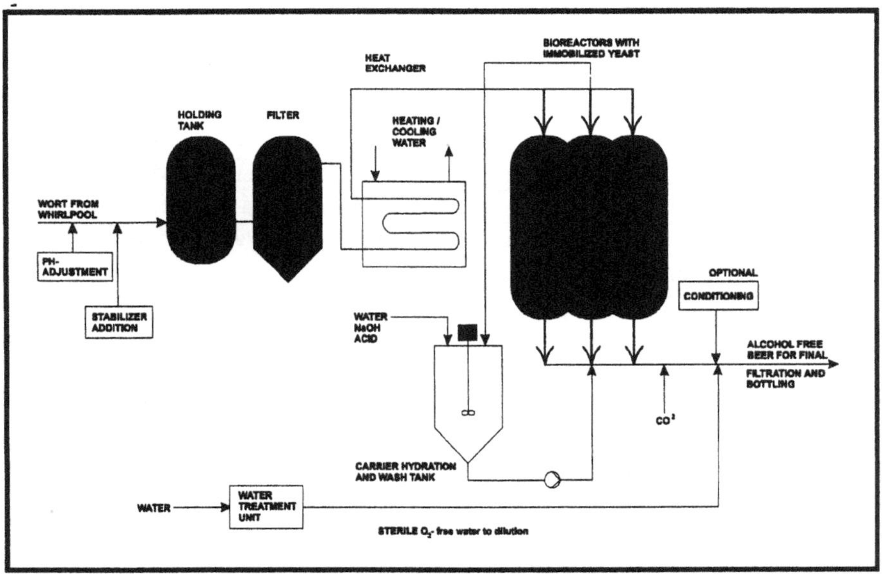

Figure 5 : Flow diagram of the 150.000 Hl/year packed bed reactor system at Bavaria Brewery (van Dieren [41]).

A major advantage of this type of carrier is that transport restrictions and diffusional limitations are minimised. This would be an ideal situation, provided that negatively charged wort components or particles are not adversely affecting the binding capacity of brewing yeasts. Accordingly wort treatment and filtration are essential to ensure efficient and controlled fermentation. An industrial scale packed bed reactor is operating at BAVARIA Brewery in the Netherlands for the production of alcohol-free beer. A reactor of 1.5 m^3 is packed with 1m^3 (400Kg) of Spezyme ® (Cultor) and operated in downflow under strictly anaerobic conditions, low temperature (0-1°C) relatively high pressure and flowrate of 20HL/h. These conditions have been shown to reduce yeast growth and cellular activity while maintaining a high reducing capacity for wort carbonyl compounds (figure 5).

The system has a production capacity of 150.000 HL beer per annum. If necessary activation steps can be introduced by circulating fresh wort under anaerobic conditions. After 5-7 months the reactor and carrier material are cleaned and sterilised. Other companies in Denmark, Australia and Spain have purchased the CULTOR technology for the production of alcohol-free beer.

6.2.2 Fluidised bed reactors

SCHOTT Engineering has proposed the use of open pored sintered glass beads as carrier material within a fluidised bed reactor for the continuous production of alcohol-free beer [5].

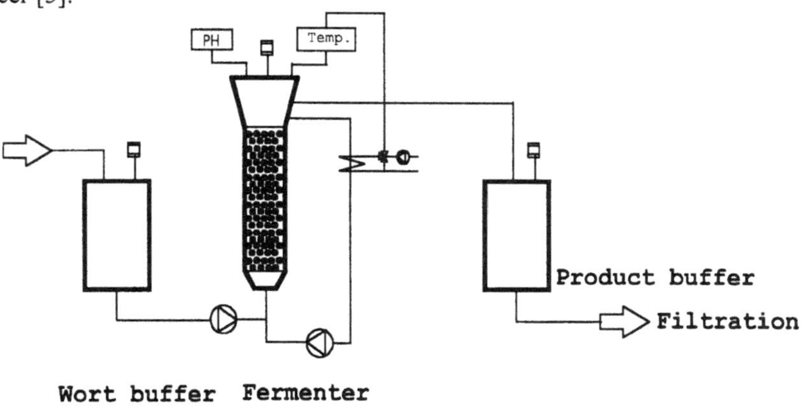

Figure 6 : Flow diagram of the Schott Engineering fluidised bed fermenter (Breitenbücher & Mistler [5]).

Siran glass beads have been reported to have several advantages for cell immobilisation:

- large active surface area up to 0.2 m^2/g
- controllable pore sizes
- controllable pore volume
- open pore volume of more than 95%
- possibility of varying pore diameter between 10 and 400 μm
- biologically and chemically stable
- easy to clean, reusable and sterilisable with steam
- not compatible
- neutral in taste, food approved

The system is suited for long term operation and alcohol concentrations in the final product can be adjusted by regulating residence time and/or temperature accordingly.

Brewery BECK & Co have tested a 60 L SCHOTT fluidised bed reactor capable of producing 8HL/day of alcohol-free beer (figure 6) [4]. Wort at 0°C. is used and contact time with the immobilised carrier is adjusted in order to keep the alcohol level below 0.05% by volume.

6.2.3 Gas lift loop reactor

Gas lift multichannel loop reactor systems for the continuous production of low alcohol and alcohol-free beers have been developed by MEURA-DELTA.

The immobilising carrier is constructed of sintered silicon carbide particles into a highly porous cylindrical module with 19 internal channels. This 900 mm matrix with an outer diameter of 25 mm and 2.5 mm internal diameter channels, has a void volume of 30%.

Laboratory, bench and pilot scale bioreactors were constructed and extensively studied under different feeding regimes with single or multi-stage bioreactor configurations allowing the continuous production of alcohol-free beer [38]. A one-stage bench scale immobilised cell reactor is presented in figure 7, and has a total hold-up volume of 1 litre.

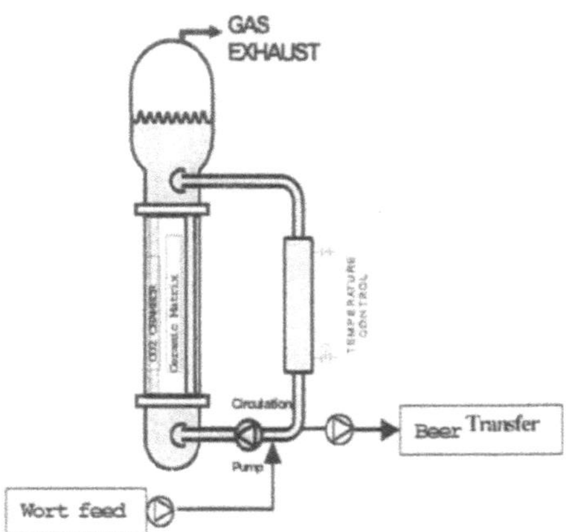

Figure 7 : Schematic layout of the Meura Delta gas lift multichannel loop reactor (van de Winkel et al. *[38]).*

The silicon carbide matrix is installed between two seals creating a closed external chamber. During fermentation a carbon dioxide pressure is build up, draining this external chamber. The excess of CO_2 escapes through the porous matrix generating small gas bubbles resulting in an upward flow through the internal channels. The gas-lift loop reactor can operate using either natural circulation, or with a circulation pump with an internal circulation rate of 15 renewals per hour for optimal mass transfer. Both the porous matrix and down-comer are equipped with cooling jackets for operating at steady-state temperature regimes. The scale up capability was evaluated under industrial conditions for over 12 months. Single stage and two-stage modular pilot scale reactors with 7 or 19 matrices and up to 1.2 hl reactor volume were operative for 4 months without any contamination when strict sterile wort feeding was maintained.

The main objectives were the production of an alcohol-free beer with specifications of 0.05, 0.1 or 0.5% alcohol by volume and sufficient removal of wort carbonyl compounds. Results indicate that the residence time appeared to be a major factor in determining residual aldehyde levels. In contrast to the ethanol production, aldehyde reduction was not additive in the two stages. An average of 85% of the total reduction occurred in the first stage. The reduction in the second stage is much lower as a result of the lower residual aldehyde concentration flowing through the second stage.

From these experiments it was concluded that the scaling-up is reproducible and predictable. Only one stage is necessary for alcohol-free beer production at an optimal temperature of 10°C with minimal aeration levels below 1 mg dissolved oxygen/l.

6.3. IMMOBILISED YEAST SYSTEMS FOR CONTINUOUS FLAVOUR MATURATION OF BEER

Packed bed reactors employing immobilised yeast on DEAE cellulose or on open pore sintered glass beads are now commercially used for continuous maturation.

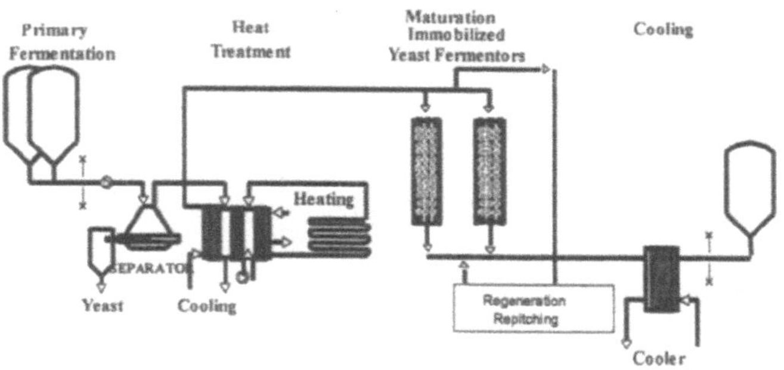

Figure 8 : Flow scheme of Cultor's two-stage heat/immobilised continuous maturation system (adapted from Pajunen & Gronquist [33]).

6.3.1. DEAE cellulose carrier (Spezyme ®)

The first successful industrial process utilising immobilised yeast in the brewing industry has been developed by the finish Co. CULTOR in association with SYNEBRYCHOFF brewery and the German engineering firm TUCHENHAGEN. The current status of the continuous maturation process has been described by Pajunen *et al.* [31;32]. Their objectives in initiating this technology were increased and more flexible volumetric productivity, as well as reducing storage needs. The process parameters were focused on the application of immobilised cells to reach final attenuation and sufficiently reduce diacetyl following an unchanged free-cell primary fermentation.

The equipment (figure 8) includes a hermetic centrifuge, a regenerative plate heat exchanger with a variable volume holding cell, immobilised yeast reactors, and a beer cooler. All of the equipment is constructed out of stainless steel, with automatic valves and flow controls.

In this system, residual yeast in the beer at the end of primary fermentation is centrifuged out to prevent the development of utilised yeast flavours during subsequent processing. Removal of yeast cells (less than 10^4 per millilitre) and proteins has been found to be critical with respect to the risk of fouling of the heat exchangers and clogging of the packed beer reactors.

The heat treatment (7 minutes at 90°C) employed by the investigators at Sinebrychoff enables all the α-acetolactate to be converted to diacetyl. This heat treatment is applied to the beer prior to the immobilised yeast beer downflow to prevent flow channelling frequently recorded in upflow operations.

The contact time of 2 h has proven to be close to the optimum for completing fermentation of the fermentable extract and achieving diacetyl levels below the taste threshold value (0.05 mg/l). The Sinebrychoff maturation system described above has not demonstrated any detectable differences in beer flavour for beer matured using traditional technology. Total lagering time was reduced from 10 to 14 days to 2 to 3 hours. No new synthesis of α-acetolactate was observed in the bioreactor. The only definable process parameter seen to influence beer quality was the length of the heat treatment time. When this time was increased to 20 min. the hydroxymethylfurfural level slightly increased.

An industrial version (1 million hl/y) of this system is now used at the SYNEBRYCHOFF KERAVA BREWERY in Finland [33]. After the heat treatment there are four reactors with a maximum total flow of 140hl/h, which corresponds to the centrifuge capacity and the time needed for emptying of the fermenters for primary fermentation rapidly. The structure of the immobilised system is modular, allowing 2, 3 or 4 fermenter operations.

Operational advantages of the immobilised yeast system are:

- Short lagtime before steady state
- Standby possible
- Easy and rapid startup
- Immobilisation and regeneration in the reactor
- Modular operation possible
- Flexibility compared with filtration

As a result of the successful industrial exploitation of the two-stage heat/immobilised cell maturation process COMPANHIA CERVEJARIA BRAHMA from Brazil decided in 1993 to test the Cultor Immo-System on an industrial scale with three 10 m³ reactors [30].

In Brahma's standard « unitank » process, thermal decarboxylation of acetolactate was achieved immediately after yeast collection and centrifugation to remove the yeast in suspension. Within less than 20 days of the start-up, results allowed to conclude that the

Cultor system had the feasibility to complete the secondary fermentation in 2 hours instead of the 12 days required in the traditional process.

Major problems were the difficulty to reach the desired degree of attenuation and the higher pH values as compared to the regular products. In contrast foam stability and flavour shelf life were significantly better. Moreover, preliminary results from the taste panel were rather in favour of the immobilised cell treated beer. Unfortunately, further evaluation is not available because of company restructuring and reappraisal of priorities.

6.3.2. Sintered glass bead carrier (Siran ®)

The ALFA LAVAL Co in association with SCHOTT ENGINEERING have developed a maturation process similar to that of Cultor employing the Siran glass beads of Schott within a packed bed reactor.

Laboratory scale tests demonstrated the suitability of using porous glass beads [11;16]. As a result the HARTWALL BREWING Co from Finland decided to build a 400.000 hl/year maturation plant in 1991 which became operational in 1992 (figure 9).

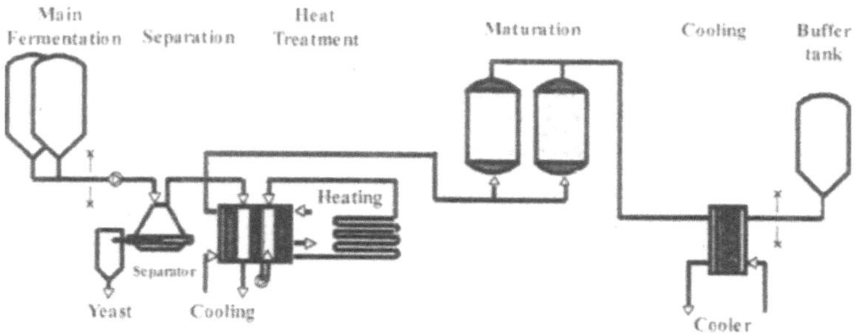

Figure 9 : The Alfa Laval two-stage heat/immobilised continuous maturation system (adapted from Hyttinen et al. [11]).

Two bioreactors with a bed volume of 2.5 m³ are used. A more even flow, preventing plugging and channelling, was obtained by changing the flow direction from a downwards to an upwards flow. After centrifugation the « Green Beer » was heated to 80°C for 10 minutes to complete the conversion of α-acetolactate into diacetyl and acetoin. Flavour comparisons have not revealed any differences between the batch and continuous maturation processes.

7. Concluding remarks

The different advanced immobilised yeast cell bioreactor systems described in this review have been assessed as to their potential for application in the brewing industry. Clearly, many alternatives superior to the conventional batch technology exist.
Reactor concept is an important consideration in immobilised cell system design, although often dictated by the type of application, the support chosen and the desired mass transfer requirements.

The reactors most commonly used for the production of beer with immobilised cells include packed bed, fluidised bed, gas lift draft tube and loop reactors.

Among these only packed bed reactors employing immobilised yeast on DEAE cellulose or porous glass beads have been operated at large scale for continuous maturation as well as for the production of alcohol-free beer and low alcohol beers. In terms of economic evaluation, major advantages are considerable savings in time and plant requirements. Mixed particles reactors including fluidised bed and gas/liquid circulating fermenters have been found more advantageous for primary fermentation. Much progress has been made and many of the processes under development have been advanced to the point of pilot plant testing.

Reduced capital and operation costs are expected. However, these predictions need to be supported by a term of large-scale trials before the potentialities and limitations of continuous immobilised yeast systems can be fully assessed. This is an essential step for allowing economic and quality evaluations to be made on a real basis.

Now the challenge is to view the subject of immobilised cell technology less experimentally and more positively as a permanent feature of future activity in the brewing industry.

References

1. Andries, M., van Beveren, P.C., Goffin, O.& Masschelein, C.A. (1995) Design of a multipurpose immobilised yeast bioreactor for application in the brewing process, *Proceedings European Brewery Convention Symposium: Immobilised Yeast Applications in the Brewing Industry*, Espoo, Finland, pp. 134-143.
2. Andries, M., van Beveren, P.C., Goffin, O.& Masschelein C.A (1996) Design and application of an immobilised loop bioreactor for continuous beer fermentation, in *Immobilised Cells: Basics and Applications* (R.H.Wijfels, R.M. Buitelaar, C. Bucke & J. Tramper, eds) Amsterdam: Elsevier Science, pp. 672-678.
3. Andries, M., Derdelinckx, G., Ione, K.G., Delvaux, F., van Beveren, P.C., & Masschelein, C.A. (1997) Zeolites as catalysts for the cold and direct conversion of acetolactate into acetoin, *Proceedings of the European Brewery Convention Congress*, Maastricht, pp. 477-484.
4. Avaisidis, A., Wandrey, Ch., Eils, H.G. & Katzke, M. (1991) Continuous fermentation of alcohol-free beer with immobilised yeast in fluidised bed reactors, *Proceedings of the European Brewery Convention*, Lisbon, pp. 569-578.
5. Breitenbücher, K. & Mistler, M. (1995) Fluidised bed fermenters for the continuous production of non-alcoholic beer with open-pore sintered glass carriers, *European Brewery Convention Symposium: Immobilised Yeast Applications in the Brewing Industry*, Espoo, Finland, pp. 77-88.
6. Cop, J., Dijon, D., Iserentant, D. & Masschelein, C.A. (1989) Reactor design optimisation with a view to the improvement of amino acid utilisation and flavour development in calcium alginate brewing yeast fermentations, *Proceedings of the European Brewery Convention*, Zürich, pp. 315-322.

7. Curin, J., Pardonova, B., Polednikova, M., Sedova, H. & Kahler, M. (1987) Beer production with immobilised yeast, *Proceedings of the European Brewery Conventions Congress*, Madrid, pp. 433-440.
8. Debourg, A., Laurent, M., Goossens, E., Borremans, E., & Masschelein, C.A. (1994) Wort aldehyde reduction of free and immobilised yeast systems, *Journal of the American Society of Brewing Chemists*, **52**, 100-105.
9. Dulieu, C., Boivin, P., Malanda, M., Dautzenberg, H. & Poncelet, D. (1997) Encapsulation of α-acetolactate decarboxylase to avoid diacetyl formation, *Proceedings of the European Brewery Convention Congress*, Maastricht, pp. 455-460.
10. Grönquist, A. Pajunen, E. & Ranta, B. (1989) Secondary fermentation with immobilised yeast-industrial scale, *Proceedings of the European Brewery Convention Congress*, Zürich, pp339-346.
11. Hyttinen, I., Kronlöf, J. & Hartwall, P. (1995), Use of porous glass at Harwall brewery in the maturation of beer immobilised yeast, *European Brewery Convention Symposium: Immobilised Yeast Applications in The Brewing Industry*, Espoo, Finland, pp. 55-61.
12. Inoue, T., Murayama, H., Kajino, K., Kamiya, T., Mitsui, S. & Mawatari, M. (1991) Direct spontaneous conversion of acetolactate into non-diacetyl substance in fermenting wort, *Proceedings of the European Brewery Convention Congress*, Lisbon, pp. 369-376.
13. Inoue, T. (1995) Development of a two-stage immobilised yeast fermentation system for continuous beer brewing, *Proceedings of the European Brewery Convention Congress*, Brussels, pp. 25-36.
14. Jones, M. & Pierce, J.S. (1964) Absorption of amino acids from wort by yeasts, *Journal of the Institute of Brewing*, **70**, 307-315.
15. Krikilion, Ph., Andries, M., Goffin, O., Van Beveren, P.C. & Masschelein, C.A. (1993) Optimal matrix and reactor design for high gravity fermentations with immobilised yeast, *Proceedings of the European Brewery Convention Congress*, Brussels, pp. 419-426.
16. Kronlöf, J., Linko, M. & Pajunen, E. (1995) Primary fermentations with a two-stage packed system – Pilot scale experiences, *European Brewery Convention Symposium: Immobilised Yeast Applications in the Brewery Industry*, pp. 118-124.
17. Masschelein, C.A., Carlier, A., Ramos-Jeunehomme, C. & Abe, I. (1985) The effect of immobilisation on yeast physiology and beer quality in continuous and discontinuous systems, *Proceedings of the European Brewery Convention Congress*, Helsinki, pp. 339-346.
18. Masschelein, C.A. & Ramos-Jeunehomme, C. (1985) The potential of alginate immobilised yeast in brewery fermentations , *Proceedings of the European Brewery Convention Congress*, Johannesburg, pp. 392-399.
19. Masschelein, C.A. (1987) New fermentation methods, *Proceedings of the European Brewery Convention Congress*, Madrid, pp. 209-220.
20. Masschelein, C.A. (1989) Recent and future developments of fermentation technology and fermenter design in brewing, in *Biotechnology Applications in Beverage Production*, chapter 7, (Cantarelli, C. & Lanzarini, G., Eds) Elsevier Applied Sciences, pp. 41-48.
21. Masschelein, C.A. (1990) Brewing into the 1990s, *Proceedings of the 21st Convention of the Institute of Brewing*, Australia and New Zealand Section, pp. 6-14.
22. Masschelein, C.A. (1990) Novel fermentation systems: their influence on yeast metabolism and beer flavour, *Proceedings of the Third Aviemore Conference on Malting, Brewing and Distilling*, pp. 103-116.
23. Masschelein, C.A., Ryder, D.S. & Simon, J.P. (1994) Immobilised cell technology in beer production, *Critical Reviews in Biotechnology*, **14**, 155-177.
24. Masschelein, C.A. (1997) A realistic view on the role of research in the brewing industry today, *Journal of the Institute of Brewing*, **103**, 103-112.
25. Mensour, N., Margartitis, A., Briens, C.L. Pilkington, H. & Russell, I. (1995) Gas lift systems for immobilised cell fermentations, *European Brewery Convention Symposium: Immobilised Yeast Applications in the Brewery Industry*, Espoo, Finland, pp. 125-132.
26. Mensour, N., Margartitis, A., Briens, C.L. Pilkington, H. & Russell, I. (1996) Application of immobilised cell fermentations, in *Immobilised Cells: Basics and Applications* (R.H. Wijfels, R.M. Buitelaar, C., Bucke & J. Tramper Eds.) Amsterdam: Elsevier Science, pp. 661-671.
27. Mensour, N., Margartitis, A., Briens, C.L. Pilkington, H. & Russell, I. (1997) New developments in the brewing industry using immobilised yeast cell bioreactor systems, *Journal of the Institute of Brewing*, **103**, 363-370.

28. Nakanishi, H., Onaka, T., Inoue, T. & Kubo, S. (1985) A new immobilised yeast reactor system for rapid production of beer, *Proceedings of the European Brewery Convention Congress*, Helsinki, 1985, pp. 331-338.
29. Norton, S., Neufield, R. & Poncelet, D.J.C.M., *Canadian Patent Application*, 2, 133, 789, 1994
30. Nothaft, A. (1995) The start-up of an immobilised yeast system for secondary fermentation at Brahma, *European Brewery Convention Symposium: Immobilised Yeast Applications in the Brewery Industry*, Espoo, Finland, pp. 41-49.
31. Pajunen, E., Mäkinen, V., & Gisler, R. (1987) Secondary fermentation with immobilised yeast, *Proceedings of the European Brewery Convention Congress*, Madrid, pp. 441-448.
32. Pajunen, E., Gröndquist, A. & Ranta, B. (1991) Immobilised yeast reactor application in continuous secondary fermentation in industrial scale operation, *Proceedings of the European Brewery Convention Congress*, Lisbon, pp. 361-368.
33. Pajunen, E. & Gröndquist, A. (1994) Immobilised yeast fermenters for continuous lager beer maturation, *Proceedings of the Institute of Brewing, 23rd Convention Asia Pacific Section*, Sidney, pp. 101-103.
34. Pilkington, H., Margaritis, A., Mensour, N. & Russell, I. (1998) Fundamentals of immobilised yeast cells for continuous beer fermentation: a review, *Journal of the Institute of Brewing*, **104**, 19-31.
35. Ryder, D.S. & Masschelein, C.A. (1985) The growth process of brewing yeast and the biotechnological challenge, *Journal of the American Society of Brewing Chemists*, **43**, 66-75.
36. Sane, H., Kondo, K., Fujii, T., Shimuzu, F., Tanaka, J. & Inoue, T., (1987) Fermentation properties of brewer's yeast having α-acetolactate decarboxylase gene, *Proceedings of the European Brewery Convention Congress*, Madrid, pp. 545-552.
37. Van den Berg, R., Harteveld, P.A. & Martens, F.B. (1983) Diacetyl reducing activity in brewer's yeast, *Proceedings of the European Brewery Convention Congress*, London, pp. 497-504.
38. Van de Winkel, L., van Beveren, P.C. & Masschelein, C.A. (1991) The application of an immobilised yeast loop reactor to the continuous production of alcohol-free beer, *Proceedings of the European Brewery Convention Congress*, Lisbon, pp. 577-584.
39. Van de Winkel, L., Van Beveren, P.C., Borremans, E., Goossens, E. & Masschelein, C.A. (1993) High performance immobilised yeast reactor design for continuous beer fermentation, *Proceedings of the European Brewery Convention Congress*, Oslo, pp. 307-314.
40. Van de Winkel, L., McMurrough, I., Evers, G., Van Beveren, P.C. & Masschelein, C.A. (1995) Pilot-scale evaluation of Silicon carbide immobilised yeast systems for continuous alcohol-free beer production, *European Brewery Convention Symposium: Immobilised Yeast Applications in the Brewery Industry*, Espoo, Finland, pp. 90-98.
41. Van Dieren, B. (1995) Yeast metabolism and the production of alcohol-free beer, *European Brewery Convention Symposium: Immobilised Yeast Applications in the Brewery Industry*, Espoo, Finland, pp. 66-71.
42. White, F.H. & Portno, A.D. (1978) Continuous fermentation by immobilised brewer's yeast, *Journal of the institute of Brewing*, **74**, 228-233.

IMMOBILIZED YEAST BIOREACTOR SYSTEMS FOR BREWING – RECENT ACHIEVEMENTS

VIKTOR A. NEDOVIC[1], BOJANA OBRADOVIC[2], IDA LESKOSEK-CUKALOVIC[1], GORDANA VUNJAK-NOVAKOVIC[3]
[1]Institute of Food Technology and Biochemistry, Faculty of Agriculture, University of Belgrade, Nemanjina 6, PO Box 127, 11081 Belgrade-Zemun, Yugoslavia; [2]Department of Chemical Engineering, Faculty of Technology, University of Belgrade, Karnegijeva 4, 11000 Belgrade, Yugoslavia; [3]Department of Chemical Engineering, MIT, E25-342, 45 Carleton Street, Cambridge, MA 02139, USA

During recent years, immobilised yeast technology has gained increasing attention in the brewing industry. As a result of extensive research immobilised yeast technology is nowadays a well established technology for beer maturation and alcohol-free and low-alcohol beer production. However, in primary fermentation the situation is more complex and this process is still under scrutiny on the lab or pilot level. This paper describes several favourable immobilised cell - bioreactor systems for primary beer fermentation with particular emphasis on alginate-yeast microbeads - internal loop gas-lift bioreactor system, introduced into this field by our research group.

1. Immobilised cell systems in biotechnology

Whole cell immobilisation has been defined as "the physical confinement or localisation of intact cells to a certain defined region of space with preservation of some desired catalytic activity" (Karel *et al.*, 1985). Many microorganisms have a capability to adhere to different kinds of surfaces in nature what enables close vicinity to nutrients and food supply. Therefore, we can say that these biological systems are immobilised in their natural state. Furthermore, many biotechnological processes could benefit by immobilisation of biocatalysts. Immobilisation offers many potential advantages over free cell systems, such as higher cell densities and cell loads, increased volumetric productivity, shorter overall reaction times, smaller fermenter sizes which may lower capital costs, reuse of the same biocatalysts for prolonged periods of time due to constant cell regeneration, development of continuous processes which may be performed beyond the nominal washout rate, improved substrate utilisation, reduced risk for microbial contamination, simplified process design, constant product quality, improved tolerance to end products and protection of cells. Above all, immobilised cell technology results in much faster fermentation rates as compared to the existing free

M. Hofman and P. Thonart (eds.), Engineering and Manufacturing for Biotechnology, 277–292.

cell fermentations. For these reasons there is a growing interest in using immobilised cell systems for different fermentation processes, such as well-known beer (Nedovic *et al.*, 1997a, Pilkington *et al.*, 1998), wine (Divies *et al.*, 1994; Yokotsuka *et al.*, 1997) and cider (Durieux *et al.*, 1998; Nedovic *et al.*, 2000a) fermentations.

2. Applications of immobilised yeast systems in brewing

Application of immobilised cell systems in brewing industry has been investigated for the last twenty years. So far, there are several industrial applications of such systems for the secondary fermentation of beer (Pajunen *et al.*, 1991), and for the production of alcohol-free and low-alcohol beer (Lommi, 1990), but the immobilised cell systems have not been yet successfully applied for primary beer fermentation on industrial level. The main reason for this is wide flavour variation in final beers, which were produced solely by immobilised cells. It was reported that the insufficient free amino nitrogen consumption by immobilised yeast cells, coupled with mass transfer restrictions and reduced cell growth in the immobilised conditions, cause an unbalanced flavour profile of final beer (Hayes *et al.*, 1991). Amino acid metabolism in yeast is closely linked to production of flavour compounds such as vicinal diketones, higher alcohols, esters, organic acids and sulphur compounds. Thus, the way to increase free amino nitrogen consumption and consequently improve beer quality is to increase growth of the cells on one side, and to minimise the internal and external mass transfer resistances in the system on the other.

Internal mass transfer relates to transfer of substrates and products within the carrier, *i.e.* through the polymeric carrier matrix and dense aggregates of immobilised cells inside the carrier. Internal mass transfer resistances are governed by nature of cell immobilisation, size, texture, and porosity of cell carriers. External mass transfer includes the exchange of nutrients and products between the cell carrier surface and the surrounding medium and it is determined by the bioreactor design and flow pattern. Therefore, two key parameters in immobilised cell systems are the choice of the cell carrier and the bioreactor design.

2.1. CELL CARRIERS AND IMMOBILIZATION METHODS

Main purpose of immobilisation is to retain high cell concentrations within "a certain defined region of space" such as a bioreactor and increase volumetric productivity of a system. Two interrelated directions in research regarding immobilisation can be distinguished: immobilisation techniques and support materials.

Immobilisation techniques can be divided into four major groups based on physical mechanisms of immobilisation (Pilkington *et al.*, 1998): adsorption to a pre-formed carrier, physical entrapment within a porous matrix, self aggregation in flocs and containment of cells behind a barrier (Figure 1). For the immobilisation of yeast cells for beer fermentation, the first two techniques gained most attention.

2.1.1. Adsorption to a pre-formed carrier

The earliest type of cell immobilisation is based on cell adsorption onto external surfaces of solid carriers. Cells can be attached by Van der Waals forces, electrostatic interactions between oppositely charged surfaces, covalent bonding and physical entrapment in the pores. Generally this is the most gentle immobilisation technique as it is carried out by recirculation of cell suspension through a packed bed of carriers, or by mixing a suspension of cells and carriers.

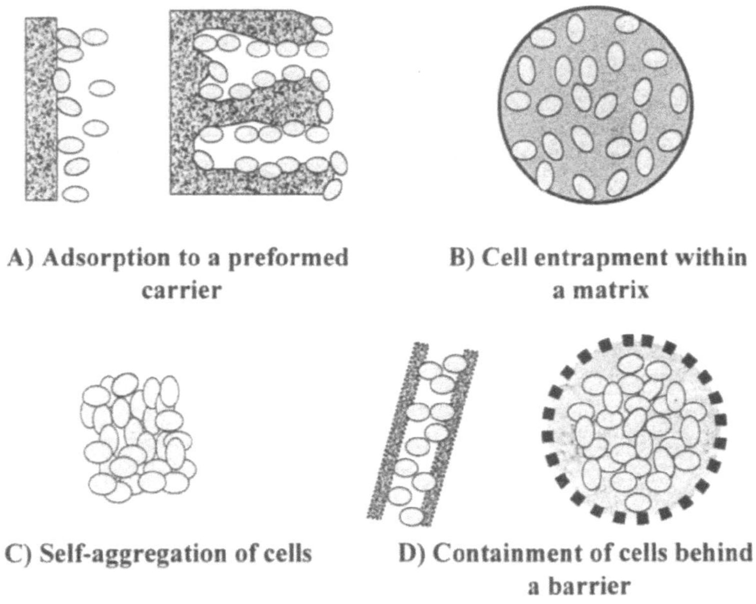

Figure 1. Principal methods for cell immobilisation

For the adsorption of yeast cells various materials proved suitable, which can be divided in two general types: materials with the yeast cells restricted to the external surfaces only, and materials with pores large enough to allow cell adsorption inside the material (Figure 1A). First group includes DEAE-cellulose and wood chips while the second group includes a variety of glass, ceramic, and synthetic materials. Depending on material, different shapes of cell carriers were applied: non-uniform granules in the case of DEAE-cellulose, spherical porous beads in the case of glass, chitosan and ceramics, porous multichannel rods in the case of silicon carbide, sponge-like porous cubes, and chips or blocks made of wood (Pajunen, *et al.*, 1991; Pajunen, 1996; Lommi, 1990; Inoue, 1995; Yamaushi *et al.*, 1994a,b; Yamauchi and Kashihara, 1996; Masschelein and Andries, 1996; Breitenbucher and Mistler, 1996; Shindo *et al.*, 1994b; Umemoto *et al.*, 1998; Scott and O'Reilly, 1995; Kronlöf *et al.*, 1996; Kronlöf and Virkajarvi, 1999, Tata *et al.*, 1999).

In all cases carriers provided mechanically good support and direct contact of cells and substrate, minimising mass transfer limitations. Additional advantage of this immobilisation method can be carrier regeneration and its repeated use. However, in these carriers cells are not protected from the surrounding medium and may be affected by sudden changes in flow, pH, and temperature. The second drawback could be high cell leakage from these carriers coupled with relatively low cell loading in these systems.

2.1.2. Cell entrapment

Cell immobilisation by entrapment is based on low porosity of matrix that at the same time retains cells within the carrier and provides metabolite diffusion (Figure 1B). Alginate, *kappa*-carrageenan and pectate gels in shape of spheres were mostly used as matrix materials for yeast immobilisation (White and Portno, 1978; Pardonova *et al.*, 1982; Hsu and Bernstein, 1985; Onaka *et al.*, 1985; Curin *et al.*, 1987; Nedovic *et al.*, 1993, 1996, 1997a; Shindo *et al.*, 1994a; Domeny *et al.*, 1996; Mensour *et al.*, 1996a,b; Pilkington *et al.*, 1999).

Two general techniques were developed for the formation of beads loaded with cells: extrusion and emulsification. The extrusion method is based on discharge of cell suspensions into a hardening solution where the gel spheres are formed and hardened. The emulsification method is based on formation of "water in oil" emulsions and solidification of droplets containing suspended cells.

Main advantage of the cell entrapment method is attainment of extremely high cell loading providing high fermentation rates. However, in some cases cell proliferation and activity can be limited by low mass transfer rates within the matrices.

2.1.3. Self-aggregation

Self-aggregation of cells (Figure 1C) can be natural or artificially induced by crosslinking agents. Natural self-aggregation of yeast cells was investigated during '60ties for continuos beer fermentation (Klopper *et al.*, 1965; Stratton *et al.*, 1994) and more recently for beer maturation (Mafra *et al.*, 1997). This technique is based on the use of highly concentrated suspensions of flocculent yeast strains. Although this is the simplest and least expensive immobilisation method it is the most sensitive to the changes in the operating conditions. In addition, there is a high risk of cell wash-out from the system. Another disadvantage could be diffusion limitations in the case of larger flocs.

2.1.4. Containment of cells behind a barrier

In this cell immobilisation technique cells are confined to a space bounded by a semipermeable barrier or immobilised within a membrane (Figure 1D). There are very little data on yeast immobilisation for beer fermentation by this method. Shindo and Kamimura (1990) reported good mechanical properties and relatively high activity of hollow polyvinylalcohol beads loaded with yeast cells. Still, the procedure for preparation of these beads is relatively complex.

2.2. BIOREACTOR DESIGN

Several promising bioreactor concepts for immobilised cell systems have been developed over the last decade related to the choice of the cell carrier (Figure 2).

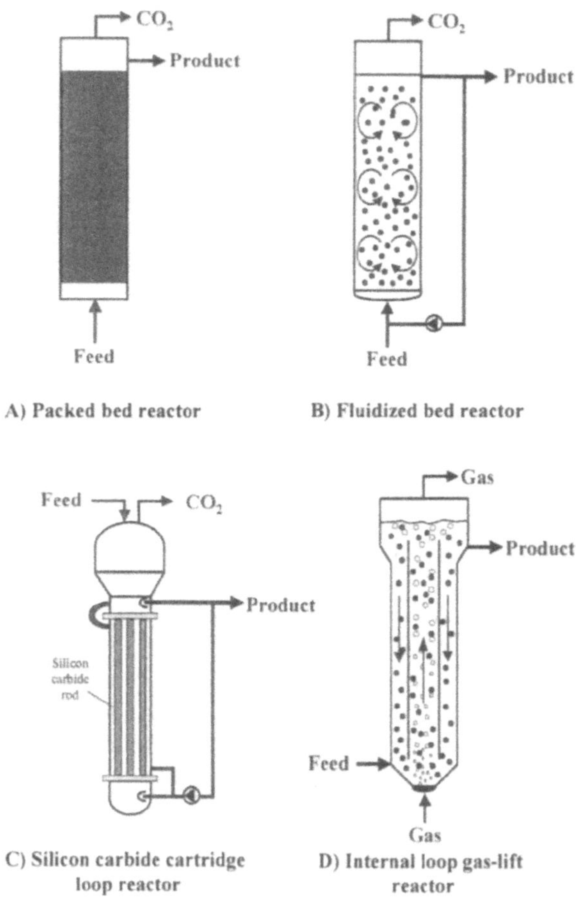

Figure 2. Bioreactor configurations for beer fermentation with immobilised yeast

2.2.1. Packed bed reactor

Packed bed reactors were mostly used for preformed carriers with cells adsorbed on the external surfaces. This type of reactor was widely investigated for beer fermentation since one of the earliest applications of immobilised yeast in this field (Narziss and Hellich, 1971) until recently developed pilot and semi-industrial plants (Kronlöf and Virkajarvi, 1999; Inoue, 1995). Packed bad reactors contain cell carriers densely packed in a column through which the feed medium (wort) is pumped (Figure 2A). This type of reactor is characterised by a simple design without moving parts providing static conditions for cell cultivation. In this way, cell leakage from carriers with cells adsorbed on external surfaces is minimised. On the other hand, lack of mixing in this type of reactor results in nonuniform feed and temperature distributions, channelling and mass transfer limitations. As a consequence, free amino nitrogen consumption was reportedly low during continuous beer fermentation (Pardonova *et al.*, 1982; Taidi, 1995).

One of the ways to avoid channelling and improve heat-transfer recently proposed can be forced circulation of fermenting beer (Andersen *et al.*, 1999; Tata *et al.*, 1999). Nitrogen consumption and beer flavour can be improved in multistage systems based on packed bed reactors. First stage in these systems was designed for aerobic fermentation, which favours amino acid consumption and is followed by packed bed reactors for anaerobic fermentation. In Kirin three-stage reactor system the first stage is carried out in a chemostat, followed by a series of packed bed reactors (Yamauchi and Kashihara, 1996). In a two stage system developed by Kronlöf *et al.* (1996, 1999) aerobic and anaerobic fermentations are carried out consecutively in two packed bed reactors in series. In both multistage systems, final concentration of free amino nitrogen was significantly lowered as compared to one stage packed bed reactor.

2.2.2. Fluidised bed reactor

Fluidised bed reactors were used for carriers with cells adsorbed inside the carrier, made either of glass or ceramics, and carriers with entrapped cells (Shindo *et al.*, 1994b; Umemoto *et al.*, 1998; Breitenbucher and Mistler, 1996; Tata *et al.*, 1999). In this type of reactor the upward flowrate of feed medium is high enough to provide fluidisation of carriers resulting in improved mixing properties and medium distribution as compared to packed bed reactors (Figure 2B). However, in the fluidised state collisions of particles can induce carrier abrasion and damage. In addition, fluidisation of glass and ceramic carriers may require high medium flowrates that could result in high pumping costs and cell leakage. On the other hand, in the case of low density carriers, fluidisation may require low flowrates at which the mass transfer rates could be too low for practical applications.

2.2.3. Silicon carbide cartridge loop reactor

Silicon carbide cartridge loop reactor was developed in an attempt to overcome problems inherent to packed and fluidised bed reactors (Van de Winkel, 1995; Masschelein and Andries, 1996; Tata *et al.*, 1999). It consists of silicon carbide multichannel rods (60% void volume) seeded with yeast cells and perfused in parallel with recirculating feed medium (Figure 2C). Cells are grown statically and nutrient supply is provided by medium circulation through the 2-mm diameter channels. The reactor is characterised by simple design, which can be easily scaled up. Disadvantages are still relatively high cost of silicon carbide matrices and lower cell growth and specific productivity in this system as compared to free cells (Masschelein and Andries, 1996).

2.2.4. Internal loop gas-lift reactor

Our research group (Nedovic *et al* 1993) successfully introduced three-phase internal loop gas-lift bioreactor in beer fermentation experimental studies. This reactor retains the advantages of fluidised beds, such as high loading of solids and good mass transfer properties and it is particularly suitable for applications with low density carriers (Vunjak *et al.*, 1992,1998). Mixing is established by circulation of liquid and solid phases providing operations at higher liquid flowrates and consequently better mass transfer rates as compared to fluidised bed reactor (Figure 2D). The absence of mechanical agitation creates a relatively low shear environment, which makes these reactors ideally suitable for the application of shear sensitive cells and solid matrices. Low-density alginate and carrageenan gel particles are typically used in three-phase gas-lift reactors as carriers of yeast cells in beer fermentations (Nedovic *et al.*, 1993, 1996, 1997a; Mensour *et al.*, 1996a,b; Vunjak *et al.*, 1998). Other important characteristics of gas-lift bioreactors are simple construction, low risk of contamination, easy adjustment and control of the operational parameters, and simple capacity enlargement (Nedovic *et al.*, 1996,1997b; Mensour *et al.*, 1996a; Mafra *et al.*, 1997).

3. Alginate-gas-lift bioreactor system

Alginate - gas-lift bioreactor system is based on the use of alginate beads as carriers for yeast cells and a three-phase internal loop gas-lift bioreactor. This system was investigated for primary beer fermentation in batch and continuous processes (Nedovic *et al.*, 1996; Nedovic *et al.*, 1997a; Vunjak *et al.*, 1998).

3.1. ALGINATE MICROBEADS LOADED WITH YEAST CELLS

Alginate has been investigated for a long time for yeast cell immobilisation and beer fermentation. However, several major problems restricted its wider use. Weak mechanical properties make alginate unsuitable for use in packed bed reactors. In addition, only systems for lab scale production of alginate beads with diameters 2 - 4 mm were available. Such beads showed significant diffusion limitations, which reduced cell growth and activity.

Installation of an internal loop gas lift bioreactor with alginate beads for beer fermentation provided new interest for this type of yeast carrier (Nedovic *et al.*, 1993). In parallel, several promising techniques and systems were developed for large scale production of gel-type polymer microbeads over the last decade (Poncelet *et al.*, 1997; Brandenberger and Widmer, 1998; Dulieu *et al.*, 1999; Prüsse *et al.*, 1999). Reduction of bead diameter significantly decreased internal mass transfer limitations.

In the case of alginate, microbeads are produced by extrusion of sodium alginate suspension of yeast cells into calcium chloride solution. Sodium ions are immediately replaced with calcium ions and gel beads are formed and solidified due to calcium alginate insolubility in water. Three extrusion techniques gained special attention (Figure 3).

Resonance method uses vibration applied at a constant frequency to liquid jet resulting in jet break-up into small uniform droplets in the range of 0.2 to 1.5 mm in diameter (Figure 3A). Vibration can be applied to the nozzle or to the liquid reservoir. This method can be easily scaled-up and systems for laboratory and pilot plant scales are currently commercially available (Marison *et al.*, 1997; Brandenberger and Widmer, 1998).

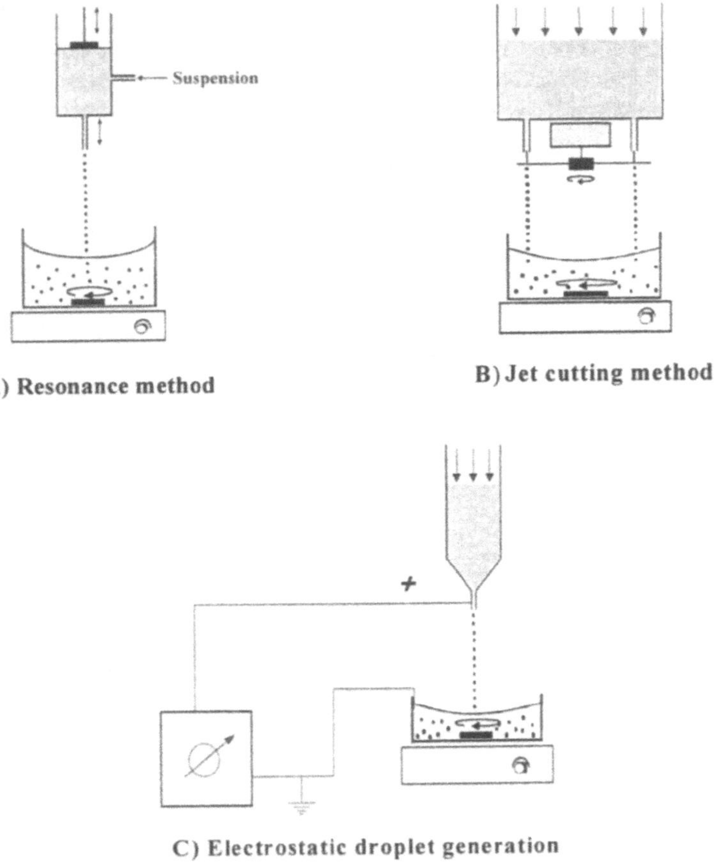

Figure 3. Extrusion techniques for cell immobilisation

Jet cutting method uses mechanical forces to break-up a liquid jet by rotating cutting wires (Figure 3B). Jet is cut into cylindrical segments that attain spherical shape with diameters in the range 0.2 to 2 mm while falling into hardening solution. This technique is commercially available for alginate bead production at industrial scale (Prüsse *et al.*, 1998; Prüsse *et al.*, 1999; Wittlich *et al.*, 1999-2000).

Electrostatic droplet generation uses electrostatic forces to disrupt a liquid surface at the capillary/needle tip forming a charged stream of small droplets (Figure 3C). In this way the liquid is exposed to an electric field, inducing the electrical charge on the liquid surface and a repulsive outside-directed force. This leads to production of uniform small-diameter microbeads down to 0.20 mm in diameter (Bugarski *et al.*, 1994; Goosen *et al.*, 1997; Melvik *et al.*, 1999; Nedovic *et al.*, 1999b). Presently only a lab scale device with one needle and a power supply of 0-10 kV is commercially available (Melvik *et al.*, 1999) although scale-up of this system in multi-needle devices with 10 or 20 needles were made recently by several research groups (Gåserød, 1998; Bugarski *et al.*, 1994).

All these techniques provide controlled production of alginate microbeads of 0.2 - 1 mm in diameter with narrow size distributions. Optimal bead diameter for yeast cell immobilisation and application in beer fermentation has yet to be determined. In alginate beads with diameters of about 3 mm cell distribution was nonuniform with dense aggregates close to the surface (around 0.25 mm in depth) implying nutrient limitations (Nedovic, 1999a). Consistently, cell growth in these beads in early phases of cultivation expressed exponential kinetics common for free yeast cells while at later times cell growth was reduced as compared to free cell systems. Alginate microbeads with diameters in the range 0.5 to 1 mm have provided high cell growth without apparent nutrient limitations in a short term cultivation (Nedovic *et al.*, 2000b). Applications of even smaller microbeads may be limited by increased cell leakage.

3.2. INTERNAL LOOP GAS-LIFT BIOREACTOR

Internal loop gas-lift reactor consists of a column with a coaxially placed internal tube. Liquid phase can be batch or continually supplied usually at the bottom of the column. Solid phase consists of beads suspended in the liquid phase. Gas phase is generally introduced at the bottom of the column into the tube that results in different bulk densities in the tube and the annular region and produces liquid and solid circulation around the tube. The tube is called riser since it contains gas-liquid-solid upflow while the annular region is called "downcomer" containing mainly liquid and solids downflow. For beer fermentation gas phase can be nitrogen or mixture of CO_2 and air, liquid phase is plant wort of 11 - 12 % extract and solid phase are yeast carriers, comprising 15 – 40 % v/v of rector volume (Nedovic *et al.*, 1993, 1996, 1997a; Mensour *et al.*, 1996a,b; Pilkington *et al.*, 1999). Internal loop gas-lift reactor provides easy temperature control by outer thermal jackets. Liquid and solid circulation may induce increased foaming which is usually solved by addition of antifoaming agents.

3.3. BEER FERMENTATION IN ALGINATE-GAS-LIFT BIOREACTOR SYSTEM

Alginate - gas-lift bioreactor system (Figure 4) was successfully applied for batch and continuous primary beer fermentation for up to 3 months at laboratory scale (Nedovic *et al.*, 1993, 1996, 1997a; Vunjak *et al.*, 1998). Alginate beads were shown in repeated batch fermentations to be suitable carriers for yeast cells. Beads with diameters in the range 1 - 2.5 mm loaded with brewing yeast (*Saccharomyces uvarum*) at concentration of about $2x10^9$ cells/ml showed no significant change in cell activity and viability over a 3 month period (Nedovic *et al.*, 1997a).

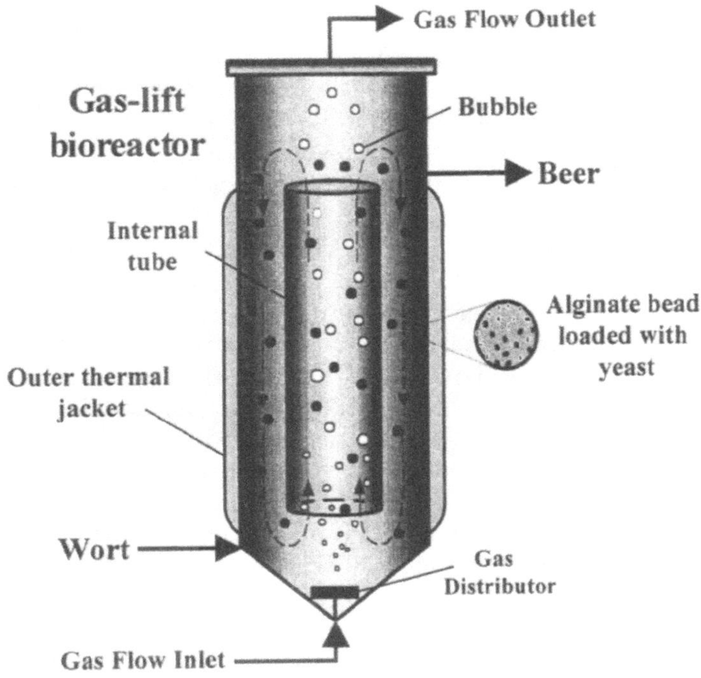

Figure 4. Alginate – gas-lift bioreactor system. Operating conditions: 4L working volume in continuous or repeated batch operations; 15 °C operating temperature; 15 - 25 % v/v fraction of solids; solid phase were alginate beads 1 – 2.5 mm in diameter loaded with yeast at $2x10^9$ cells/ml of alginate

Alginate - gas-lift bioreactor system provided significant fermentation time reduction as compared to traditional beer fermentation, which requires about 7 days (Figure 5). Primary beer fermentation took 12 to 18 h, depending on operating conditions and desired beer attenuation. The observed reduction of fermentation time can be attributed to high cell loading and efficient bulk mixing in the liquid phase which resulted in rapid mass transfer between the liquid and cell carriers.

High degree of mixing in internal loop gas-lift reactors was shown to be a consequence of high recirculation of liquid and solid phases in a riser-downcomer part and liquid backmixing in the upper section of the reactor (Obradovic *et al.*, 1994; Nedovic *et al.*, 1997b; Vunjak *et al.*, 1998). Flow in riser and downcomer was described as close to plug flow with low degree of mixing while it was considerably turbulent in the upper section (above the riser) where gas bubbles are disengaged and liquid flow direction is reverted from riser to downcomer. Overall mixing properties of the reactor were found to be mainly determined by superficial gas velocity which was shown to govern both circulation rate and flow pattern (Vunjak *et al.*, 1992; Obradovic *et al.*,

1994; Nedovic *et al.*, 1997b). As the superficial gas velocity increased, mixing time rapidly decreased to a certain limit after which no further decrease was observed (Figure 6). This limiting superficial gas velocity corresponds to transition of flow regime where gas starts recirculating together with liquid and solid phases at approximately constant circulation rate.

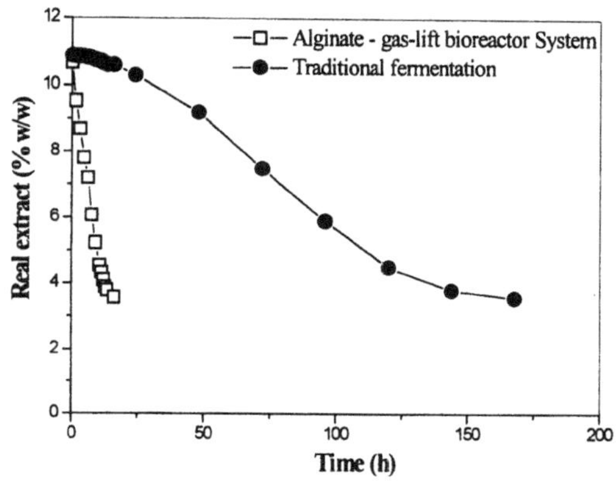

Figure 5. Extract uptake in: alginate – gas-lift bioreactor system, and traditional fermentation process

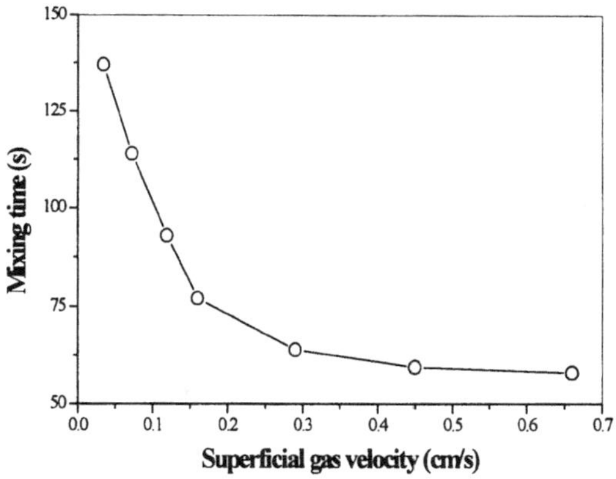

Figure 6. Mixing time in the alginate - gas-lift bioreactor system as a function of superficial gas velocity at 15 % v/v fraction of solids

Superficial gas velocity determined also the fermentation rate and free amino nitrogen utilisation, which are dependent on mixing and increased as the superficial gas velocity increased (Figure 7). At superficial gas velocities of about 0.15 cm/s the fermentation rate reached a limiting value of 0.72 %w/w/h of extract indicating efficient external mass transfer and minimal diffusion boundary layer (Figure 7A). At these conditions, improved amino nitrogen consumption (Figure 7B) resulted in formation of higher alcohols in concentrations typical for lager type beer (Table I). The average concentrations of free amino acids and various flavour-active compounds (e.g. higher alcohols, acetate esters, acetaldehyde, total diacetyl) were comparable with those in beer produced by traditional fermentation (Table 1). Final beer produced in the alginate - gas-lift bioreactor system had desired sensory and analytical profile and could not be distinguished from beer produced by traditional process.

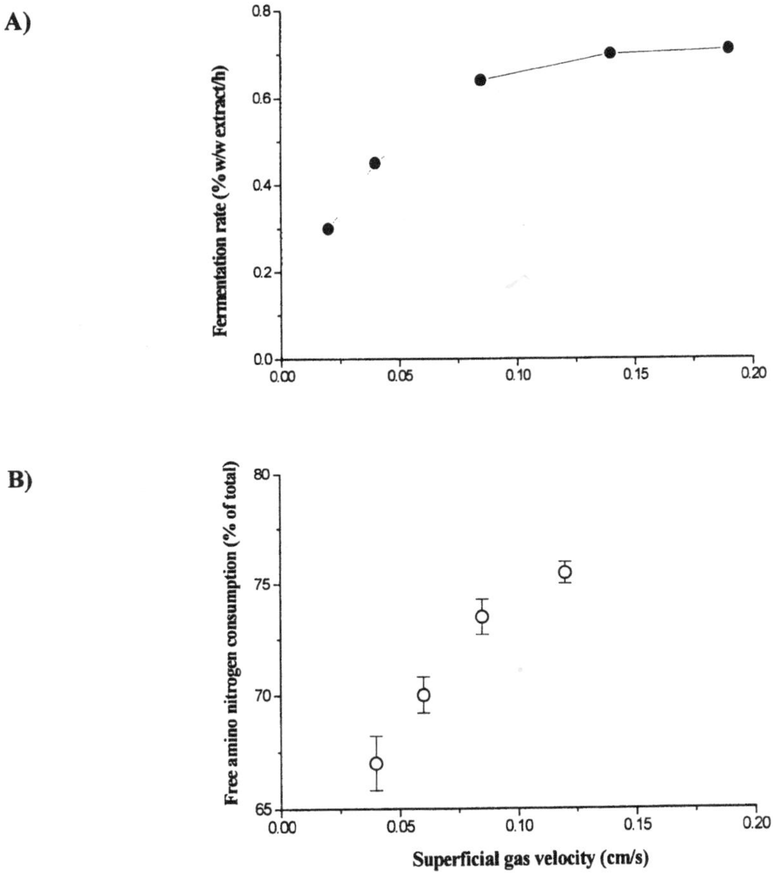

Figure 7. Fermentation in the alginate - gas-lift bioreactor system as a function of superficial gas velocity. A) Fermentation rate; B) Free amino nitrogen consumption

Table 1 Comparison of beers produced by traditional fermentation and by Alginate - Gas-lift Bioreactor System

	Traditional Brewing System	Alginate - gas-lift bioreactor system
Fermentation time	6-7 days	12-18 hours
Original extract (%w/w)	11.25	11.20
Ethanol (%w/w)	3.73	3.65
Free amino nitrogen (mg/L)	78	70
Acetaldehyde (mg/L)	4.5	6.4
Ethyl acetate (mg/L)	15.2	18.8
Isoamyl acetate (mg/L)	0.85	1.25
n – Propanol (mg/L)	14.0	19.8
Isobutanol (mg/L)	13.8	20.6
Isoamyl alcohol (2-Methil butanol+3-Metil butanol) (mg/L)	62.2	70.2
Total diacetyl (mg/L)	0.32	0.28

4. Conclusion

Immobilised cell technology offers favourable solutions for brewing industry. Main advantages are significant reduction of time and cost of beer fermentation. Several attractive concepts for yeast cell immobilisation and bioreactor designs for primary beer fermentation have been developed over the last decade. Mostly used systems are based on yeast cells adsorbed to preformed carriers or entrapped in gel matrices, and cultivated in packed and fluidised bed reactors. Internal loop gas-lift bioreactor with alginate beads loaded with yeast cells is another promising approach to process of main beer fermentation. This system has shown excellent results in repeated batch and continuos beer fermentations over up to 3 month period. During this period alginate beads loaded with yeast have shown unchangeable activity and cell viability. Fermentation time was reduced below 1 day and final beers had desired sensory and analytical profiles.

References

Andersen, K., Bergin, J., Ranta, B., and Viljava, T. (1999) New process for continuous fermentation of beer, in *Proc. 27th Congr. Eur. Brew. Conv.*, Cannes, The European Brewery Convention, Zoeterwoude, pp. 771-779

Brandenberger, H. and Widmer F. (1998) A new multinozzle encapsulation/immobilisation system to produce uniform beads of alginate, *J. Biotechnol. 63*, 73-80.

Breitenbucher, K. and Mistler, M. (1996) Fluidised bed fermenters for the continuous production of non-alcoholic beer with open-pore sintered glass carriers, in *Immobilised yeast applications in the brewing industry, Monograph XXIV*, Verlag Hans Carl Getranke-Fachverlag, Nurnberg, pp. 77-88

Bugarski, B., Qiangliang, L., Goosen, M.F.A., Poncelet, D., Neufeld, R.J. and Vunjak, G. (1994) Electrostatic droplet generation: Mechanism of polymer droplet formation, *AIChE J. 40*, 1026-1031.

Curin, J., Pardonova, B., Polednîkova, M., Sedova, H., and Kahler, M. (1987) Beer production with immobilised yeast, in *Proc. 21st Congr. Eur. Brew. Conv.*, Madrid, IRL Press, Oxford, pp. 433-439

Divies C., Cachon R., Cavin J.-F., and Prevost H. (1994) Immobilised cell technology in wine production, *Crit. Rev. Biotechnol. 14(2)*, 135-153.

Domeny, Z., Smogrovicova, D., Gemeiner, P., Malovikova, A., and Strudík, E. (1996) Calcium pectate gel to immobilize yeast for continuous beer production, in H. Dautzenberg and D. Poncelet (eds.), *Proc. Int. Workshop Bioencapsulation V*, Potsdam, Germany, P12, pp. 1-4

Dulieu, C., Poncelet, D., and Neufeld, R.J. (1999) Encapsulation and immobilisation techniques, in W.M. Kühtreiber, R.P. Lanza and W.L. Chick (eds.) *Cell Encapsulation Technology and Therapeutics*, Birkhäuser, Boston, pp. 3-17

Durieux, A., Nedovic, V., Van Nedervelde, L., Rosseels, P., Vandegans, J., Plaisant, A-M., De Baerdemacker, L., and Simon J-P (1998) Effect of Yeast and Leuconostoc oenos Immobilisation on the Formation of Flavour Components in Cider, in R.R. Price and D. Poncelet (eds.), *Proc. Int. Workshop Bioencapsulation VII: "Advances in Basic Science and Industry"*, Easton, USA.

Gåserød, O. (1998) Microcapsules of alginate-chitosan: A study of capsule formation and functional properties, *PhD thesis*, NTNU Trondheim, Norway.

Goosen, M.F.A., Al-Ghafri, A.S., Mardi, O.E., Al-Blushi, M.I.J., Al-Hajri, H.A., Mahmoud, E.S.E., and Consolacion, E.C. (1997) Electrostatic droplet generation for encapsulation of somatic tissue: Assessment of high-voltage power supply, *Biotechnol. Prog. 13*, 497-502.

Hayes S.A., Power J., and Ryder D.S. (1991) Physiology of immobilised cells and the application to brewing, *Brew. Dig. 66(11)*, 28-33.

Hsu W.-P. and Bernstein, L. (1985) A new type of bioreactor employing immobilised yeast, *J. Am. Soc. Brew. Chem. 43*, 101-103.

Inoue, T. (1995) Development of a two-stage immobilised yeast fermentation system for continuous beer brewing, in *Proc. 25th Congr. Eur. Brew. Conv.*, Brussels, IRL Press, Oxford, pp. 25-36

Karel S.F., Libicki S.B., and Robertson C.R. (1985) The immobilisation of whole cells: engineering principles, *Chem. Eng. Sci. 40(8)*, 1321-1354.

Klopper, W.J., Roberts, R.H., Royston, M.G., and Ault, R.G. (1965) Continuous fermentation in a tower fermenter, in: *Proc. 10th Congr. Eur. Brew. Conv.*, Stockholm, Elsevier Scientific Publishing Company, Amsterdam, pp 238-244

Kronlöf J., Linko, M., and Pajunen, E. (1996) Primary fermentation with a two-stage packed bed system – pilot scale experiences, in *Immobilised yeast applications in the brewing industry, Monograph XXIV*, Verlag Hans Carl Getranke-Fachverlag, Nurnberg, pp. 118-124

Kronlöf J., and Virkajarvi I. (1999) Primary fermentation with immobilised yeast, in: *Proc. 27th Congr. Eur. Brew. Conv.*, Cannes, The European Brewery Convention, Zoeterwoude, pp. 761-771

Lommi H. (1990) Immobilised yeast for maturation and alcohol-free beer, *Brew. Dist. Int. 5*, 22-23.

Mafra, I., Cruz, J.M.M., and Teixeira, J.A. (1997) Beer maturation in a continuously operating bioreactor using a flocculating brewer's yeast strain, in *Proc. 26th Congr. Eur. Brew. Conv.*, Maastricht, IRL Press, Oxford, pp. 509-517

Marison, I., Catana, E., Brandenberger, H., Heinzen, C., Kuhn, O., and Plüss, R. (1997) A novel vibrating nozzle extrusion system for application with microbial and animal cells, in F. Godia and D. Poncelet (eds.) *Proc. Int. Workshop Bioencapsulation VI: "From fundamentals to industrial applications"*, Barcelona, Spain, T1.2, pp. 1-4

Masschelein C.A., and Andries M. (1996) The Meura-Delta immobilised yeast fermenter for the continuous production of beer, *Cervesia 21(4)*, 28-33.

Melvik, J.E., Skjåk-Bræk, G., Gåserød, O., Klokk, T.I., and Skaugrud, Ø. (1999) Different parameters influencing the alginate bead size by use of an electrostatic bead generator, in G. Skjåk-Bræk and D. Poncelet (eds.), *Proc. Int. Workshop Bioencapsulation VIII: "Recent Progress in Research and Technology"*, Trondheim, Norway, O-11, pp. 1-4

Mensour N., Margaritis A., Briens C.L., Pilkington H., and Russel I. (1996) "Gas lift" systems for immobilised cell fermentations, in *Immobilised yeast applications in the brewing industry, Monograph XXIV*, Verlag Hans Carl Getranke-Fachverlag, Nurnberg, pp. 125-133

Mensour N., Margaritis A., Briens C.L., Pilkington H., and Russel I. (1996) Application of immobilised yeast cells in the brewing industry, in R.H. Wijffels, R.M. Buitelaar, C. Bucke and J. Tramper (eds.), *Immobilised cells: Basics and applications*, Elsevier Science B.V., Amsterdam, pp. 661-671

Narziss, L. and Hellich, P. (1971) Ein beitrag zur wesentlichen beschleunigung der gδrung and reifung des bieres, *Brauwelt 111*, 1491-1500.

Nedovic, V., Obradovic, B., Vunjak-Novakovic, G., and Leskosek-Cukalovic, I. (1993) Kinetics of beer fermentation with immobilised yeast cells in an internal loop air-lift bioreactor, *Hem. Ind. (in Serbian) 47*, 168-172.

Nedovic V.A., Leskosek-Cukalovic I., and Vunjak-Novakovic G. (1996) Short-time fermentation of beer in an immobilised yeast air-lift bioreactor, in J. Harvey (ed.), *Proc. 24th Conv. Inst. Brew.*, Singapore, Winetitles, Adelaide, p. 245

Nedovic, V.A., Leskosek-Cukalovic, I., Milosevic, V., and Vunjak-Novakovic, G. (1997a) Flavour formation during beer fermentation with immobilised *Saccharomyces cerevisiae* in a gas-lift bioreactor, in F. Godia and D. Poncelet (eds.) *Proc. Int. Workshop Bioencapsulation VI: "From fundamentals to industrial applications"*, Barcelona, Spain, T5.3, pp. 1-4

Nedovic, V.A., Pesic, R., Leskosek-Cukalovic, I., Laketic, D., and Vunjak-Novakovic, G (1997b) Analysis of liquid axial dispersion in an internal loop gas-lift bioreactor for beer fermentation with immobilised yeast cells, in M. Olazar and M.J.San Jose (eds.), *Proc. II European Conference on FLUIDIZATION*, The University of the Basque Country Press Service, Bilbao, pp. 627-635

Nedovic, V.A. (1999a) Immobilised cell systems in beer fermentation, *Monograph (in Serbian)*, Andrejević Foundation, Belgrade.

Nedovic, V.A., Trifunovic, O., Pesic, R., Leskosek-Cukalovic, I., Bugarski, B. (1999b) Production of microbeads containing immobilised yeast cells for continuous beer fermentation by electrostatic droplet generation, in G. Skjåk-Bræk and D, Poncelet (eds.), *Proc. Int. Workshop Bioencapsulation VIII: "Recent Progress in Research and Technology"*, Trondheim, Norway, P-4, pp. 1-5

Nedovic, V.A., Durieux, A., Van Nedervelde, L., Rosseels, P., Vandegans, J., Plaisant, A.-M., and Simon, J.-P. (2000a) Continuous cider fermentation with co-immobilised yeast and *Leuconostoc oenos* cells, *Enz. Microb. Technol. 26*, 834-839.

Nedovic, V., Obradovic, B., Leskosek-Cukalovic, I., Trifunovic, O., Pesic, R., and Bugarski, B. (2000b) Electrostatic generation of alginate microbeads loaded with brewing yeast, Submitted to *Proc. Biochem.*

Obradovic, B., Dudukovic, A., Vunjak-Novakovic, G. (1994) Local and overall mixing characteristics of the gas-liquid-solid air-lift reactor, *Ind. Eng. Chem. Res. 33*, 698-702.

Onaka, T., Nakanishi, K., Inoue, T., and Kubo, S. (1985) Beer brewing with immobilised yeast, *Bio/Technology 3*, 467-470.

Pajunen E., Gronqvist A. and Ranta B. (1991) Immobilised yest reactor application in continuous secondary fermentation in industrial scale operation, in *Proc. 23rd Congr. Eur. Brew. Conv.*, Lisbon, ILR Press, Oxford, pp. 361-368

Pajunen, E. (1996) Immobilised yeast lager beer maturation: DEAE-cellulose at Sinebrychoff, in *Immobilised yeast applications in the brewing industry, Monograph XXIV*, Verlag Hans Carl Getranke-Fachverlag, Nurnberg, pp. 24-34

Pardonova, B., Polednikova, M., Sedova, H., Kahler, M., and Ludvik, J. (1982) Biokatalysator für die bierherstellung, *Brauwissenschaft 35*, 254-258.

Pilkington H., Margaritis A., Mensour N.A., and Russell I. (1998) Fundamentals of immobilised yeast cells for continuous beer fermentation: A review, *J. Inst. Brew. 104*, 19-31.

Pilkington, P.H., Margaritis, A., Mensour, N.A., Sobczak, J., Hancock, I., Heggart, H., and Russell, I. (1999) The use of kappa-carrageenan gel to immobilize lager brewing yeast, in G. Skjåk-Bræk and D. Poncelet (eds.), *Proc. Int. Workshop Bioencapsulation VIII: "Recent Progress in Research and Technology"*, Trondheim, Norway, P-17, pp. 1-4.

Poncelet, D., Dulieu, C., and Neufeld, R.J. (1997) Fundamentals of the dispersion into encapsulation technology, in F. Godia and D. Poncelet (eds.) *Proc. Int. Workshop Bioencapsulation VI: "From fundamentals to industrial applications"*, Barcelona, Spain, T1.1, pp. 1-4

Prüsse, U., Bruske, F., Breford, J., and Vorlop, K.-D. (1998) Improvement of the jet cutting method for the preparation of spherical particles from viscous polymer solutions, *Chem. Eng. Technol. 21*, 153-157.

Prüsse, U., Dalluhn, J., Sonnenburg, J., Breford, J., and Vorlop, K.-D. (1999) Scale-up strategies for bead production by jet cutting, in G. Skjåk-Bræk and D. Poncelet (eds.), *Proc. Int. Workshop Bioencapsulation VIII: "Recent Progress in Research and Technology"*, Trondheim, Norway, O-6, pp. 1-4

Scott, J.A. and O'Reilly, A.M. (1995) Use of a flexible sponge matrix to immobilize yeast for beer fermentation, *J. Am. Soc. Brew. Chem. 53*, 67-71.

Shindo S and Kamimura, M. (1990) Immobilisation of yeast with hollow PVA gel beads, *J. Ferm. Bioeng. 70*, 232-234.

Shindo, S., Sahara H., and Koshino S. (1994a) Suppression of α-acetolactate formation in brewing with immobilised yeast, *J. Inst. Brew. 100*, 69-72.

Shindo S., Sahara H., Watanabe N., and Koshino S. (1994b) Main fermentation with immobilised yeast using fluidised-bed reactor, in *Proc. 23rd Conv. Inst. Brew.*, Sydney, pp. 109-113

Stratton, M., Campbell, S., and Banks, D. (1994) Process developments in the continuous fermentation of beer, in *Proc. 23rd Conv. Inst. Brew.*, Sydney, pp. 96-100

Tata, M., Bower, P., Bromberg, S., Duncombe, D., Fehring, J., Lau, V., Ryder, D., and Stassi, P. (1999) Immobilised yeast bioreactor systems for continuous beer fermentation, *Biotechnol. Prog. 15*, 105-113.

Taidi, B. (1995) Beer production with immobilised yeast, in *Abstracts of the 17th ISSY: "Yeast Growth and Differentiation: Biotechnological, Biochemical and Genetic Aspects"*, Edinburgh, UK, O21

Umemoto S., Mitani Y., and Shinotsuka K. (1998) Primary fermentation with immobilised yeast in a fluidised bed reactor, *MBAA Tech. Quart. 35(2)*, 58-61.

Van de Winkel, L. (1995) Design and optimisation of multipurpose immobilised yeast bioreactor system for brewery fermentations, *Cerevisia 20*, 77-80.

Vunjak-Novakovic, G., Jovanovic, G., Kundakovic, Lj., and Obradovic, B. (1992) Flow regimes and liquid mixing in a draft tube gas-liquid-solid fluidised bed, *Chem. Eng. Sci. 47*, 3451-3458.

Vunjak-Novakovic G., Vunjak N., Kundakovic Lj., Obradovic B., Nedovic V., Sajc L., and Bugarski B. (1998) Air-Lift Reactors: Research and Potential Applications, *Trends Chem. Eng. 5*, 159-172.

White, F.H. and Portno, A.D. (1978) Continuous fermentation by immobilised brewers yeast, *J. Inst. Brew. 84*, 228-230.

Wittlich, P., Jahnz, U., Prüsse, U, and Vorlop, K.-D. (1999-2000) Jet cutting method as a new process for the production of spherical beads from highly viscous polymer solution, in M. Hofman (ed.), *ECB9 CD ROM Proceedings*, Branche Belge de la Societe de Chimie Industrielle, ECB9/2762

Yamauchi Y., Kashihara T, Murayama, H., Nagara, A., Okamoto, T., and Mawatari, M. (1994a) Scaleup of immobilised yeast bioreactor for continuous fermentation of beer, *MBAA Tech. Quart. 31*, 90-94.

Yamauchi Y., Okamoto, T., M., Murayama, H., Nagara, A., Kashihara T, and Nakanishi, K. (1994b) Beer brewing using an immobilised yeast biorector, *J. Ferm. Bioeng. 78*, 443-449.

Yamauchi Y. and Kashihara T. (1996) Kirin immobilised system, in *Immobilised yeast applications in the brewing industry, Monograph XXIV*, Verlag Hans Carl Getranke-Fachverlag, Nurnberg, pp. 99-117

Yokotsuka, K., Yajima, M., and Matsudo, T. (1997) Production of bottle-fermented sparkling wine using yeast immobilised in double-layer gel beads or strands, *Am. J. Enol. Vitic. 48*, 471-481.

NEW MATRICES AND BIOENCAPSULATION PROCESSES

ULRICH JAHNZ[1], PETER WITTLICH[2], ULF PRÜSSE[2] AND KLAUS-DIETER VORLOP[2]
[1]geniaLab® BioTechnologie - Produkte und Dienstleistungen GmbH
Bundesallee 50, D-38116 Braunschweig, ulrich.jahnz@geniaLab.de
[2]FAL - Federal Agricultural Research Centre,
Institute of Technology and Biosystems Engineering
Bundesallee 50, D-38116 Braunschweig, klaus.vorlop@fal.de

Summary

A PVA-matrix is presented which is capable of gelating at room temperature resulting in lens-shaped particles (LentiKats®). Immobilisation of biocatalysts in LentiKats® is possible without significant loss of biological activity. The hydrogels are long term mechanically and chemically stable and show hardly any biodegradability.

Using the new JetCutter method uniform and monodisperse beads can be generated from high viscous fluids at large throughput. The technique is suited for technical and industrial scale.

1. Introduction

Bioencapsulation describes the process of immobilising biological catalysts by enclosing them in a stable matrix. Those biocatalysts can either be growing, resting or dead cells, purified enzymes or even enzymes from crude fermentation broths. As encapsulation always is connected to additional costs one has to show care in finding the best immobilisation technique. i.e. the technique that combines lowest costs and highest efficiency for the given application.

1.1. TECHNIQUES FOR THE IMMOBILISATION PROCESS

As described in literature immobilisation offers the possibility of increasing the efficiency of industrial biotechnological processes: Bringing microscopic particles to macroscopic structures by encapsulation often helps to protect the biocatalyst from contamination and thus allows work under non-sterile conditions. Moreover, considerably higher productivity is possible due to increased concentration of catalytic activity. Easy retention of cells or enzymes in continuously run processes is facilitated.

M. Hofman and P. Thonart (eds.), Engineering and Manufacturing for Biotechnology, 293–307.

Different immobilisation methods are demonstrated in figure 1. From the variety of available techniques, these widely used are on the one hand the adsorption and on the other hand the encapsulation of biocatalysts (Cougghlan and Kierstan, 1987; Vorlop and Klein 1985; Kennedy and Medo, 1990). Encapsulation is further classified into microencapsulation and entrapment. In contrast to adsorption, the method of encapsulation offers better protection of the biocatalyst what an important factor when immobilising sensitive cells. Here, of course, the used polymers should guarantee lowest toxicity against biocatalysts while having sufficient mechanical, chemical, and biological stability.

When enzymes have to be entrapped, cut-off of the applied immobilisation material has to be taken into consideration. Often enzymes have to be cross-linked or bound to a carrier for enlargement before they can be encapsulated successfully.

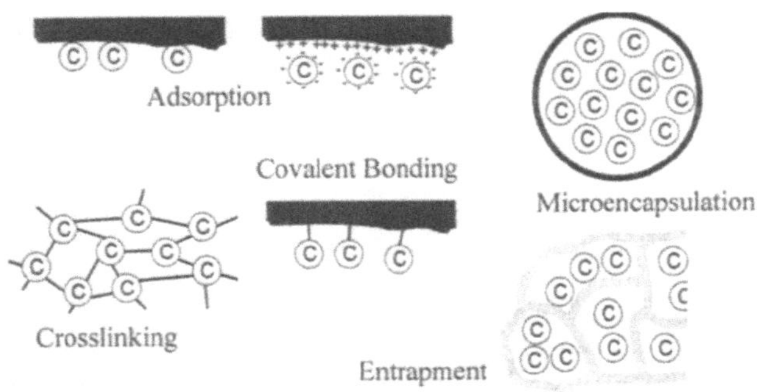

Figure 1. Immobilisation methods

1.2. SHORT OVERVIEW OF SUITABLE MATERIALS FOR ENCAPSULATION

The material used most often for immobilisation is the naturally occurring polymer alginate that can easily be solidified by means of ionotrophic gelation. Gel formation by temperature change can be used for agarose or gelatine among others. Most methods that apply biopolymers offer gentle conditions for the biocatalyst.

An increased stability and lower biodegradability of polymers can be achieved when employing synthetic gels (Leenen *et al.*, 1996; Muscat *et al.*, 1996). These can consist of gels based on chemically bound polymers like polyurethanes or acrylate-co-polymers, or on gels formed by means of hydrogen bonds like polyvinylalcohol hydrogels.

1.3. SHAPES OF PARTICLES WITH IMMOBILISED BIOCATALYSTS

The size and the shape of the immobilised biocatalyst both have a strong influence on its properties of stability, diffusion, and retention in the production process.

Spherical shapes have been used for long time since many immobilisation techniques result in beads. Furthermore this allows diffusion effects to easily be described with mathematical models which help to understand and optimise the overall process.

If beads are too large in diameter, the biocatalysts in the core of the bead probably will suffer from diffusional limitations that will result in sub-optimal specific activity of the immobilised biocatalyst. Only the outer shell of large beads will be catalytically active. Moreover large particles are susceptible to a ready deterioration caused by stirring facilities and other mechanical charges when employed in stirred reactors. In contrast too small beads cause a higher drop of pressure and often tend to clog outlet lines in continuous processes and in general evoke problems when they have to be retained.

In fact, very often the method that is applied to immobilise the catalyst, i.e. the kind of polymer and the machines used for the bead production, determines the size of the immobilised biocatalyst. A stable matrix means a high content of polymer with high viscosity in the polymer solution. Since it was not possible to work with high viscous solutions in the past, often the stability of beads was not as good as required.

Many more or less useful approaches where made to overcome the problem of producing particles from different materials. For example Tanaka *et al.* (1996) cut large blocks of polymer with encapsulated biocatalyst into cubes of about 3 mm. Apart from diffusional problem this could result in insufficient mechanical stability especially at the edges of the cubes. Use of particles was completely circumvented by application of biocatalysts entrapped in a layer that is used to coat surfaces within the reactor, e.g. inside a static mixer. This technique unfortunately results in comparatively small surface of the immobilisation layer that is connected to low specific activity. The same is true for polymers shaped like filaments or cells and enzymes enclosed in membrane systems. The latter cause even more problems due to membrane fouling effects.

Of course, the requirements for an immobilised particle depend on the type of reactor to be used. The material in a stirred reactor has to be more stable and withstand especially attrition than that used for a packed bed, but often the reactor hardware cannot be changed since it is already installed in a production line.

Finding the optimal method for immobilising biological matter for a special application requires two decisions to be made: Selecting the suitable matrix, i.e. the polymer or material for encapsulation, and finding the immobilisation device to create appropriate size, shape, and amount of particles.

2. Techniques for bead production

Several techniques have been developed to produce beads from viscous fluids. The following aspects have to be considered when evaluating an immobilisation apparatus:

- What is the range of *viscosity* that can be handled? Many polymers dissolved in water show high viscosity even at low concentrations.
- Have the beads produced a reasonable *diameter* and how narrow is the *distribution of bead diameters*? To have homogenous conditions throughout the reactor one has to apply beads as identical to each other as possible. This is especially true when

the bead is processed further, e.g. by a drying step, since beads varying in diameter will be affected differently.

- Does the process require *sterile conditions* during bead production and how can this be achieved?
- Which *production scale* of beads is possible? Immobilisation is of particular interest to enhance profitability of biotechnical processes for production of bulk-chemicals. This demands the availability of immobilised material in industrial scale. It is important from the outset to consider that a method working under laboratory conditions will have to be upscaled at some point.

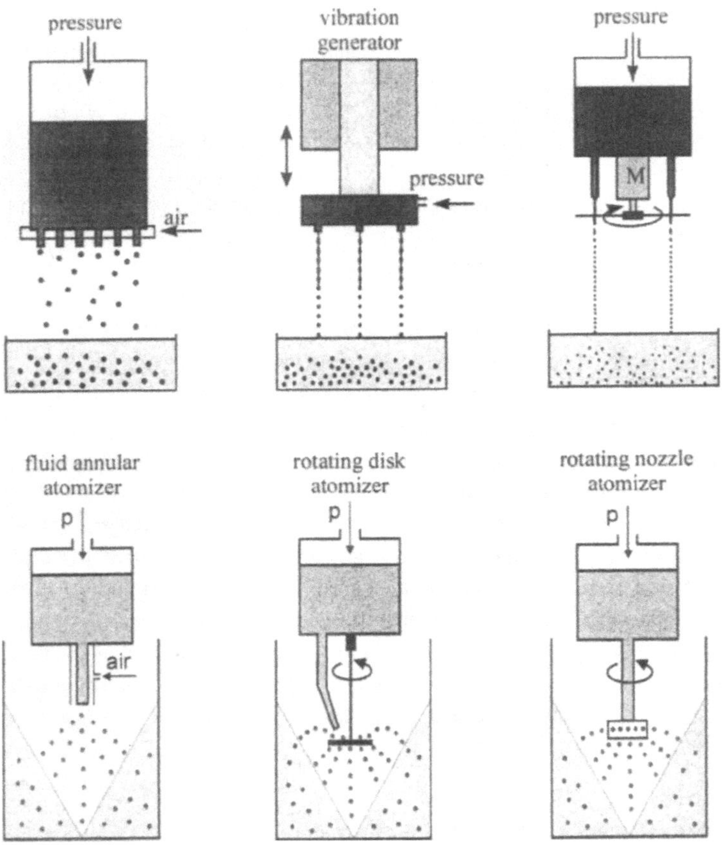

Figure 2. Devices for bead/particle-generation (blow-off, vibrator, JetCutter, different atomisers).

When a liquid flows slowly out of a vertical nozzle, the surface tension causes the formation of an orb at the tip of the nozzle until it is released due to gravitation. Since using smaller nozzles, e.g. cannulas, has only little influence on the size of the bead the detachment of the drop of liquid has to be forced to make smaller beads. Three main mechanisms are recently in use:

- a laminar flow of air surrounding the nozzle (blow off),
- a defined oscillation to break up a jet of liquid, or
- a real cutting of the liquid.

The principle of these methods is shown in the upper row of figure 2. In addition, this figure presents various atomiser principles. Atomisers were developed for large scale production of particles but they in general produce particles with a rather broad particle size distribution.

2.1. BLOW-OFF-DEVICES

As mentioned above, one possibility to release small beads is to add the force of a continuous flow of air to gravitation (Vorlop and Klein, 1983). This can be achieved by sheathing the cannula for the liquid with a larger cannula that is connected to compressed air. By applying a laminar flow of air and regulating the mass flow of the liquid, the size of the produced beads can be controlled. The described technique demands only little technical equipment but has drawbacks: the throughput is very limited even with multi-nozzle systems, and only fluids with low viscosity can be handled. Due to the outer flow of air the processed material tends to dry at the tip of the cannula and thus causes problems in long-term operation. However, the method is appropriate for most initial experiments under lab conditions.

2.2. VIBRATION

The formation of single droplets from a continuous flow of liquid can also be achieved by applying a vibration either to the outlet nozzle or to the liquid itself. The flowing liquid expresses the shape of the obtruded vibration and is laced at the troughs and thus forms individual droplets (Brandenberger and Widmer, 1997). By altering the frequency of the vibration, the size of the resulting particles can be controlled. The distribution of particle sizes is very uniform and higher throughput is often achieved by using multiple nozzles in parallel. Devices based on this principle have been commercialised by several companies. However, beads with diameters below 1 mm can only be made from fluids with viscosity of up to 300 to 500 mPa· that limits the use of the technique.

2.3. ATOMIZERS

Atomisers use high energy dissipation to form droplets from fluids. The force can either be applied by using a nozzle and compressed air, or by spraying the fluid onto a rotating disk or using similar techniques. Although these devices have a large throughput the bead diameter is only insufficiently controlled and the distribution of particle size is accordingly broad. In addition atomisers only work with low viscous fluids and thus cannot be applied for every task.

2.4. JETCUTTING

All techniques discussed so far lack the possibility of processing fluids with high viscosity and most also suffer from problems when they have to be scaled up. To counteract these problems we invented the method of JetCutting that works with fluid viscosity up to several Pa· (Vorlop and Breford, 1994).

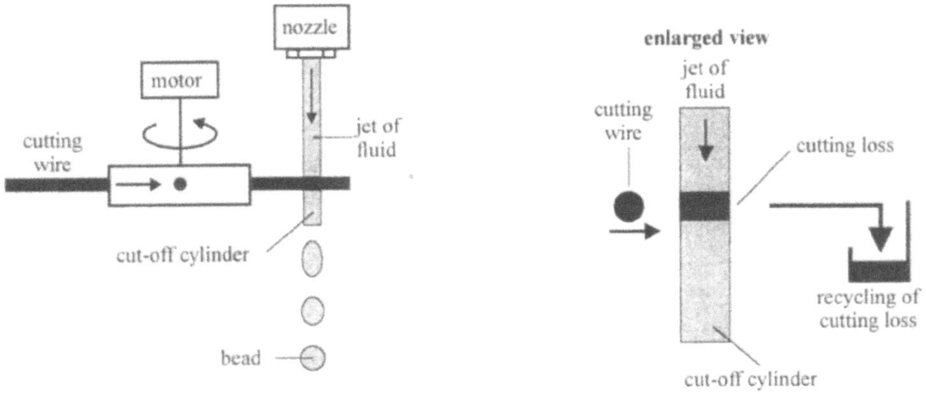

Figure 3. Schematic view of the JetCutting process.

A continuous jet of fluid is pressed out of a nozzle at high speed and cut into cylindrical segments by means of a fast moving cutting tool. This cutting tool is most often realised as a group of thin wires connected to a rotating device (see figure 3). The cylindrical segments produced by the cutting event form beads due to surface tension while they continue to fall down towards hardening.

The amount of liquid which is slung away during the cutting process depends on several parameters: On the one hand the wire should be kept as thin as possible, on the other hand it is advantageous to use rather a thin jet of fluid which is cut into longer cylinders than to work with a thicker jet divided into short segments.

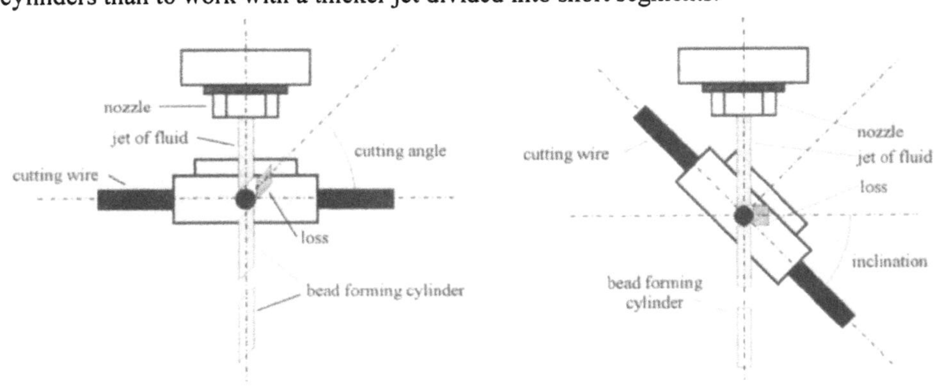

Figure 4. Comparison of horizontal and inclined cutting.

As can be seen from figure 4 it is also advisable to incline the plane of cutting in ratio to the direction of the jet, since both elements are subject to a continuous motion (Prüße *et al.*, 1998).

Only with an inclination are the resulting fragments of the jet indeed true cylinders without fraying ends at the top and bottom. These fraying tips tend to spray away and thus increase the amount of recyclable loss.

When using wires of 50 µm in diameter and setting the optimal inclination the amount of liquid that is slung away is drastically reduced and is below 2% of the processed material.

Since the velocity of the jet of fluid is constant when using a pulsation free pump, and the cutting tool rotates electronically controlled with constant speed, all segments are the same size and resulting beads are monodisperse (see figure 5). The lower limit for beads is approx. 150 µm at the moment, the upper limit is about 2.5 to 3 mm (depending on the properties of the fluid).

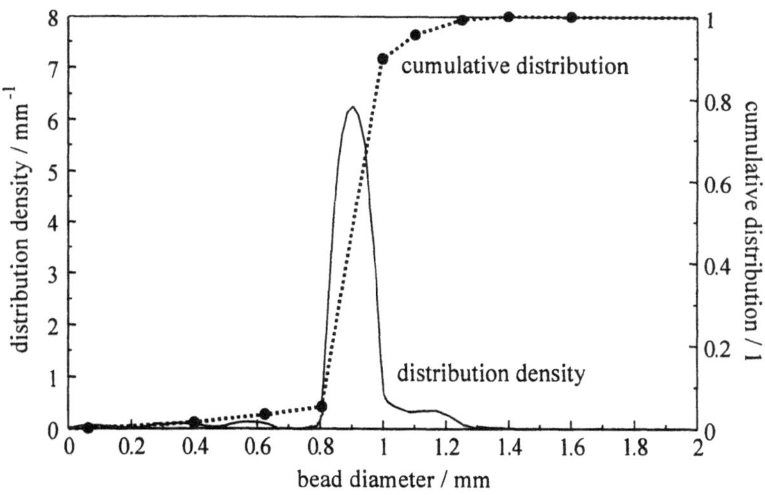

Figure 5. Size distribution for beads from 6% (w/w) PVA-solution.

The production rate by JetCutting depends on the flow rate of the liquid and the speed of the cutting tool. The upper limit at the moment is 15,000 beads per second and nozzle, but some 25,000 beads will be possible in the future. Table 1 lists the resulting throughput, i.e. the bead production rate, depending on the bead diameter and the frequency of bead generation.

Depending on the material, special precautions have to be taken to allow a smooth transition into the bath were the beads are collected. Due to the enormous number of beads produced and the high speed of the beads (up to 30 m/s), just taking a stirred bath is not enough for many applications. In these cases, equipment for giving the assimilating fluid a flow like in a geyser, a maelstrom or in a wide gutter have to be used.

Table 1. Bead production rates in kg/(h·nozzle) for fluids with a density of 1 kg/L

	Throughput in kg/(h·nozzle)		
Ø bead / mm	5000 Hz*)	10,000 Hz*)	50,000 Hz*)
0.2	0.08	0.15	0.8
0.6	2.04	4.07	20.4
1.0	9.42	18.8	94.2
1.5	31.8	63.6	318
2.0	75.4	151	754
2.5	147	295	1470
3.0	254	509	2540

*) *frequency of bead generation*

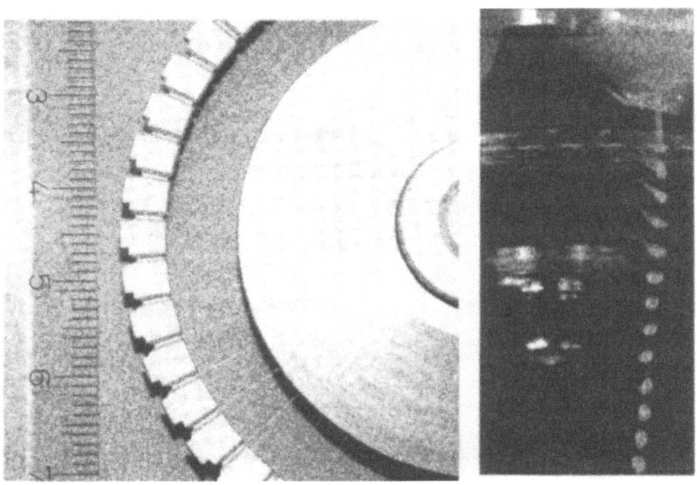

Figure 6. Detailed view of cutting tool (100 µm cutting wire) as available from genialab® high speed motion picture of cutting process.

Materials processed so far include the biopolymers alginate, chitosan, carrageenan, gelatine and synthetic materials like polyvinylalcohol and silicon. Applications were the entrapment of microorganisms or enzyme preparations, or the formulation of fragrances, vitamins, and other ingredients for food and pharmaceutical industries on a large scale. In addition, sol-gel materials for preparation of inorganic carriers, or molten waxes for entrapping pharmaceutically active substances have been tested.

3. Materials for encapsulation

Any material that can be turned into hydrogels by physicochemical means is in principle suitable for the encapsulation of biocatalysts. The formation of the gel can either be due to ion interaction or chemical bonding or can also be facilitated by thermal solidification. The polymers used can either be of natural origin or synthetic materials.

3.1. NATURAL POLYSACCHARIDES FOR IONOTROPIC GELATION

The classical material for encapsulation is the well-known alginate (Kierstan and Bucke, 1977; Vorlop and Klein, 1983; Smidsrød and Skjåk-Bræk, 1990). The major source of alginate is found in the cell walls and the intracellular spaces of brown seaweed.

The use of alginate as an immobilising agent in most applications rests in its ability to form gels that can develop and set at room temperature by interacting with calcium ions.

Due to this ionotrophic gelation, the reaction conditions are mild, and bacteria, eukaryotic microorganisms and also higher cells show excellent survival rates. Excessive research has been done on alginate, but the stability of the resulting gel is often too low when working with powerful growing cells like fungi or when the beads are used in a reactor with high shearing strain. Also, the gels are susceptible to calcium-leaching what will result in deterioration when working with phosphate or citrate buffered media. Moreover, alginate is readily biodegraded when working under non-sterile conditions and is thus not suitable for bulk-use when sterility cannot be ensured.

Other biopolymers have to be mentioned which have been tested for immobilisation purposes. Pectinate also can gelate in the presence of calcium or, depending on the type of pectinate when heated and cooled down, in the presence of sucrose (Berger and Rühlemann, 1998). Carrageenan molecules in solution aggregate to a double helix, which can be stabilised by interaction with potassium or ammonia (Chibata *et al.*, 1987). Hence it can be used with phosphate-buffered media which is advantageous compared to alginate. The mass product chitosan is recovered from deacetylation of chitin and is a polycation. Therefore it forms stable hydrogels in the presence of polyanions like e.g. tripolyphosphate (Vorlop and Klein, 1987).

3.2. SYNTHETIC HYDROGELS BY CHEMICAL REACTION

Synthetic gels can be formed by polymerisation or cross-linking of prepolymers or monomers. The disadvantage of these prepolymers is that the chemical reaction is executed in the presence of the microorganisms which causes significant losses in activity in most cases. To circumvent this drawback, efforts were taken to soften the reaction conditions. The development of polycarbamoyl sulphonate (PCS) as a replacement for polyurethanes (PU) is one example (Muscat *et al.*, 1996). By blocking the toxic isocyanate groups during the polymerisation, a less toxic prepolymer can be obtained.

In the industrially employed PEGASUS-process, different polyethylene glycol (PEG) prepolymers are used to immobilise nitrifying sludge (Emori *et al.*, 1996). The cells are protected from the cross-linking reagent during immobilisation by a macromolecular coagulant. The obtained PEG pellets consist of 3 mm blocks and show good mechanical properties. However, due to large dimension, the diffusional properties are not optimal and final specific activity of immobilised biomass is limited.

3.3. HYDROGELS FROM POLYVINYL ALCOHOL

Polyvinyl alcohols (PVA) are hydrophilic polymers whose aqueous solutions are capable of gelling on their own when stored for prolonged time. Hydrogen bonds between hydroxyl groups of neighbouring polymer chains form a non-covalent network. However, this process takes days and the gels obtained at temperatures above 0°C usually are rather weak and thus not suitable.

A different effect is obtained when PVA solution is subjected to freezing (Lozinsky and Plieva, 1998; Lozinsky, 1998). Due to phase separation during the freezing process the formation of hydrogen bonds is enhanced and the resulting hydrogel is significantly stronger. Hydrogels from PVA by this cryogelation are mechanically very stable and show more or less no abrasion when employed in stirred reactors. In gelated form PVA is hardly biodegradable and thus can be used when working under non-sterile conditions. Chemically the hydrogels can be utilised with any physiological compound since they do not dissolve. Merely by heating to above 60°C the hydrogels can be melted.

Parameters influencing the gel-strength are the degree of deacetylation of the used PVA, the chain length of the polymer, its concentration in the solution, and the rate of thawing. Usually concentrations of 7 to 15 % (w/w) of polymer with a molecular weight of about 80 to 100 kDa are used. The slower the thawing process, the more rigid the resulting gel will be. Especially the temperature range of about -15 to +5°C is crucial for a satisfactory stability and the emerging hydrogel should be thawed with only a few degrees per hour in this scope.

An alternative to the slow-thawing method for the reinforcement of the hydrogel is the multiple freezing-thawing, which is scarcely used since it consumes significant amounts of energy. In both cases the freezing often causes significant loss of microbial activity when living cells are to be immobilised.

Another known means of gelating a PVA-solution is the dripping into boric acid, but the resulting hydrogel is less stable and rather brittle.

4. LentiKats®

As mentioned above the method of cryogelation of PVA often inflicts a loss in biological activity. To counteract the stress we developed a method that allows the gelation of PVA-solutions at room temperature by means of controlled partial drying. Due to their characteristic lenticular shape the resulting particles are named LentiKats®.

4.1. DESCRIPTION OF PROPERTIES

Ready-to-use LentiKat®Liquid solution is mixed with the biocatalyst, i.e. cells or enzyme preparation, and small droplets are floored on a suitable surface. By exposing these droplets to air, the water starts to evaporate and thus leads to enhanced formation of hydrogen bonds. Once about 70% of the polymer-biocatalyst-solution have been removed, the hydrogel is stable enough and can be re-swollen in a stabilising solution before the ready LentiKats® are employed.

The particles formed by this procedure combine the advantages of both large and small beads: On the one hand side they measure about 3 to 4 mm in diameter and can be retained by established sieve technology or rapidly by settling. On the other hand they are only 200 to 400 µm thick and thus cause hardly any diffusional limitations to the enclosed biocatalysts.

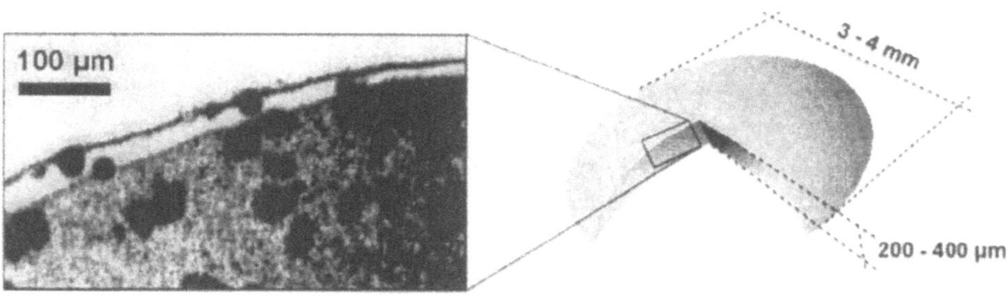

Figure 7. Schematic view and image of a LentiKat®.

Since LentiKats® are based on polyvinyl alcohol they have the same properties as discussed for the PVA cryogels. But based on the very mild encapsulation conditions, very high survival rates can be obtained. The complete immobilisation procedure takes place in less than one hour and also the stress of the partial drying is tolerable for most organisms as shown below.

4.2. PRODUCTION DEVICES FOR LAB- AND TECHNICAL SCALE

For first trials it is sufficient to form the droplets by a simple lab-syringe on a standard petri dish. Once the results show that the method is in principle compatible to the tested biocatalyst, it is advisable to change the method of preparing the LentiKats® to keep identical conditions for different experiments. The mechanical properties of the produced LentiKats® and the survival rates of entrapped microorganisms strongly depend on the time of gelation and the uniformity in size of the droplets. Using a syringe creates droplets differing in size and, even more crucial, this successive approach leads to unequally distributed times of gelation. Moreover the amount of particles which can be manually produced is very limited and even for lab-scale applications it is a tedious work.

As an improvement a new device was developed to produce more than 400 identical droplets simultaneously in one step. Based on the well-known principle of printing technology a special printer head was designed for multiple transfer of equal amounts of PVA solution to a special surface (figure 8). Handling the printer is a cyclic sequence of the following steps: Loading each tip with the same amount of polymer solution by dripping into, replacing the stock of polymer solution with a standard 145-mm-petri-dish, lowering the printer head and flooring the droplets on the petri-dish and finally exposing the fresh droplets to gelation conditions.

Figure 8. LentiKat®Printer(right), detailed view of the printer head (under) and continuous production unit (left) (photos right and under by courtesy of geniaLab®; picture on left by courtesy of BCS mega).

After appropriate time gelation is abruptly terminated by flooding with a stabilising solution. The described procedure guarantees that all particles are generated in the same way and no delays occur. This is especially important when the data of consecutively run experiments have to be compared and differences caused by the immobilisation process itself have to be precluded. Up to approx. 100 grams of identically immobilised material can be prepared in acceptable time for lab-scale applications under sterile or non-sterile conditions.

For the production of technical and industrial amounts of LentiKats®, a conveyor belt system was designed and built. Here a set of parallel nozzles drip the solution onto a continuously moving belt which runs in a drying chamber. After passing through the chamber the ready LentiKats® are scraped off and can be employed. The current machine produces technical amounts but can easily be scaled up since the principle of dripping, drying and scraping off is well established in various parts of the food industry and appropriate machinery is available for all scales.

4.3. EXAMPLES FOR APPLICATIONS OF LENTIKATS®

The benefiting influence of avoiding extreme temperature conditions in contrast to the cryogelated PVA-hydrogels was shown by immobilising a mixed culture of sensitive nitrifying bacteria (*Nitrosomonas europaea* and *Nitrobacter winogradskyi*). The initial activity of biomass could be raised from below 1 % (entrapped by freezing-thawing method, -20°C) to 75 % for immobilisation in LentiKats® compared to suspended cells of nitrifiers.

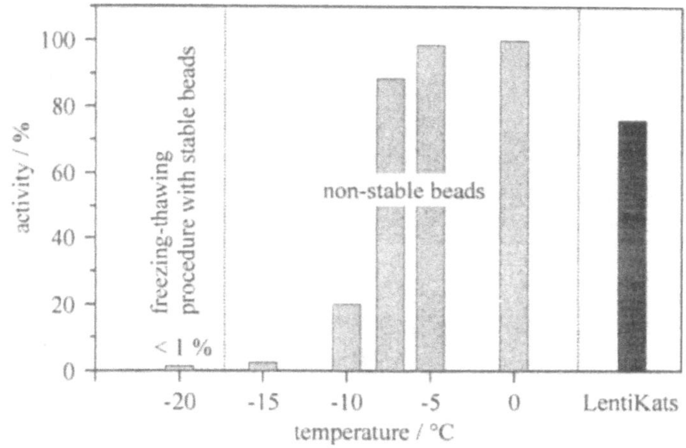

Figure 9. Comparison of PVA beads made by freezing-thawing procedure and LentiKats®; beads made in the temperature range of –15 to 0°C are not stable but were made to investigate inactivation effects.

As another example of successful immobilisation in LentiKats®, the anaerobic bacterium *Clostridium butyricum* producing 1,3-propanediol from glycerol was encapsulated. The optimal initial biomass concentration was $2\cdot10^7$ cells/ml regarding the final hydrogel. LentiKats® with this concentration showed the highest activity.

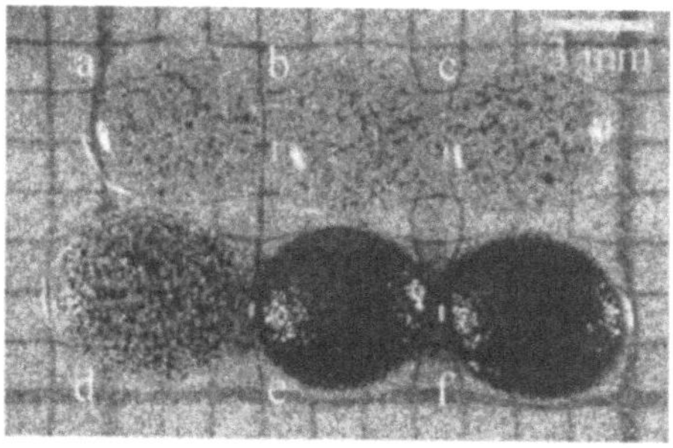

Figure 10. Determination of optimal biomass content; initial concentrations in cells per ml were $a=1\cdot10^5$, $b=5\cdot10^5$, $c=2\cdot10^6$, $d=1\cdot10^7$, $e=5\cdot10^8$, and $f=5\cdot10^8$, respectively (photographs were taken after incubation until steady state was reached, cells are stained)

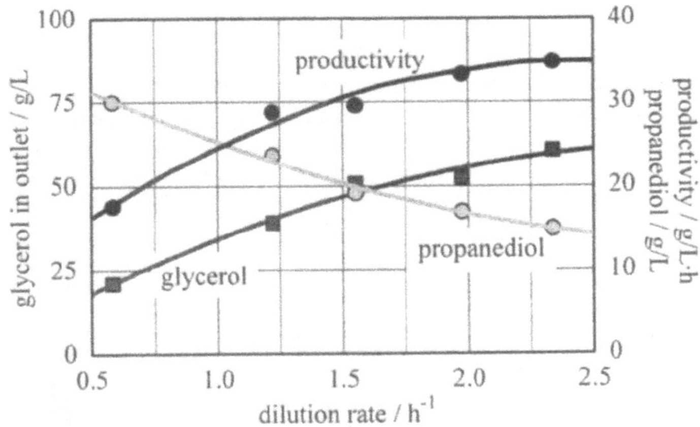

Figure 11. Continuous operation with C. butyricum entrapped in LentiKats® (inlet: approx. 90 g/L raw glycerol, 20% (v/v) immobilised catalyst, 38°C, pH 7.8, non-sterile conditions).

After 10 weeks continuous operation under non-sterile conditions in a stirred 0.5-L-fermenter the LentiKats® showed no breakup. A productivity of 35 g/(lh) propanediol from raw glycerol was reached. This is the highest productivity known for this process so far (Wittlich *et al.*, 2000).

5. Conclusions

In this paper we presented two techniques, both suitable for the encapsulation of biological catalysts. The *JetCutting* method produces monodisperse beads from highly viscous fluids. Due to its enormous production capacity it is applicable not only for lab- and technical scale but is a new device for industrial production of beads.

LentiKats® provide a new matrix for immobilisation, which has extraordinary properties and allows encapsulation at low costs. Due to mild conditions, e.g. no harsh chemicals and the use of room-temperature for gelation, even sensitive organisms show high rates of survival as was shown with nitrifying bacteria. The improved shape of LentiKats® reduces diffusion limitations and leads to very high values for specific activity and productivity.

References

Berger, R., Rühlemann, I. (1988) Stable ionotrophic gel for cell immobilisation using high molecular weight pectic acid, *Acta Biotechnol.*, 8, 401-405.

Brandenberger, H., Widmer, F. (1997) A new multinozzle encapsulation/immobilisation system, in Godia, F., Poncelet, D. (eds.) *Proceedings of the International Workshop Bioencapsulation VI*, Barcelona, poster 9.

Chibata, I., Tosa, T., Sato, T., Takata, I. (1987) Immobilisation of cells in carrageenan, *Methods Enzymol.*, 135, 189-198.

Cougghlan, M.P., Kierstan M.P.J. (1987) Preparation and applications of immobilised microorganisms: a survey of recent reports, *J. Microbiol. Methods*, 8, 51-90.

Emori, H., Mikawa, K., Hamaya, M., Yamaguche, T., Tanaka, K., Takeshima, T. (1996) PEGASUS Innovative biological nitrogen removal process using entrapped nitrifiers, in Wijffels, R.H., Buitelaar, R.M., Bucke, C. Tramper, J. (eds.) *Immobilised Cells: Basics and Applications*, Elsevier Sciences B.V.

Kennedy, J.F., Melo, E.H.M (1990) Immobilised enzymes and cells, *Chem. Eng. Prog.*, 86, 81-89.

Kierstan, M., Bucke, C. (1977) The immobilisation of microbial cells, subcellular organelles, and enzymes in calcium alginate gels, *Biotechnol Bioeng* 19 (3), 387-97.

Leenen, E.J.T.M., Dos Santos, V.A.P.M., Tramper J., Wijffels, R.H. (1996) Characteristics and selection criteria of support materials for immobilisation of nitrifying bacteria, in Wijffels *et al.* (eds.) *Immobilised Cells: Basics and Applications*, Elsevier Sciences B.V., 205-212.

Lozinsky, V.I. (1998) Cryotropic gelation of poly(vinyl alcohol), *Russian Chemical Reviews*, English Edition., 67-7, 573-586.

Lozinsky, V.I., Plieva, F.M. (1998) Poly(vinyl alcohol) cryogels employed as matrices for cell immobilisation. D. Overview of recent research and developments, *Enzyme Microbiol Technol.*, 23, 227-242.

Muscat, A., Prüße, U., Vorlop, K.-D. (1996) Stable support materials for the immobilisation of viable cells, in Wijffels, R.H., Buitelaar, R.M., Bucke, C. Tramper, J. (eds.) *Immobilised Cells: Basics and Applications*, Elsevier Sciences B.V., 55-61.

Prüße, U., Bruske, F., Breford, J. and Vorlop K.-D. (1998). Improvement of the Jet Cutting method for the production of spherical particles from viscous polymer solutions, *Chem. Eng. Technol.* 21, 153-157.

Smidsrød, V., Skjåk-Bræk, G. (1990) Alginate as immobilisation matrix for cells, *Trends Biotechnol.*, 8, 71-78.

Tanaka, K., Sumino, T., Nakamura, H., Ogasawara, T. and Emori, H. (1996) Application of nitrification by cells immobilised in polyethylene glycol, in Wijffels, R.H. *et al.* (eds.) *Immobilised Cells: Basics and Applications*, Elsevier Sciences, 622-632.

Vorlop, K.-D., Klein, J. (1983) New developments in the field of cell immobilisation: formation of biocatalysts by ionotrophic gelation, in Lafferty, R.M. (editor) *Enzyme Technology*, Springer Berlin, 219-235.

Vorlop, K.-D., Breford, J. (1994). German Patent *DE 4424998.*

Vorlop, K.-D., Klein, J. (1985) Immobilisation techniques, in Moo-Young (editor) *Cells in Comprehensive Biotechnology 2,* Pergamon Press, 203-224.

Vorlop, K.-D., Klein, J. (1987) Entrapment of microbial cells in chitosan, *Methods Enzymol.*, 135, 259-268.

Wittlich, P., Schlieker, M., Willke, T., Vorlop, K.-D. (2000) Leistungssteigerung biotechnischer Prozesse durch neuartige Immobilisierungsmethoden für Biokatalysatoren am Beispiel der 1,3-Propandiol-

PART VII
DOWNSTREAM PROCESSING

INDUSTRIAL DOWNSTREAM PROCESSING

MADS LAUSTSEN
Novo Nordisk A/S

1. Introduction.

Industrial downstream challenges and solutions are highly dependent on the application of the product in question. The downstream processing field does cover both high price and high purity products like pharmaceuticals and fine chemicals, as it covers low cost production of bulk proteins like detergent enzymes.

To present a clear picture of the state of the art within industrial downstream processing, in which all relevant techniques and problem solutions have been discussed is therefore not an easy task. Further, where focus in public research is publication of ideas and results, knowledge and developed techniques in the industries will normally be held confidential where possible.

Indeed industry is often quite reluctant to disclose details regarding production processes and strategies as in many cases small differences in production cost or in development speed can make the difference between success or failure.

A drawback of these confidentiality issues is at times that alignment of public research with industrial needs is difficult.

This paper discusses some of the more important aspects of current downstream processing in industrial scale and future challenges for the industries. As a special focus some of the main differences between pharmaceutical production and bulk enzyme production are highlighted.

2. General aspects connected to downstream processing.

2.1. INTELLECTUAL PROPERTY RIGHTS.

Patent filing has for years been used extensively for protecting products, production organisms, potential protein engineering sites together with cloning and expression tools. Industrial fermentation and purification processes have previously been less protected as processes and developed techniques have been held confidential. However, in the later years a larger number of patent applications have also been filed within the

M. Hofman and P. Thonart (eds.), Engineering and Manufacturing for Biotechnology, 311–324.

pharmaceutical downstream field. Also patent filing related to bulk enzyme production has started increasing. A major reason for this is the importance of freedom of operation, as even production processes that could have been running for years potentially can be hit by an issued patent. Infringement of third parts valid patent is something that can be very problematic. For pharmaceuticals it is indeed a must that the process developed is not violating patent rights as such processes are very troublesome to change. For enzyme production customers often want guaranty that the enzyme product and the process it has been produced by do not violate any third parts patent rights.

Patent filing within fermentation and recovery/purification processes is therefore very much done for safe guarding current and future freedom of operation.

As patent filing has been strengthened within the field, one can obtain information about industrial downstream processes and focus by looking into new patent applications from the industries.

2.2. PUBLIC RESEARCH IN DOWNSTREAM PROCESSING.

Research in the downstream field possesses some extra difficulties for universities and other public research centres as relevant research often needs to be performed with industrial relevant process streams - fermentation broth or partially purified product. Such process streams, or even information about them, will often not be handed out from companies. Further, for especially primary separation trials broth needs to be fresh why trials have to be performed close to the production site. Because of these problems it is often easier for many companies to do research and development for process streams of which large quantities of production relevant product is needed, like for primary separation and crystallisation trials.

Much of the public research is and has been dedicated to chromatographic purification - the low volume, high purity area with fine chemical and pharmaceutical purification as the target. There are several good reasons for that. It is obviously of significant importance to the public health that industries are capable of producing safe and pure pharmaceuticals. With increased sensibility of quality analysis, increasing purity demands have followed, these setting higher demands to purification processes. Further, as industrially relevant chromatography trials can be performed in bench scale with a minimum of process liquid requirement, this research is relatively easy to perform for public institutes.

2.3. QUALITY.

Quality issues play a central role in process development, as it does in the final production processes. Where in the health care field quality is controlled through pharmaceutical GMP with strict demands to process validation and process documentation, the enzyme field has other less strict control systems. Production of food and feed enzymes has to comply with Food GMP regulations. Here with a clear focus on ensuring sanitary processing. Food enzyme production under Kosher regulations will further be restricted on raw materials and process conditions. Production of technical enzymes is on the other hand not subjected to these regulations. However, for both food and feed enzyme production and for technical enzyme production the ISO 9000 system is a natural platform for ensuring quality and efficient production. As bulk enzyme production is performed in large scale environmental issues have special focus, why it is likely that many of the companies in this field will adopt the ISO 14000 system as a means of reducing the environmental impact from the processing.

2.4. UPSTREAM PROCESS.

The upstream process, though not being the focus for this paper, does deserve some comments, as it is the starting point for the first part of the downstream process.

Typically, pharmaceutical products are fermented by either micro-organisms (often yeast or E. coli) or by mammalian cells. Tissue extractions still exist, but are being substituted by fermentation processes where possible. Lately production in transgenic animals like goats and pigs has become an interesting alternative to mammalian cell production with significantly improved product expression in the transgenic milk compared to titres in mammalian cell cultures.

Production in transgenic plants might in the future become an attractive alternative to the current fermentation systems for both the pharmaceutical and the enzyme industries. Production of Albumin in Tobacco plants being one example of the potentials in this technology.

In short micro-organisms like yeast and E. coli, can be used for production of smaller and less complex molecules, often with reasonable yield, where the more complex proteins typically need to be produced in mammalian cells or alternatively in transgenic animals.

Besides product concentration differences the different matrix bases for the production systems set some different requirements to the subsequent recovery processes. For recovery of milk from transgenic animals milking and fat removal steps obviously need to be incorporated, potentially together with a casein precipitation step.

Downstream processes based on both mammalian cells and transgenic animals need to be supplemented with a virus inactivation/removal step. Furthermore animals and raw materials derived from animals (e. g. serum) need to be controlled carefully for safeguarding against TSE problems.

Production of pharmaceuticals in micro-organisms can be done as soluble product expressed in e. g. yeast and E. coli or as insoluble inclusion bodies in E. coli. Inclusion body production with the advantage of yielding very good fermentation yield (g/litre level), but with the down side of a complication of the downstream process with a not always easy solid/solid separation of inclusion bodies from the rest of the fermentation solids together with necessary denaturation and renaturation steps.

In the enzyme business micro-organisms are the dominant production method with Bacillus and Aspergillus being the most widely used strains.

As industrial enzymes normally are of microbial origin and as they in most cases are smaller less complex and stable proteins, fermentation titres in this industry are in most cases measured in the multiple gram/litre level.

In this industry focus is on extracellular production systems. However, for a not insignificant part extracellular expression has not been possible, why intracellular products from time to time have to be handled. The production of Glucose Isomerase being a larger scale example of a successful intracellular enzyme product.

3. Pharmaceutical production.

3.1. GENERAL DOWNSTREAM ISSUES.

In pharmaceutical industries the main premises and demands to a downstream process are in overall terms set by relatively low production volume and high quality demands for the product. In details some of the more dominant requirements are:

High product quality with respect to:

- host protein
- DNA
- product aggregation
- virus safety
- micro heterogeneity

General process requirements:

- fast to develop
- good scalability from bench scale to production scale
- easy to operate
- easy to validate

General production premises:

- low to moderate fermentation titres (compared to enzyme production)
- seldom strict demands on yield or cost
- sanitary and validated facility
- validated cleaning

Downstream processes can be characterised by two parts: - recovery and purification. Recovery being the part where the product is separated from the biomass (primary separation) and concentrated by a membrane process or a catcher step where the product is bound to a chromatographic matrix and eluted in high concentration. The recovered product is purified by a process normally consisting of 3-5 chromatographic steps.

3.2. RECOVERY.

3.2.1. Primary separation.

In the pharmaceutical industry centrifugation, filtration and microfiltration are the standard primary separation techniques. Where centrifugation and filtration are the old well functioning and robust techniques microfiltration has gained some foothold the later years as this technique offers closed sanitary processing, this being indeed attractive for pharmaceutical processes. It is therefore likely with the continued development within microfiltration membranes and hardware that more microfiltration processes will be developed in the future.

Aqueous two phase extraction as well as fluidised/expanded bed chromatography have been proposed as alternative techniques for primary separation.

Aqueous two phase extraction was for years expected to have some significant production relevance especially to solving problems regarding extraction of product from homogenised fermentation broth. However, the technique has never gained the expected success. The reasons for this are; large use of chemicals, some of them difficult to get rid of in the wastewater, and the fact that the conventional techniques actually perform quite well and often with less environmental problems.

The expanded bed technique has on the other hand been developed into an alternative to the established centrifugation, filtration and microfiltration processes. This in particular for processes where solid liquid separation problems are difficult to solve with the conventional techniques.

3.2.2. Intracellular products.

For processes based on intracellular expression of soluble product cell opening is needed. Many laboratory techniques exist; however, in production scale almost only pearl milling or homogenisation is used. These techniques are by far the most efficient in an industrial environment, even though they create some downstream difficulties. The breakage does liberate, not only the product but it also generates colloids difficult to remove with traditional solid/liquid techniques. Flocculation is here one technique used to facilitate colloid removal. If a clear supernatant/filtrate is not possible to obtain, it is however, not always crucial even though a rule of thumb for loading on columns teaches that one should ensure loading being free from particulate matter. Successful industrial processes do exist where high colloid containing liquids derived from homogenised broth are loaded onto catcher columns. Using relatively ridged matrixes packed in compressed columns typically solves the channelling problems one could expect when loading with such high particulate matter. Axial compression columns and also the radial flow columns are typical means for solving these problems.

Processes based on inclusion bodies are recovery wise significantly more complicated. Besides the cell opening the inclusion bodies need to be separated from the rest of the fermentation solids. To that controlled centrifugation is applied - decanters well suited for such separations. However, the critical step in this process is dissolution of the inclusion bodies and in particular the renaturation of the product.

These steps require high levels of chemicals for the solubilisation and necessitate quite extensive dilutions for the renaturation. Over time several alternative approaches have been suggested like reversed micelles and electrodialysis, however, so far large chemical use and dilution are still the production realities.

Even though processes based on inclusion bodies recovery wise are more difficult than processes based on fermentations producing soluble product, the high fermentation titres obtainable by this method often makes inclusion body production the preferred choice - production of insulin being an example.

3.2.3. Concentration.

The common approach in pharmaceutical industry is binding of the product obtained in the primary separation to a chromatographic column with subsequent elution in high concentration. Ion exchange columns are often used for this step but also hydrophobic columns and affinity can be applied this early in the process as long necessary care is taken towards hydrophobic fermentation constituents like foam controlling agents.

For larger productions ultrafiltration has become an attractive alternative for protein concentration, as this method is inexpensive and as it in combination with diafiltration can produce a high concentrated product in a well-defined salt concentration well suited for subsequent chromatography.

3.2.4. Precipitation/crystallisation.

In some cases it can be advantageous or even necessary to de-couple the purification process from the recovery. Especially in cases in which prolonged storage of the recovered product is performed prior to purification - e.g. when the purification is performed on a different location than the recovery or when the recovered product is split into several purification batches. A de-coupling of the recovery and purification will also allow for pooling several recovered batches into one batch prior to purification.

Freezing of the recovered product is one way to de-couple recovery from the purification. However, precipitation or crystallisation does offer the advantages of high product stability and product concentration. Especially crystallisation does produce a very pure and stable product very well suited for chromatographic purification. By including crystallisation as the last step in the recovery > 90 % purity can be obtained for the recovered product.

3.3 PURIFICATION.

3.3.1. Chromatographic principles.

The use of HIC and of affinity chromatography has often been complicated from large batch to batch variations in capacity and efficiency, especially if used early in the process as catcher steps. In most cases varying levels of foam control agents carried down through the process causes these problems. However, simple adsorption and/or filtration procedures efficiently solve such problems. For adsorption especially activated carbon and hydrophobic water treatment resins can be applied with success where removal by filtration is enhanced by the use of foam controlling agents with low solubility in the product stream. Control of foam controlling agents therefore opens the way for efficient robust HIC, reverse phase and affinity steps early in the purification trail. Another benefit from such control is often enhanced filtration and membrane processes.

For development of pharmaceutical processes many standard techniques are available for setting up and running chromatographic columns. Equipment set-up and procedures for running efficient loading and elution are therefore not the big challenges.

The set-up of columns and loading/elution principles applied are the relatively simple batch type processes with standard columns and gradient or step elution. The more complex methods like continuous chromatography are not adapted and will most likely not gain a significant implementation, as fast development and simple processing are a necessity.

With use of ion exchange, HIC, gel filtration and some times reverse phase chromatography product quality demands for host protein and DNA are normally achievable without too much development and optimisation. However, most products or process liquids do possess some characteristics that should be accounted for when developing the process. Proteolysis, aggregation, solubility, stability and in particular micro heterogeneity are the potential problematic parameters.

With continuing developments in analytical techniques and increased demands to product purity separation tools for controlling the micro heterogeneity will be of increasing importance. E.g. with ion exchange chromatography it is practically impossible in large scale to separate protein species with less that 0.1 pH unit difference in isoelectric point.

Besides precisely optimised HIC, reverse phase, ion exchange and affinity chromatography matrixes industry (and product quality) will unquestionably benefit from increased research focus related to the development of new and more efficient techniques for resolving micro heterogeneity separation problems. It is here of value to focus on increased protein chemistry knowledge with the goal of avoiding micro heterogeneity in the first part secondly to ensure stability of product through production process and in subsequent storage.

3.3.2. Matrix quality.

An ongoing debate is about the strategy for using high performance matrixes early or late in processes. Some processes do perform well when only based on low-pressure matrixes. However, in many cases high performance matrixes are needed for obtaining the necessary purification efficiency. In most cases quality of matrixes will change through the process with the initial chromatographic steps being performed with traditional low-pressure matrixes, or even with inexpensive water treatment resin types, where the later steps are equipped with medium or high pressure matrixes. A current trend is to move the higher performance matrixes up earlier in the processes, this however setting stricter demands to the initial recovery steps. The decision regarding where in the process to use higher performance matrixes is in reality a case to case evaluation of pros and cons - very much based on specific company traditions.

3.4. RESEARCH AND DEVELOPMENT OF PARTICULAR INTEREST FOR PHARMACEUTICAL DOWNSTREAM PROCESSING.

The continuing strives for refining matrixes with improved stability, flow characteristics and new improved ligands will also in the coming years be beneficial for the downstream operations. In particular new and improved affinity matrixes must be expected to bring some further advantages to the field. Some of the more interesting affinity systems commercialised being systems based on dyes as ligands, as these are robust and have chemical stability allowing for standard CIP procedures.

For processes based on inclusion bodies the product Solubilisation and the renaturation/refolding steps require large amounts of chemicals and process water. Development of low cost environmentally friendly techniques for these process steps will therefore have positive impact on both process economy and environment.

With increased focus on producing products with small/controlled micro heterogeneity research and development within new high performance purification technologies is of continuing interest for the industry as is improved analytical tools for analysis of micro heterogeneity.

The transfer of the extreme power of analytical electrophoretic techniques into production relevant processing has long been hoped for. Many such larger scale systems have failed in the past. However, published results with membrane based electrophoresis systems show that membrane based large scale electrophoresis might have some interesting potentials as a new inexpensive high resolution technique well suited for resolving difficult micro heterogeneity separation problems [1, 2, 3].

Protein chemistry related to micro heterogeneity and product stability does continue to be of significant importance to the downstream field. Also of significant importance is research in protein interactions with soluble components as well as protein - surface interactions. Increased knowledge in these fields will be of clear value for the downstream processing field.

4. Enzyme production.

4.1. GENERAL DOWNSTREAM ISSUES RELATED TO ENZYME PRODUCTION.

Where in pharmaceutical production process development speed, process validation and in particular high purity are essential elements; a somewhat different set of process conditions/requirements is essential for the enzyme producer. In the enzyme industry the highest possible purity of a product is not the target, as product purity does depend heavily on the application for the product and of specified requirements from customers. Where one might expect higher purity requirements for food and feed enzymes relative to enzymes for industrial use this is not always the case. Many food and feed enzymes are enzyme complexes with many necessary enzyme activities. In this case the purification task is to recover the pool of enzymes and potentially free it from non-proteinaceous matter. On the other hand in some industrial applications like enzymes for detergents significant purity requirements can be set by customers with in some cases > 95 % enzyme purity. Another difference to pharmaceutical production is scale of production, as pharmaceutical production often only is counted in kilo of active ingredient, whereas bulk enzyme production for some products can be several ton/day of the active enzyme. This sets extra focus on efficient use of resources and on environmentally friendly processing.

Bulk enzymes are often commodity products with sales prices several orders of magnitudes less than sales prices for pharmaceutical products. Production cost and capacity utilisation are therefore of crucial importance for the enzyme producer, as even small differences in process economy between companies can make the difference between success or failure.

A way to obtain low production cost is to ensure high throughput in the production facility, why high capacity equipment with low downtime between batches is used.
As an enzyme producer needs to produce several different products in the same facility by smaller or longer production campaigns, flexibility is another necessary feature for an enzyme production.

Nearly in all cases fermentation titres in the enzyme industry need to be counted in multiple gram/litre to be competitive. The recovery processes therefore have to be able to handle not only large fermentation volumes, but also high product concentrations. A condition quite different from the pharmaceutical industry in which processes are targeted smaller fermentation volumes with lower product concentrations.

The very high concentrations of active enzyme in the process liquids further stress working environment issues. It is therefore important to secure the operators in the plants against process dust and aerosols, as such exposure might lead to allergenic problems. Therefore good ventilation and enclosed processing are of importance to the field.
The most important characteristics of bulk enzyme processing can be summarised to:

- high yield
- low cost
- high capacity
- robustness
- environmentally friendliness
- high flexibility
- enclosed operations
- capability of producing highly concentrated product with necessary purity.

The downstream production of bulk enzymes can be split into three parts:

- harvest (primary separation)
- concentration
- purification

4.2. HARVEST.

The often most important step in the whole process is the primary separation where the enzyme is separated from the fermentation solids. Here yields of at least 90 % are often needed, this in a high capacity process generating a minimal amount of cell sludge. In many countries the cell sludge from the production is spread on fields as a fertiliser. It is therefore important that only very environmentally friendly processing aids are being used for the separation. Moreover the separation needs to produce a process liquid well suited for subsequent concentration.

The common production organisms for industrial enzymes are for bacteria Bacillus strains and for fungi Aspergillus strains. For fungi primary separation is easy with often no pre treatment requirement for efficient separation on filters or centrifuges. For bacteria however, flocculation is widely used. Both inorganic salts and polymeric flocculants are used for this purpose. An optimised flocculation procedure can not only ensure an efficient high capacity primary separation, but also ease the subsequent concentration and in some cases even make further purification unnecessary.

For primary separations the equipment types of choice are drumfilters and continuos centrifuges or decanters.

For harvest of intracellular enzymes cell opening is primarily performed with pearl milling or homogenisation. The colloids hereby generated might - like for pharmaceutical production - impact the subsequent processing negatively. However, as flocculation is a standard tool for the enzyme industry, these problems can often be fully resolved by optimised flocculation procedures.

4.3. CONCENTRATION.

The concentration needs to be low cost and at the same time able to concentrate to high dry matter concentrations. As some process liquids can have a tendency to form precipitates during the concentration, the process equipment also needs to be able to handle solid containing liquids. For the concentration step vacuum evaporation and ultrafiltration are the unit operations of choice. As evaporation energy wise is the more expensive unit operation and at the same time has the disadvantages, unlike membrane concentration, that it also concentrates smaller components in the liquid like peptides and inorganic salts, membrane concentration has today become the preferred concentration method.

Over the years quite some focus has been allocated to the development of efficient ultra filtration systems with improved hardware and with membranes more resistant to fouling as the result.

On the membrane part industry is looking for hydrophilic, mechanical and chemical stable membranes with narrow pore distributions.

Hardware wise many different systems are available; spiral, hollow fibre and plate and frame systems. All of the systems have industrial relevance. The optimal system to choose for a specific process does however depend on viscosity and solid concentrations in the process liquid, as not all the systems handle these issues equally well. As a rule of thumb one can expect the most energy efficient systems to be the most susceptible to viscosity and precipitation problems.

4.4. PURIFICATION.

For purification of enzymes precipitation with salt or organic solvent is one of the older standard procedures. For organic precipitation acetone and alcohol have been the preferred solvents where salt precipitation has been performed with $(NH_4)_2SO_4$ or Na_2SO_4. Na_2SO_4 actually being the preferred salt as it is both more efficient than $(NH_4)_2SO_4$ and as it can be reused by crystallisation at low temperature.

Precipitation techniques can also be applied as a concentration tool, as the precipitated enzyme can be formulated directly without prior solubilisation or it can be dissolved in fairly concentrated form.

However, as these techniques require large amounts of chemicals, they possess some difficult solid liquid separations and can have a negative environmental impact, organic precipitation and salt precipitation are avoided today when possible.

Adsorption with activated carbon, bentonite or water treatment resins are also older standard methods, this is in particular the case for activated carbon.

As described for primary separation in many cases also the flocculation step does result in a considerable purification, why this together with ultrafiltration and potential diafiltration often can result in the necessary product purity [4].

Removal of unwanted enzyme activities can be difficult as most low cost protein purification techniques are not sufficiently efficient for these separations. Such activities will often be tried removed by mutation of the production organism. However, if not successful selective inactivation of the unwanted activity with high temperature and/or extreme pH might be a useful technique [5]. Continuous processing has opened up for increasing use of such extreme conditions.

Chromatographic procedures as used in the pharmaceutical productions are in nearly all cases too costly for bulk enzyme production. In some instances however, unwanted by-products - including colour, nucleic acids and foam controlling agents are adsorbed on ion exchange or hydrophobic columns. Such procedures are in most cases though avoided as they increase cost and complicate the process. If however adsorption steps are necessary for impurity removal water treatment resins, activated carbon or bentonite will be the preferred choices - biotech matrixes as used in pharmaceutical industries are too expensive and not necessarily more efficient for these applications.

In the later years various crystallisation techniques have been adapted to industrial scale. Where crystallisation techniques few years back only were applied on very pure solutions developments have shown it possible to develop efficient industrial crystallisation processes even on very impure protein solutions. Successful large scale crystallisations have been reported for a broad range of different crystallisation conditions; - low conductivity, organic solvents, water soluble polymers and various salts as crystallisation agents [6, 7, 8, 9].

The advantage of including a crystallisation step in an industrial process is clearly a high purity of the final product with a quality almost independent of upstream variations. A further advantage of a crystallisation process is the possibility for developing high strength product formulations with savings in formulation chemicals and transport cost as an economic benefit.

Unquestionably crystallisation is today's most efficient low cost protein purification technique for bulk protein production.

4.5. FUTURE CHALLENGES CONNECTED TO DOWNSTREAM PROCESSING OF BULK ENZYMES.

Environmental issues are getting increasingly more important why focus on reuse of process liquids and chemicals like permeates, rinse water, mother liquors and CIP will have high priority in process optimisations. Also reduction in solid waste is of importance, here reduction of filterpad and filter-aid use is at focus. One way is to develop continuos processes with short process time (hours instead of days), as this reduces microbial growth and potential precipitation problems. For efficient filter-aid reduction microfiltration has often been proposed as an alternative technique for primary separation as it is a closed and environmentally friendly process. Where microfiltration has obtained some foothold in the pharmaceutical industries, successes within bulk protein production have been limited. Microfiltration is here up against some quite effective and well-established operations. It has further to operate with fermentations having both high biomass and viscosity but also having a high tendency for membrane fouling. However, the newest developments in membranes and in hardware have already today made microfiltration processes economic feasible for a number of bulk processes. Especially developments within the newer high shear systems together with improved membrane quality will be beneficial for the industry.

A field of large impact for processing economy is the ability to ensure sanitary processing, this most dominant for food and feed enzymes. Efficient CIP procedures together with sanitary equipment design are here essential. To obtain necessary germ reductions processes need to include germ reduction steps, this normally done by dead-end germ filtration. To save on filtration cost and for avoiding reprocessing caused by high germ counts fast total viable count analysis is valuable. Such analyses are possible to perform within hours, but no real online analysis is available. Developments within particle counting in the < 10 μm area would therefore be of significant value.

An alternative approach to protein concentration and purification is selective precipitation. Even though these techniques are not currently applied to enzyme production interesting results on yield, purification and stabilisation have been presented with selective precipitation with e.g. detergents and dyes being quite efficient agents [10, 11]. It is therefore likely that selective precipitation will obtain some implementation in enzyme purification and formulation.

A problem new to the enzyme industry is enzyme solubility. During resent years, industrial fermentations and production organisms have been optimised to yield high enzyme titres. Work has also been performed to improve on the enzyme purity in the fermentation for facilitation of the downstream process. These developments have in some instances had the consequence that the product does precipitate or crystallise spontaneously in the fermentation or during the downstream process. Harvest of partly insoluble enzyme is a significant complication to standard processing, as the process might need to recover the soluble fraction as well as the insoluble fraction. In some, but not all instances fast processing and addition of components preventing precipitation or crystallisation can solve these problems. Substrate or substrate analogues are examples of such components.

The continuing developments within production organisms and fermentation technology [12] with higher product yield and cleaner background will in the years to come accelerate these process problems. Where in resent years the ability for an enzyme to crystallise was seen with great pleasure, in the coming years spontaneous enzyme precipitation and crystallisation will bring unpleasant challenges to the field.

5. Summary.

Pharmaceutical and bulk enzyme productions differ significantly in relation to process requirements and solutions.

Where pharmaceutical downstream processing is centred on chromatographic purification, bulk enzyme production is based on the techniques known from the chemical industries with flocculation, membrane concentration, adsorption, precipitation and crystallisation as the dominant technologies.

Pharmaceutical production benefits from the continuing research in refining chromatography based systems. Many well designed columns, matrixes and procedures therefore exist, making removal of host impurities a not too difficult matter. However, as sensitivity of impurity detection increases and focus increases for producing safe pharmaceuticals with little/controlled micro heterogeneity research and development within the high-resolution area will be of significant value. High-resolution chromatographic systems therefore still deserve focused research and development, as do new high-resolution technologies. Also focus on analytical tools, protein chemistry and on protein interactions with surfaces and soluble components will be of clear value.

For bulk enzymes low production cost, high capacity and environmental friendliness are very important parameters. As price competition within the field can be fierce, much refinement is going on to ensure efficient low cost production. However, a hard challenge enzyme producers is going to handle in the near future - besides general cost and waste reductions, is efficient processing of enzymes partially precipitated or crystallised in the fermentation broth. Research within enzyme solubility and crystallisation will therefore be of value for the enzyme field. Research and development within enclosed, sanitary and environmentally friendly technologies do also have significant value for the enzyme producer.

References.

1. Wenisch, E., Schneider P., Hansen S. A., Rezzonico R. and Righetti P. G. (1993) Isoelectric focusing in a multicompartment electrolyser with zwitterionic membranes, exemplified by purification of glucoamylase, *Journal of Biochemical and Biophysical Methods 27*, 199-213

2. Horvath Z. S., Corthals G. L., Wrigley C. W. and Margolis J (1994) Multifunctional apparatus for electrokinetic processing of proteins, *Electrophoresis 15*, 968-971

3. Laustsen M. (1993) Methods and systems for high molecular weight electrodialysis, U. S. Patent Application 08/176,037
4. Nielsen N. (1993) Method for purification of an aqueous enzyme solution, WO 94/01537
5. Laustsen M. and Nielsson S. (1997) Selective Inactivation of Enzyme Activities, PCT/DK96/00489
6. Becker, Nathaniel, T.; Braunstein, Edit, L.; Gros, Ernst, H.; Fewkes, Robert; Heng, Meng, H. (1996) Crystalline cellulase and method for producing same, WO 97/15660
7. Gros, Ernst, H. and Cunefare, Lerry, L. (1997) Crystalline protease and method for producing same, WO 97/33983
8. Nilsson B. M., Laustsen M. and Pahle C. (1998) Separation of proteins EP 0691 982
9. Nilsson B. M., Laustsen M. and Rancke-Madsen A. (1998) Separation of proteins US 5,728,559
10. Lovrien R. and Matolis D. (1996) Hard and soft sulphate and sulfonate anions in protein precipitation-biorecognition 211th ACS National Meeting New Orleans, LA
11. Becker N. T. and Anderson K. A. (1998) Recovery of proteins by precipitation using lignosulfonates, WO 98/30580
12. De Laat, Wilhelmus, Theodorus, Antonius, Maria; Preusting, Johannes, Cornelis, Gerardus; Koekman, Bertus, Pieter. (1998) Fermentative production of valuable compounds on an industrial scale using chemically defined media, WO 98/37179

SEPARATION OF α-LACTALBUMIN AND β-LACTOGLOBULIN BY PREPARATIVE CHROMATOGRAPHY USING SIMULATED MOVING BEDS

S.L. LUCENA, P.T.V. ROSA, L.T. FURLAN AND C.C. SANTANA
DPB/FEQ/UNICAMP, C.P. 6066, CEP 13083-970 Campinas, SP, Brazil

Abstract

Simulated moving bed (SMB) is an important separation process that uses a series of columns of adsorption connected being formed a circuit that is divided in, for example, four zones defined by two entrances (feeding and desorbant) and two exits (extract and rafinate). Those entrances and exits are periodically moved in the direction of the liquid flow simulating a countercurrent movement with the phase solid adsorbent. In this work, experimental data for equilibrium isotherms in an ion-exchange resin were coupled to mathematical formulation leading to a computational routine developed for the estimation of the concentration profiles in the extract and in the rafinate of constituted by the proteins α-lactalbumin and β-lactoglobulin in a binary mixture. Those proteins are present in cheese whey with concentrations of the order of 1,5 g/L to 3,0 g/L, respectively, and, in high levels of concentrations, exhibit application in the veterinary medicine and as supplement for of culture of cells media, being generally residues of milk processing industries. Chromatographic profiles obtained for a SMB with twelve columns indicates conditions that meets the complete separation of the two components from the mixture.

1. Introduction

The products of biotechnological origin possess great diversity and are generally present in fermentation broths and of cultures of cells in low concentrations. Diluted systems coupled to considerable amounts of chemical species that interfere in the recovery processes, concentration and final purification turn those difficult and onerous tasks, answering in most of the processes for the largest part of the costs of biomolecules production. In the special case of the proteins the percentages of the purification costs reaches values of the order of 60% in relation to the total costs of the production process, could located in the range of up to 80 to 90% for original fermentation products of recombinant DNA (Blanch and Clark, 1997). The

M. Hofman and P. Thonart (eds.), Engineering and Manufacturing for Biotechnology, 325–337.

conventional liquid chromatographic (Ganetsos and Barker, 1993). The largest disadvantages of the usual chromatographic separations lasts in the discontinuity of the process and in the dilution of the product. It is a well-known fact in the operations of adsorption that continuous systems in which the phase solid is contacted in the direction opposed the one of the flowing phase that the profile of mass transfer stays stationary and the adsorbent is used in a more efficient way.

Purified proteins derived from milk products like whey are acquiring important applications in medicine, veterinary, as food functional products and for cells media culture. In many countries, whey from milk products like cheese are discarded in mainstreams, causing pollution problems. As restrictions on the discharge of pollutants increase, new strategies for waste treatment must be found. Future initiatives to improve recovery of by-products will require additional research to investigate possible uses for valuable waste stream components and to develop cost-effective techniques for their recovery (Beszedits,1982). A typical composition of a cheese whey is shown in Table 1.

Table 1: Proteins main composition of cheese whey from bovine milk (Source: McKenzey,1970)

Protein	Concentration (g/l)	Molecular Mass (kDa)	Isoeletric point
β-lactoglobulin	3 to 4	18.4	5.2
α-lactalbumin	1.5	14.2	4.7 to 5.1
BSA	0.3 to 0.6	69	4.9
IgG, IgA, IgM	0.6 to 0.9	150 to 900	5.8 to 7.3
lactoperoxidase	0.06	78	9.6
lactoferrin	0.5	78	8.0

Increasingly opportunities exist for the commercial extraction of these bioproducts even in diluted solutions. The utilisation of chromatographic processes based on ion-exchange resins coupled to chemical engineering principles is one example of process that can be applied to the design , scale-up and optimisation of large scale systems (Carrére, 1993). A fundamental understanding of adsorption processes as well as of novel equipment configurations enables foresee the fractionation of individual proteins in economic basis. The search of preparative methods to separate and purify fragile products like proteins is an important issue connected to the increasing demand and higher throughput of proteins in the biotechnological industry. The utilisation of large scale expanded bed columns and the search of continuous methods of separation are examples of trying to meet some of the preparative techniques requirements. Continuous processes using Simulated Moving Beds (SMB) as a chromatographic procedure was utilised by Huang et al. (1986) who demonstrated the feasibility of SMB as a chromatographic procedure to purify enzymes. Gottschlich et al. (1997) applied an SMB system to purify α-chymotrypsin on immobilised soybean trypsin inhibitor, based therefore on biospecific affinity phenomena. This paper addresses the utilisation of adsorption kinetics and equilibrium data in ion-exchange adsorbents connected to a linear model for SMB performance A model binary system produced artificially by using α-lactalbumin and β-lactoglobulin was chosen to demonstrate the feasibility of separation of these major bovine milk serum components.

2. Basic concepts of processes of separation with simulated moving beds

In Figure 1 two different processes of adsorption are presented in a countercurrent mode of operation. The well-known process called true moving bed (TMB) allows the obtaining of a continuous operation, distinctly of the classic chromatographic elution process. On the other hand, due to the difficulties of implementing the circulation of the solids, efforts have been driven to develop processes that maintain the advantages of the countercurrent operation but that avoid the circulation of the solids. In most of the innovations in that sense the movement of the solids is obtained by periodic changes in the feeding and discharge in a system of multiple columns resulting the outline of well-known process as simulated moving bed (SMB).

Since 1964, continuous chromatographic systems have been used mainly in industrial scale in the petrochemical industries (Processes SORBEX and PAREX, developed by United Oil Products), and of processing of sugar (Barker and Abusabah, 1985). The technology of SMB has also been studied and applied the products of fine chemistry (Ganetsos and Barker, 1993) and of biotechnological origin (Yamamoto et al., 1992), especially in enantiomers separation. SMB presents economic advantages over other chromatographic systems for several reasons: it is a continuous process and it allows to separate starting from a similar composed mixture, allowing high productions and low solvent consumption. In general in that system type the volume of requested adsorbent is approximately 25% of the requested in batch chromatography (Gottschlich et al., 1996).

As depicted in the Figure 1, SMB uses a series of columns of adsorption (eight columns or twelve columns, for example) with an appropriate adsorbent. The columns are connected to recipients that contain the feeding and the eluent and that receive the currents of exit of the product through lines controlled by a group of valves of multiple positions. That group of controlling valves allows that they are alternate, in regular intervals of time, the points of entrance of the feeding, of the eluent and of the exit currents. The system changes therefore the positions between the entrance points and exit, simulating the countercurrent flow.

From the point of view of the operational variables, the project of SMB is relatively complex because it involves at least ten specific parameters to know: diameter of the columns, four lengths of separation zones, four flowing currents and a average velocity associated to the control of the opening of the valves of multiple positions. LMS is usually used for a mixture that contains two similar products, of the which it is attempted the separation. The use of SMB in the separation of multicomponent mixtures is not still very well known (Ganetsos and Barker, 1993). The main claim of this separation method consists in its ability to separate mixtures of difficult resolution and for products of high added value .

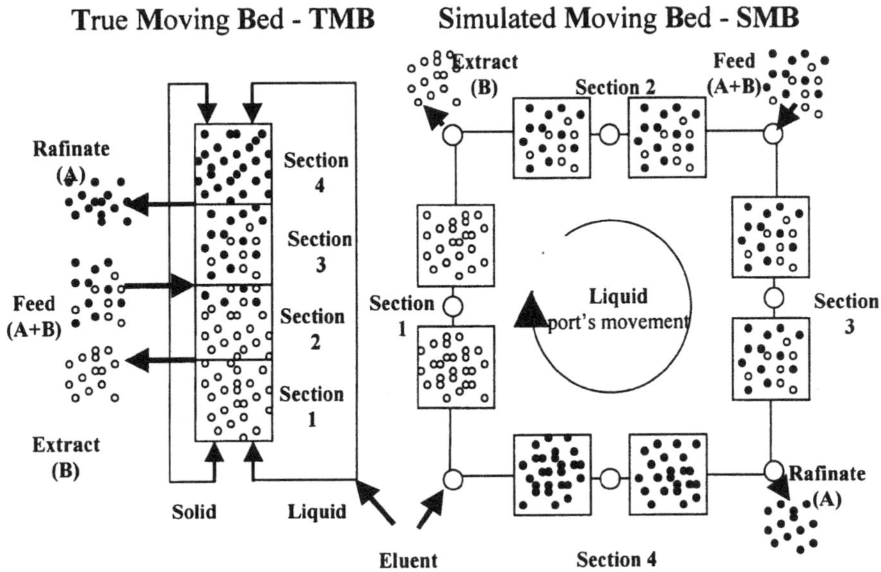

Figure 1: Schematic representations of the true moving bed (to the left) and of the simulated moving bed (to the right)

3. Mathematical formulation

The equation of the mass balance for the solute in the flowing phase is given by the equation of the rate as deduced by Blanch and Clark (1997):

$$\varepsilon D_{xx}\frac{\partial^2 C}{\partial z^2} - u_o\frac{\partial C}{\partial z} - \varepsilon\frac{\partial C}{\partial t} - (1-\varepsilon)\frac{\partial q}{\partial t} = 0 \qquad (1)$$

where C is the concentration of the component i in the liquid phase, q is the concentration of the component i in the surface of the particles, z is the axial position, and ε it is the bed porosity, uo it is the superficial speed and D_{xx} is the coefficient of axial dispersion.

The terms of the equation (1) represents the axial dispersion, the convection in the bed, the accumulation in the liquid and accumulation in the particles, respectively. If it be considered that there are no resistances for the mass transfer of the adsorbed component in the film of the particle and in the pores of the resins, the concentrations

of the material to be adsorbed in the solid and in the liquid will be directly related for the isotherm of adsorption of the material.
The boundary conditions for the system are given for:

$$C = 0 \ \text{ em } t = 0 \ ; \ \forall z \tag{2}$$

$$u_0 C_F = u_0 C_0 - \varepsilon D_{xx} \left.\frac{\partial C}{\partial z}\right|_{z=0} \ ; \ \forall t \tag{3}$$

$$\left.\frac{\partial C}{\partial z}\right|_{z=L} = 0 \ ; \ \forall t \tag{4}$$

where C_F is the concentration of the component i in the feeding and C_0 is the concentration of the component i in the beginning of the bed and u_0 is the superficial liquid velocity. Boundary condition described by equation (3) is known as a Dankwerts type condition (Dankwerts, 1953) and takes in account the effect of axial dispersion at the column entrance.

3.1. APPLICATION FOR A COLUMN OF ADSORPTION

As pointed out above, the value of the concentration of the solute in the solid (q) it is related with the concentration of the solute in the liquid through the isotherms of adsorption. Dimensionless forms previous equations with the application of classical forms of isotherms for the concentration q of the component i in the surface of the particles are presented in Appendix. Equations (5) and (6) presents the equations in the dimensionless form obtained starting from the equation (1) for isotherms of adsorption of the linear type and Langmuir type, while the equations (7) and (8) represents the equations for the isotherm of adsorption of competitive Langmuir for two components. The set of equations were solved for the study of adsorption in one column using the method of the orthogonal collocation method for the axial position and the Runge-Kutta adaptive method with control step (Press et al., 1992) for the time.

3.2. APPLICATION TO THE SIMULATED MOVING BED

The simulated moving bed was considered as a group of interlinked columns of adsorption as shown in Figure 1. It was considered that the moving bed is formed by 3 columns in each section, resulting in a total of 12 columns for the system. The operation is started with the whole system filled with the solvent and, therefore, the concentrations of the components in all the columns were considered as being the same and equal to zero in this period. As the columns are connected in series, the liquid flowrate is constant in each section of the bed. The twelve columns were jointly simulated and the feed concentration of the columns are different from the feeding of the system. The eluent was considered to be pure, without proteins. In the feeding columns and at the point of the solvent introduction the concentrations were obtained as described in equations (5) and (6):

$$FC_F = fr_3 C_{col7,out} + fr_2 C_{col6,in} \tag{9}$$

$$fr_1 C_{col1,out} = fr_4 C_{col12,in} \tag{10}$$

where fr1, fr2, fr3 and fr4 are the flow rates in the sections 1,2,3 and 4, and Ccol7,out and Ccol1,out are the concentrations in the exit of the columns 7 and 1.

Equations (1) through (4) are applied for each column and integrated along the time. When switching time is reached, the positions of the currents of the system were rotated, passing for the columns to the left. The concentration profiles in each column was maintained constant during the change of the currents. This procedure was used up to reach the required final number of rotations.

4. Adsorption isotherms

The isotherms were obtained to 25 °C after determination of the kinetics of adsorption so that the time to reach the equilibrium was known. About 4 ml of the solutions of proteins with different concentrations were incubated in a rotative agitator with a certain amount of the resin (about 20 mg) in syringes of 5 ml. After 45 minutes (time of equilibrium) the concentrations in the supernatant were measured in an spectrophotometer to 280 nm being taken as reference a calibration curve. Using a mass balance, the concentrations of proteins adsorbed in the resin could be ascertained. The concentrations of proteins in the supernatant and in the resin were then adjusted being used the model of Langmuir, as shown in Figure 2 , while Figure 3 depicts the linear part of the experimentally obtained isotherms.

5. Results and discussion

5.1. INDIVIDUAL COLUMN OF ADSORPTION

The simulated moving bed consists basically of a group of interlinked individual columns of adsorption in a recurrent way, and the material that leaves a column it is fed in an adjacent column. The understanding of the behaviour of an adsorption column is of vital importance for the understanding of the operation of the simulated moving bed. Several factors can alter the form of the profile of concentration of a certain solute in the exit of a column as function of the time (breakthrough curves). Among the main variables we can mention the superficial velocity of the liquid inside the column, height and diameter of the bed of resins, diameter of the particles, coefficient of axial dispersion, concentration of the solute in the feeding and the mathematical description of the adsorption isotherm for the solute in the particular resin. Some important forms of isotherms of adsorption of proteins in resins are the linear , of Langmuir and of competitive Langmuir. It can be observed in the equations (5) the (8) that the variation of the dimensionless concentration with the dimensionless time is not altered by the concentration of feeding of the solute for the case of the linear isotherm, even so this variation is function of the feeding concentration for the case of the isotherms of Langmuir. Thus, for low values of solute concentration (values of small K_iC_i if compared with 1), practically the phenomenon of competition of adsorption is not observed and the linear isotherm represents well the process of adsorption. With the increase of the solute concentration, the linear isotherm doesn't represent more the process of adsorption and the models of Langmuir and of competitive Langmuir describes better the equilibrium. Figure 4 presents the breakthrough curve obtained through the simulation of the equations (5) the (8) together with the initial condition (2) and with the boundary conditions (3) and (4) for the adsorption of two solutes that are fed in the column of adsorption with a such concentration that the value of K_iC_i is not small if compared with the unit.

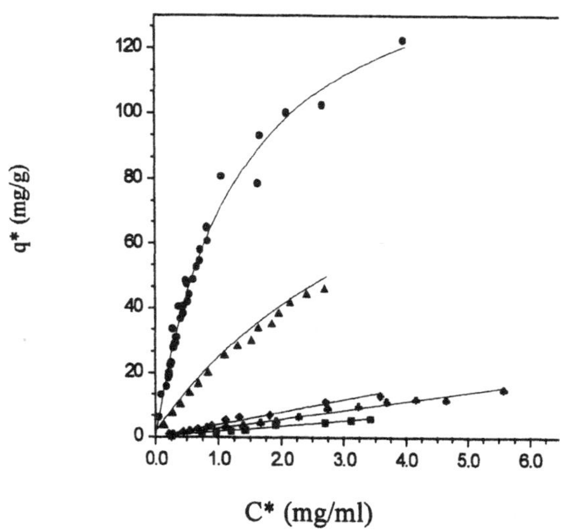

Figure 2: Adsorption isotherms for α-lactalbumin and β-lactoglobulin in Accell Plus QMA at 25 °C. Buffer: pH 6.5 - 20 mM Bis-Tris. Temperature: 25°C ● β-Lg 125 mM NaCl, q = 159.67C*/(1.29+C*). ▲ β-Lg 160 mM NaCl, q* = 123.81C*/(4.36+C*). ◆ β-Lg 250 mM NaCl, q* = 3.925C*. ✚ β-Lg 160 mM NaCl, q* = 2.819C*. ■ β-Lg 250 mM NaCl, q* = 1.746C*.*

Symbols shown in the figure are results of the simulation process. The competition of the two solutes in adsorption of the resins results in a smaller liquid rate of adsorption, resulting in smaller breakthrough times. In this value of feeding concentration the term of KC cannot be disregarded in the Langmuir equation, because there is a great variation among the profiles obtained by this model of adsorption and for the linear model. The choice among the model of Langmuir and of competitive Langmuir depends on the system under study. In the case of affinity resins that presents non-specific adsorption, the model of Langmuir can be used . Considering the case of resins that don't have great specificity for a certain solute, the isotherm of competitive Langmuir should be used since it is possible to determine the adsorption characteristics of the competitive species.

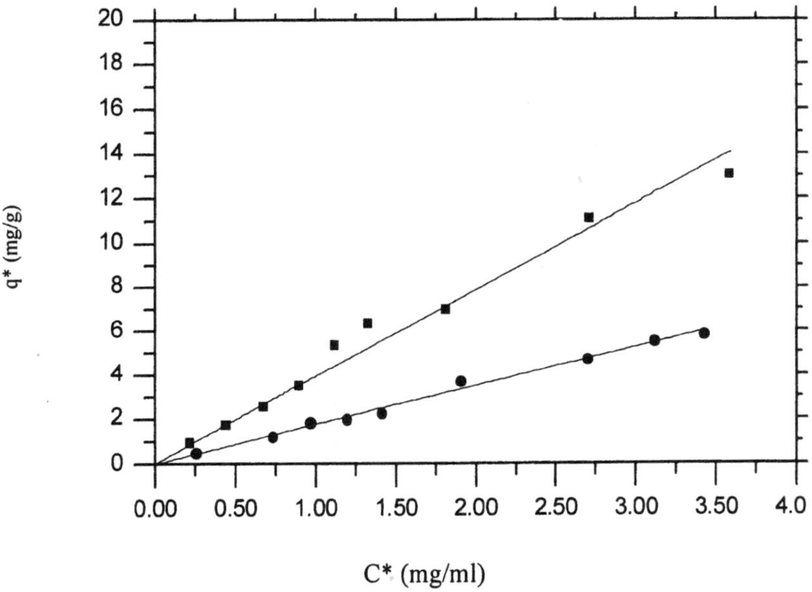

Figure 3: Linear parts of isotherms for α-lactalbumin and β-lactoglobulin in Accell Plus QMA at 25 °C. Buffer: pH 6.5 - 20 mM Bis-Tris / 250 mM NaCl. Temperature: 25°C. ■ β-Lg, $q^* = 3.925C^*$, $R2 = 0.992$. ● α-La, $q^* = 1.740C^*$, $R2 = 0.990$.

5.2. SIMULATED MOVING BED

In a SMB, the compound that interacts more intensely with the adsorbent will move in the solid phase direction and will be recovered in the extract stream. Conversely, the material that exhibits lower tendency to accumulate on the solid particles will be carried on by the liquid flow and will be recovered in the raffinate port. The β-lactoglobulin and α-lactalbumin proteins have different affinities to the ionic exchange resin in the experimental conditions tested, as can be observed in Figure 3. Thus, as β-lactoglobulin has higher affinity to the resin than α-lactalbumin it should be expected when a mixture of these two proteins are been separated in a SMB that β-lactoglobulin will be present at the extract stream. In order to examine the performance of the SMB to separate mixtures of β-lactoglobulin and α-lactalbumin, equations (2) to (5) and (9) to (10) were simulated. The dynamic values of the proteins concentration in the extract and raffinate obtained by simulation can be observed in Figure 5. In the conditions tested, the steady state is obtained after 180 minutes and the products in the rafinate and in the extract are practically pure. The β-lactoglobulin is recovered preferentially in the extract and is obtained in more diluted form than the α-lactalbumin due to their higher affinity to the resin that implies in a higher desorbant flow rate to remove them to the solid phase. We can also notice in Figure 5 that the protein concentrations in the exit streams of the SMB are dependent to the switching time.

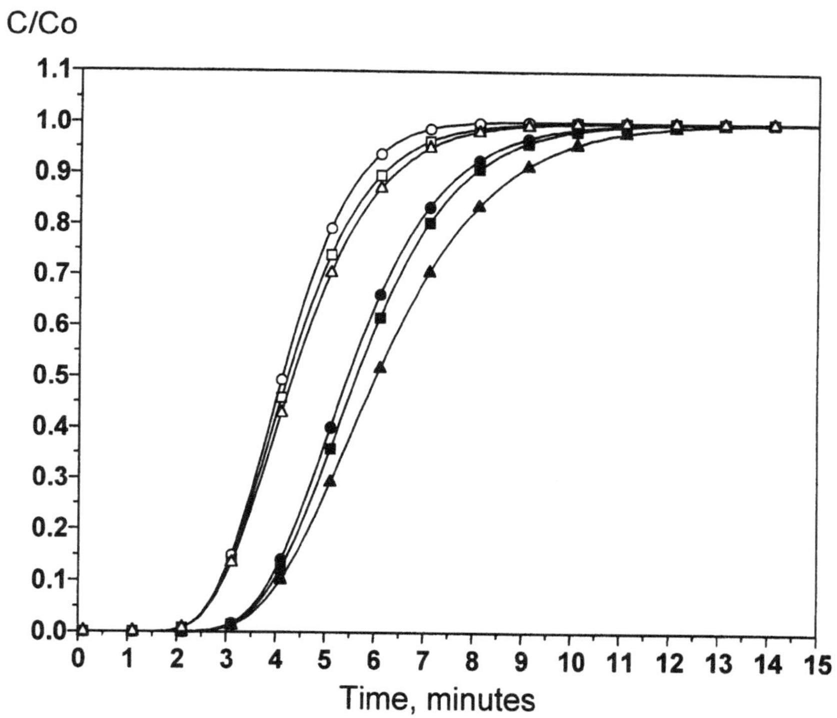

Figure 4: Breakthrough curves in a column, simulated for several isotherms of adsorption. Used conditions: L = 10 cm, dc = 1,6 cm, Co = 0,05 g/ml, Pe = 20, F = 1 ml/min. ■ *Non-Competitive Langmuir (K1=3,0 ml/g: qm =0,3 g/ml),* □ *Non-Competitive Langmuir (K1=1,5 ml/g: qm =0,3 g/ml),* ● *Competitive Langmuir (K1=3,0 ml/g: qm =0,3 g/ml),* O *Competitive Langmuir (K1=1,5 ml/g: qm =0,3 g/ml),* ▲*Linear (1,3 g/ml x 3 ml/g = 0,9),* Δ *Linear (1,3 g/ml x 3 ml/g = 0,9),*

At the time that the SMB reaches the steady state, there will be a well-established concentration profile inside the columns. Figure 6 shows the stationary concentration profile in the SMB for the separation of the mixture of β-lactoglobulin and α-lactalbumin using the conditions presented in the Figure 5 legend.

We can observe in Figure 6 that the performance of the SMB is affected considerably by the switching time. In order to verify the influence of this parameter on the separation of β-lactoglobulin and α-lactalbumin, the equations (2) to (5) and (9) to (10) were simulated for several values of switching time keeping constant the values of the flow rates in the sections. The results obtained by simulation are presented in Figure 7. We can observe in this figure that there is a range of switching time values (from 11.75 to 13.50 minutes) where the raffinate and the extract are completely pure. For switching times lower than 11.75 minutes, the α-lactalbumin is not completely removed from the solid phase in the section III and will be eluted at section IV, contaminating the extract. For switching times greater than 13.50 minutes, there is some desorption of β-lactoglobulin from section II that will contaminate the raffinate. The range of switching time were the separation is complete is not the same where the concentrations of the compounds in the exit streams are at the maximum values. Thus, someone should decide if purity or concentration is the main purpose of the separation in order to set the switching time.

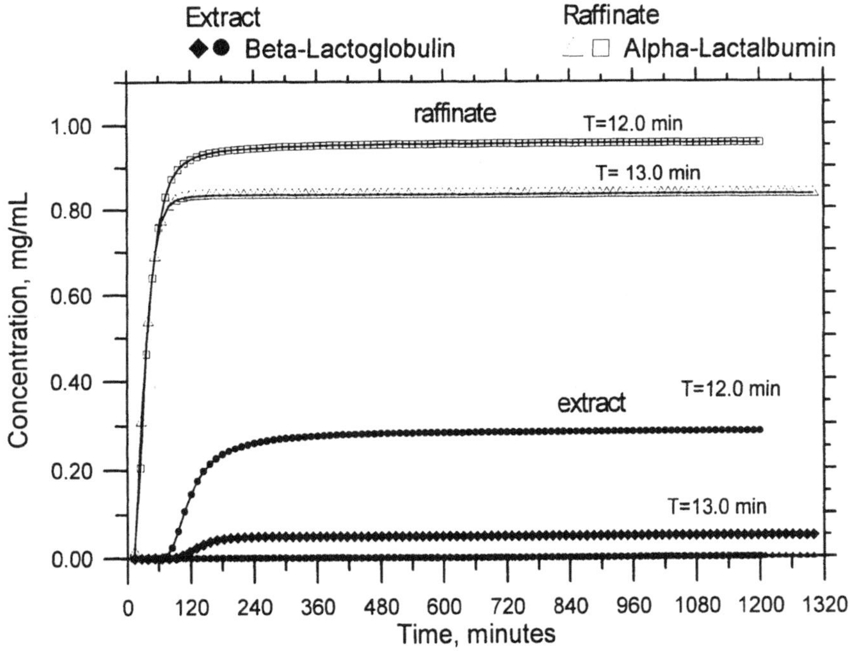

Figure 5: Concentration profile in simulated moving bed as a function of time

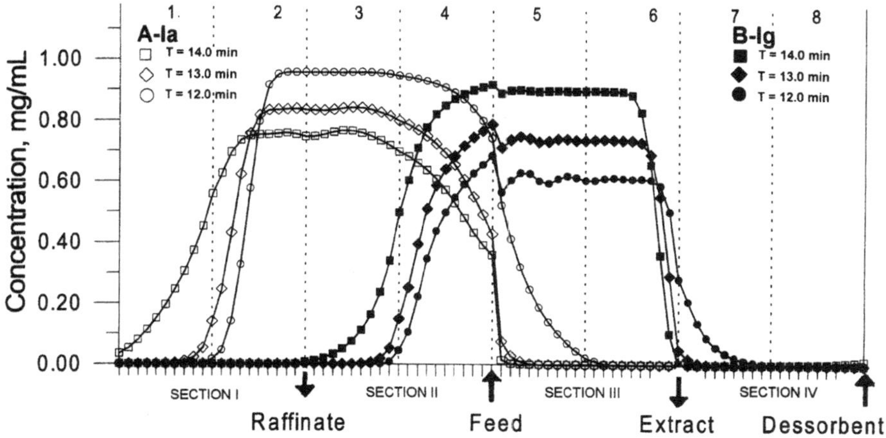

Figure 6: Concentration profile as a function of zone position

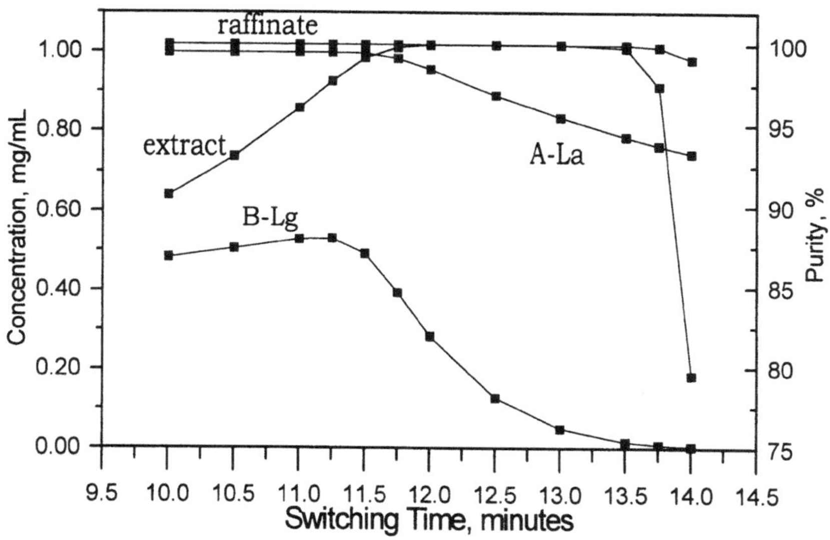

Figure 7: Concentration profile as a function of switching time

6. Conclusions

Adsorption isotherms for the proteins α-lactalbumin and β-lactoglobulin were experimentally obtained. The developed mathematical models are capable to predict the breakthrough curves for adsorption columns operating separately for three types of isotherms of adsorption. The mathematical model developed for the simulated moving bed represents the performance of the system on separating individual components from a mixed feed. The computational routine for the simulated moving bed for compositions was solved for the case of an isolated adsorption column, as presented in the Figure 4. The parameters of the isotherms experimentally obtained indicates that the proteins are strongly adsorbed in the studied conditions (high values of K_i what would imply in a very slow movement of the proteins along the adsorbed system. The parameters of adsorption isotherms can be modified through the increase of the ionic force or of the pH of the media, turning possible the use of this technique for the purification of those proteins.

References

Barker, P.E. and Abusabah, E.K.E. (1985), The separation of Synthetic Mixtures of Glucose and Fructose and also Inverted Sucrose Feedstocks Using Countercurrent Chromatographic Techniques, Chromatographia, vol. 20, no. 1, 9-12.

Beszedits, S. (1982) , Protein recovery from Food Processing Wastewater, B&L Information Services.

Blanch, H.B. and Clark, D. S. (1997), Biochemical Engineering, Chapter 6: Product Recovery, Marcel Dekker Inc., New York.

Carrére, H., Extraction des Proteines du Lactoserum par Chromatographie d' Échange d' íons en Lit Fluidisé, Thése de Doctorat, Institute Polytechnique de Toulouse, 1993.

Dankwerts, P.V., Chemical Engineering Science, (1953), 2, 1-12 .

Ganetsos, G. and Barker, P.E., Eds. (1994), Preparative and Production Scale Chromatography, Marcel Dekker, Inc., New York.

Gottschlich, N., Weidgen, S., Kasche, V. (1996). Continuous biospecific affinity purification of enzymes by simulated moving-bed chromatography: theoretical description and experimental results, Journal of Chromatography , 719, 267-274.

Gottschlich, N. and Kasche, V., (1997). Purification of monoclonal antibodies by simulated moving-bed chromatography, Journal of Chromatography , 765, 201-206.

Huang, S.Y., Lin, C.K., Chang, W.H. and Lee, W.S., (1986) Enzyme purification and concentration by simulated moving bed chromatography: in the experimental study, Chemical Engineering Communications, Vol. 45, 291-309.

Press, W.H., Teukolsky, S.A., Vetterling, W.T. and FlanneryY, B.P. (1992), Numerical Recipes in Fortran, Cambridge University Press.

Yamamoto, S., Nakanishi, K. It is Matsuno, R. (1988),. Ion Exchange Chromatography of Proteins, Marcel Dekker, Inc., New York.

Appendix

Dimensionless forms of Equation (1) for several isotherms

Linear Isotherm ($q = KC$)

$$\frac{\partial \Phi}{\partial \tau} = \frac{T_N u_o}{L[\varepsilon + (1-\varepsilon)K]}\left(\frac{1}{Pe}\frac{\partial^2 \Phi}{\partial x^2} - \frac{\partial \Phi}{\partial x}\right) \tag{5}$$

Langmuir Isotherm ($q = \dfrac{q_m KC}{1+KC}$)

$$\frac{\partial \Phi}{\partial \tau} = \frac{T_N \ u_o}{L}\frac{(1+K_1 C_F \Phi)^2}{\left[\varepsilon(1+K_1 C_F \Phi)^2 + (1-\varepsilon) q_m K_1\right]}\left(\frac{1}{Pe}\frac{\partial^2 \Phi}{\partial x^2} - \frac{\partial \Phi}{\partial x}\right) \tag{6}$$

Langmuir Competitive Isotherm ($q_i = \dfrac{q_m K_i C_i}{1+K_1 C_1 + K_2 C_2}$)

$$\frac{\partial \Phi_1}{\partial \tau} = \frac{\dfrac{C_{F1}}{C_{F2}}\left\{\left[\varepsilon + A q_m K_2 (1+K_1 C_{F1}\Phi_1)\right]\dfrac{\partial \Phi_2}{\partial \tau} - \dfrac{T_N u_o}{L}\left(\dfrac{1}{Pe}\dfrac{\partial^2 \Phi_2}{\partial x^2} - \dfrac{\partial \Phi_2}{\partial x}\right)\right\}}{q_m K_1 K_2 A C_{F2}\Phi_2} \tag{7}$$

$$\frac{\partial \Phi_2}{\partial \tau} = \frac{\dfrac{T_N u_o}{L}\left\{q_m K_1 K_2 C_{F1}\Phi_2 A\left(\dfrac{1}{Pe}\dfrac{\partial^2 \Phi_1}{\partial x^2} - \dfrac{\partial \Phi_1}{\partial x}\right) + S_1\left(\dfrac{1}{Pe}\dfrac{\partial^2 \Phi_2}{\partial x^2} - \dfrac{\partial \Phi_2}{\partial x}\right)\right\}}{S_1 S_2 - (q_m K_1 K_2 A)^2 C_{F1} C_{F2}\Phi_1 \Phi_2} \tag{8}$$

where: $S_1 = \varepsilon + A q_m K_1 (1 + K_2 C_{F2} \Phi_2)$

$S_2 = \varepsilon + A q_m K_2 (1 + K_1 C_{F1} \Phi_1)$

In the above equations the dimensionless parameters: $\Phi = C/CF$, $\tau = t/TN$, $Pe = Lu_o/\varepsilon D_{xx}$, $x = z/L$, TN is an arbitrary time, L is the length of the column, K is the constant of the linear isotherm, K_i it association constant and q_m is the maximum adsorption capacity of the resin

HIGH-SPEED PECTIC ENZYME FRACTIONATION BY IMMOBILISED METAL ION AFFINITY MEMBRANES

SILVIA ANDREA CAMPERI, MARIANO GRASSELLI[1] AND OSVALDO CASCONE
Cátedra de Microbiología Industrial y Biotecnología. Facultad de Farmacia y Bioquímica. UBA. (1113) Junín 956, Buenos Aires, Argentina.
[1] Departamento de Ciencia y Tecnología. Universidad Nacional de Quilmes. Roque Sáenz Peña 180, (1876) Bernal, Prov. Buenos Aires, Argentina. E-mail: scamperi@ffyb.uba.ar- Fax: 54-11-4508-3645

Abstract

Immobilised metal ion affinity polysulphone hollow-fibre membranes with a high capacity for protein adsorption were prepared and their application for commercial pectic enzyme fractionation was studied. The pass-through fraction containing pectin lyase (PL) is useful for fruit-juice clarification without methanol production on account of pectin-esterase (PE) being retained by the IDA-Cu^{2+} membrane.

1. Introduction

Commercial preparations of pectic enzymes normally contain a mixture of depolymerising (pectin lyase, PL, and polygalacturonase, PG) and de-esterifying (pectin-esterase, PE) enzymes (Rombouts and Pilnik, 1980).

The use of PL alone, instead of the combination of PE and PG for fruit juice clarification, prevents the release of methanol in the juice, thus constituting a potential health hazard in non-concentrated juices (Szajer and Szajer, 1982). Moreover, the volatile ester content, responsible for the specific aroma of various fruits, is not damaged (Alaña *et al.*, 1989). Furthermore, the use of PG and PE-containing enzyme complexes decreases fruit juice stability because of the coagulating processes caused by the interaction of the de-esterified pectin derivatives with the endogenous Ca^{2+}. For these reasons and in order to utilise pectinase activities more rationally, there is a current need for purification of commercial pectinase preparations to allow more specific and controllable effects (Alaña *et al.*, 1989).

M. Hofman and P. Thonart (eds.), Engineering and Manufacturing for Biotechnology, 339–349.

We reported a fractionation method to separate PL from PE and PG to obtain a fraction which will not produce methanol during juice clarification (Navarro del Cañizo *et. al.*, 1994, Camperi *et al.,* 1996). This is based on immobilised metal ion affinity chromatography (IMAC), a protein purification method that exploits the affinity of surface functional groups, mainly histidines, towards transition metals (Porath et al., 1975, Hemdan et al., 1989). IMAC is a good option in preparative protein purification, taking into account its high yields, and ligand economy and stability (Arnold, 1991). Due to beaded soft gels are utilised as supports, the main drawback of this fractionation scheme is that low flow rates must be used to prevent gel deformation and allow mass transfer. Also, the sample must be clarified before loading onto the column.

Membrane (internal pore diameter between 0.1 and 1 μm) is a good alternative to macroporous beads as separation based on membranes is characterised by the absence of pore diffusion, which is the main transport resistance in conventional column chromatography using porous particles. Proteins are directly transported by convection to the affinity ligand onto the inner surface of the through-pores of the membrane thus making adsorption rates faster. Additionally, membrane chromatography can overcome the high operating pressure and low adsorption rate, the typical disadvantages of bead-based chromatography (Brandt *et al.*, 1988; Roper and Lightfoot, 1995; Thömmes and Kula, 1995). Furthermore, it has been proposed that, unlike conventional column chromatography, solutions containing debris or solid particles can be processed by cross-flow microfiltration in chromatographic membranes without a previous clarification treatment (Kroner *et al.*, 1992).

Saito *et al.* (1989) developed a new type of affinity hollow-fibre membrane by radiation-induced co-grafting of a cross-linking agent with the reactive monomer. Grafting is a useful method for chemical modification of existing polymers. In this way, a higher degree of chemical modification of chromatographic supports can be obtained, thus meaning a greater amount of reacting sites for ligand attachment to the support (Mueller-Schulte and Daschek, 1995).

A number of different module configurations (hollow fibres, spiral-wound cartridges, flat-sheet membranes, etc.) are available on the market. A hollow-fibre membrane is superior to a flat-sheet membrane because of its high surface area/volume ratio (Yamagishi *et al.* 1991) and the easy scale-up of the chromatography by simple bundling of numbers of hollow fibres (Kubota *et al.*, 1997).

Polysulphone is a suitable membrane polymer because of its good film-forming properties and its resistance to thermal and biological degradation. Its heat stability allows performing chemical modifications without impairing its performance. We made tentacle cation-exchange hollow-fiber membranes of high capacity for proteins from epoxy-activated microfiltration polysulphone membranes (Camperi *et al.*, 1999).

Here we report a simple and economical method for pectinase fractionation, based on affinity chromatography, using a cartridge of Cu(II)-iminodiacetic (IDA) as the immobilised ligand on polysulphone membranes.

2. Materials and methods

2.1. ENZYMES AND REAGENTS

Polysulphone hollow-fibre microfiltration membranes, kindly donated by A/G Technology Co., Needham, Massachusetts, USA were epoxy-activated by Innovatec S.A., Buenos Aires, Argentina. They had a nominal 0.65 μm internal pore diameter and a nominal 80% porosity. The inner and outer diameters were 0.75 and 1.25 mm respectively.

Chicken egg lysozyme, horse skeletal muscle myoglobin, haemoglobin and citric pectin were from Sigma Chemical Co., St. Louis, USA. L-histidine hydrochloride was from BDH Chemicals Ltd., England. Bioconcentrated Plus, Biocon, Ireland was utilised as a pectic enzyme source.

All other reagents were AR grade.

2.2. HISTIDINE, LYSOZYME, MYOGLOBIN AND HAEMOGLOBIN CONCENTRATION MEASUREMENTS

Histidine solution concentration was determined by measuring their absorbance at 220 nm, lysozyme and myoglobin at 280 nm, and haemoglobin at 430 nm.

2.3. PECTIC ENZYME ASSAY

PL was assayed by monitoring the increase in absorbance at 235 nm as described by Albersheim (1966). One PL unit is defined as the amount of enzyme that causes a rise in absorbance of 1.0 per min, at 235 nm.

PE activity was determined by monitoring the decrease in absorbance of bromocresol green at 617 nm due to carboxyl groups being released in pectin according to Vilariño *et al.* (1993). One PE unit is the amount of enzyme required to release 1 μEq of carboxyl groups per min.

2.4. CHELATING HOLLOW FIBRE SYNTHESIS

Iminodiacetic acid (IDA) was immobilised onto the epoxy-activated membranes by suspending the fibres in 1M IDA-2Na in dimethyl sulphoxide (DMSO)/water (1:1) (Yamagishi *et al.* 1991). The reaction was performed at 80°C for 24 h. In order to hydrolyse the remaining epoxy groups, the fibres were then immersed in 0.5 M sulphuric acid for 2 h at 80°C. After washing the fibres with water, they were immersed in 0.5 M $CuSO_4$. Three hours later they were washed again with distilled water. Figure 1 shows a scheme of the chelating hollow-fiber membranes.

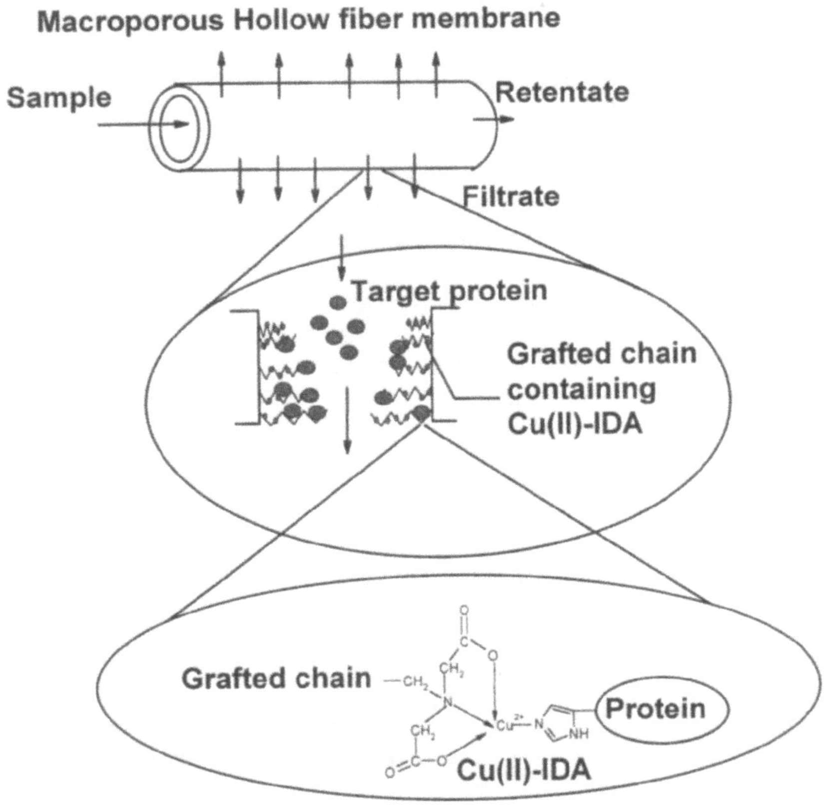

Figure 1. Scheme of the chelating hollow-fibre membranes

2.5. PURE WATER FLUX DETERMINATION FOR A SINGLE CHELATING HOLLOW FIBRE.

It was determined with a dead-end constant pressure apparatus as described by Yamagishi *et al.* (1991). An 8-cm long hollow fibre was positioned in a U-shaped configuration and pure water was forced to permeate outwards at a constant filtration pressure of 1 bar, in the dead-end mode. Space velocity (SV) was calculated as the flow rate divided by the membrane volume.

2.6. MEASUREMENT OF THE AMOUNT OF IDA INTRODUCED

The amount of IDA introduced was determined from the measurement of the copper saturation capacity assuming a stochiometric ratio. Copper content was measured spectrophotometrically by soaking the fibres with 0.1 M EDTA, pH 7.5, for 3 h and comparing the absorbance of the supernatant at 715 nm with that of 0.1 M EDTA with Cu(II) at various concentrations (Wuenschell *et al.*, 1991).

2.7. ADSORPTION ISOTHERMS MEASUREMENT

The adsorption isotherms for histidine, lysozyme, myoglobin, haemoglobin and pectin-esterase binding to IDA membranes were measured basically as described by Chase (1984). A 20 mM sodium phosphate buffer, pH 7.0, 0.25 M NaCl was used as the adsorption buffer. Small pieces of IDA membrane were put into tubes (approximately 10 μl membrane volume in each one) containing increasing amounts of each adsorbate, in a final volume of 1.5 ml. The suspension was stirred gently for 24 h at 20°C to allow the system to reach its equilibrium. Each adsorbate solution was then removed and its concentration or activity at equilibrium (c*) was determined as indicated above. The equilibrium concentration or activity of adsorbate bound to the membrane per unit of total membrane volume (q*), was calculated as the total amount of adsorbate present at the beginning of the experiment less the amount still in the soluble phase at equilibrium.

Values for the dissociation constant (Kd) and the maximum adsorption capacity (qm) were determined according to Chase (1984) and are given as the mean $\pm$ SE.

2.8. ASSEMBLING A HOLLOW-FIBRE MEMBRANE MODULE

A/G Technology Co., Needham, Massachusetts, USA donated the module cartridge. The cartridge had four openings: two on the lumen side and two on the shell side. Ten chelating hollow fibres, 8 cm long, were put into the cartridge in parallel and plugged at both ends using epoxy resin. The effective membrane length was 6.5 cm (total volume, 0.408 ml).

2.9. BREAKTHROUGH CURVES FOR PE AND PL ADSORPTION

The sample was a solution of Biocon Bioconcentrated Plus 23 mg/ml in a 20 mM sodium phosphate buffer, pH 7.0, 0.25 M NaCl containing 600.3 U/ml of PE and 289 U/ml of PL. It was pumped at a SV of 5 min^{-1} through the cartridge in the dead-end flow mode. The lumen side was used as an inlet and the shell side as an outlet for the permeate. The outlet of the cartridge was monitored for PE and PL activity in all the fractions collected. Figure 2 shows a schematic diagram of the system utilised for breakthrough curves measurement.

2.10. UTILISATION OF THE CU(II)IDA-CARTRIDGE FOR PECTIC ENZYME FRACTIONATION

5 ml solution of Biocon Bioconcentrated Plus 23 mg/ml in 20 mM sodium phosphate buffer, pH 7.0, 250 mM NaCl was pumped through the cartridge in the dead-end flow mode. PE was eluted with 0.1 M EDTA, pH 7.0.

The activity of PE and PL was measured in the washing and in eluate solutions.

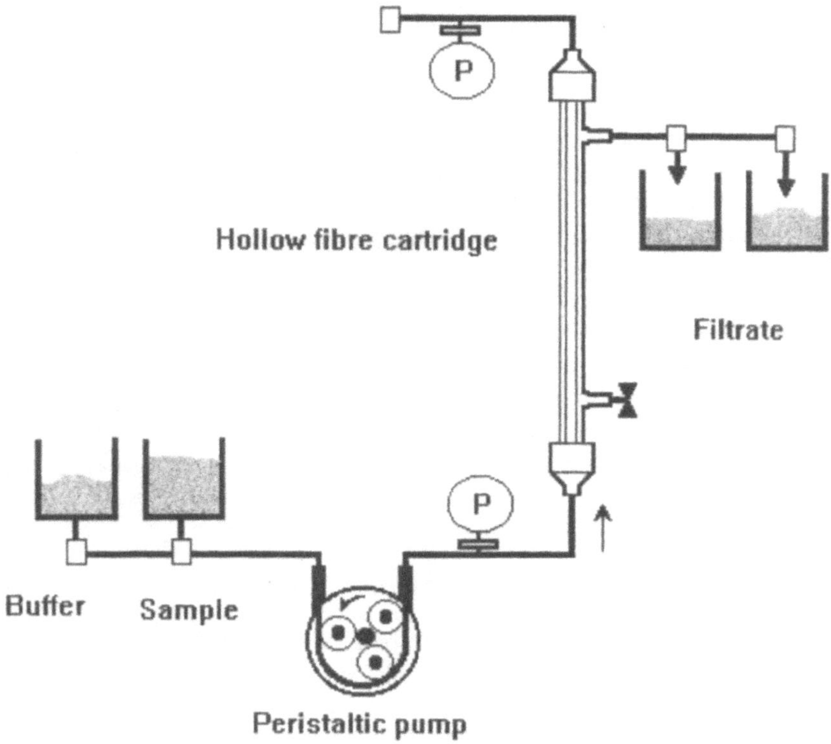

Figure 2. Schematic diagram of the system utilised for breakthrough curves measurement

3. Results and discussion

3.1. CHROMATOGRAPHIC CHARACTERISATION OF THE DERIVATISED MEMBRANES

IDA was immobilised on epoxy-activated PSU membranes. The Cu(II)IDA-membranes had a copper saturation capacity of 60 μmol/ml and a pure water SV of 234 min^{-1} at a filtration pressure of 1 bar.

Non-selective adsorption of biomolecules on the derivatised hollow fibres was assessed through histidine and lysozyme adsorption onto the IDA-membranes in the absence of copper. A negligible biomolecule adsorption was observed.

The adsorption isotherms for histidine, lysozyme, myoglobin and haemoglobin binding to Cu(II)IDA membranes showed a good fit of experimental data to a Langmuir-type isotherm as is indicated in Figure 3. Table 1 shows the calculated maximum adsorption capacity (qm) and the values for the dissociation constant (Kd) for the pure adsorbates.

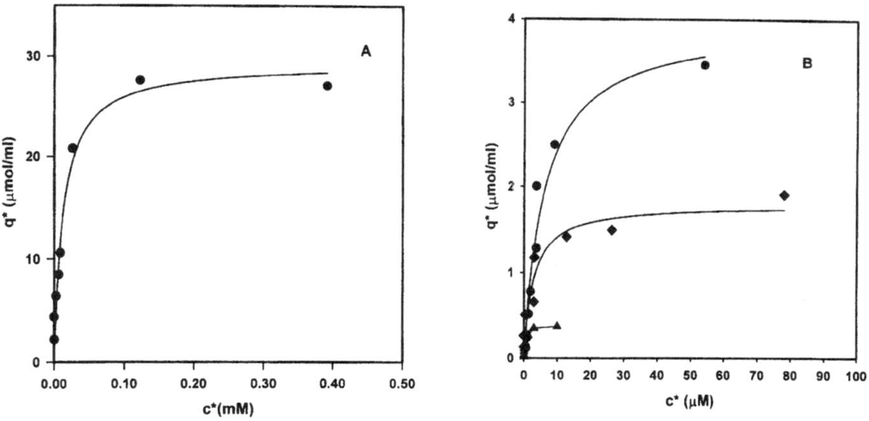

Figure 3. Adsorption isotherm for the binding of (A) histidine and (B) pure proteins (lysozyme ●, myoglobin ◆ and haemoglobin ▲) to the Cu(II)IDA immobilised onto hollow-fibre membranes

Table 1. Maximum adsorption capacity (qm) and dissociation constant (Kd) for pure adsorbates.

Adsorbate	qm (µmol/ml)	Kd (µM)	Molecular weight (kDa)	Surface Histidines
Histidine	29±2	13±3	-	-
Lysozyme	4.0±0.3	10±1	14.6	1
Myoglobin	1.8±0.2	2.8±0.5	18.8	4
Haemoglobin	0.38±0.01	0.41±0.06	68.0	24

The molar ratio of histidine adsorbed to Cu(II)-IDA was approximately 0.5 thus indicating that only 50% of the immobilised ligand was accessible to histidine. This reduced capacity was attributed to inaccessibility of the chelated copper ions and steric hindrance as well.

Lysozyme saturation capacity (qm) was 4.0 µmoles/ml, similar to those of the commercially available chelating gels (6.8 to 7.5 and 3.8 to 4.1 µmol of lysozyme per ml of Chelating Sepharose Fast Flow and TSK Gel Chelate respectively).

The accessibility of the ligand immobilised onto the membrane for interaction with proteins decreased with the rise in molecular weight of the protein - a result of the molecular-sifting properties of the membrane. On the other hand, the adsorption of proteins to IDA membranes could block the access to other copper sites thus explaining why the IDA to protein molar ratio increased with the protein molecular weight (Belew, *et al.*, 1987).

Table 1 shows that the Kd values decrease with the number of surface histidines. Many chromatographic studies on protein and peptides demonstrated that retention in Me(II)-IDA adsorbents is dictated primarily by the availability of histidyl residues (Arnold, 1991).

The adsorption isotherm of the PE binding showed a qm of 8000 U/ml and a Kd value of 20.3 U/ml.

3.2. PROPERTIES OF THE HOLLOW-FIBRE MEMBRANE MODULE

The SV for the module was the same as that of a single hollow fibre. This result indicates an advantage of membrane chromatography over conventional bead-packed columns: scale-up of the former does not require a high-pressure pump whereas scale-up of the latter does so, unless a thin column with a large diameter, i.e., the equivalent of a functional porous membrane, is used.

The chemical stability of the IDA-membrane cartridge was examined by repeated adsorption and elution cycles of lysozyme. The capacity of the module to adsorbed lysozyme stayed constant thus evidencing chemical stability of the ligand.

3.3. BREAKTHROUGH CURVES FOR PE AND PL ADSORPTION

Figure 4 shows the breakthrough curve for the adsorption of PE and PL working at a SV of 5 min^{-1} and with an input stream adsorbate concentration of 600.3 U/ml of PE and 289 U/ml of PL. The dynamic capacity of the column under these conditions was 7500 PE U/ml. PL was not adsorbed by the chromatographic membrane.

3.4. UTILISATION OF THE IDA-CARTRIDGE FOR PECTIC ENZYME FRACTIONATION

In order to test the usefulness of the cartridge for pectic enzyme fractionation, the hollow-fibre cartridge (0.408 ml) was loaded with 3000 U of PE and 1445 U of PL (5 ml of Biocon Bioconcentrated Plus 23 mg/ml) at a SV of 5 min^{-1}.

Figure 5 shows the pattern obtained. 99 per cent of the PE activity was retained by the chromatographic matrix and eluted quantitatively with 0.1 M EDTA, pH 7.0, thus indicating that the fractionation procedure can be successfully scaled-up. The time of the fractionation process, 10 min was far shorter than when working with chelating soft gel: 50 min (Camperi *et al.*, 1996) where lower flow rates must be used to allow mass transfer. The better hydrodynamic properties of the membranes resulted in an enormous saving of time and a higher productivity: 750 PE U/ml.min compared with that previously obtained working with chelating soft gels: 52 U/ml.min (Camperi *et al.*, 1996).

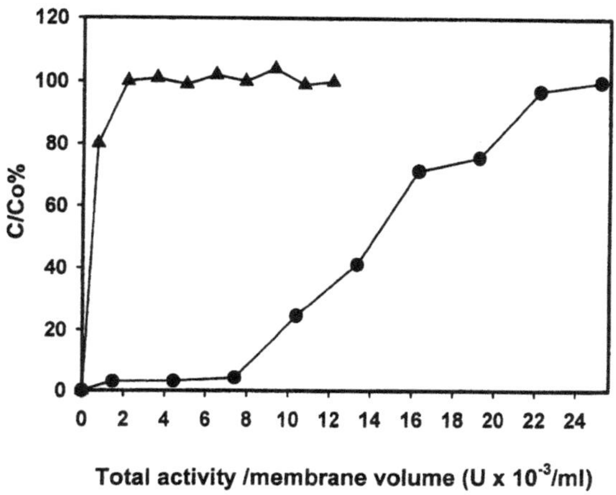

Figure 4. Breakthrough curve for the adsorption of PE (●) and PL (▲) working at a SV of 5 min^{-1} and with input stream adsorbate concentration of 600.3 U/ml of PE and 289 U/ml of PL.

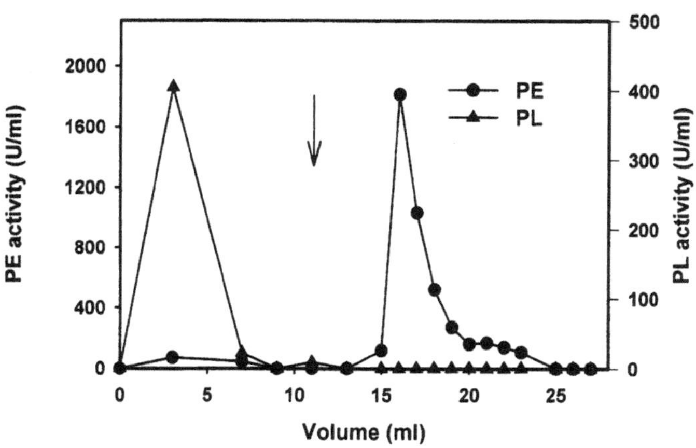

Figure 5. Elution pattern of Biocon Bioconcentrated Plus chromatographed on the Cu(II)-IDA hollow-fibre cartridge. Details in Materials and Methods. The arrow indicates the buffer change.

4. Conclusions

The high capacity of the membrane cartridge for PE and its excellent hydrodynamic properties allows a very fast fractionation of a commercial pectic enzyme preparation at a low operating pressure. The fraction passing through - containing all the PL activity loaded onto the cartridge - can be used directly to clarify fruit juice without production of methanol.

Acknowledgements

This work was supported by grants from the Universidad de Buenos Aires, the Consejo Nacional de Investigaciones Científicas y Técnicas de la República Argentina (CONICET) and the Agencia Nacional de Promoción Científica y Tecnológica.
M.G. and O.C. are career researchers of the CONICET.

References

Alaña, A., Gabilondo, A., Hernando, F., Moragues, M.D., Domínguez, J.B., Llama, M.J. and Serra, J.L. (1989) Pectin lyase production by a *Penicillium italicum* strain, *Appl. Envir. Microb. 55*, 1612-1616.

Albersheim, P. (1966) Pectinlyase from fungi, in: Neufeld, E.F. and Ginsburg, V. (eds.), *Methods in Enzymology*, Academic Press, New York, vol. 8, pp.628-631.

Arnold, F.H. (1991) Metal-affinity separations: a new dimension in protein processing, *Bio/Technology 9*, 151-155.

Belew, M., Yip, T.T., Anderson, L. and Porath, J. (1987) Interaction of proteins with immobilised Cu^{2+}. Quantitation and equilibrium constants by frontal analysis, *J. Chromatog. 403*, 197-206.

Brandt, S., Goffe, R.A., Kessler, S.B., O'Connor, J.L. and Zale, S.E. (1988) Membrane-based affinity technology for commercial scale purification, *Bio/Technology 6*, 779-782.

Camperi, S.A, Auday, R.B, Navarro del Cañizo, A.A.and Cascone, O. (1996) Study of variables involved in fungal pectic enzyme fractionation by immobilised metal ion affinity chromatography, *Process Biochem. 31*, 81-87.

Camperi, S.A., Navarro del Cañizo, A.A., Wolman, F.J., Smolko, E.E., Cascone, O. and Grasselli, M. (1999) Protein adsorption onto tentacle cation-exchange hollow-fiber membranes, *Biotechnol. Prog. 15*, 500-505.

Chase, H.A. (1984) Prediction of the performance of preparative affinity chromatography, *J. Chromatog. 297*, 179-202.

Hemdan, E.S., Zhao, Y., Sulkowski, E. and Porath, J. (1989) Surface topography of histidine residues: a facile probe by immobilised metal ion affinity chromatography, *Proc. Nat. Acad. Sci. USA 86*, 1811-1815.

Kroner, K.H, Krause, S. and Deckwer, W.D. (1992) Cross-flow anwendung von affinitätsmembranen zur primärseparation von proteinen, *Bioforum 12*, 455-458.

Kubota, N., Konno, Y., Saito, K., Sugita, K., Watanabe, K., Sugo, T. (1997) Comparison of protein adsorption onto porous hollow-fiber membrane and gel bead-packed bed, *J. Chromatog. 782*, 159-165.

Mueller-Schulte, D. and Daschek, W. (1995) Application of radiation grafted media for lectin affinity separation and urease immobilisation: a novel approach to tumour therapy and renal disease diagnosis, *Radiat. Phys. Chem. 46*, 1043-1047.

Navarro del Cañizo, A.A., Hours, R.A., Miranda, M.V. and Cascone, O. (1994) Fractionation of fungal pectic enzymes by immobilised metal ion affinity chromatography, *J. Sci. Food Agric. 64*, 527-531.

Porath, J., Carlsson, J., Olsson, I. and Belfrage, G. (1975) Metal chelate affinity chromatography, a new approach to protein fractionation, *Nature 258*, 598-599.

Rombouts, F.M. and Pilnik, W. (1980) Pectic Enzymes, in: Rose AH (ed.) *Economic Microbiology*, Academic Press, London, vol. 5, pp. 228-282.

Roper, D.K. and Lightfoot, E.N. (1995) Separation of biomolecules using adsorptive membranes, *J. Chromatog. A*., 702, 3-26.

Saito, K., Ito, M., Yamagishi, H., Furusaki, S., Sugo, T. and Okamoto, J. (1989) Novel hollow fiber membrane for removal of metal ion during permeation: preparation by radiation-induced co-grafting of a cross-linked agent with reactive monomer, *Ind. Eng. Chem. Res. 28*, 1808-1812.

Szajer, I. and Szajer, Cz.(1982) Clarification of apple juices by pectin lyase from *Penicillium paxilli*, *Biotechnol. Lett. 4*, 553-556.

Thömmes, J, and Kula, M.R. (1995) Membrane Chromatography- An integrative concept in the downstream processing of proteins, *Biotechnol. Prog. 11*, 357-366.

Vilariño, C., Del Giorgio, J.F., Hours, R.A. and Cascone, O. (1993) Spectrophotometric method for fungal pectin-esterase activity determination, *Lebensm Wiss u Technol*. 26, 107-110.

Wuenschell, G. E., Wen, E., Todd, R., Shhek, D. and Arnold, F. H. (1991) Aqueous two-phase metal affinity extraction of heam proteins. *J. Chromatog.* 543, 345-354.

Yamagishi, H., Saito, K., Furusaki, S., Sugo, T. and Ishigaki, Y. (1991) Introduction of high-density chelating group into a porous membrane without lowering the flux. *Ind. Eng. Chem. Res.* 30, 2235-2237.

PART VIII
ECONOMIC FINALITIES

ECONOMIC BENEFITS OF THE APPLICATION OF BIOTECHNOLOGY - EXAMPLES

MARLENE ETSCHMANN, PETER GEBHART AND DIETER SELL
DECHEMA e.V., Theodor-Heuss-Allee 25, D-60486 Frankfurt am Main, Germany, fax: ++49 69 75 64 388, e-mail: sell@dechema.de

Summary

Biotechnological processes or process steps can substitute traditionally applied techniques. Of all the arguments to be considered when choosing a biotechnical process, improving the process economy is the most important one.

Different areas of application for biotechnology have been examined and different process alternatives (biotechnological vs. conventional processes) compared from an economic point of view.

Overview

Biotechnological methods are increasingly being used to substitute chemical processes in a wide range of industries. This even affects sectors where at first sight it may seem surprising to find biotechnology at work, for example in textile finishing or in pulp and paper production. Biotechnological methods may have innumerable advantages if looked at from a researcher's point of view, but to be applied in practice they have to meet one stipulation: they have to be cheaper than the conventional process. Environmental advantages are often only regarded as a pleasing side effect. A change in thinking still has to take place to the effect that costs for environmental protection should be considered during process design. Production integrated measures can reduce environmental costs dramatically compared to conventional end-of-pipe techniques, which are added to existing processes.

In the following, five examples are presented which demonstrate that it is economically and ecologically advantageous to replace a traditional process by one with one or more biotechnological steps. The first two, the production of 7-aminocephalosporanic acid and stone washing of jeans, are well known and were assessed some time ago. The next two, production of riboflavin and biopulping are

M. Hofman and P. Thonart (eds.), Engineering and Manufacturing for Biotechnology, 353–360.

currently being implemented on an industrial scale. The third, bleach cleanup in textile finishing, has been applied for several years, but was investigated and calculated only recently by the authors.

1. Production of 7-aminocephalosporanic acid

7-aminocephalosporanic acid (7-ACA) is a pharmaceutical chemical, a key product for most semisynthetic cephalosporin antibiotics. Hoechst Marion Roussell, which uses it to produce several antibiotics, has developed a process based on biochemical catalysts. In the two-stage enzymatic synthesis D-alpha-amino acid oxidase and glutaryl amidase are used to form 7-ACA. The reactions are carried out at room temperature in an aqueous solution. In contrast to the chemical process no chlorinated hydrocarbons, toxic auxiliaries or heavy-metal salts are needed.

This is considerably more environmentally friendly than the chemical process formerly applied and reduces the percentage of process costs used for environmental protection (including waste incineration, purification of waste water and waste gas) from 21% to 1%. The absolute environmental protection costs are thus reduced by 90% per tonne of 7-ACA. (Wiesner et al. 1995, OECD 1998).

Table 1: Comparison of chemical and enzymatic production of 7-ACA

waste per t 7-ACA	chemical process	enzymatic process
waste to be incinerated	31 t	1 t
zinc (as ZnH_4PO_4)	1.8 t	-
waste water	0.1 t COD	1.7 t COD
waste gas emissions	7.5 kg	1 kg

2. Stonewashing of jeans

Many of the 70 million jeans sold in Europe every year are stonewashed. This means that they are subjected to a washing process which locally abrades the indigo dyestuff from the cotton yarn and thus produces the desired look. In the past the abrasion effect on the garments originated from pumice stone. Nowadays there is an increasing tendency to use cellulase enzymes or a combination of pumice and enzymes. The Institute for Applied Environmental Economics, The Hague, performed a life cycle assessment (LCA) to compare these three methods with respect to their environmental economic costs. The results show that enzymatic stone washing, also known as "bio-stoning", has numerous advantages, most of which are based on the simple fact that no stones are involved. Thus, there is

- no need to remove pumice fragments from the garments
- no need to landfill pumice sludge
- less machine damage
- lower maintenance costs

to name just a few of the benefits.

Bio-stoning scores best in the comparison of environmental economic costs, however European textile finishers prefer to stone wash with a mix of enzymes and pumice as only the stones create the effect desired by the European consumer (Kothuis and Schelleman 1996).

Table 2: Comparison of different stonewashing methods

Environmental economic costs for 3 washing methods (in ECU)	pumice	mix	cellulase
Air	8.31	5.57	4.13
Water	28.10	20.16	16.37
Waste	2.01	1.26	0.62
Total	8.42	26.99	21.12

3. Production of riboflavin

Riboflavin, or Vitamin B_2, is produced on a scale of several thousand tons per year. It is used as a vitamin in feed, food and pharmaceutical applications. Because of its yellow colour, it can also be used as a food colorant.

For large-scale applications, riboflavin is produced by a combination of chemical and fermentation processes. First, ribose is obtained by fermentation, then it is converted into riboflavin by a multistep chemical process. This procedure has been continuously improved in the last decades, with the principle still dating back to the 1930s. Only in the 1990s did a fundamental change take place, insofar as bacterial strains were developed which directly transform glucose to riboflavin. These strains were produced by a combination of classical mutation-selection and molecular biology methods. The quality of the riboflavin produced is equivalent or even slightly superior to riboflavin from chemical synthesis. The difference in the processes lies in the environmental impact: Biotechnological production uses almost exclusively renewable raw materials. The use of organic solvents and other chemical substances can be reduced, air emissions and waste are decreased by 36%. 25% less energy is used compared to the conventional process. These figures were determined by F. Hoffmann-La Roche in the preliminary stages of the building of a commercial-scale riboflavin fermentation plant at Grenzach, Germany. Verification of the figures is anxiously awaited when the facility comes on stream in mid-2000 (Loon van, et. al. 1996, Eggersdorfer et al. 1996, Bretzel et al. 1999).

4. Biopulping

Pulp and paper production is an industrial section where the idea of production integrated biotechnology took root relatively early. The first studies on the application of fungi for wood delignification date back to the 1950s. Significant energy savings by fungal treatment in mechanical pulping were first demonstrated in the 1970s, but it took some twenty years more to achieve industrial application.

To make paper, the wood fibres which are "glued" together by lignin have to be separated from each other. This can be done by chemical degradation and removal of the lignin (chemical pulping) or by physically tearing the fibres apart (mechanical pulping).

About 25% of the world's wood pulp production is produced by mechanical pulping, which has a high yield, but is energy-intensive.

In two research projects, Biopulping Consortium I and II, joint research groups from US universities and industry evaluated the technical and commercial feasibility of using a fungal pretreatment with mechanical pulping to save energy and/or improve paper quality. The assumption that fungal pretreatment would have less environmental impact than chemical pretreatment proved to be right (Akhtar et al. 1995). The fungi alter the wood cell walls, soften the chips and thus substantially reduce the electric energy needs for pulping. The paper quality increases and 30% electric energy can be saved by inoculating the wood chips (Scott 1998). Many strains of fungi were studied, of which *Ophiostoma piliferum* proved to be one of the most efficient. It is available under the product name Cartapip®, marketed by Agrasol Inc., Charlotte, North Carolina, USA.

5. Bleach cleanup

The textile finishing industry is characterised by high consumption of energy and resources and time-consuming production processes. For these reasons production-integrated biotechnological processes could make a considerable contribution to conserving energy and water, reducing emissions and to shortening the processes and consequently the throughput time.

As a rule, process innovations that "only" relieve the environment are not sufficient incentive to companies to modify their operations. They are at most desirable by-products. Only economic advantages convince decision-makers in companies to apply ecologically advantageous, innovative processes. The textile finishing industry differs from other branches in that it is scarcely possible to offer unrivalled products and new or significantly improved quality. Thus, the substitution of a process by one that is economically advantageous can make a considerable contribution towards consolidating or improving the position of a company with regard to the competition.

Nowadays hydrogen peroxide is generally used for bleaching textiles. To achieve high quality in the ensuing dyeing process it is necessary to remove bleach residues as completely as possible from the textile.

In conventional processes residual peroxide is removed by repeatedly (at least twice) rinsing the textile in hot water. This method is not only energy and water-intensive, but cannot guarantee the complete removal of residual hydrogen peroxide that is required.

For this reason an enzymatic process was developed that may now be considered to be established and that has been extensively, but by no means exhaustively, applied in the textile finishing industry. With this new, biotechnological process only one high-temperature rinse is necessary (at 80-95°C, depending on the type of fabric) after the oxidative bleaching. The catalase enzyme is added to the next rinse and allowed to react for approx. 15 minutes at 30-40°C. In the example studied the compound "KAPPAZYM AP-Neu" (Kapp Chemie GmbH, Miehlen, Germany) was applied. The enzyme degrades residual peroxide into water and oxygen. Then the necessary consecutive steps can be started. The results are of significantly higher quality compared with the conventional process.

As one rinse is omitted, both the water and energy consumption and the process duration are reduced.

5.1. MATERIALS AND METHODS

An analysis of the bleach cleanup process during textile finishing was performed at Windel Textil GmbH & Co (Bielefeld, Germany), a medium sized textile finishing company, in 1998.

5.1.1. Selection of the production plant

As there are numerous, only slightly differing bleaching processes, a representative process had to be found for each of the two machine types used, namely the beam and the jet dyeing machines. The beam-dyeing machine derives its name from the fact that the textile roll is wrapped around beam-shaped metal cylinders made of perforated steel. The cylinder with the material is then inserted into the machine.

With the jet-dyeing machine, the textile is transported through the dyeing liquid by the action of jets. Thus the mechanical impact on the fabric is minimised.

In the period under study, May to July 1998, 355 bleaching processes were carried out using the catalase KAPPAZYM AP-Neu. 281 processes, which is approx. 80% of the total, were carried out on the beam dyeing machines, the remaining 20% (74 processes) on the jet dyeing machines.

The calculations shown were made for a beam dyeing machine with a capacity of 5,800 l of liquid and an average load of 226 kg of knitted fabric containing cotton. The jet dyeing machine chosen holds 157 kg of material and 3,000 l of liquid per run.

5.1.2. The process

Figure 1 shows the sequence of a representative bleaching process in a beam dye. After oxidative bleaching with H_2O_2, residual peroxide has to be removed so as not to interfere with subsequent steps. The traditional way to clean up bleach is by twice rinsing with hot water. In the new process "Kappazym AP-Neu" is added after the first rinse. This compound contains the catalase enzyme, which converts any remaining peroxide into water and oxygen. Thus, the second rinsing cycle can be omitted (grey area in Fig. 1), as the liquid is clean enough to start the next process step, reductive bleaching.

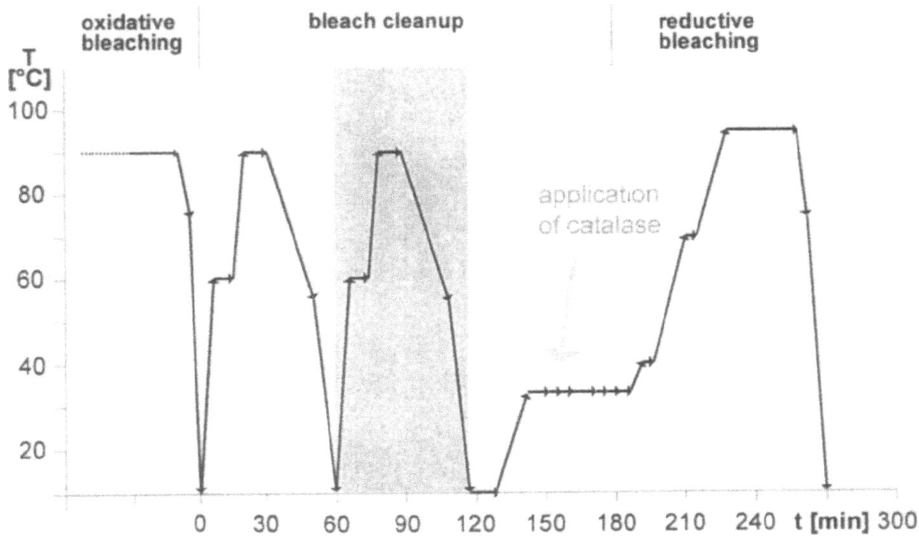

Figure 1: Sequence of a bleaching process in a beam-dyeing machine

5.1.3. Economic analysis

On account of the technical specifications of the two processes, it was not necessary to make any changes to the textile finishing plant that required investments. Only the programmes for the machine control system had to be modified. To determine the economic differences between the old and the new process, a complete analysis of the two was performed. In this, the fixed and proportional costs taken from the current calculation, the current cost for chemicals and resources and the current municipal prices for water and energy were used. Comparability of the economic results over a longer time span was optimised by basing the analysis on a year's production.

5.2. RESULTS

For data protection reasons no absolute figures can be given, but the costs of the new process are given in relation to the traditional process in Table 3.

Besides hydrogen peroxide a whole range of other chemicals are used in bleaching blended fabrics, e.g. stabilisers, common salt and fabric-protective agents. Steam is required to heat the water, cooling water to cool it. The term process water means water that comes into direct contact with the textiles.

Table 3 : Comparison of process costs according to type of machine

	Beam dyeing machine			Jet dyeing machine		
Costs for:	Savings	Percentage of total costs	Weighted savings	Savings	Percentage of total costs	Weighted savings
chemicals	+7.18	34.33%	+2.47%	+10.73	17.11%	+1.84%
steam	-14.36	4.81%	-0.69%	-9.30	3.67%	-0.34%
cooling water	-16.66	9.99%	-1.66%	-19.48	12.73%	-2.48%
process water	-19.97	20.06%	-4.01%	-19.97	14.95%	-2.99%
other finishing costs	-9.77	30.82%	-3.01%	-8.98	51.54%	-4.63%
Total		100.00%	-6.90%		100.00%	-8.60%

The enzymatic process can achieve savings both with the beam dyeing and the jet dyeing machines. Admittedly costs for chemicals rise by 7% and 11% respectively. This is due to the additional cost of the enzyme. In all other areas the costs drop, some by up to 20%. Moreover it should be borne in mind that the individual cost factors represent different percentages of total costs. As one rinsing cycle is omitted completely, the reduction in process water costs is particularly striking. The enzyme application reduce the bleaching process by one hour. Thus, costs such as those for labour, machinery and electricity, among others, are also reduced. Theses costs are included in "other finishing costs" and contribute substantially to the savings.

The enzymatic process finally turns out to be about 7% and 8% cheaper respectively than the traditional process.

The substitution of multiple rinsing by enzyme application in the bleach-cleanup process at Windel Textil GmbH & Co produced the following advantages:

- natural resources are saved by the reduction in water consumption (as a rinsing agent as well as a coolant for the whole process) and steam (as the source of process energy)
- the environmental impact is reduced both by the decrease in use of resources and the lower production of wastewater
- the process is significantly cheaper for the company. Depending on the machine type used and the fabric to be treated, costs were reduced by 7 to 8%
- none of the machinery had to be modified

These results demonstrate that without big technical or financial investment even a small change in a production process can lead to a significant decrease in use of resources and environmental impact. It is also an effective means of improving a company's competitiveness.

6. Conclusions

Biotechnological processes offer many opportunities for application in different industrial areas. However, an awareness of these possibilities is frequently lacking. This may – at least in part – have to do with the public perception of biotechnology. Most of the time biotechnology is portrayed as a universal technology for fighting human diseases and world hunger, whereas biotechnology in sectors other than pharmaceutical, medical or agricultural rarely hits the headlines.

Entrepreneurs perceive environmental protection negatively as a cost factor. According to studies carried out for the European Commission in 1995, 77% of all entrepreneurs asked stated that environmental protection based on legislation measures resulted in increased costs. It has to be borne in mind that the majority of investments regarding environmental protection are still made for end-of-pipe or add-on technologies, which will never be productivity factors. The situation is different for preventive and integrated measures. They can create competitive advantages and even decrease operational costs (Heiden 1999). As shown previously, biotechnology can be integrated into industrial processes with benefits both from an economical and from an ecological point of view. This applies to many industrial fields, if executives would only recognise biological processes as an equally good alternative. An analysis, which shows their economic benefits, may be a convincing argument for a decision in favour of a biotechnical process.

References

Akhtar M., Kirk T.K., Blanchette R.A. (1995) Biopulping: An overview of consortia research. In: Srebotnik, E., Messner, K. (eds) Proceedings of the 6th Int. Conference on Biotechnology in the Pulp and Paper Industry: Advances in Applied and Fundamental Research. Facultas Universitätsverlag. Wien.

Bretzel W., Schurter W., Ludwig ,. B., Kupfer E., Doswald S., Pfister M., Loon van A. P. G. M. (1999). Commercial riboflavin production by recombinant *Bacillus subtilis*: down-stream processing and comparison of the composition of riboflavin produced by fermentation or chemical synthesis. Journal of Industrial Microbiology & Biotechnology 22, 19-26.

Eggerdorfer M. et al. (1996). Vitamins. In: Elvers, B., Hawkins, S. (eds) Ullmann's Encyclopedia of Industrial Chemistry, Vol A 27 443-613.

Heiden S.(1999). Integrierter Umweltschutz – Biotechnologie auf neuen Wegen. In: Industrielle Nutzung von Biokatalysatoren. Heiden S., Bock A.-K., Antranikian G. (eds). Erich Schmidt. Berlin. 3-26.

Kothuis B., Schelleman F. (1996) Environmental Economic Comparison of Biotechnology with Traditional Alternatives. Institute for Applied Environmental Enonomics (TME), The Hague.

Loon van A. P. G. M., Hohmann H.-P., Bretzel W., Hümbelin M., Pfister M. (1996). Development of a Fermentation Process for the Manufacture of Riboflavin. Chimia 50 (9), 410-412.

OECD (1998) Biotechnology for clean industrial products and processes. 83.

Scott G. M., Akhtar M., Lentz M. J., Kirk K. T., Swaney R. (1998) New technology for papermaking: commercializing biopulping. Tappi Journal 81 (11), 220-225.

Wiesner J. et al. (1995) Production-Integrated Environmental Protection. In: Ullmann's Encyclopedia of Industrial Chemistry, Vol B 8 213-309

ENZYME STABILITY AND STABILISATION : APPLICATIONS AND CASE STUDIES.

DR. GUIDO A. DRAGO* AND DR. TIM D. GIBSON.
Applied Enzyme Technology Ltd, 175 Woodhouse Lane, Leeds, LS2 3AR, UK. Telephone: +44-113-233-3030 FAX: +44-113-233-2593 e-mail G.A.Drago@leeds.ac.uk. (Corresponding author)*

Summary

The stabilisation of enzymes is of great importance in many applications. The two main types of stability may be defined as: 1) Storage or Shelf Stability and 2) Operational Stability. The first relates to the stability of enzymes when stored as a dehydrated preparation, a solution or immobilised and is particularly concerned with retention of activity over time. The second generally relates to the retention of activity of an enzyme when in use. Both types of enzyme stability will be discussed using case studies from the analytical field (alkaline phosphatase, alcohol oxidase, acetylcholine-esterase and a recombinant luciferase) and enzyme based biosterilisers (peroxidases). The introduction of an electrophoretic technique for predicting protein-polymer interactions will be described. In addition, stabilisation using covalent immobilisation of pre-stabilised enzyme complexes will be presented using glucose oxidase as an example and a brief discussion on the likely factors influencing stability of enzymes is included.

1. Introduction

The stabilisation of enzymes is of great importance in a variety of applications. Enzymes are used in the fields of biocatalysis, analytical chemistry, food processing, environmental treatment, detergent manufacture, biosensor production for medical diagnostics and other measuring applications, to name but a few. In all these areas the retention of the biological activity of the enzyme molecule is paramount, and this depends on stabilisation of the biological structure of the enzyme. In most cases, the actual mechanism of stabilisation of enzymes is a little understood phenomena. As enzyme structures are solved, reaction mechanisms are understood and the mechanisms of protein folding and deactivation are worked out, the mechanisms of enzyme

M. Hofman and P. Thonart (eds.), Engineering and Manufacturing for Biotechnology, 361–376.

stabilisation are also becoming more fully understood. With the combination of expertise in the field of enzyme stabilisation and the advances in the methods used for predicting enzyme structural changes during the degradation processes, it is becoming increasingly easy to predict how to stabilise an enzyme for a specific industrial application.

Protein engineering can be a useful tool for increasing the stability of certain enzymes, for example luciferase, bacterial proteases (e.g. savinase - used in detergents) and carbohydrases, for example β -glucosidase. Of course this is only possible so long as structural data is available for the enzyme under examination. The engineering of an enzyme structure can also lead to instability that is not always by design. The practicality of carrying out such time consuming studies on altering enzyme structure in order to improve stability is a matter of some debate. On the one hand one can generate an enzyme with all the attributes required of its application, on the other one cannot always predict the result of a particular mutation. In any case, even when an enzyme has been stabilised in this fashion, practical application of traditional techniques, such as granulation in the presence of stabilising additives are still applied. One very important point is to define what is meant by the term 'enzyme stability'.

The two main types of stability may be defined as:

- Storage or Shelf Stability
- Operational Stability.

The first relates to the stability of enzymes when stored as a dehydrated preparation, a solution or as an immobilised preparation and is particularly concerned with retention of activity over time. Clearly such considerations are extremely relevant for enzyme producers from the point of manufacture of their products to the supply of the end users. The shelf life of enzyme based products generally depends on the stability of the enzyme in this context. The second generally relates to the retention of activity of an enzyme when in use. This is important for systems using enzymes for biocatalysis or biotransformations and analytical monitoring systems. The retention of activity in this context is often measured in terms of a half-life or T_{50} (where the deactivation does not follow first order kinetics), referring to the time taken for the amount of enzyme activity to fall to half its original value.

Both types of enzyme stability will be discussed in this paper, using case studies from the analytical field (alkaline phosphatase, alcohol oxidase, acetylcholine-esterase, a recombinant luciferase and enzyme based biosterilisers (peroxidases). In all cases the stabilisation technique described has been restricted to the use of additives to modify the microenvironment of the enzyme under investigation. No other technique will be described in detail, except covalent immobilisation of pre-stabilised enzyme complexes. Initial trials using this system, (marketed as the PolyEnz™ Process by AET Ltd) have indicated significantly higher levels of thermal stability for immobilised biocatalysts, glucose oxidase being used as a model enzyme [1]. The potential to produce operationally stable biocatalysts for use in the biosynthesis / biotransformation field will be discussed. Also a brief discussion of the likely factors influencing stability of enzymes has been included.

2. Materials and methods

Alcohol oxidase from *Hansenula polymorpha* was prepared in house by the method of Gibson [2]. Acetylcholine-esterase (930 units mg^{-1}, true cholinesterase, type III from electric eel E.C.3.1.1.7) and β-galactosidase were purchased from Sigma. Horseradish peroxidase (HRP-4), glucose oxidase (GO-3) and bovine glutamate dehydrogenase (GLDH) were purchased from Biozyme Ltd. Recombinant luciferase was a gift from Celsis Ltd. The Pyrococcus furiosus glutamate dehydrogenase and β-glucosidase were a kind gift from Dr. Serve Kenyan at Wageningen Agricultural University, Holland. The polyelectrolytes DEAE-dextran and dextran sulphate were obtained from Amersham Pharmacia Biotech (Uppsala, Sweden) and Gafquat 755N, Gafquat HS-100 were obtained from ISP (Europe) Ltd, Guildford, Surrey, UK.

The techniques used for the stabilisation of enzymes described in this abstract are varied. Detailed protocols are described in the sited literature, except in certain instances as described below.

Alkaline phosphatase was stabilised as part of a UK government award (SMART Award). The actual formulations are available from Applied Enzyme Technology Ltd under license.

Alcohol oxidase and acetylcholine-esterase biosensors were prepared by carbodiimide immobilisation of the enzymes onto carbon electrodes, which were then treated by dip coating in a mixture of stabilisers. The sequence of transducer activation, enzyme immobilisation and subsequent stabilisation using polyelectrolyte-protein complex formation are described in Gibson *et al.* [3, 4] and Rippeth *et al.* [5].

Solutions of HRP-4 were stabilised using admixtures of polyelectrolytes, polyalcohols and a specific buffer composition (Patent Application Pending). The actual formulation is available from Applied Enzyme Technology Ltd under license.

Solution stabilisation of recombinant luciferase was carried out by incubation of the enzyme with combinations of polyelectrolytes and polyalcohols in an commercial buffer supplied with the enzyme, the composition of which was unknown. The thermal degradation was carried out using the same techniques as described in Gibson [6] and Pierce *et al.* [7].

Immobilised glucose oxidase and pre-stabilised glucose oxidase-complexes were carried out by the procedure of Appleton *et al.* [1]. In all cases the biological activity of the enzyme was used as the main parameter to determine the stabilisation effect. Other techniques to determine any molecular and structural modifications occurring have been utilised. These include gel electrophoresis, circular dichroism, fluorescence and turbidimetric measurements. These methods have used to ascertain protein stability, especially where no simple method is available to directly measure biological activity, however in practical terms the results obtained from activity assays are usually sufficient.

The methods used to ascertain the effect of the stabilisers, generally focus upon thermostability as a suitable parameter for the demonstration of enzyme activity retention. This is a well accepted technique, provided the correct controls are evaluated and true long term stability studies are carried out in real time to corroborate short term, elevated temperature degradation's [8, 9]. The evaluation of biosensor shelf life using the standard techniques described for pharmaceutical protein shelf-life estimations at elevated temperatures has recently been published [10]. Dry stabilisation studies were investigated by incubation of the dehydrated enzyme preparations or biosensors over freshly dried silica gel as desiccant. Early studies used a single temperature as an indication of the stabilisation effect, usually 37°C whereas the later experimental procedures reported in McAteer *et al.* [10] used a series of different temperatures. The dehydrated preparations were assayed for residual enzyme activity at selected time points throughout the incubation period and the results usually depicted as a time course of enzyme activity retention (Arrhenius plot).

Solution stabilisation studies were carried out using an elevated temperature method, where the activity of the enzyme under investigation decayed to half the original activity within 15 - 20 minutes, this is described in detail in Pierce *et al.* [7]. Each enzyme exhibits a characteristic deactivation response to temperature, which is dependant on the buffer used, the pH and the molarity of the solution. The control values of any particular enzyme system were determined using a solution of native enzyme dissolved in a defined buffer system and the effects of potentially stabilising additives were determined by comparison to the control deactivation profiles observed.

The electrophoretic separation of protein-polyelectrolyte complexes was carried out using standard polyacrylamide isoelectric focusing. Samples were pre-incubated in polyelectrolytes for 30-60 minutes prior to electrorhetic separation. Focusing was carried out for 2.5 hours at 1500v. The gel was subsequently fixed for 15 minutes in 5% sulphosalicylic acid and 10% trichloroacetic acid, rinsed with destaining solution (30% methanol, 10% acetic acid, 60% distilled water). The gel was subsequently stained with Coomassie blue for 10 minutes and destained until the background staining was low and the bands appear easily distinguishable. The gel was then dried down onto Gel Bond PAG film (Pharmacia Biotech.).

3. Results

3.1. ALKALINE PHOSPHATASE SOLUTION STABILITY: ENZYME SOURCE AND BUFFER PARAMETERS

The source of the enzyme can be critical to the native stability of the enzyme and the ability of additives to further stabilise the enzyme in question. This is the case with alkaline phosphatase. Alkaline phosphatase isolated from bovine sources has a different stability profile compared with that isolated from bacteria (*Bacillus species)*. The stabilisation effects observed are more dramatic in the case of the bacterial enzyme compared with the bovine enzyme, figure 1.

Alkaline phosphatase is presented as an example of the effect of buffer choice in the apparent stability of an enzyme in solution, figures 2 and 3. It can be seen from these results that the buffer used can have an effect on the stability of an enzyme, when the pH and concentration are kept constant, figure 2. In this case the most significant effect is seen using HEPES buffer, which is clearly incompatible with this enzyme. In figure 3, the effect of pH is clearly apparent. The thermostability of the enzyme increases as the pH is reduced. These type of considerations are often extremely important in downstream processing of enzymes, where the wrong choice of buffer or pH can deactivate the enzyme being purified.

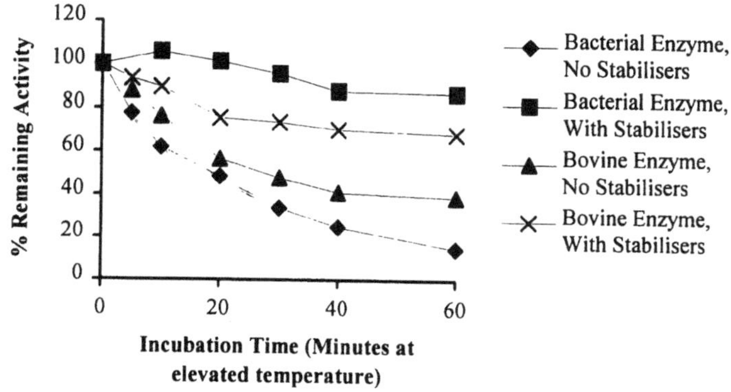

Figure 1. The thermostabilisation of both bovine and bacterial alkaline phosphatases is shown above. The activity increases for the bacterial enzyme is proportionately higher than the bovine enzyme, when the proprietary stabilisation system (developed at AET) is used. Temperatures for denaturation are 60.2°C for the bovine enzyme and 62°C for the bacterial enzyme, in Bis-Tris/HCl buffer, pH 6.5.

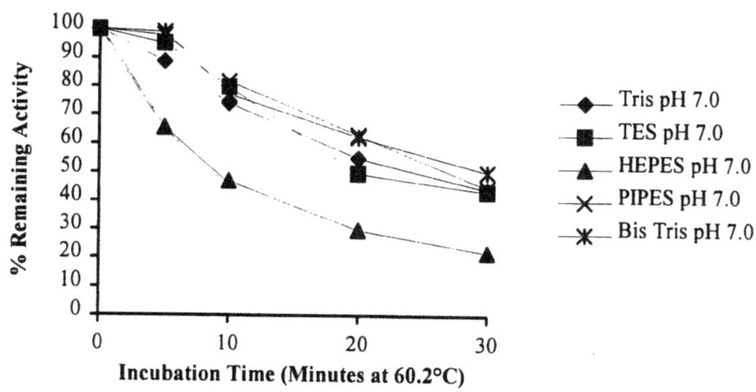

Figure 2. The effect of buffer type is shown here in figure 1. Bovine alkaline phosphatase shows different solution stability profiles depending on the buffer used. The pH and the temperature was held constant (pH changes with respect to temperature were corrected). It can be seen that HEPES buffer destabilises the enzyme, whereas all other buffers maintain roughly the same activity.

In our experience similar buffer specificity has been obtained with several other proteins including antibody conjugate solutions. The preference for Tris/HCl buffer systems over phosphate buffers by antibody conjugates in solution is measured by increased stability in the presence of Tris buffer. Another case in point is the purification protocol for alcohol oxidase. The maintenance of the pH during ammonium sulphate precipitation is vital in order to maintain the high activity of the enzyme preparation, since denaturation occurs if the pH falls below 6.5. Alcohol Oxidase is also extremely sensitive to high levels of NaCl while other proteins such as Glucose Oxidase exhibit increased solution stability at high NaCl concentrations (unpublished results, AET Ltd.).

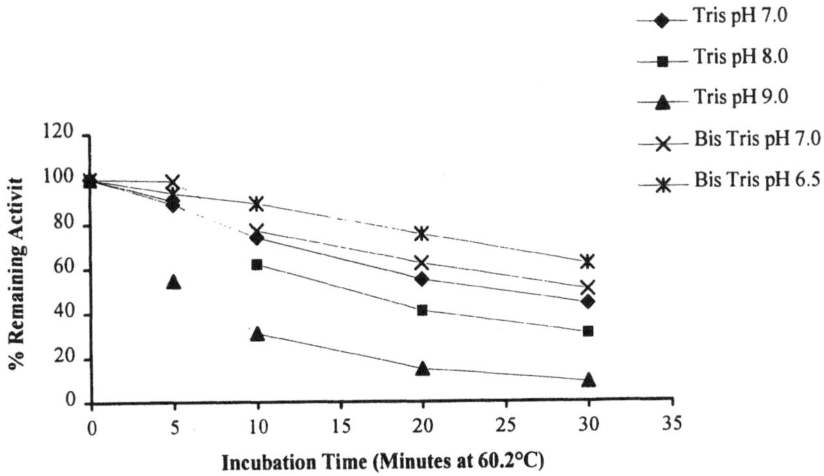

Figure 3. The pH effect in Tris/HCl and Bis-Tris/HCl buffers is shown above for bovine alkaline phosphatase. The more alkaline the pH, the more unstable the enzyme becomes, indicating that choice of pH is vital to promote stability in solution.

3.2. HORSERADISH PEROXIDASE STABILITY IN SOLUTION

Enzyme stability in solution is often difficult due to the multiple interactions and the presence of large amounts of water. The examples of HRP stability in solution for application in biosterilisers have been carried out in conjunction with a client manufacturing such reagents. Polyelectrolyte based formulations have been successfully used to stabilise dilute solutions of HRP (36μg.ml^{-1}) in the absence of any other protein. Full activity has been maintained for 6 months at 40°C, figure 4. The stability of peroxidases in solution is suitable for many application areas including environmentally friendly biosterilisers, where the enzyme is used as a means of generating bactericidal molecules 'in situ'. Peroxidases are also one of the main enzyme labels for detection of antibody based diagnostics (immunoassays), nucleic acid based diagnostics and basic enzyme assay protocols. Many different types of peroxidase - antibody conjugate have been stabilised using polyelectrolyte-based formulations and the performance of immunoassays has been improved as a result. The enzyme kinetics of the peroxidase label and the efficiency of the binding reaction between the antibody and its antigen does not seem to be affected (unpublished results, AET Ltd.).

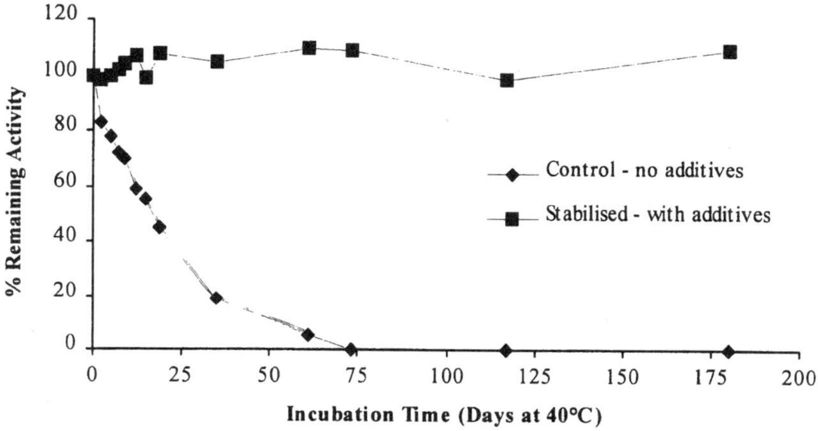

Figure 4. The solution stability of horseradish peroxidase is shown using a proprietary stabilisation system formulated specifically for peroxidases by AET Ltd. The enzyme concentration was 36μg. ml^{-1} with no other carrier protein being added. This type of stabilisation has applications in biosterilisers and immunoassay reagents, where peroxidases are used as the enzyme label.

3.3. ALCOHOL OXIDASE DRY STABILITY : ALCOHOL BIOSENSORS

Alcohol oxidase is a multisubunit enzyme (8 subunits) [11] that is rather unstable. However, certain patents and publications have shown that the enzyme can be stabilised for different application areas [12-18]. It is active in organic solvents provided the enzyme is hydrated [19]. It is also used in commercially available biosensors for the measurement of alcohol from numerous sources [20]. The dry stabilisation of the native enzyme isolated from the methylotrophic yeast, *Hansenula polymorpha,* reveals an extremely stable enzyme when stored in combinations of polyelectrolytes and polyalcohols. Full activity is retained for over 5 months at 37°C and 67% activity retained after 12 months at this temperature, indicating that the microenvironment of the enzyme is very important for 3D structural stability. The unstabilised enzyme loses all activity within 1 month at 37°C.

Table 1. Stability of Alcohol Oxidase Based Biosensors

Stabilisers Used \ Days Incubation at 37°C	0	2	7	14	21	35
	% Remaining Activity Relative to Initial Amperometric Response					
No Stabilisers pH 7.0	100	75	15	3	-	-
DEAE-Dextran 1%w/v pH 7.0	100	40	17	2	-	-
Lactitol 10%w/v pH 7.0	100	80	46	38	19	-
0.5% w/v DEAE Dextran + 10% w/v Lactitol pH 6.0	100	98	99	93	94	86
0.5% w/v DEAE Dextran + 10% w/v Lactitol pH 7.0	100	101	96	89	78	82
0.5% w/v DEAE Dextran + 10% w/v Lactitol pH 8.0	100	87	86	76	61	58
Stabilisers Used \ Weeks Incubation of Dehydrated Sensors.	**0**	**1**	**4**	**8**	**16**	**24**
	% Remaining Activity Relative to Initial Amperometric Response					
0.5% w/v DEAE Dextran + 10% w/v Lactitol pH 7.0 Temperature: 4°C	100	109	102	97	93	82
0.5% w/v DEAE Dextran + 10% w/v Lactitol pH 7.0 Temperature: 22°C	100	103	98.5	88.5	81.5	66

The examples given for alcohol oxidase based biosensors show significant differences in the storage stability of the enzyme when the additives are included, see table 1. The sensors maintain 86% of original activity after 35 days at 37°C in the presence of stabilisers in a pH dependent fashion, with pH 6 giving significantly better stability than pH 7 or pH 8. Longer term studies, 24 weeks at room temperature (22°C) and 4°C, resulted in sensors that retained 66% and 82% activity respectively. Unstabilised control sensors manufactured without stabilisers lost 97% activity within the first 2 weeks.

3.4. ACETYLCHOLINEESTERASE STABILITY AND BIOSENSORS

Acetylcholine-esterase (AChE) is an enzyme that degrades acetylcholine at synaptic junctions and is an ester hydrolase by nature. The 3D structure of AChE is well defined and exhibits a strong dipole as exhibited by the distribution of charged residues on the surface of the protein [21] (see figure 5).

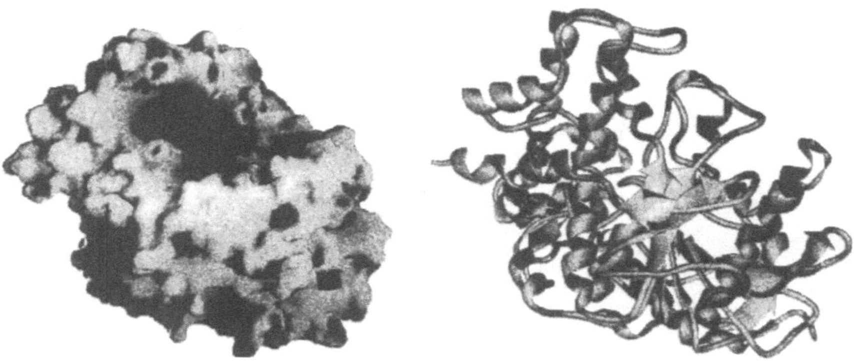

Figure 5. The electrostatic surface potential and the ribbon structure of acetylcholine-esterase orientated in the same conformation is shown. The main points to note are that the active site is embedded in the middle of the negative face (coloured red) of the protein, whilst the majority of the lysine residues are located at the back and sides. Immobilisation was carried out using carbodiimide coupling to lysine and therefore the orientation would tend to project the active site out into the bulk media. Stabilisation carried out using negative or positive polyelectrolytes affects the electrochemical response of biosensors made using this enzyme due to the overall surface charge produced.

A rational structural approach has been clearly demonstrated with the AChE biosensors produced by Rippeth *et al.* [5]. Stabilisation of the immobilised enzyme with polyelectrolyte combinations shows a distinct difference to that of the soluble enzyme dehydrated from solution (table 2). A possible explanation for this difference in stability, is thought to correspond to the orientation of the enzyme onto the surface of the transducer. Immobilisation onto the pre-activated carbon transducer, utilises covalent coupling to the amino groups, most of which are situated on the back face of the enzyme opposite to the active site, figure 5. This leaves the negative face of the enzyme containing the active site exposed to the bulk solution, creating a surface that in molecular terms is negatively charged. Interaction between a positively charged (cationic) polyelectrolyte and the immobilised enzyme will tend to favour an electrostatic attraction between the negative face of the enzyme and the positive charges of the polymer backbone, leading to complex formation with an excess of positive charges. The substrate used to measure enzyme activity is acetylthiocholine, which is positively charged and thus effectively partitioned away from the active site. This would explain the reduction in response between sensors stabilised by lactitol or trehalose alone and when a cationic polyelectrolyte is added. If a polyelectrolyte of the opposite charge (anionic) is used, the interaction between the enzyme structure and the negative charges of the polymer would tend to create a more negative microenvironment and therefore effectively partition the positively charged substrate towards the enzyme and thereby enhancing substrate binding at the active site. Referral to table 2 shows that adding anionic polymer and lactitol as stabilisers increases the sensor responses dramatically (unpublished results). The actual stabilisation factor for these biosensors seems to be a combination of stabilisation of the enzyme structure by:

- immobilisation
- differently charged additive combinations used as stabilisers.

Table 2. Stability of Native Acetylcholine-esterase and Acetylcholine-esterase Biosensors

Stabilisers Used	**Native Acetylcholine-esterase Activity DAbs.min^{-1} and (percentage activity)**			**Acetylcholine-esterase Biosensors Current Response in nA and (percentage activity)**	
	Before Drying	After Drying	43 Days at 37°C	Freshly Prepared	43 Days at 37°C
No Stabilisers	0.255 (100)	0.030 (11.8)	0.007 (2.8)	275 (100)	168 (61.1)
10% w/v Lactitol (Polyalcohol)	0.255 (100)	0.010 (3.9)	0.020 (7.8)	574 (100)	527 (91.8)
0.5% w/v DEAE Dextran + 10% w/v Lactitol (Cationic Polyelectrolyte)	0.260 (100)	0.120 (46.2)	0.110 (42.3)	Not Done	Not done
0.5% w/v Gafquat 755N + 10% w/v Lactitol (Cationic Polyelectrolyte)	0.240 (100)	0.225 (93.8)	0.220 (91.7)	335 (100)	312 (93.1)
0.5% w/v Dextran sulphate + 10% w/v Lactitol (Anionic Polyelectrolyte)	0.365 (100)	0.083 (22.7)	0.110 (30.1)	947 (100)	1056 (111.5)
Separate Control for the anionic polyelectrolyte stabiliser.	0.365 (100)	0.008 (2.2)	0.010 (2.7)	-	-
10% w/v Trehalose (Polyalcohol)	0.278 (100)	0.073 (26.3)	0.070 (25.2)	532 (100)	458 (83.6)

3.5. RECOMBINANT LUCIFERASE STABILITY IN SOLUTION

Samples of recombinant luciferase were a generous gift from Celsis Ltd. The main point to note with this enzyme is that it is very labile in solution, with rapid denaturation occurring at a temperature of 25°C for the native enzyme. Various recombinant enzymes have been produced by a number of companies (Celsis, Promega, etc.) in order to i) secure a commercial supply of the enzyme from a bacterial source and ii) to attempt to improve the thermostability of the enzyme for use in assay protocols. The recombinant enzyme obtained, exhibits a half life of 14 minutes at a temperature of 45°C, which is a distinct improvement on the wild type enzyme. Addition of polyelectrolyte stabiliser combinations increased the thermostability even further to give a half-lives of 26.7 and 33.6 minutes, depending on the formulation used, figure 6.

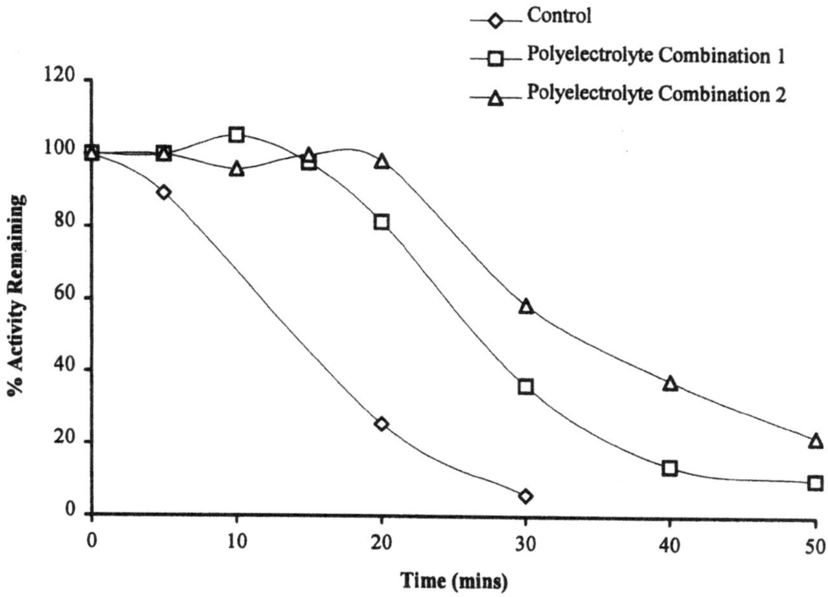

Figure 6. Solution stabilisation of recombinant enzymes is shown by this example. Native luciferase is very unstable, beginning to lose activity at around 25°C. Protein engineering has improved this to give a half-life of 14 minutes at 45°C in this example. Addition of cationic polyelectrolyte stabiliser combinations improves this to give half-lives of 26.7 and 33.6 minutes respectively.

3.6. IMMOBILISED GLUCOSE OXIDASE : PRE-STABILISED COMPLEXES

Following on from covalent immobilisation of enzymes to produce biosensors, the chemical industry has expanded the use of enzymes in the areas of synthesis and biotransformation very rapidly over the last few years. The biocatalysis field has been the subject of intense activity with developments such as cross linked enzyme crystals (CLEC's) marketed by Altus Biologics Inc. and rational design of protein stability, presented at the Enzyme Stabilisation Conference in Leeds, 1998 by Prof. Klibanov contributing to the growth of this area. The long term operational stability of biocatalysts is an area that is vitally important in commercial terms and is often the deciding factor as to whether a process can be made viable using a biocatalyst as opposed to the more traditional chemical processes. Each enzyme acts as an individual in most cases and what works for one type of enzyme rarely works for another. Also the problem is compounded in that the activity of the enzyme is often compromised during the immobilisation process, leaving an enzyme that may be stable, but with a significantly lower amount of residual activity. Yet another aspect of biosynthesis and biotransformations is the need to have a high level of shelf-stability of the immobilised biocatalyst. Manufacture of a large batch of biocatalyst is more cost effective, therefore there is a great need for shelf stability. Ideally, biocatalysts would be stored in the same fashion as chemical catalysts, with the minimal requirement for specialist storage conditions.

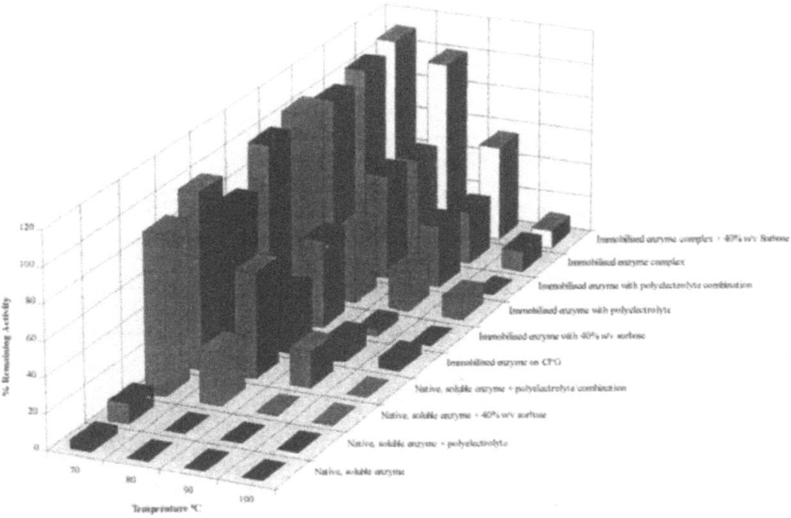

Figure 7. The formation of an immobilised, thermostabilised glucose oxidase using combinations of polyelectrolytes and sorbose is shown in the chart above. Native soluble enzyme has been included as a comparison. Each series of bars represents the residual activity of an enzyme preparation, stabilised as labelled on the z-axis after being subjected to 20 minutes incubation at the respective temperatures shown. The combinations of polyelectrolyte and sorbose give the highest levels of thermostabilisation at temperatures up to 90°C, with the immobilised enzyme complex giving the maximum activity retention.

Covalent immobilisation of glucose oxidase has been described in many cases [22, 23], and in the past this enzyme has been immobilised by direct covalent attachment to silanised controlled pore glass to produce flow injection assay systems for fermentation monitoring [2]. Recently the immobilisation of glucose oxidase - polyelectrolyte complexes has been carried out onto a silanised glass support and the thermal stability was compared to the native immobilised enzyme on the same support [1]. The results show that the immobilised complex is more stable than the native immobilised enzyme. Although the addition of stabilisers after immobilisation of the native enzyme improve stability, this is not to the same level as observed for the immobilised enzyme stabiliser complex. Incorporation of a polyalcohol into the buffer solution bathing the immobilised biocatalyst further improves the thermal stability of the immobilised enzyme. This improvement in stability can be clearly seen in figure 7, where a composite series of results for the activity retention after 20 minutes exposure to the temperatures shown is depicted. The highest level of stability is obtained upon immobilisation of the glucose oxidase - polyelectrolyte complex with sorbose in the solution.

3.7. DETECTION OF PROTEIN-POLYELECTROLYTE COMPLEXES BY ISOELECTRIC FOCUSING

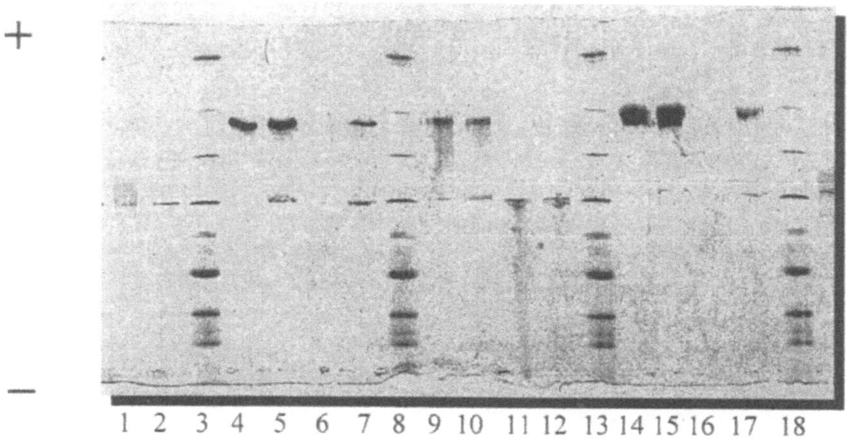

Lane 1 Bovine liver GDH
Lane 2 L-GDH + 0.01% Dextran sulphate
Lane 3 IEF markers
Lane 4 PF-GDH
Lane 5 PF-GDH + 0.01% Dextran sulphate
Lane 6 PF-GDH + 0.01% DEAE Dextran
Lane 7 PF-GDH + 0.01% Gafquat HS100
Lane 8 IEF markers
Lane 9 β-Galactosidase
Lane 10 β-Gal + 0.01% Dextran sulphate
Lane 11 β-Gal + 0.01% DEAE Dextran
Lane 12 β-Gal + 0.01% Gafquat HS100
Lane 13 IEF markers
Lane 14 β-Glucosidase
Lane 15 β-Gluc + 0.01% Dextran sulphate
Lane 16 β-Gluc + 0.01% DEAE Dextran
Lane 17 β-Gluc + 0.01% Gafquat HS100
Lane 18 IEF markers

Figure 8. Study of protein-polyelectrolyte complexes by isoelectric focusing

During the stabilisation process do polyelectrolytes interact with enzymes in a predictable fashion and can this be measured? At AET Ltd. we have utilised gel electrophoresis to examine the interactions between proteins and polymers and use this technique to predict a specific formulation for the stabilisation enzymes. Figure 8 clearly shows how electrophoresis can be used to measure these interactions. Due to the extremely large size of the polymers, interaction between the polymer and protein is detected as the retardation of the enzyme in the gel matrix. In this example 4 different enzymes are screened using 3 different polymers. Two are cationic (positively charged) DEAE dextran and Gafquat HS100 and one anionic (negatively charged) dextran sulphate. The *Pyrococcus furiosus* glutamate dehydrogenase (Pf GLDH), β-galactosidase and *Pyrococcus furiosus* β-glucosidase all have acidic isoelectric points making their overall charge negative. As one would predict, they are retarded by DEAE dextran the most positively charged of the polymers tested (lanes 6, 11 and 16) and to a lesser extent Gafquat HS-100 (lanes 7,12 and 17). Bovine GLDH that has an isoelectric point (pI) of 7.3 is retarded by dextran sulphate (lane 2). Due to is neutral pI it does in fact bind all the above polymers to varying degrees (data not shown). Not only does this technique help us to predict interactions between proteins and polymer molecules but has also been used to determine the affinity of polymer binding. This data has been used by AET Ltd. to stabilise enzymes at the optimal polymer concentration. The ability to predict the amount of polymer needed to stabilise an enzyme allows the costs of stabilisation to be cut considerably. This also helps reduce the possibility of stabilising polymers interfering with the final application of the stabilised enzyme.

4. Discussion and conclusions

Based on the many examples from the literature and the experiences of the research carried out at AET Ltd. it is suggested that one of the major contributory factors involved in influencing the stability of native, soluble enzymes is the relative water activity at the enzyme (or protein) surface, whether the enzyme is dissolved, immobilised or suspended (as is the case for organic solvents). This factor is likely to be the single most important parameter in promotion of structural stability of enzymes and relates to the immediate microenvironment around the protein structure. In adding polyhydric alcohols to aqueous solutions, the bulk and surface water activity is modified relative to the absolute concentration of the additive and it is well known that stabilisation effects observed are dependent on the concentration of polyalcohol present [24, 25]. The retention of enzyme activity in organic solvents where the water activity is carefully controlled is nothing short of remarkable [26]. When used in combination with additives promoting electrostatic interactions (polyelectrolytes) or surface chemical interactions leading to immobilisation or crosslinking, the efficacy is usually enhanced significantly, indicating synergy of action at the surface of the enzyme structure leading to increased stability of the enzyme. The solvation characteristics of an enzyme are likely to be changed significantly when associated into a protein - polyelectrolyte complex and as such will interact differently with polyalcohol solutions leading to elevated stability. It is thought that the surrounding environment of the enzyme and the surface interactions occurring are likely to be just as important as the actual amino acid sequence and corresponding secondary and tertiary structure of the enzyme molecule. Of course some additives such as metal ions are directly related to enzyme structure and as such are not strictly surface interactions. Addition of dilute solutions of metal salts e.g. magnesium, zinc often stabilise proteins to a high degree and act synergistically with polyelectrolyte combinations [17]. Observations with the recombinant luciferase, where point mutation of a single amino acid produced a rise in the thermal denaturation temperature of 14°C, indicated further stabilisation of the enzyme activity could conferred by modification of the surrounding environment. The half-life of deactivation of the recombinant itself being doubled when polyelectrolyte additive combinations were present.

The production of stable enzyme based systems for a multitude of applications can be realised by the use of many different techniques and procedures. In practical terms, the inclusion of soluble additives and immobilisation techniques outnumber the more sophisticated molecular engineering approach. However this is becoming much more commonplace, especially in the production of recombinant enzymes. Where protein engineering works to confer stability, the resulting stable mutant recombinant enzyme produced is usually incorporated in a process in exactly the same manner as the native enzyme. This being especially true in the production of detergents, where the manufacturing plant and processes are already in place. Personal communications with industrial enzymologists indicate that the inclusion of additives is just as effective in elevating stability of enzymes for industrial applications and does not carry the sheer amount of time and effort (with the associated costs) as a protein-engineering regime. In our case it has been found the enzyme stability can be significantly elevated by using novel polyelectrolyte stabiliser combinations that have been developed over the last 10 years. Both shelf-stability and operational stability have been improved, with increases in shelf life being demonstrated for well over 30 different enzymes (unpublished results, AET Ltd.). This indicates that the methodology is relatively generic in nature and can be adapted for many application areas. The molecular mechanisms of stabilisation are currently under investigation, with the aim of being able to predict the type of stabilisers needed for specific enzymes. By using more sophisticated techniques such as circular dichroism, fluorescence spectroscopy, Differential scanning calorimetry, electrophoretic techniques, analytical centrifugation and electron microscopy. Data accumulated from such experiments will help us to understand more about how proteins denature at the molecular level and ultimately enable us to stabilise enzymes in a more predictable fashion.

Acknowledgement

AET Ltd works closely with academics at the University of Leeds and other Universities in this field of enzyme stabilisation.

References

1. Appleton B, Gibson T D and Woodward J R. (1997) High Temperature Stabilisation of Immobilised Glucose Oxidase: Potential Applications in Biosensors, *Sensors and Actuators **B. 43,*** 65-69.
2. Gibson T D. (1991) PhD Thesis, *Stabilised Enzyme Based Diagnostic Systems,* Biotechnology Unit, University of Leeds.
3. Gibson T D, Hulbert J N, Parker S M, Woodward J R and Higgins I J. (1992) Extended Shelf Life of Enzyme Based Biosensors using a Novel Stabilisation System, *Biosensors and Bioelectronics 7,* 701-708.
4. Gibson T D, Pierce B L J, Hulbert J N and Gillespie S. (1996) Improvements in the Stability Characteristics of Biosensors using Protein-Polyelectrolyte Complexes, *Sensors and Actuators B 33,* 13-18.
5. Rippeth J J, Gibson T D, Hart J P, Hartley I C and Nelson G. (1997) Flow-Injection Detector Incorporating a Screen Printed Disposable Amperometric Biosensor for Monitoring Organophosphate Pesticides, *The Analyst 122,* 1425-1429.

6. Gibson T D. (1996) Protein Stabilisation using Additives Based on Multiple Electrostatic Interactions. In F Brown. (ed.) *'New Approaches to Stabilisation of Vaccines Potency'. Dev. Biol. Stand* ***87,*** 207-217, Karger.
7. Pierce B L J, Gibson T D and Bunnel P. (1998). A Preliminary Model and Evidence for the Mechanism of Stabilisation of Analytical Enzymes in Aqueous Solution by Polyelectrolytes and Sugar Derivatives, in A.O.Scott, (ed.), *Biosensors for Food Analysis,* The Tetley Group Limited, Greenford, UK, The Royal Society of Chemistry, pp54 - 60
8. Kirkwood T.B.L. (1977) Predicting the Stability of Biological Standards and Products, *Biometrics* ***33,*** 736-742.
9. Remmele R L, Nightlinger N S, Srinivasan S, Gombotz W R. (1998) Interleukin-1 receptor (IL-1R) liquid formulation development using differential scanning calorimetry, *Pharm. Research* ***15,*** 200-208.
10. McAteer K, Simpson C E, Gibson T D, Gueguen S, Boujtita M and El Murr N. (1999) Proposed Model for Shelf Life Prediction of Stabilised Commercial Enzyme-Based Systems and Biosensors, *J. Mol.Catal. B: Enzymatic. 7,* 47-56.
11. Woodward J.R. (1990) Biochemistry and Applications of Alcohol Oxidase from Methylotrophic Yeasts, in Codd G.A., (ed.), *Advances in Autotrophic Microbiology and One-Carbon Metabolism,* Kluwer Academic Publishers, Dordrecht, Vol 1.
12. Phillips R.C. (1985) Colorimetric Ethanol Analysis Method and Test Device, Eur. Patent Application 0,133,481. A1.
13. Hopkins T.R. (1988). Stabilised Alcohol Oxidase Compositions and Method for Producing Same, U.S. Patent 4,729,956
14. Gibson T D and Woodward J R. (1989) Enzyme Stabilisation, PCT/GB89/01346.
15. Gibson T D and Woodward J R. (1991) Enzyme Stabilisation , PCT/GB91/00443.
16. Gibson T D, Higgins I J and Woodward J R. (1992) The Stabilisation of Analytical Enzymes using a Novel Polymer-Carbohydrate System and the Production of a Stabilised, Single Reagent for Alcohol Analysis, *The Analyst 117,* 1293-1297.
17. Gibson T D, Hulbert J N and Woodward J R. (1993) Preservation of Shelf Life of Enzyme Based Analytical Systems using a Combination of Sugars and Sugar Alcohols with Cationic Polymers or Zinc Ions, *Anal. Chim. Acta. 279,* 185-192.
18. Gibson T D, Hulbert J N, Pierce B and Webster J I. (1993) The Stabilisation of Analytical Enzymes using Polyelectrolytes and Sugar Derivatives, in W J J van den Tweel, A Harder and R M Buitelaar (eds.), *Stability and Stabilisation of Enzymes,* Elsevier, pp337-346.
19. Zaks A and Klibanov A M. (1988) The Effect of Water on Enzyme Action in Organic Media, *J. Biol. Chem.* ***263,*** 8017 - 8021.
20. Woodward J R, Gibson T D and Spokane R B. (1993) Construction and Function of Alcohol Oxidase Based Biosensors, *Abst. Papers. Amer. Chem. Soc. 205,* 5.
21. Ripoll D R, Faerman C H, Axelsen P H, Silman I and Sussman J L. (1993) An electrostatic mechanism for substrate guidance down the aromatic gorge of acetylcholine-esterase, *Proc. Nat. Acad. Sci. USA* ***90,*** 5128 - 5132.
22. Hossain, M M. and Do, D D. (1985) Fundamental studies of glucose oxidase immobilisation on controlled pore glass, *Biotech. & Bioeng. 27,* 842-851.
23. Scouten, W.H., Luong, J.H.T. and Brown, R.S. (1995) Enzyme or protein immobilisation techniques for application in biosensor design, *Trends in Biotechnology 13,* 178-185.
24. Monsan, P. and Combes, D. (1988) Enzyme stabilisation by immobilisation, *Methods in Enzymology 137,* 584-598.
25. Back J.F, Oakenfull D. and Smith M.B. (1979) Increased Thermal Stability of Proteins in the Presence of Sugars and Polyols, *Biochem. 18,* 5191-5196.
26. Zaks A. and Klibanov A M. (1984) Enzymatic Catalysis in Organic Media at 100°C, *Science 224,* 1249 - 1251.

IMPROVEMENTS OF ENZYME STABILITY AND SPECIFICITY BY GENETIC ENGINEERING

M. POHL AND M.-R. KULA
Institute of Enzyme Technology Heinrich-Heine Universität Düsseldorf, D-52426 Jülich, Germany

1. Introduction

In former time, if an enzyme with improved properties was needed, a different organism producing the desired enzyme usually was screened. This is still a possible approach, but it is labour intensive and relies on chance only. In screening enzyme producers rather than enzymes it is difficult to design the experiments to differentiate enzyme properties in the primary microbial screen. The development of genetic engineering techniques has opened up more direct routes to alter enzyme properties by changing the amino acid sequence of a given enzyme and screen for improved variants. Changing the amino acid sequence is accomplished by altering the DNA sequence of the structural gene of the enzyme. This alteration can be done in a random way allowing for errors in the transcription of the gene, or by site-directed mutagenesis, replacing a selected amino acid residue by one or several proteinogenic amino acids. The latter approach requires structural knowledge as well as some information on structure/function correlations. We used site-directed mutagenesis successfully in two projects to improve enzyme properties in catalysts of considerable interest in the production of fine chemicals, formate dehydrogenase (FDH) and pyruvate decarboxylase (PDC), respectively. In the following we review the strategy and results obtained. For the experimental details the reader is referred to the original literature referenced.

2. Results

2.1. FORMATE DEHYDROGENASE

Formate dehydrogenase (EC 1.2.1.2) is produced by microorganisms able to utilise methanol as carbon and energy source. We have worked intensively with FDH from *Candida boidinii* [1] exploiting the enzyme for *in situ* regeneration of NADH in

M. Hofman and P. Thonart (eds.), Engineering and Manufacturing for Biotechnology, 377–382.

coenzyme dependent reductions [2]. Formate is a safe and cheap hydrogen source and the reaction has a favourable equilibrium in the conversion of formate to CO_2. This approach is the method of choice for coupled enzymatic processes e.g. in the industrial production of L-*tert* leucine by reductive amination [3, 4].

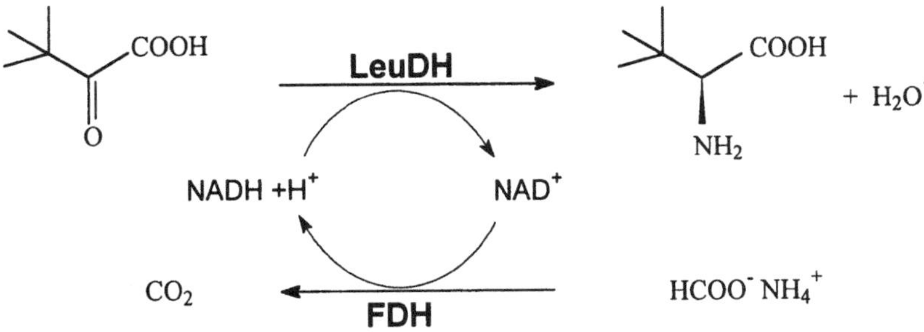

*Figure 1: Coupled enzymatic process for the synthesis of L-*tert*-leucine by reductive amination of tri-methyl pyruvate using leucine dehydrogenase (LeuDH). Regeneration of NADH is performed by FDH from C. boidinii [1-4].*

Continuous enzyme production with *C. boidinii* in 200-L scale [5] has been described as well as an efficient isolation procedure [6]. Since the enzyme was readily available from the wild type yeast it has been cloned only recently from *C. boidinii* [7, 8]. FDH is also of high interest with regard to the enzyme mechanism of hydride transfer in dehydrogenase catalysed reactions, which is studied mainly by V. Tishkov and V. Popov's groups in Moscow. From their work the high resolution X-ray structure of the bacterial FDH from *Pseudomonas spec.* is available [9].

The FDH from methylotrophs are highly homologous [10], they do not contain metal ions and have a rather low specific activity compared to other dehydrogenases of only 6-8 U/mg in homogeneous preparations. It would be very interesting to increase the reaction rate, but this appeared a risky project based on the limited information available. Increasing the stability of FDH would, in the end, achieve a similar goal to reduce the amount of enzyme protein necessary to produce a unit mass of product. If the product specific enzyme consumption rate is taken arbitrary as 1000 U/kg, this translates for an enzyme of low intrinsic activity such as FDH into ≈ 100 mg/kg, while for an enzyme like phenylalanine dehydrogenase (≈ 3000 U/mg) it means only 0,3 mg/kg. Increasing the stability is therefore of high economic significance for a low activity enzyme. In case of FDH we knew from previous work in purification and enzyme catalysis that *C. boidinii* FDH is susceptible to oxidative deactivation. The amino acid sequence of *C. boidinii* FDH contains a total of two cysteine residues in position 23 and 262. Using the crystal structure of *Pseudomonas* FDH as a model, both cysteines are likely to be accessible at the surface of the protein, but placed too far apart to form intramolecular disulfide bridges in the native molecule. Thishkov *et al.* had reported that replacement of cysteine in the bacterial FDH by serine or methione had dramatically increased the resistance of *Pseudomonas* FDH against treatment with Hg^{2+}-ions [11]. We decided therefore to replace Cys23 and Cys262 in *C. boidinii* FDH by small aliphatic amino acids using the appropriate genetic techniques and analyse for resistance of the mutants generated against oxidative stress. The latter was intensified using 10 μM Cu^{2+} at pH 7.5. In Fig 2 it is shown, that exchanging Cys23 for serine improved the stability considerably. Replacing both cysteines in the sequence further increased the half-life of the double mutant. On the other side the amino acid exchange had no measurable effect on the kinetic properties of FDH, the reaction rates and K_M values are identical within the error margin for the analytical procedures employed [7, 12]. The temperature stability of the muteins was slightly reduced compared to the wild type [7, 12]. A similar observation was reported by Tishkov *et al.* [11] for the bacterial FDH mutein.

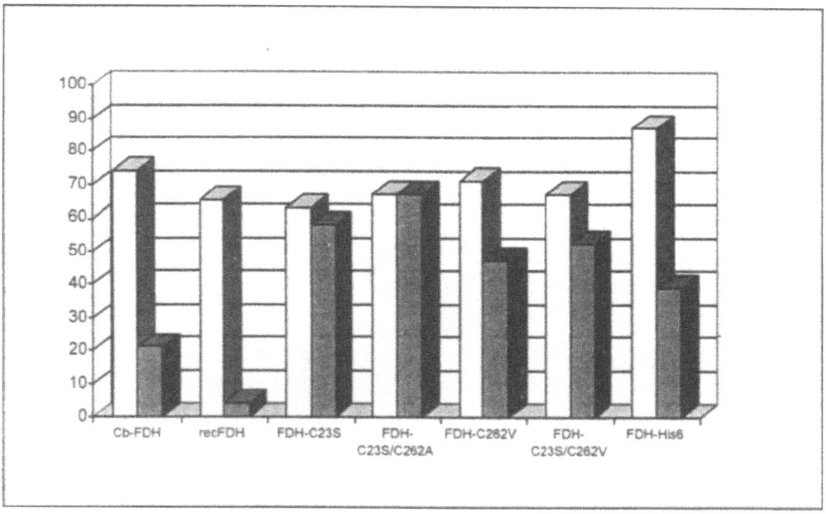

Figure 2: Residual activities (black columns) of FDH isolated from C. boidinii (CbFDH), recombinant FDH (recFHD), and various muteins after incubation in potassium phosphate buffer, pH 7.5, containing 10 μM $CuCl_2$ for 15 h at 25°C, relative to an untreated sample (white columns). FDH-His6 refers to recombinant wild type FDH elongated with a hexahistidine tag at the C-terminus.

The new FDH variants have been produced in *E. coli* at 30-L scale. More than 50,000 U have been isolated for application studies currently in progress at Degussa-Hüls. A final conclusion can be reached only after these trials have been completed and evaluated, but we expect a substantial reduction of product specific enzyme consumption due to the increased stability of the catalyst.

2.2. PYRUVATE DECARBOXYLASE

The oldest biotransformation process still in operation today utilises "fermenting yeast" to achieve a stereospecific acyloin condensation-type carboligation between benzaldehyde and acetaldehyde yielding (*R*)-phenylacetylcarbinol ((*R*)-PAC), the key intermediate in the industrial ephedrine synthesis ((1*R*), (2*S*)-ephedrine) [13]. Acetaldehyde is supplied by the yeast culture during the fermentation of glucose *via* decarboxylation of pyruvate by pyruvate decarboxylase (PDC, EC 4.1.1.1). The same enzyme is also responsible for the carboligation of acetaldehyde with benzaldehyde. The catalytic mechanism of PDC is well-studied [14]. In Fig. 3 the general concept is illustrated. PDC needs thiamine diphosphate (ThDP) as an essential coenzyme. In the course of the reaction an active intermediate is the carbanion-enamine which is termed "active aldehyde". Protonation of this intermediate yielding hydroxyethyl-ThDP, followed by release of acetaldehyde is the reaction path during decarboxylation, which constitutes the main reaction of PDC. During carboligation, which is a side reaction of the enzyme, the "active aldehyde" reacts with a second aldehyde (here: benzaldehyde) yielding a chiral acyloin.

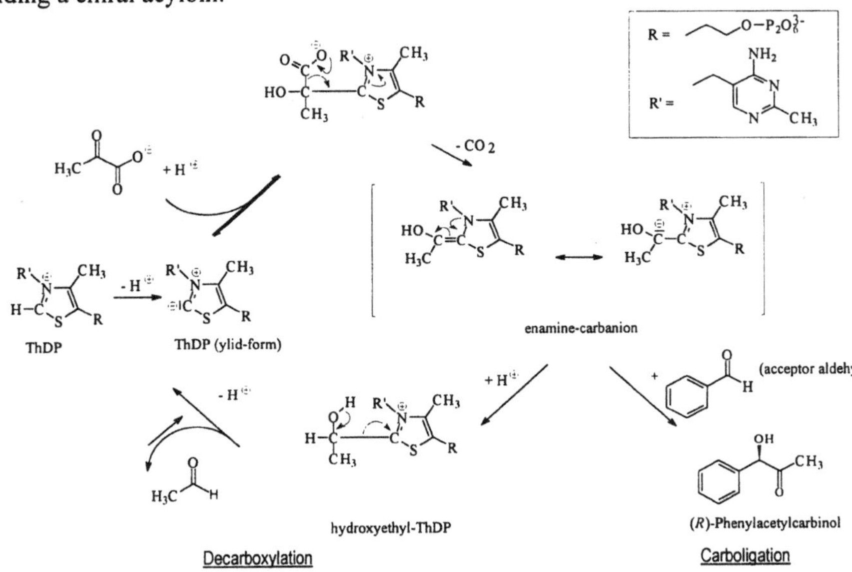

Figure 3: Reaction paths of Pyruvatedecarboxylase. The "active aldehyde" represents the intermediate for the decarboxylation and the carboligation reaction.

The biotransformation using whole cells is not without problems, which mainly arise from competing reactions e.g. the reduction of benzaldehyde by the fermenting yeast and the toxicity of the aromatic aldehyde for the cells. In principle, these problems could be by-passed using an enzymatic process for the production of (*R*)-PAC. If such an approach is taken glucose can no longer be used as a source of pyruvate. Since the enzymatic reaction is reversible the formation of the "active aldehyde" (Fig. 3) does not essentially require decarboxylation of pyruvate and is also formed upon addition of acetaldehyde to ThDP *via* hydroxyethyl-ThDP.

Since PDC isolated from yeast is very unstable [14], we employed the significantly more stable enzyme from *Zymomonas mobilis* for our studies. Unfortunately, the carboligase activity of the bacterial PDC is by a factor of 5 lower compared to the yeast enzyme. The goal of our studies was to increase the carboligase activity of PDC from *Z. mobilis* by site-directed mutagenesis, thereby keeping the good stability of the enzyme. Since the 3-dimensional structure of PDC from *Z. mobilis* [15] has only resently been available, we used the crystal structure of PDC from yeast as a model to identify amino acids in the active site. PDC is a homotetramere and the active site is formed at the interface of two subunits, the overall architecture can be viewed as a dimer of dimers. The substrates approach the active site through a channel like cleft between the subunits. Homology modeling indicated that tryptophan 392 found in the bacterial enzyme protrudes in the channel thereby reducing the free path and effectively blocking access of benzaldehyde to the active site in *Z. mobilis*. By contrast, an alanine residue is found in the equivalent position of yeast PDC. Replacing Trp392 in PDC from *Z. mobilis* by alanine resulted in a catalyst with about 50% of the wild type decarboxylase activity but unaltered K_M for pyruvate (1.0 mM) and a 3 fold increase in the carboligase activity leading to (*R*)-PAC under optimal conditions [16, 17]. It was found that acetaldehyde inhibited the wild type and the mutant enzyme at rather low concentrations in the mM range in an apparently irreversible reaction [16]. Acetaldehyde in the reaction mixture cannot be avoided, it is either the starting material or it is produced in the decarboxylation reaction from pyruvate. If pyruvate is used as a substrate and care is taken to reduce acetaldehyde *in situ* to ethanol by a coupled enzymatic process, (*R*)-PAC of high enantiomeric purity >98% ee) could be continuously produced [16, 17]. However, this method requires the use of at least 3 different enzymes, which makes this process only useful in small scale.

To generate a better catalyst two goals have to be achieved: *i.* a higher tolerance against acetaldehyde and *ii.* an improved carboligase activity. For this purpose further alterations at position 392 were evaluated. Replacement of tryptophan by methione, isoleucine, penylalanine or valine resulted in stable variants with good activity. Among these the Trp392Met and Trp392Ile muteins exhibited the highest carboligase activity with 2.5 or 2.6 U/mg, respectively, representing a five fold increase compared to the wild type enzyme [14,18]. Replacement by amino acids with small side chains like glycine and alanine decreases the stability of the tetramer, charged amino acids at position 392 such as histidine or glutamate lead to very unstable enzymes, which loose activity rapidly by dissociation of coenzyme and loss of quaternary structure. Obviously Trp392 serves not only as a barrier in the substrate channel but contributes significantly to PDC stability. The hydrophobic aliphatic amino acids methionine or isoleucine appear to have the critical dimensions and contribute in a similar way as tryptophan to stabilize the quaternary structure. Unexpected was the observation that the muteins Trp392Met, Trp392Ile exhibit a much better tolerance against acetaldehyde than the wild type PDC [19]. The underlying reasons for this result are unclear and cannot be derived in a rational way from our present knowledge about PDC. This finding is a nice illustration of the present limits of a rational protein design as a substitution of Trp for Ile would not be considered *a priori* useful in a strategy to decrease the sensitivity of a protein towards inactivation by acetaldehyde. One may have found such a mutation by a random mutagenesis and screening. In this particular case we found it by chance and have now a catalyst available, which has carboligase activity similar to yeast PDC but with a much higher operational stability. Since the muteins Trp392Ile, Trp392Met of the *Z. mobilis* PDC maintain high stereoselectivity with ee values ≥ 98% with respect to the formation of (*R*)-PAC, these catalysts are presently further evaluated for application in an enzyme catalysed process.

3. Conclusion

The results discussed as numerous other examples in the literature clearly illustrate the potential of modern techniques to alter properties of a given enzyme to adapt it to other reaction conditions and to improve performance. Taylor made enzymes will find increased application as catalysts in the production of fine chemicals.

Acknowledgement

We are indebted to our collaborators and students Drs. H. Bruhn, K. Mesch, G. Goetz, H. Slusarczyk and Dipl.-Biol. S. Felber for their enthusiastic and dedicated work. Our research was in part supported by the "Deutsche Forschungsgemeinschaft" (Grant Ku 188/5-1-3, Po 558/1-1) and SFB 380, by BMBF Project Az 03D0031C7 and Degussa AG, Hanau, and by BASF AG, Ludwigshafen.

References

1. Schütte, H., Flossdorf, J. Sahm, H. and Kula, M.-R. (1976). Eur. J. Biochem. **62**, 151-160
2. Kula, M.-.R. and Wandrey, C. (1987). Methods Enzymol. **136**, 9-21
3. Bommarius, A.S., Drauz, K., Hummel, W., Kula, M.-R. and Wandrey, C. (1994). Biocatalysis **10**, 37-47
4. Bommarius, A.S.,Schwarm, M., Stingl, K., Kottenhahn, M. and Hutmacher, K. (1995) Tetrahedron: Asymmetry **6**, 2851-2888
5. Weuster-Botz, D., Paschold, H., Striegel, B., Gieren, H., Kula, M.-R. and Wandrey, C. (1994). Chem. Eng. Technol. **17**, 131-137
6. Kroner, K.H., Schütte, H., Stach, W. and Kula, M.-R. (1982). J. Chem. Tech. Biotechnol. **32**, 130-137
7. Slusarczyk, H. (1997) Stabilisierung der NAD-abhängigen Formiatdehydrogenase aus *Candida boidinii* mittels gerichteter Mutagenese, Dissertation Universität Düsseldorf
8. Sakai, Y., Murdanoto, A.P., Konishi, T., Iwamatsu, A.and Kato, N. (1997). J. Bacteriol. **179** (14), 4480-4485
9. Lamzin, V.S., Dauter, Z., Popov, V.O., Harutyunyan, E.H. and Wilson, K.S. (1994). J. Mol. Biol. **236**, 759-785
10. Popov, V.O. and Lamzin, V.S. (1994). Biochem. J. **130**, 625-643
11. Tishkov, V.I., Galkin, A.G., Marchenko, G.N., Egorova, O.A., Sheluho, D.V., Kulakova, L.B., Dementieva, L.A. and Egorov, A.M. (1993b). Biochem. Biophys. Res. Commun., **192**, 976-981
12. Slusarczyk, H, Pohl, M and Kula, M.-R. (1997) Deutsche Patentanmeldung No. 197 53 350.7
13. Hildebrandt, G. and Klavehn, W. (1932) Deutsches Reichspatent Nr. 548 459.
14. Pohl, M. (1997). Adv. Biochem. Eng. Biotechnol., 58, 16-43
15. Dobritzsch, D., König, S., Schneider, G. and Lu, G.G. (1998). J. Biol. Chem., 273, 20196-20204
16. Bruhn, H., Pohl, M., Grötzinger, J. and Kula, M.R. (1995). Eur. J. Biochem., 234, 650-655
17. Bruhn, H., Pohl, M., Mesch, K. and Kula, M.R. (1995) Deutsche Patentanmeldung 195 23 269. 0-41.
18. Iding, H., Siegert, P., Mesch, K. and Pohl, M. (1998). Biochim. Biophys. Acta, 1385, 307-322
19. M. Pohl, K. Mesch, H. Iding, G. Goetz, M.-R. Kula, M. Breuer und H. Hauer (1997) Verfahren zur Herstellung enantiomerenreiner Phenylacetylcarbinole aus Acetaldehyd und Benzaldehyden in Gegenwart von Pyruvatdecarboxylasen aus Zymomonas. Deutsche Patentanmeldung, 19736104.8

AN APPROACH TO DESICCATION-TOLERANT BACTERIA IN STARTER CULTURE PRODUCTION

WEEKERS F. [1], JACQUES P. [2],
MERGEAY M. [3] AND THONART P. [1,2]
(1) University of Liege. Walloon Center for Industrial Biology, Bat. B 40 - 4000 Sart-Tilman, Liege. Belgium.
(2) Agricultural University of Gembloux. Bio-industries Unit, Passage des Déportés, 2 - 5030 Gembloux, Belgium.
(3) SCK/CEN. Boerentang, 200 - B2400 Mol. Belgium.

1. Introduction

Water is necessary for life on Earth. At the cell scale, water serves as solvent for organic and inorganic solutes, metabolites and as substrate for, or product of metabolic activity. In addition, it is well established that water plays an important role in maintenance of structural and functional integrity of biological membranes and macromolecules (Crowe et al., 1987). Nevertheless, a number of organisms are able to survive almost complete desiccation : a phenomenon known as anhydrobiosis (Crowe et al., 1992).

Desiccation-tolerance is not a property of "normal growth" conditions, but rather an ability to survive adverse hydration conditions. During the dry period, the cells are not active, but in a dormant state, ready to resume activity when the hydration conditions are favourable again (Kaprelyants et al., 1993). These anhydrobiotic organisms (plant seeds, soil dwelling rotifers, crustacean cysts, nematodes, bacterial spores,...) attain a desiccated state very resistant to ionising radiation, heat, UV radiation and may persist in the dry form for decades (Aguilera and Karel, 1997; Brown, 1976; Potts, 1994; Jawad et al., 1998). These properties may be used as indirect indicators of desiccation tolerance (Sanders and Maxcy, 1979; Weekers et al., 1999a). Some bacterial species are also able to withstand desiccation without formation of differentiated forms through the accumulation of "protective components" such as disaccharides.

The drying resistance of bacteria is relevant for the conservation of starter cultures and has important economical consequences (Rapoport and Beker, 1987). The preservation of bacteria in a desiccated form has a main advantage on a frozen form : a lower cost in storage and transport. The disadvantages of dried cultures are considerable loss of activity and poor shell-life under uncontrolled conditions (Lievense and Van't

M. Hofman and P. Thonart (eds.), Engineering and Manufacturing for Biotechnology, 383–398.

Riet, 1993; Lievense and Van't Riet, 1994). For these reasons bacteria can be selected according to their desiccation-tolerance. In this strategy, desiccation tolerance is used as the main selective pressure and is considered as the most important property for the production of starter cultures. Therefore products such as a starter culture for the bioremediation of xenobiotic-contaminated soils are stored in a dry form (Weekers et al., 1999a; Weekers et al., 1998; Weekers et al., 1996). The xenobiotic catabolic activity may be introduced afterward into the drought-tolerant strains by the mean of natural conjugation in order to broaden the potential applications of the product (Weekers et al., 1999b).

In this paper, the new approach of desiccation-tolerant bacteria selection is described. Mechanisms of damages to the cells due to desiccation and adaptations towards drought-tolerance are proposed for undifferentiated cells. The strains selected according to the new strategy serve as examples and a comparison is made between the desiccation-tolerant strains, some sensitive strains and a drought-tolerant reference, *Deinococcus radiodurans* (ATCC13939) on a desiccation-tolerance point of view. The importance of drought-tolerance for technological applications of bacteria is emphasized.

2. Selection of desiccation-tolerant bacteria

Bacterial strains were isolated according to their desiccation-tolerance. Their potential technological applications such as growth on xenobiotic compounds were looked at afterward and were improved with plasmids. Soil bacteria were isolated from dried xenobiotic-polluted soil samples according to their desiccation-tolerance and were compared to reference strains chosen with equivalent technological properties i.e. the ability to decompose recalcitrant xenobiotic compounds (table 1). *Deinococcus radiodurans*, ATCC13939 served as a drought-tolerant reference. *Deinococcaceae* developed a very effective DNA repair ability what provides them resistance to ionising radiation. Mattimore and Battista (1995) have shown that *D. radiodurans* was also resistant to desiccation, since functions necessary to survive desiccation were also necessary to survive ionising radiation.

Under different drying conditions with or without protective additives and with different drying technologies, the bacteria selected from desiccated soil samples exhibited a better tolerance to desiccation than the references. Whatever the drying technique was, the survival ratio between the soil strains and the references was 4 to 65 folds higher. The survival was as good as the one of *Deinococcus radiodurans* with most techniques, but was lower after storage in the dry form. The difference in behaviour among the strains may arise from variations in sensitivity of the different targets of the desiccation damages or from the mechanisms of drought-tolerance that each strain utilises.

Table 1 : Survival of the strains to various desiccation processes : freeze-drying (w. or w/o. trehalose, 0,5%), slow drying and storage in reduced water activity (a_W) conditions. nd : not determined

			Survival to desiccation processes			
Strains	**Identification**	**Description**	**freeze-rying**	**freeze-drying (w. trehalose)**	**slow drying**	**storage at low a_w.**
T902	*Rhodococcus erythropolis*	Isolated from desiccated soil, growth on diesel oil	26%	30%	66%	12%
TF1	*Acinetobacter johnsonii*	Isolated from desiccated soil, growth on diesel oil	6%	16%	10%	6%
TF7	*Micrococcus luteus*	Isolated from desiccated soil, growth on diesel oil	2%	8%	5%	4%
ATCC 13939	*Deinococcus radiodurans*	desiccation-tolerant reference	16%	nd	42%	40%
LB400	*Pseudomonas sp.*	Growth on PCB	0.1%	nd	0.3%	0.03%
LH168	*Acinetobacter calcoaceticus*	Growth on oil	0.2%	nd	0.3%	0.03%
A5.1.	*Alcaligenes eutrophus*	growth on 4-chlorobiphenyl	0.02%	0.1%	0.6%	<0.01%
SK15	*Arthrobacter sp.*	Growth on biphenyl	0.36%	nd	nd	nd
LB126	*Sphingomonas sp.*	Growth on fluorene	0.05%	nd	0.2%	<0.01%
LH240	*Pseudomonas sp.*	Growth on oil	<0.01%	nd	0.8%	0.8%
AEX5	*Alcaligenes eutrophus*	Growth on chlorobiphenyls and chlorobenzoates	0.1%	nd	0.9%	<0.01%
PaW1	*P. putida*	Growth on toluene	0.09%	nd	0.2%	0.1%
GpO1	*P. oleovorans*	Growth on oil	0.01%	nd	0.3%	0.03%

3. Targets of desiccation damages and the proposed mechanisms responsible for dessication tolerance

Desiccation-damage targets were identified : the phospholipid bilayer membranes, the nucleic acids and the proteins. This list includes all the major cell components.

3.1. MEMBRANES

Membrane damages have been identified as responsible for most of the loss of viability during desiccation. To monitor the loss of membrane specific permeability during dehydration-rehydration cycles, the level of lactate dehydrogenase (LDH) activity in the supernatant after rehydration of the desiccated cells was measured and compared to the level of activity of a completely lysed sample. It serves as an indication of the leakage of the content of the cells due to phospholipid bilayer disruption (Castro et al., 1997; Weekers et al., 1999a). The effect of the water activity of the culture medium used prior to desiccation-rehydration cycles was measured (table 2).

Table 2 : Level of membrane lysis (%) as measured by LDH activity in the supernatant after drying and after storage at low a_w of the cells grown in different a_w conditions.

	$a_w = 0.98$		$a_w = 0.96$		$a_w = 0.94$		$a_w = 0.92$	
	slow drying	storage at low a_w	slow drying	storage at low a_w	slow drying	storage at low a_w	slow drying	storage at low a_w
ATCC 13939	55	56	49	50	40	40	No growth	No growth
T902	25	28	23	25	24	23	19	20
TF1	92	92	88	88	79	80	76	74

The level of membrane lysis of *Deinococcus radiodurans* after slow drying (see table 2) was equivalent to the cell mortality during dehydration (see table 1). Membrane leakage would be the only cause of cell death in *Deinococcus radiodurans*. This result is related to the ability of *Deinococcaceae* to repair, upon rehydration, the DNA damages caused by the desiccation (Battista, 1997). In comparison, some desiccated-soil-strains have a level of membrane lysis that does not account for the total mortality fraction. Other mechanisms of desiccation damage are involved in cell death.

Membrane lysis measurements do not evolve with storage time. This type of damages occurs only during the drying (or rehydration) time. Cells grown in a medium with reduced a_w, undergo less membrane lysis (Chen and Alexander, 1973; Weekers et al., 1999a).

3.1.1. Membrane desiccation-damage mechanisms

Membranes are mainly composed of phospholipids with membrane-proteins held in association by hydrophobic forces. Even for purified phospholipids there are several possible structures. The lamellar bilayer, in gel or liquid crystalline phase, and the hexagonal phase are the more frequent. The organisation of the phospholipid-water system is mainly dependent on composition, temperature, hydration state of the bilayer and on ionic strength and pH of the surrounding medium.

In water and at physiological temperatures, the polar head groups of the phospholipids are hydrated (about 10 molecules of water per phospholipid head). The water molecules spatially separate the polar head groups. When water is removed, the head groups get closer together. The packing, in turn, increase the opportunity for the hydrocarbon chains to interact. As a result, the temperature at which the chain melts to form the liquid crystalline phase (T_m) increase (Crowe and Crowe, 1982; Crowe et al., 1993a; Crowe et al., 1993b).

Thus, when phospholipid bilayers are dried, their phase transition temperature, T_m, increases, which, in turn, makes them undergo a phase transition from liquid crystalline phase to gel phase even when kept at a constant room temperature (figure 1). The hexagonal phase is usually not reached during drying because it is localised in high temperature region (> 80°C).

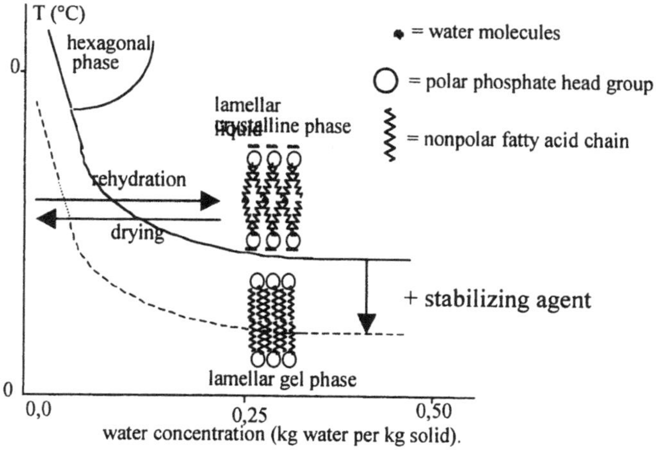

Figure 1 : Phase diagram of a simple phospholipid-water system. Variation of T_m of the bilayer with water content (adapted from Lievense and Van't Riet, 1993)

For example, palmitoyloleoyl phosphatidylcholine (POPC) has a T_m in water of -7°C. In the dry state, T_m reaches +60°C. Thus, when dried at room temperature, the phospholipid bilayer goes through a phase transition. Dry baker's yeast (*Saccharomyces cerevisiae*) packages always state that rehydration should be operated in warm water (about 40°C). It has been established that T_m for membrane phospholipids in these dry yeast cells is 35-40°C. In the hydrated cells, the T_m is about 10°C. Thus, if the dry cells are placed in water at temperatures below 40°C, their membranes undergo a phase transition during rehydration (Gelinas et al., 1989).

Unfortunately, phospholipid bilayers, as they undergo phase transitions, are known to become transiently leaky (Chapman, 1994; Crowe et al., 1989; Linders et al., 1997) In addition, biological membranes consist of a mixture of phospholipids. Each type enters the gel phase at a different temperature and hydration state thus at different times leading to segregation of the different phospholipids during drying. This separation is called 'lateral phase separation' and is considered an important mechanism in damaging biological membranes during dehydration. It becomes then important to prevent such phase transition during drying.

3.1.2. Role of disaccharides in membrane tolerance to desiccation

In the presence of disaccharides such as trehalose and sucrose the melt temperature of the phospholipidic bilayers is lowered (Crowe et al., 1987; Goodrich et al., 1991). This phenomenon enables the drying of biological membrane systems without going through a phase transition, avoiding, in turn, leakage of the content of the bilayer membrane system (Crowe et al., 1988; Crowe et al., 1985; Hoekstra et al., 1992; Strauss et al., 1986). The phospholipid bilayer is in liquid crystalline phase, even in the dry state and at room temperature.

The sugar molecules replace the water shell around the polar phosphate groups acting as 'spacers' (Crowe et al., 1993) (figure 2). To be effective at protecting the membranes, the disaccharides must be present on both sides of the lipidic bilayer (Eleutherio et al., 1993). That implies that if the cell produces trehalose, it must exit the cell to protect the outer side and if trehalose is added in a formulation of a dry starter culture to protect the cells of desiccation damage, it must enter the cells.

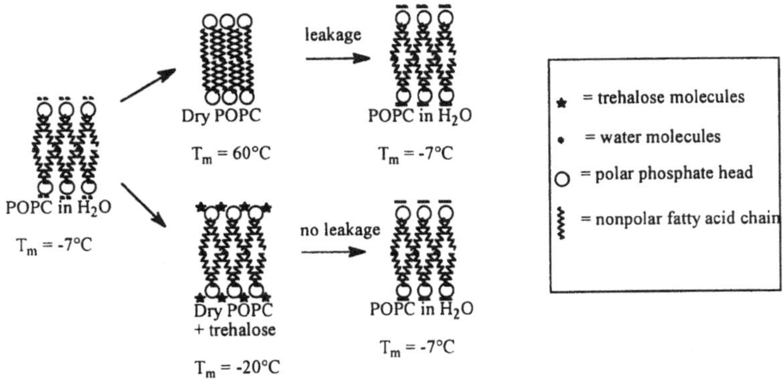

Figure 2 : Role of disaccharides in stabilisation of phospholipid bilayer during dehydration (adapted from Aguilera and Karel, 1997)

3.2. PROTEINS

In consideration of adaptation of microorganisms to extreme conditions (temperature, pH and pressure), it is generally assumed that the protein evolution is driven toward the achievement of optimum function rather than maximum stability. Adaptation to desiccation can be viewed quite differently for one main reason : a desiccated cell does not grow and the time the cell remains desiccated may represent the largest part of the 'life' of the cell and of its composing proteins. Unless desiccation-tolerant cells accumulate proteins that serve some structural or protective role (and no evidence for that has been forthcoming), the consideration of protein function in a desiccated cell is irrelevant. However, the question of optimal function might be critical at the time the cell emerge from desiccation. Since there is no evidence that proteins from desiccation-tolerant bacteria are more stable than the equivalent proteins of their desiccation-sensitive counterparts, we must take into account other mechanisms of proteins stabilisation.

3.2.1. The anhydrobiotic cell and a water replacement hypothesis

According to the *preferential exclusion hypothesis*, when some cells are submitted to an osmotic stress they are know to produce 'compatible solutes' which act as stabilisers for the proteins by being preferentially excluded from their direct vicinity. Such an exclusion is thermodynamically not favourable, but if the proteins were 'unfolded' from their native configuration, they would expose an even greater surface to the solute what would be even more unfavourable (Levine and Slade, 1992). As a consequence, the presence of these solutes stabilises the proteins. However, the anhydrobiotic cell is characterised by a far lower water content than a cell submitted to an osmotic stress or than a cryotolerant cell in presence of extracellular icc. The preferential exclusion hypothesis does not hold at these low moisture contents, but only in intermediate moisture systems (Crowe et al., 1993).

Some desiccation-tolerant prokaryotes accumulate large amounts (up to 20% of the dry weight) of either or both of the disaccharides trehalose and sucrose. The observations led to the conclusions that they were efficient at protecting enzymes during both freeze-drying and air-drying. However, as the preferential exclusion hypothesis does not hold, they do not act as compatible solutes.

A *water replacement hypothesis* was developed to account for the protective effect of these polyhydroxyl compounds in the desiccated systems (Clegg et al., 1982; Crowe et al., 1993a; Crowe et al., 1993b). Essentially, the hypothesis is that the compounds, such as trehalose, replace the shell of water around the macromolecules, circumventing damaging effect during drying. The expression 'water replacement' may also be applied to the role of trehalose in stabilising the lipidic bilayer systems of the membranes (Leslie et al., 1995).

3.2.2. Vitrification of the cytoplasm as mechanism of tolerance to desiccation

There is some controversy whether the exclusion hypothesis is the explanation of the stabilising effect of the disaccharides on the cells or not. Some groups of researchers involved in the food industry claim that the most important property of saccharides relevant for the protection of anhydrobiotic cells is their ability to form a vitreous (glassy) phase (Crowe et al., 1997; Slade and Levine, 1991).

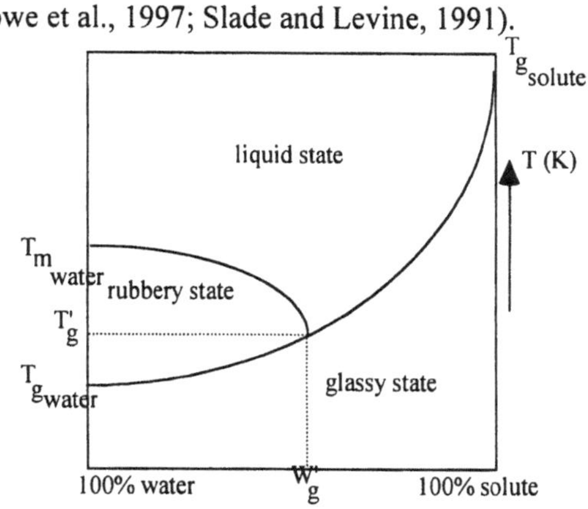

Figure 3 : 'Dynamic' phase diagram of a glass forming solute in water.

In intermediate moisture systems, such as bacterial cells, most physical and chemical processes (with the exception of free radical reactions) are under kinetic control, i.e., they are diffusion limited. The living organisms may be in a stationary state, but not in thermodynamic equilibrium. In a simple system composed of two components, a solute in water, a continuum of hydration states can be achieved from a pure solute to a solution of infinite dilution. Each hydration state has a characteristic temperature that defines the point of kinetic change in physical state : the glass transition temperature. This transition occurs between a metastable glassy solid that is capable of supporting its own weight against gravity to a rubbery viscous fluid that can flow in real time. At temperatures below this glass transition temperature, $T<T_g$, diffusion-limited processes are inhibited by an extremely high viscosity and water is virtually unavailable (figure 3).

Water acts as a plasticiser : the net effect of increasing the water content, W, is equivalent to the net effect of increasing the temperature. The viscosity of the system decreases. T'_g is an invariant point on the continuum curve of T_g and represents the state-specific subzero T_g of the maximally freeze-concentrated, amorphous solute/unfrozen water matrix surrounding the ice-crystals in a frozen solution. T'_g corresponds to, and is determined by, the intersection of the glass curve and the non-equilibrium extension of the equilibrium liquidus curve for the T_m of ice. This solute-specific point defines the composition of the glass that contains the maximum practical amount of plasticising moisture (W'_g) (Levine and Slade, 1992).

Several solutes may be accumulated by anhydrobiotic organisms causing the vitrification of the cytoplasm under physiological conditions. The vitrified cells are stabilised during conservation in the dry state. A water system alone cannot be in the vitrified phase under physiological conditions, the glass transition temperature of pure water being −137°C. When compounds such as trehalose, sucrose or polyhydroxy-compounds are added to water, they raise the T_g of the system, which stabilises the vitrified phase under 'normal' conservation conditions (figure 4). Sucrose has a T_g of 67°C and trehalose has a T_g of 79°C.

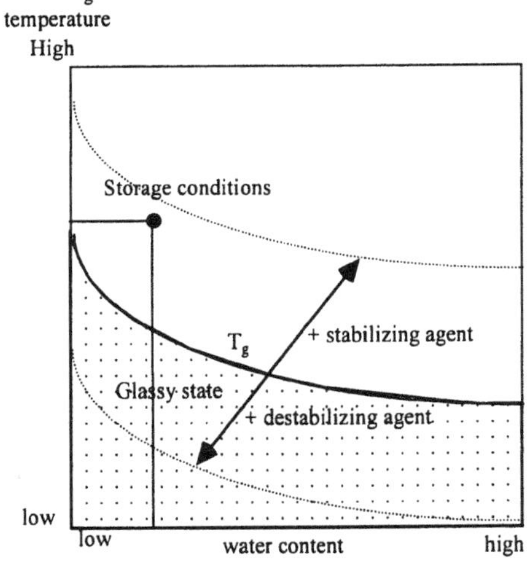

Figure 4 : Effect of (de)stabilising solutes on T_g of a desiccated product (adapted from Lievense and Van't Riet, 1993)

Vitrification stabilises anhydrobiotic systems by virtually stopping the rate of all chemical reactions, which were under diffusion limitations (including the reactions of degradation of biological materials during low-moisture preservation). Vitrification also inhibits water loss by reducing diffusion rates to the free surface, prevents fusion of vesicles during dehydration and stops solute leakage during rehydration.

Trehalose permits water content as high as 2 water molecules per glucose ring while still in the glassy state (i.e. up to 17 weight % of water) at ambient conditions.

3.3. NUCLEIC ACIDS

Both DNA and RNA are targets of desiccation damages. In large parts, the damages reflect the accumulation of mutations during the time there is no cell growth i.e. desiccation (Potts, 1994). The mechanisms of repair are unlikely to operate in air-dried cells and these damages become manifest only upon rehydration. Damage to DNA in the dry form may arise through chemical modifications (alkylation, oxidation), cross-linking (between protein and DNA), base removal such as depurination, or ionising or non-ionising radiations. As opposed to the damage to the membranes, DNA damages continuously accumulate during the time of dry storage. The control of the conditions of storage are essential for the preservation of dry starter cultures.

A

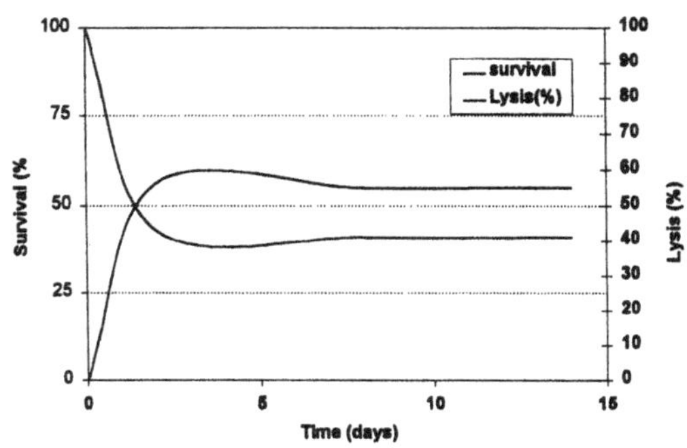

B

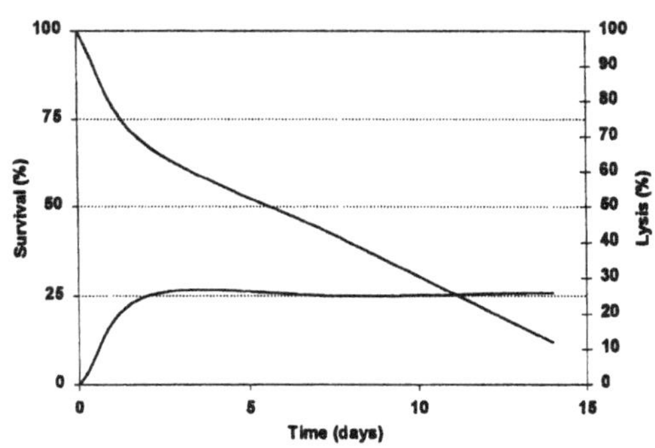

Figure 5 : Evolution of survival (%) and membrane lysis (%) during storage of D. radiodurans (A) and Rhodococcus erythropolis (B)

The survival value of *Deinococcus radiodurans* is stable during storage, because the major damage (i.e. membrane alterations) occurs during the drying process itself and later does not evolve anymore (figure 5A). *Deinococcus radiodurans* accumulates single- and double-strand breaks during the time of desiccation, but *Deinococcaceae* can tolerate massive DNA damages because, upon rehydration, they have the ability to efficiently repair them. As opposed to *Deinococcus*, the storage in the dry form of the drought-tolerant *Rhodococcus* strain results in a continuous decrease of the survival although the membrane injuries are not evolving anymore (figure 5B). *Rhodococcus* strains have not been shown to be able to erase all DNA damages upon rehydration. The accumulation of DNA damages are probably responsible for the steady decrease of survival although in the experiments with *Rhodococcus*, damages to the DNA are not dissociated from the damages to the proteins.

3.3.1. Mechanisms of tolerance to DNA damages during desiccation

Deinococcus radiodurans has a unique mechanism of tolerance to nucleic acid damages occurring during desiccation or irradiation as discussed before. For a good review about it see Battista's paper (Battista, 1997).

Some terrestrial cyanobacteria utilise another technique to prevent DNA damages : the accumulation of photoprotective pigments with broad UV absorption spectrum. They play a role in the radiation tolerance of the dry cells.

Vitrification may play an important role in slowing down the rates of destruction of the nucleic acids, by impeding the diffusion of the reactive species (excepted, as noted before, for the free-radical reactions).

3.3.2. UV irradiation as a tool for the selection of drought-tolerant bacteria

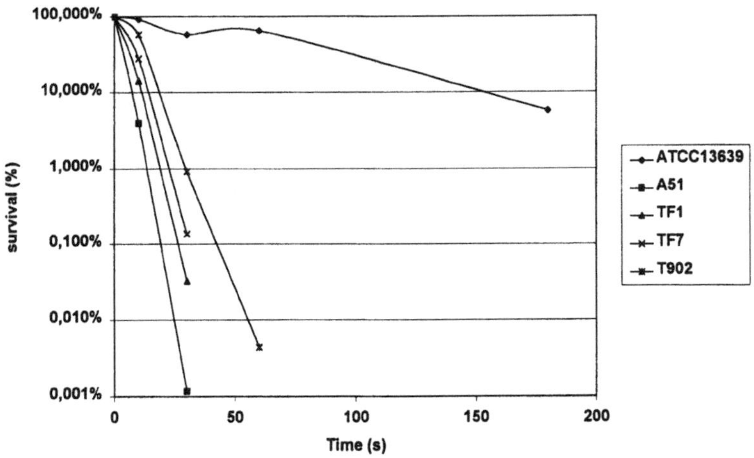

Figure 6 : Log of the percentage of survival of the strains (see table1) vs. duration of UV radiation (254 nm)

Desiccation-tolerance is viewed as an important factor for the use of living microorganisms in technological processes. Therefore quick selection procedures for drought-tolerant bacteria were set-up (Weekers et al., 1999a). As functions necessary to survive ionising radiation are also necessary to survive desiccation, UV radiation tolerance measured by exposure to UV radiation (254 nm) served as a tool to quantify the drought-tolerance of a strain. This measurement only accounts for the DNA damages and not for the damages to the membranes.

As DNA damages are the main consequence of irradiation, the *Deinococcus* strain is notably more resistant than the other strains. The behaviour of the drought-tolerant group of strains is different from the sensitive ones modelled by $A_{5.1}$. The survival of the former is up to 1000 folds higher than the latter (figure 6) after 30 seconds of irradiation.

4. Factors influencing survival

When used as a preservation technique, desiccation may be optimised toward a higher level of survival of the final dry product. Several factors influence the final survival and each of these factors may be optimised.

4.1. BACTERIAL SPECIES

Bacterial species cannot usually be chosen for a given use. The nature of the microorganism is dictated by the properties required by the technological applications one will make of it. However, when selecting bacteria according to their desiccation tolerance as a first criteria before looking for biodegradation properties (or any other activity) among the drought-tolerant selected strains, the selection is made on the species basis. The required activity may be transferred to the strain in a second step with genes borne on a plasmid such as catabolic plasmid. With this strategy, the selection is made according to the bacterial species. As a rule of thumb, one can roughly distinguish between sensitive and less sensitive vegetative bacterial cells, by distinguishing between Gram-negative and Gram-positive bacteria, respectively. With the method used for the selection of drought-tolerant strains from a desiccated soil, 20 strains out of 26 collected were gram-positive (77%). This ratio is in agreement with the general tendency of higher resistance for the gram-positive bacteria. The difference in membrane behaviour during desiccation is mainly responsible for the observed difference in desiccation-tolerance between Gram positive and Gram negative bacteria.

4.2. GROWTH CONDITIONS

The growth conditions and growth phase of the bacterial cell is important for their survival (Gelinas et al., 1989; Labuza et al., 1972; Linders et al., 1997; Siegele and Kolter, 1992).

- The growth conditions will influence the membrane composition, which in turn, influences the membrane phase transition, as discussed before.

- The degree of saturation of the phospholipids of the membrane (influenced by the aeration of the culture) is related to the stability of the membrane. Higher degree of saturation would correlate to a higher survival.
- Generally, the highest dehydration resistance is found for bacteria harvested in the stationary growth phase.
- Most protective compounds are produced during the stationary phase, as yeasts producing trehalose. It is better to wait until that time to collect and dry them.

4.3. PROTECTIVE ADDITIVES

Research on the protective effect of various additives is abundant (Combes et al., 1990; De Cordt et al., 1994; De Valdez et al., 1985; De Winder et al., 1989; Gianfreda et al., 1991; Graber and Combes, 1989; Izutsu et al., 1994; Roser, 1991). It is probably because -and it has been acknowledged- the use of additives is the most fruitful strategy for obtaining optimal survival after drying. Positive effect has been reported for sugars, polyalcohols, glycerol, carboxylic acids, milk, skim milk, culture medium, proteins, amino acids, polymers, metallic cations and salts. The effective protective effect of each additive is species-specific.

The interactions of some of these additives with membranes were discussed earlier and the hypothesis of the water replacement or the glass formation theory were explained. Some additives may also act as anti-oxidant or encapsulating agents isolating the cells from the lethal effect of oxygen and oxygen related species.

4.4. CELL CONCENTRATION

It is generally reported that higher concentrations in the suspension to be dried give higher survival. An explanation to this effect could be the release of intracellular compounds of damaged cells that could protect other cells.

4.5. DRYING GAS, RATE AND EXTEND

When oxidation of cellular compounds plays an important role in reducing survival, nitrogen can advantageously replace air in air-drying processes.

Drying rate, or the rate of water activity reduction, has also an effect on the survival (Antheunisse and Arkesteijn-Dijksman, 1979; Gervais et al., 1992; Poirier et al., 1996). If the rate of drying is too rapid, there is no time for adaptation mechanisms to take place (such as accumulation of protective compounds) and the survival is low. On the other hand, if the drying rate is too slow, the cells are submitted to an environment of unfavourable water activity for longer periods of time, which is also unfavourable. There is an optimal intermediate drying rate, which balances these two opposite effects. The fluidised bed drying technique with warm air (45°C) allows drying rates that respect best this balance. Drying rate effect in given conditions can be predicted by the reduction of a_w in solution by addition of glycerol or any other solute reducing the water activity of the solution.

The survival of bacteria is undoubtedly related to the final water concentration, due to the dehydration inactivation. A low water concentration is necessary to obtain storage stability and, therefore, an optimum has to be found between survival after drying and stability during storage. The use of protective additives also influences this effect. With desiccation-tolerant soil strains, the long-term conservation is the same with a residual a_w of 0.29 and 0.17. It means that once the stability of the dry product is guaranteed, it is not necessary to decrease a_w to lower values, because survival at low water activity is smaller than survival at higher values.

4.6. REHYDRATION

Rehydration conditions, such as temperature, composition, osmolality, rehydration medium or rate, can significantly influence survival. Some authors regard rehydration as the most important step since the damages to membranes happen more during rehydration than dehydration. The rehydration temperature directly correlate with the membrane phase transition theory.

4.7. STABILITY DURING STORAGE

In the preservation of commercial bacterial starter cultures, low inactivation rate during storage is as important as high viability after drying. High survival yields must be obtained but they are not sufficient. To enter the commercial chain, the half-life of the dry product must be at least equivalent to the time necessary for the product to reach the final user. Storage stability is increased by decreasing temperature. The presence of oxygen in the storage atmosphere is detrimental (Mary et al., 1993). The inactivation during storage was related to the formation of radicals in the presence of oxygen. As possible radical reaction, fatty acid oxidation and DNA damage have been reported. Light is also expected to be detrimental and storage in the dark is recommended. The glass formation theory is also relevant for the storage of biological products. Many authors report a maximum water concentration below which the cells have to be stored (Lievense and Van't Riet, 1994; Scott, 1956). Since the diffusion coefficient in the glassy state are very small, diffusion-limited reactions become undetectable. To reach the glassy state, low temperature and low moisture content are usually necessary, but one can achieve the glass state by adding compounds that bring the glass transition curve to higher temperature closer to ambient. Starch hydrolysis products such as maltodextrins are efficient at stabilising products by their glass forming properties. Radical reactions are not diffusion limited and will thus not be reduced by the glassy state. However, the rate of diffusion of oxygen into the product will be slowed and this will decrease the rate of production of radicals.

5. Conclusions

The study of desiccation tolerance of cells requires the application of a judicious mix of biophysics, structural biochemistry, and molecular ecology to the study of the whole cells and their purified components. The membrane lysis is responsible for most of the mortality during desiccation. It does not evolve during storage of the desiccated product. It is possible to reduce or prevent the phase transition of the membranes with protecting compounds and growth conditions affect the desiccation-tolerance of the microorganisms. The conditions of storage of desiccated biological materials must be controlled because the nucleic acid damages accumulate during the time of desiccation.

The complexity of the desiccation-tolerance phenomena is related to the complexity of a cell and to the multiplicity of its components. There is not a universal additive that will protect all cells from all desiccation damages, nor there are techniques and conditions that will allow best survival and storage preservation in any case. Each species is a different case.

Quick selection techniques such as resistance to UV radiation exposure can be used to select desiccation-tolerant strains for their technological application.

Acknowledgements

F. Weekers is a recipient of a FRIA (Fonds pour la Formation à la Recherche dans l'Industrie et l'Agriculture) grant.

References

Aguilera, J., and Karel, M. (1997). Preservation of biological material under desiccation. Critical Review in Food Science and Nutrition *37*, 287-309.

Antheunisse, J., and Arkesteijn-Dijksman, L. (1979). Rate of drying and the survival of microorganisms. Antonie Van Leeuwenhoek *45*, 177-184.

Battista, J. (1997). Against all odds : the survival strategies of Deinococcus radiodurans. Annual Review of Microbiology *51*, 203-224.

Brown, A. (1976). Microbial water stress. Bacteriol. rev. *54*, 803-846.

Castro, H. P., Teixeira, P., and Kirby, R. (1997). Evidence of membrane damage in Lactobacillus bulgaricus following freeze-drying. Journal of Applied Microbiology *82*, 87-94.

Chapman, D. (1994). The role of water in biomembranes. Journal of Food Engineering *22*, 367-380.

Chen, M., and Alexander, M. (1973). Survival of soil bacteria during prolonged desiccation. Soil Biology and Biochemistry *5*, 213-221.

Clegg, J., Seitz, P., Seitz, W., and Hazelwood, C. (1982). Cellular responses to extreme water loss : the water replacement hypothesis. Cryobiology *19*, 306-316.

Combes, D., Graber, M., and Ye, W. (1990). Stabilising effect of polyhydric alcohols. Annals of the New-York Academy of Sciences *613*, 559-563.

Crowe, J., and Crowe, L. (1982). Induction of anhydrobiosis : membrane changes during drying. Cryobiology *19*, 317-328.

Crowe, J., Crowe, L., Carpenter, J., Rudolph, J., Winstrom, A., Spargo, C., and Anchordoguy, T. (1988). Interaction of sugars with membranes. Biochimica et Biophysica Acta *947*, 367-384.

Crowe, J., Hoekstra, F., and Crowe, L. (1989). Membrane phase transition are responsible for imbibitional damage in dry pollen. Proceeding of the National Academy of Sciences of USA *86*, 520-523.

Crowe, J. H., Crowe, L. M., and Carpenter, J. F. (1993a). Preserving dry biomaterials : the water replacement hypothesis, part 1. Biopharm, 28-34.

Crowe, J. H., Crowe, L. M., and Carpenter, J. F. (1993b). Preserving dry biomaterials : the water replacement hypothesis, part 2. Biopharm, 40-43.

Crowe, J. H., Crowe, L. M., Carpenter, J. F., and Wistrom, C. A. (1987). Stabilisation of dry phospholipid bilayers and proteins by sugars. Biochemistry Journal *242*, 1-10.

Crowe, J. H., Hoekstra, F. A., and L.M., C. (1992). Anhydrobiosis. Annual Review of Physiology *54*, 579 - 599.

Crowe, J. H., Oliver, A. E., Hoekstra, F. A., and Crowe, L. M. (1997). Stabilisation of dry membranes by mixtures of hydroxyethyl starch and glucose : the role of vitrification. Cryobiology *35*, 20-30.

Crowe, L. M., Crowe, J. H., Womersley, A. R. C., and Appel, L. (1985). Preservation of freeze-dried liposomes by trehalose. Archives of Biochemistry and Biophysics *242*, 240-247.

De Cordt, S., Hendrickx, M., Maesmans, G., and Tobback, P. (1994). The influence of polyalcohols and carbohydrates on the thermostability of a-amylase. Biotechnology and bioengineering *43*, 107-114.

De Valdez, G., De Giori, G., De Ruiz Holgado, A., and G., Oliver. (1985). Effect of drying medium on residual moisture content and viability of freeze-dried lactic acid bacteria. Applied and Environmental Microbiology *49*, 413-415.

De Winder, B., Matthijs, H., and Mur, L. (1989). The role of water retaining substrata on the photosynthetic response of three drought tolerant phototropic microorganisms isolated from a terrestrial habitat. Archives of Microbiology *152*, 458-462.

Eleutherio, E., De Araujo, P., and Panek, A. (1993). Role of trehalose carrier in dehydratation resistance of *Saccharomyces cerevisiae*. Biochimica et Biophysica Acta *1156*, 263-266.

Gelinas, P., Fiset, G., Leduy, A., and Goulet, J. (1989). Effect of growth conditions and trehalose content on cryotolerance of baker's yeast in frozen dough. Applied and Environmental Microbiology *55*, 2453-2459.

Gervais, P., Maréchal, P., and Molin, P. (1992). Effects of kinetics of osmotic pressure variation on yeast viability. Biotechnology and Bioengineering *40*, 1435-1439.

Gianfreda, L., Toscano, G., Pirozzi, D., and Greco, G. (1991). The effect of sorbitol on acid phosphatase deactivation. Biotechnology and Bioengineering *38*, 1153-1158.

Goodrich, R., Crowe, J., Crowe, L., and Baldeschwieler, J. (1991). Alteration in membrane surfaces induced by attachment of carbohydrates. Biochemistry *30*, 5313-5318.

Graber, M., and Combes, D. (1989). Effect of polyols on fungal alpha-amylase thermostability. Enzyme and Microbial Technology *11*, 673-677.

Hoekstra, F., Crowe, J., Crowe, L., Van Roekel, T., and Vermeer, E. (1992). Do phospholipids and sucrose determine membrane phase transitions in dehydration pollen species? Plant Cell Environment *15*, 601-606.

Izutsu, K., Yoshioka, S., and Terao, T. (1994). Effect of mannitol crystallinity on the stabilisation of enzymes during freeze-drying. Chemical and Pharmaceutical Bulletin *42*, 5-8.

Jawad, A., Snelling, A., Heritage, J., and Hawkey, P. (1998). Exceptional desiccation tolerance of *Acinetobacter radioresistens*. Journal of Hospital Infection *39*, 235-240.

Kaprelyants, A., Gottschal, J., and Kell, D. (1993). Dormancy in non-sporulating bacteria. FEMS Microbiology Reviews *104*, 271-286.

Labuza, T., Jones, K., Sinskey, A., Gomez, R., Wilson, S., and Miller, B. (1972). Effect of drying conditions on cell viability and functional properties of single-cell protein. Journal of Food Science *37*, 103-107.

Leslie, S., Israeli, E., Lighthart, B., Crowe, J., and Crowe, L. (1995). Trehalose and sucrose protect both membranes and proteins in intact bacteria during drying. Applied and Environmental Microbiology *61*, 3592-3597.

Levine, H., and Slade, L. (1992). Another view of trehalose for drying and stabilising biological material. Biopharm, 36-40.

Lievense, L., and Van't Riet, K. (1993). Convection Drying of Bacteria I. The drying process, In *Advances in Biochemical Engineering/Biotechnol*ogy, *A. Fiechter*, ed. (Berlin: Springer-Verlag Berlin Heidelberg) 50, 45-63

Lievense, L., and Van't Riet, K. (1994). Convection Drying of Bacteria II. Factor influencing survival., In *Advances in Biochemical Engineering/Biotechnol*ogy, *A. Fiechter*, ed. (Berlin: Springer-Verlag Berlin Heidelberg) 51, 71-89

Linders, L., Wolkers, W., Hoekstra, F., and Van't Riet, K. (1997). Effect of added carbohydrates on membrane phase behaviour of dried *Lactobacillus plantarum*. Cryobiology *35*, 31-40.

Mary, P., Moschetto, N., and Taillez, R. (1993). Production and survival during storage of spray-dried Bradyrhizobium japonicum cell concentrates. Journal of Applied Bacteriology *74*, 340-344.

Mattimore, V., and Battista, J. (1995). Radioresistance of *Deinococcus radiodurans* : function necessary to survive ionising radiation are also necessary to survive prolonged desiccation. Journal of Bacteriology *178*, 633-637.

Poirier, I., Maréchal, P., and Gervais, P. (1996). Water stress and bacterial survival : influence of the kinetics of water potential variation. Med. Fac. Landbouww. Univ. Gent *61/4a*, 1559-1564.

Potts, M. (1994). Desiccation tolerance of prokaryotes. Microbiological reviews *58*, 755-805.

Rapoport, A., and Beker, M. (1987). Conservation of Yeasts by dehydration. In *Advances in Biochemical Engineering/ Biotechnol*ogy, *A. Fiechter*, ed. (Berlin: Springer-Verlag Berlin Heidelberg), *35*, 127-171.

Roser, B. (1991). Trehalose drying : a novel replacement for freeze-drying. Biopharm *4*, 47-53.

Sanders, S., and Maxcy, R. (1979). Isolation of radiation-resistant bacteria without exposure to irradiation. Applied and Environmental Microbiology *38*, 436-439.

Scott, W. (1956). The effect of residual water on the survival of dried bacteria during storage. Journal of General Microbiology *19*, 624-633.

Siegele, D., and Kolter, R. (1992). Life After Log. Journal of Bacteriology *174*, 345-348.

Slade, L., and Levine, H. (1991). Beyond water activity : recent advances based on an alternative approach to the assessment of food quality and safety. Critical Review in Food Science and Nutrition *30*, 115-360.

Strauss, G., Schurtenberger, P., and Hausser, H. (1986). The interaction of saccharides with lipid bilayer vesicles : stabilisation during freeze-thawing and freeze-drying. Biochimica et Biophysica Acta *858*, 169-180.

Weekers, F., Jacques, P., Springael, D., Mergeay, M., Diels, L., and Thonart, P. (1999a). Desiccation-tolerance : an advantage for the biotechnological applications of bacteria. In *9th European Congress on Biotechnology*, M. Hofman, ed. (Brussels).

Weekers, F., Jacques, P., Springael, D., Mergeay, M., Diels, L., and Thonart, P. (1999b). Improving the catabolic functions in desiccation-tolerant soil bacteria. Applied Biochemistry and Biotechnology *77-79*, 251-266.

Weekers, F., Jacques, P., Springael, D., Mergeay, M., Diels, L., and Thonart, P. (1998). Effect of drying on bioremediation bacteria properties. Applied Biochemistry and Biotechnology *70-72*, 311-322.

Weekers, F., Jacques, P., Springael, D., Mergeay, M., Diels, L., and Thonart, P. (1996). Selection of hydrocarbon degrading bacteria with drought resistance characteristics. Med. Fac. Landbouww. Univ. Gent ***61/4b***, 2161-2164.

BIOTECHNOLOGICAL RESEARCH AND THE DAIRY INDUSTRY:

A Functional Interaction

HEIKE NEUBAUER AND BEAT MOLLET
Nestlé Research Center
Nestec Ltd Vers-chez-les-Blanc
P.O. Box 44 CH-1000 Lausanne 26 Switzerland

Abstract

The application of biotechnological processes in the manufacturing of dairy products has a long tradition. Microorganisms already occurring in food, like lactic acid bacteria, yeast's and moulds, are used in order to preserve perishable foodstuffs and to improve the flavour, colour, texture and digestibility. This article gives an overview of the role and development of lactic acid bacteria as industrial starter cultures for dairy fermentations, and illustrates the use of modern biotechnological techniques for the development of new starter strains. Due to the increasing knowledge about the biochemistry, fermentation technology and nutritional aspects of traditional fermentation processes and due to the advances in modern microbiology techniques, including molecular biology and genetic engineering, new starter strains can be specifically selected among a large number of natural strains or designed by genetic engineering to meet the product requirements.

1. Introduction

1.1. THE HISTORY OF LACTIC ACID BACTERIA

Lactic acid bacteria are typically involved in a large number of spontaneous food fermentations. Reference to such products is already documented in archaic texts from Uruk/Warka (Iraq) dated around 3200 B.C (Nissen *et al.* 1991). Beer brewed by the Babylonians and exported to Egypt around 3000 B.C. was most likely the product of alcoholic and lactic fermentations. Present day sorghum, maize and millet beers in Africa possess similar features in which the lactic fermentation plays a key role in

M. Hofman and P. Thonart (eds.), Engineering and Manufacturing for Biotechnology, 399–412.

acceptability and microbiological safety in tropical climates (Haggblade and Holzapfel, 1989).
Since a long time, these interactions of lactic acid bacteria with foods have attracted the attention of scientists and the first research was done on lactic acid fermentation by Pasteur in 1857, followed by the first isolation of a pure bacterial culture, *Bacterium lactis*, by Lister in 1873. In 1890, Weigmann in Kiel and Storch in Copenhagen introduced, almost simultaneously, the use of starter cultures for cheese and sour milk production. This opened the way for the industrialisation of fermentations with lactic acid bacteria. At about the same time, Elie Metchnikoff at the Pasteur Institute in Paris and other scientists realised the similarity between the food fermenting bacteria and some of the inhabitants of the human intestinal flora and proposed their use in the diet due to health and life prolonging properties. Hence, yoghurt products started to gain popularity in Europe. The development of fruit and flavoured yoghurt in the 1950s helped this product to become even more popular in the western world. Today, a multitude of different fermented dairy and non-dairy foods is commercialised worldwide and the consumer demand especially for fresh refrigerated dairy products is still growing considerably.

2. Classification of lactic acid bacteria

The natural microbial diversity of lactic acid bacteria bears great potential for various applications in food technology and biotechnology. Hence, there is a strong need to develop culture collections, and to classify or categorise the bacteria according to their natural habitat, general properties, their present or past use in foods, and safety issues relevant to man and the environment. The taxonomic classification helps to answer questions like: which bacterial species have a long safety record in food products and are suited for human consumption? And, are there any lactic acid bacteria involved in or known as causative agent for human disease or infections, and hence not suited for consumption? Furthermore, it allows comparisons to be made with bacteria found in (or used for) fermented food and those colonising the human intestinal flora further substantiating the early findings of Metchnikoff.

2.1. THE GROUP OF LACTIC ACID BACTERIA

According to Orla-Jensen (1919), lactic acid bacteria are Gram-positive, non-motile, non-sporeforming bacteria that ferment carbohydrates and higher alcohols to produce lactic acid as the major end product. They comprise different genera and more than 100 different species (Orla-Jensen, 1919; Kandler and Weiss, 1986). Lactic acid bacteria are evolutionary very dispersed comprising microorganisms with different morphologies (coccoidal and rod-shaped), with different optimal growth temperatures (mesophilic and thermophilic conditions), and with different major fermentation pathways. Lactose may be taken up via the phosphoenolpyruvate-sugar phosphotransferase system (PTS) or as free lactose, and glucose may be fermented via the glycolytic (homofermentative) or the pentose phosphate heterofermentative pathways.

2.2. CLASSICAL BACTERIAL TAXONOMY COMBINED WITH MOLECULAR BIOLOGY

The classical approach to bacterial taxonomy was based on morphological, physiological and biochemical features. Nowadays other characteristics of the cells such as cell wall composition and protein fingerprinting by analysis of the total soluble cytoplasmatic proteins, isoprenoid quinones are included. With the evolution of molecular biology other taxonomic tools became available, such as mol% G+C content of the DNA, pulsed field gel electrophoresis (PFGE), random amplification of polymorphic DNA-PCR (RAPD-PCR), Insertion Sequence (IS) typing, DNA:DNA hybridisation studies and structures and sequence of ribosomal RNA (rRNA). These new tools are especially useful for the identification of lactic acid bacteria which cannot be reliably differentiated by the classical methods, as shown for the species of the *Lactobacillus acidophilus* group, comprising of *Lb. acidophilus* sensu stricto[1], *Lb. amylovorus, Lb. crispatus, Lb. gallinarum, Lb. gasseri,* and *Lb. johnsonii* (Schleifer *et al.*, 1995; Klein *et al.*, 1998). A reliable and fast identification of these organisms is currently only possible with the help of molecular biological methods.

The implementation of molecular biological methods has led to significant changes in the taxonomy of lactic acid bacteria (Schleifer, 1987; Schleifer *et al.*, 1995; Stiles and Holzapfel, 1997; Klein *et al.*, 1998). It has been proposed that the taxonomy and physiology of lactic acid bacteria can only be understood by combining the morphological, biochemical and physiological characteristics with the molecular-based and genomic techniques (Klein *et al.*, 1998). However, the classification of lactic acid bacteria is still under investigation and not without some controversy on the definition of the boundaries between some genera and species (for reviews see: Stiles and Holzapfel, 1997; Axelsson, 1993; Pot et al., 1994).

2.3. ISOLATION OF NEW STRAINS OF LACTIC ACID BACTERIA

Lactic acid bacteria are generally found in fermented foods, but also in the gastrointestinal microflora. It is from these habitats that they can be isolated for the use as starter cultures in controlled fermentation processes. An example of the isolation and identification of a new species from cottage industry or spontaneously fermented foods is the new *Lactobacillus* species, *Lb. panis*, named after the latin word for bread since it was first isolated from rye sourdough (Wiese *et al.*, 1996). Such species can potentially be further developed into new starter strains for fermentation processes.

However, a major drawback in the isolation of new strains is that many species cannot be cultured *in vitro* at all. It has been speculated that at present only 10 to 50% of the bacteria from the gastrointestinal tract can be cultured in the laboratory (McFarlene *et al.*, 1994; Amann *et al.*, 1995; Langendijk *et al.*, 1995; Wilson *et al.*, 1996). Furthermore, only a fraction of the total culturable and only a few unculturable species are known. The development of novel molecular typing techniques based on DNA technology, by which bacteria can be identified without cultivation, now at least

1 The unconventional abbreviations *Lb.* and *Lc.* are used in this text to avoid confusion between the genera *Lactobacillus* and *Lactococcus*, respectively.

partly overcomes this problem and has greatly facilitated the retrieval of new species (Amann *et al.,* 1990; 1995; Zoetendal *et al.,* 1998). At present, several thousand 16S rRNA sequences of different bacteria are available in genetic databases (Angert *et al.,* 1990; Snel *et al.,* 1994). However, the lack of cultivability restricts their physiological analysis and use in the food technology.

3. Lactic acid bacteria as starter cultures

As outlined above, different members of the lactic acid bacteria are applied as starter cultures for the production of a great variety of foods such as fermented cheese, milk, bread, wine, pickles and meat (Table 1). Important prerequisite for lactic acid bacteria strains used in industrial application are that they must be cultivable and stable on an industrial scale and through the manufacturing processes and storage conditions of the food product over the many years of industrial use. This is especially important for probiotic strains that need to retain their metabolic activity in order to exert the desirable health beneficial effects. Using modern molecular-based tools, industrial food-microbiology laboratories are able to characterise and monitor the genetic stability and activity of their starter strains.

3.1. THE ROLE OF LACTIC ACID BACTERIA IN THE FERMENTATION OF MILK

Lactic acid bacteria need to fulfil specific requirements for the transformation of the raw material, e.g. milk, to the final product. These requirements are different depending on the nature of the raw material, the desired end product and the final quality demand. The role of the lactic acid bacteria in food fermentations can be summarised as: i) generation of flavour, ii) texturing capacities, iii) microbial preservation of the raw material and iv) probiotic, health beneficial properties (Table 2).

The numerous metabolic pathways involved in these functions vary widely. Flavour compounds in dairy products such as yoghurt often result from hydrolysis of milk proteins and lipids and subsequent metabolism of the products, but are equally generated from the bacterial carbon metabolism. The main end product of sugar metabolism, pyruvate, plays a central role since it is converted to metabolites such as lactic acid, acetaldehyde, diacetyl and acetoin which are important for flavour. At the same time, the pH drop caused by the production of lactic acid and other organic acids is responsible for the precipitation of casein and thus necessary for the development of the typical yoghurt texture. Extracellular polysaccharides produced by several lactic acid bacteria have an important impact on texture and viscosity of fermented milk products. Lactic acid bacteria generally inhibit the growth of spoilage and pathogenic bacteria due to the production of lactic acid and 'natural preservatives' such as organic acids, hydrogen peroxide and antibacterial peptides, i.e. bacteriocins (Ray and Daeschel, 1992).

Table 1. Different strains of lactic acid bacteria and their use in fermented foods

In fermented dairy products (yoghurt, probiotic products, cheese, butter, sour milk)	*Streptococcus thermophilus* *Lactococcus lactis* *Lactobacillus acidophilus* *Lactobacillus delbrueckii* ssp. *bulgaricus* *Lactobacillus helveticus* *Lactobacillus casei* *Leuconostoc mesenteroides*
In fermented vegetables (wine, mixed pickles)	*Lactobacillus plantarum* *Lactobacillus fermentum* *Leuconostoc sp.* *Oenococcus oenos*
In fermented meat products (salami, crude ham)	*Lactobacillus curvatus* *Lactobacillus sake* *Pediococcus acidilactici* *Carnobacteria piscicola*
In fermented cereals (sourdough, silage, soy sauce)	*Lactobacillus sanfrancisco* *Lactobacillus panis* *Lactobacillus plantarum* *Lactobacillus fermentum* *Pediococcus acidilactici* *Tetragenococcus halophilus*

Table 2. The role of lactic acid bacteria in food

Preservation:	Acidification
	Antibacterial substances
Flavour:	Primary and secondary metabolic end-products
	Enzymatic activities
Texture:	Bacterial cell-mass
	Polymerising enzymes
	Exopolysaccharides
Health benefits:	Adhesive protection
	Pathogen exclusion
	Immunomodulation

3.2. THE NEW AGE OF STRAIN AND PRODUCT DEVELOPMENT

Several of the early biotechnological processes are still in use, although they are applied today under well-controlled conditions on an industrial scale. To obtain fermentation products of a reproducible and high quality, present large-scale fermentations are initiated by the addition of well-defined lactic acid bacteria starter cultures. Over the last decades, more and more starter cultures were developed for specific product and quality ranges which led to an increased competition in the classical yoghurt market between food companies, local dairies and co-operatives. Therefore the necessity for food companies to further distinguish their products by superior quality and taste, and to offer innovative new products satisfying the consumer needs is increasing. Hence, food companies started to develop proprietary starter cultures to improve their own fermented dairy products.

In earlier times, new starter cultures were selected by extensive and time consuming screening of large numbers of natural lactic acid bacteria and suitable strains combined by simple trial and error. Today, technological aspects like fermentation and storage conditions, organoleptic influences on taste and texture, as well as health and nutritional aspects of the products and the starter cultures are becoming more comprehensible in molecular terms. Modern microbiology techniques in analytical biochemistry, fermentation technology, and molecular biology allow a more efficient and specific screening of strains or spontaneously occurring mutants in order to identify the one(s) exactly suited for a special purpose. Furthermore, the molecular techniques can also be applied for appropriate genetic improvement of a given natural strain.

4. Improved starter strains – case studies

In the following sections, some results and practical applications in the field of fresh fermented dairy products, including some examples from our own institute, will be presented.

4.1. SELECTION OF NATURALLY IMPROVED STRAINS

4.1.1. Mild, shelf-stable yoghurt

Yoghurt results from the growth association between *Streptococcus thermophilus* and *Lactobacillus bulgaricus*. Both organisms grow in milk where they ferment lactose to lactate, lowering the pH of the product. Upon refrigerated storage of the completed yoghurt for several days (in the supermarkets or at the consumer's home), the pH may drop further. This so-called post-acidification leads to a gradually increasing acid and bitter taste of the yoghurt, thus degrading the initial organoleptic quality of the product.

S. thermophilus on its own ferments milk into a mild but flavourless product. It is *Lb. bulgaricus* which mostly contributes to the typical yoghurt flavour, and lowers the pH to values below 4.2 (Oberman, 1985). The approach to limit post-acidification and still produce the yoghurt flavour was to regulate growth and maintenance of *Lb. bulgaricus* by controlling its energy metabolism. Hence, *Lb. bulgaricus* starter strains were screened for the presence of spontaneous Lac minus mutants, having no or reduced residual β-galactosidase activity. Mutants were identified having genetic deletions within the β-galactosidase gene (*lacZ*) or expanding beyond the *lac* region, thereby inactivating a further gene vital for growth in milk, encoding the cell wall bound proteinase (Mollet and Delley, 1990; Germond *et al.*, 1995). Such Lac minus and Lac Prt minus mutants were not able to grow in milk as single-strain cultures without the supplementation of glucose and peptones. However, if grown in mixed cultures with a lactose fermenting *S. thermophilus* strain, Lac minus *Lb. bulgaricus* strains were able to grow despite the absence of glucose. Hence, *S. thermophilus* provides the mutant partner strains with the energy necessary to propagate in milk. Once the fermentation process has been terminated, growth and lactose metabolism of *S. thermophilus* and *Lb. bulgaricus* cease, resulting in a non post-acidifying yoghurt product. Such products have been shown to keep their mild taste and organoleptic properties for more than 6 months stored at 4°C (Mollet, 1996).

4.1.2. Probiotics, bacteria with health beneficial properties

The term 'probiotic' is used to describe living micro-organisms that are administered to man or animal to improve the integrity of the intestinal microflora and thereby maintain and improve the health status of the consumer (Fuller, 1989). In the last years, the manufacture and marketing of probiotic products for human consumption has increased worldwide. Probiotic bacteria need to fulfil certain biological requirements like the survival of the harsh acidic conditions in the stomach, the resistance to the toxicity of the gastric conditions, digestive enzymes, bile salts, local immune mechanisms and interactions with other bacteria. These probiotic properties are not valid for all lactic acid bacteria or all strains of a given species (for review see: Salminen *et al.*, 1993; Brassart *et al.*, 1994). However, they are most likely to be found among strains of *Lb. acidophilus* and *Lb. casei* or the human *Bifidobacterium* species.

The most important desirable effects of the probiotics (reviewed by Salminen *et al.*, 1996; Tannock, 1997; Holzapfel *et al.*, 1998; Kasper, 1998; Vaughan and Mollet, 1999) are activity against pathogens, stabilisation of the gut ecosystem, strengthening the gut mucosal barrier, immune modulation, anti-mutagenic and anti-tumourigenic activity, anti-allergic effects and nutritional advantages. The reported effects of well-characterised probiotic strains are shown in Table 3. Numerous studies analysed the effectiveness of probiotics in enhancing the resistance against certain intestinal tract infections (DeSimone *et al.*, 1988; Paubert-Braquet *et al.*, 1995; Boudraa *et al.*, 1990) and in the treatment of different types of intestinal disorders. Upon ingestion of selected strains the duration or severeness of gastro-enteritis was significantly reduced (Isolauri *et al.*, 1994; Holzapfel *et al.*, 1998; Kasper, 1998). These effects were suggested to result from a modulation of the immune response by particular strains of lactic acid bacteria (Kaila *et al.*, 1992; Majamaa *et al.*, 1995; Tannock, 1997; Hamann *et al.*, 1998) but also from antagonistic activities of the strain against pathogens. These antagonistic activities include competitive exclusion, interbacterial aggregation, or production of antimicrobial substances such as organic acids, hydrogen peroxide and/or bacteriocins (Klaenhammer, 1993; Lindgren and Dobrogosz, 1990; Spencer and Chesson, 1994).

Attachment to the human intestinal epithelial cells is considered to be an important factor. Adherence could be reconstructed *in vitro* on cultivated human intestinal cell lines (HT-29 and Caco-2) (Conway *et al.*, 1987; Elo *et al.*, 1991; Chauvière *et al.*, 1992; Coconnier *et al.*, 1992; Bernet *et al.*, 1994). Recent papers report that the microflora in the small intestine can influence the expression of epithelial surface structures, which may serve as receptors for attachment of other microorganisms. (Bry et al., 1996; Umesaki et al 1997). Of special interest are observations on the protective effects of some lactic acid bacteria against carcinogenesis. There is evidence that selected strains may reduce colon cancer risk by influencing the concentration of short-chain fatty acids and ammonia, by binding, inhibiting or inactivating of mutagens, and by reducing the activity of carcinogen-generating enzymes in the gut (Kasper, 1998; Tannock, 1997; Holzapfel *et al.*, 1998).

Table 3. Reported effects of probiotic bacterial strains

Strain	Reported effects in clinical studies
Lactobacillus johnsonii La1	transient persistence in the gut; immune enhancement; vaccine adjuvant; adherence to intestinal cells; reduction of gastric inflammation and infection by *Helicobacter pylori*; reduction of *Clostridium perfringens* in the gut flora; balancing of the intestinal microflora
Lactobacillus acidophilus NCFB 1748	lowering of the activity of faecal enzymes; decrease of faecal mutagenicity; prevention of radiotherapy-related diarrhoea; treatment of constipation
Lactobacillus GG (ATCC 53013)	adjuvant; prevention of antibiotic-associated diarrhoea; reduction of rotavirus diarrhoea; treatment of relapsing *Clostridium difficile* diarrhoea; prevention of acute diarrhoea; alleviation of Crohn's disease; antagonism against cariogenic bacteria
Lactobacillus casei Shirota	immune enhancement; balancing of intestinal bacteria; prevention of intestinal disturbances; reduction of rotavirus diarrhoea; faecal enzyme reduction; positive effects on superficial bladder cancer
Lactobacillus gasseri (ADH)	reduction of faecal enzyme activity; survival in the intestinal tract
Bifidobacterium lactis (Bb12)	reduction of rotavirus and other viral diarrhoeas; transient persistence in the gut of infants; balancing of the intestinal microflora
Lactobacillus reuteri	colonisation of the intestinal tract; mainly tested in animal studies, without effect on faecal enzyme activity

The identification of genetic determinants important for the probiotic properties will permit rapid screening for new effective strains. Today, probiotic bacteria are mostly used in fermented milk products. In the future, new probiotic functional foods will most likely also include infant formulae, fruit juices, fermented soy and cereal-based products (Lee and Salminen, 1995).

4.2. THE GENETIC ENGINEERING APPROACH

4.2.1. Texture producing strains

Over the past ten years, a certain consumer trend in the yoghurt market has moved towards 'light' products with low or no fat content. Increasing thickening properties of the yoghurt are required to compensate for the lower fat content. This could be achieved by adding stabilisers, gums, pectins or starch. However, the use of exocellular polysaccharide ("EPS") producing bacteria for the fermentation process satisfies much better the market need for "natural" products and the legal definition of yoghurt in some European countries. The use of EPS producing strains increases the viscosity of yoghurt and decreases susceptibility to syneresis. A significant characteristic, and problem for the yoghurt industry, is the instability of this "slimy" property of the starter strains. Often, the spontaneous loss of the EPS producing ability of lactic acid bacteria has been related to the instability of plasmid encoded genes. However, this seems not to be true for the thermophilic bacteria *Lb. bulgaricus* and *S. thermophilus* which often do not contain such plasmids (for reviews see: Cerning, 1990, 1994).

The structure of several EPS produced by thermophilic lactic acid bacteria has been determined and characterised as heteropolysaccharides (Doco *et al.*, 1990; Gruter *et al.*, 1993; Yamamoto *et al.*, 1994). The genetic basis of the biosynthesis and secretion of these polysaccharides is being elucidated. By taking advantage of an agar plate assay containing ruthenium-red, colonies of EPS and non-EPS producing yoghurt bacteria can be differentiated and genetically analysed (Stingele and Mollet, 1995). The genes involved in synthesis and secretion of EPS were identified from ropy *S. thermophilus, Lactococcus lactis* and *Lb. bulgaricus* starter strains (Stingele *et al.*, 1996; De Vos, 1996; Lamothe *et al.*, 1999). It now becomes feasible to genetically transfer the texturing character of those strains to other, industrially important starter strains, and to modify production and the biochemical structure of these EPS by genetic engineering.

In fact, a genetic transfer of a ropy phenotype from *S. thermophilus* to a non-ropy *Lc. lactis* has already been demonstrated. The 13 genes of the EPS gene cluster of *S. thermophilus* Sfi6 were successfully transferred to a heterologous *Lc. lactis* strain which became capable of producing an EPS (Stingele *et al.*, 1999). However, the primary structure of the newly produced EPS from *Lc. lactis* was different from the one of the original strain: the GalNAc was substituted by a Gal residue and α-1,6-branched Gal was missing in the repeating unit. It seems that *Lc. lactis* did not provide all necessary nucleotide sugars for the biosynthesis, thus forcing some flexibility to some of the transferred glycosyltransferase specificities. Depending on the genetic background in which a EPS cluster is expressed, this degree of flexibility adds to the potential of creating new exocellular polysaccharide structures.

4.2.2. Novel flavour producing strains

Lactic acid bacteria convert lactose to lactic acid what results in acidification and preservation of the raw material milk. Depending on the starter strains and process technology used, different products are obtained varying in flavour. It would now be interesting to evaluate and develop novel starter strains to add new flavours to the fermented products.

A lot of progress has been achieved in understanding the lactose metabolism of different lactic acid bacteria and the generation of flavours related to the secondary end-products deviating from this pathway. One of the best-studied species is *Lc. lactis*, which displays a simple fermentation metabolism where lactose or glucose is mainly converted to L-lactate (de Vos and Vaughan, 1994). The last step in this metabolism is conducted by the enzyme L-lactate dehydrogenase, which converts pyruvate to L-lactate. Genetic engineering was applied to modify the metabolic pathway of a *Lc. lactis* strain to produce L-alanine, instead of L-lactate (Hols *et al.* 1999). To re-route the conversion of pyruvate to L-alanine, the L-alanine dehydrogenase gene (*alaDH*) of a lactic acid bacteria unrelated *Bacillus* species has been cloned and overexpressed in *Lc. lactis*. To achieve controlled expression of *alaDH*, the gene was linked to a nisin inducible, food-grade promoter system (de Ruyter *et al.* 1996; Kuipers *et al.* 1997). It was therefore possible to redirect ca. 1/3 of the total carbon flux of *Lc. lactis* to the generation and secretion of alanine instead of L-lactate. Inactivation of the L-lactate dehydrogenase as well as the alanine racemase genes finally resulted in a strain converting lactose stoichiometrically to L-alanine. The efficiency of this bioconversion process was reported to reach 99% (Hols *et al.* 1999).
The use of such a *Lc. lactis* strain as a biocatalyst for stereospecific conversion of lactose to alanine opens interesting perspectives for producing amino acids for applications in foods. This example shows a quite complex genetic construct in *Lc. lactis*, importing and expressing a gene, the *alaDH*, not naturally found in lactic acid bacteria. It also suggests the potential of using trans-species genetic engineering in developing novel starter strains for fermentation processes, demonstrating real product innovations.

5. Outlook and Conclusions

In earlier times, once the importance of lactic acid bacteria had been recognised, starter strains were selected by trial and error. Today, the food industry disposes of a modern microbiology, including analytical biochemistry and genetical tools, immunological and cell-biological tests to specifically screen and analyse hundreds or thousands of natural lactic acid bacterial strains. The progress in molecular biology and genetic engineering will further broaden the possibilities of using lactic acid bacteria in food and may allow in the future to improve existing products and to develop novel products and applications.

The above mentioned examples illustrate the enormous potential that lactic acid bacteria bear for today's and tomorrow's applications in different domains of the food industry. They will continue to play an important role for the fermentation processes of a variety of different food products by contributing to their conservation, flavour development, texture and health beneficial properties. They will also be used increasingly as a natural source for food ingredients and additives to non-fermented products to accomplish for example antimicrobial, texturing or probiotic purposes.

Acknowledgements

We thank Anne Constable for critically reading this manuscript.

References

Amann, R.I., Binder, B.J., Olsen, R.J., Chrisholm, S.W., Devereux, R. and Stahl, D.A (1990) Combination of 16S rRNA-targeted oligonucleotide probes with flow cytometry for analysing mixed microbial populations, *Appl. Env. Microbiol. 56,* 1919-1925

Amann, R.I., Ludwig, W., Schleifer, K.H. (1995) Phylogenetic identification and in situ detection of individual microbial cells without cultivation, *Microbiol. Rev. 59,* 143-169

Angert, E.R., Clements, K.D. and Pace, N.R. (1993) The largest bacterium, *Nature 362*, 239-241

Axelsson, LT. (1993) Lactic acid bacteria: classification and physiology, in S. Salminen and A. von Wright (eds.) *Lactic acid bacteria*, Marcel Dekker Inc., New York, pp. 1-63.

Bernet, M.F., Brassart, D., Neeser, J.R. and Servin, A.L. (1994) *Lactobacillus acidophilus* LA1 binds to cultured human intestinal cell lines and inhibits cell attachment and cell invasion by enterovirulent bacteria, *Gut 35*, 483-489.

Boudraa, G., Touhani, M., Pochart, P., Soltana, R., Mary, J.Y. and Desjeux, J.F. (1990) Effect of feeding yoghurt versus milk in children with persistent diarrhoea, *J. Pediatr. Gastroenterol. Nutr. 11,* 509-512

Brassart, D., Neeser, J.-R., Michetti, P. and Servin, A.L. (1994) The selection on dairy bacterial strains with probiotic properties based on their adhesion to human intestinal epithelial cells, *Proceedings Lactic 94*

Bry, L., Falk, P.G., Midtvedt, T. and Gordon, J.I. (1996) A model of host-microbial interactions in an open mammalian ecosystem. *Science, 273,* 1380-1383

Cerning, J. (1990) Exocellular polysaccharides produced by lactic acid bacteria, *FEMS Microbiol. Rev. 87,* 113-130.

Cerning, J. (1994) Polysaccharides exocellulaires produits par les bactéries lactiques, in H. de Roissart and F.M. Luquet (eds.) *Bactéries lactiques, aspects fondamentaux et technologiques*, Vol. 1, Lorica, Uriage, France, pp. 309-329.

Chauvière, G., Coconnier, M.-H., Kerneis, S., Fourniat, J. and Servin, A.L. (1992) Adhesion of human *Lactobacillus acidophilus* strain LB to human enterocyte-like Caco-2 cells, *J. Gen. Microbiol. 138,* 1689-1696.

Coconnier, M.-H., Klaenhammer, T.R., Kerneis, S., Bernet, M.F. and Servin, A.L. (1992) Protein-mediated adhesion of *Lactobacillus acidophilus* BG2FO4 on human enterocyte and mucus-secreting cell lines in culture, *Appl. Environ. Microbiol. 58*, 2034-2039.

Conway, P., Goldin, B.R. and Gorbach, S.L. (1987) Survival of lactic acid bacteria in the human stomach and adhesion to intestinal cells, *J. Dairy Sci. 70*, 1-12.

de Ruyter, P.G.G.A., Kuipers, O.P. and de Vos, W.M. (1996) Controlled gene expression systems *Lactococcus lactis* with the food-grade inducer nisin, *Appl. Environ. Microbiol. 62,* 3662-3667

DeSimone, C., Tzantzoglou, S., Baldinelli, L., Di Fabio, S., Bianchi-Salvatori, B., Jirilli, E., and Vesely, R. (1988) Enhancement of host resistance against *Salmonella typhimurium* infection by a diet supplemented with yoghurt, *Immunopharmacol. Immunotoxicol. 10,* 399-415

de Vos, W.M., Vaughan, E.E. (1994) Genetics of lactose utilisation in lactic acid bacteria, *FEMS Microbiol. Rev. 15*, 217-237.

de Vos, W.M. (1996) Metabolic engineering of sugar catabolism in lactic acid bacteria, *Antonie van Leeuwenhoek 70*, 223-242.

Doco, T., Wieruszeski, J.-M., and Fournet, B. (1990) Structure of an exocellular polysaccharide produced by *Streptococcus thermophilus, Carbohyd. Res. 198*, 313-321.

Elo, S., Saxelin, M. and Salminen, S. (1991) Attachment of *Lactobacillus casei* strain GG to human colon carcinoma cell line Caco-2: comparison with other dairy strains, *Lett. Appl. Microbiol. 13*, 154-156.

Fuller, R. (1989) Probiotics in man and animals *J. Appl. Bacteriol. 66,* 365-78

Germond, J.-E., Lapierre, L., Delley, M. and Mollet, B. (1995) A new mobile genetic element in *Lactobacillus delbrueckii* subsp. *Bulgaricus, Mol. Gen. Genet. 248*, 407-416.

Gruter, M., Leeflang, B.R., Kuiper, J., Kamerling, J.P. and Vliegenthart F.G. (1993) Structural characterisation of the exopolysaccharide produced by *Lactobacillus delbrueckii* subspecies *bulgaricus* rr grown in skimmed milk, *Carbohyd. Res. 239*, 209-226.

Haggblade, S. and Holzapfel, W.H. (1989) Industrialisation of Africa's indigenous beer brewing, in K.H. Steinkraus (ed.), *Industrialisation of indigenous fermented foods* Marcel Dekker, New York, pp. 191-283.

Hamann, L., EL-Samalouti, V., Ulmer, A.J., Flad, H.-D. and Rietschel, E.Th. (1998) Components of gut bacteria as immunomodulators, *Int. J. Food Microbiol. 41*, 141-154

Hols, P., Kleerebezem, M., Schank, A.N., Ferain, T., Hugenholtz, J., Delcour, J., and de Vos, W.M. (1999) Conversion of *Lactococcus lactis* from homolactic to homoalanine fermentation through metabolic engineering, *Nature Biotechnology, 17*, 588-592.

Holzapfel, W.H., Haberer, P., Snel, J., Schillinger, U. and Huis in't Veld, J.H.J. (1998) Overview of gut flora and probiotics, *Int. J. Food Microbiol. 41*, 85-101

Isolauri, E., Kaila, M., Mykkanen, H., Ling, W.H., and Salminen, S. (1994) Oral bacteriotherapy for viral gastroenteritis, *Dig. Dis. Sci. 39*, 2595-2600

Kaila, M., Isolauri, E., Soppi, E., Virtanen, E., Laine, S. and Arvilommi, H. (1992) Enhancement of the circulating antibody secreting cell response in human diarrhoea by a human *Lactobacillus* strain, *Pediatr. Res. 32*, 141-144.

Kandler, O. and Weiss, N. (1986) Regular, non-sporulating gram-positive rods, in P.H.A. Sneath, N.S. Mair, M.E. Sharpe and J.G. Holt (eds.) *Bergey's manual of systematic bacteriology*, vol. 2. The Williams & Wilkins Co., Baltimore, pp. 1208-1260.

Kasper, H. (1998) Protection against gastrointestinal diseases – present facts and future developments, *Int. J. Food Microbiol. 41*, 127-131

Klaenhammer, T.R. (1993) Genetics of bacteriocins produced by lactic acid bacteria, *FEMS Microbiol. Rev. 12*, 39-86.

Klein, G., Pack, A., Bonaparte, C. and Reuter, G. (1998) Taxonomy and physiology of probiotic lactic acid bacteria, *Int. J. Food Microbiol. 41*, 103-125.

Kuipers, O.P., de Ruyter, P.G.G.A., Kleerebezem, M., and de Vos, W.M. (1997) Controlled overproduction of proteins by lactic acid bacteria, *Trends in Biotechnol. 15*, 135-140

Lamothe, G.T., Stingele, F. , Neeser, J.-R., Desachy, P. and Mollet, B. (1998) Genetic analysis of exopolysaccharide (EPS) production in *Lactobacillus delbrueckii* subsp. *bulgaricus*, Abstracts of the ASM Conference on Streptococcal Genetics, Vichy, France, p.67.

Langendijk, P.S., Schut, F, Jansen, G.J., Raangs, G.C., Kamphuis, G.R., Wilkinson, M.H.F. and Welling, G.W. (1995) Quantitative fluorescence in situ hybridisation of *Bifidobacterium* spp. with genus-specific 16S rRNA-targeted probes and its application in faecal samples, *Appl. Env. Microbiol. 61*, 3069-3075

Lee, Y.-K. and Salminen, S. (1995) The coming of age of probiotics, *Trends in Food Science & Technology 6*, 241-244.

Lindgren, S.E. and Dobrogosz, W.J. (1990) Antagonistic activities of lactic acid bacteria in food and feed fermentations, *FEMS Microbiol. Rev. 87*, 149-164

Majamaa, H., Isolauri, E., Saxelin, M., and Vesikari, T. (1995) Lactic acid bacteria in the treatment of acute rotavirus gastroenteritis, *J. Pediatr. Gastroenterol. Nutr. 20*, 333-338

McFarlene, G.T. and Gibson, G.R. (1994) Metabolic activities of the normal colonic microflora, in S.A.W. Gibson (ed.), *Human health: contribution of microorganisms*, Springer, Frankfurt,Germany, pp. 17-38

Mollet, B. (1996) New technologies in fermented milk, *Cerevisia, 21*, 63-65.

Mollet, B. and Delley, M. (1990) Spontaneous deletion formation within the ß-galactosidase gene of *Lactobacillus bulgaricus, J. Bacteriol. 172*, 5670-5676.

Nissen, H.J., Damerow, P. and Eglund, R.K. (1991) *Frühe Schrift und Techniken der Wirtschaftsverwaltung im alten Vorderen Orient – Informationsspeicherung und – verarbeitung vor 5000 Jahren,* 2. ed. Verlag Franzbecker, Bad Salzdetfurth.

Oberman, H. (1985) Fermented milks, in B.J.B. Wood (ed.) *Microbiology of fermented foods*, vol. 1, Elsevier applied science publishers, New York, pp. 167-194.

Orla-Jensen, S. (1919) *The lactic acid bacteria*, Host & Son, Copenhagen.

Paubert-Braquet, M. Gan, X.H., Gaudichon, C. Hedef, N., Serikoff, A., Bouly, C. Bonavida, B. and Braquet, P. (1995) Enhancement of host resistance against *Salmonella typhimurium* in mice fed a diet supplemented with yoghurt or milk fermented with various *Lactobacillus casei* strains, *Int. J. Immunother. XI*, 153-161

Pot, B., Ludwig, W., Kersters, K. and Schleifer, K.H. (1994) Taxonomy of lactic acid bacteria, in L. De Vuyst and E.J. Vandamme (eds.) *Bacteriocins of lactic acid bacteria: microbiology, genetics and applications*, Elsevier Science Publishers Ltd., Amsterdam, pp. 13-90.

Ray, B. and Daeschel, M.A. (1992) *Food Biopreservatives of Microbial Origin*, CRC Press, Boca Raton, FL.

Salminen, S., Deighton, M. and Gorbach, S. (1993) Lactic acid bacteria in health and disease, in S. Salminen and A. von Wright (eds.) *Lactic acid bacteria*, Marcel Dekker Inc., New York, pp. 199-225.

Salminen, S., Isolauri, E. and Salminen, E. (1996) Clinical uses of probiotics for stabilising the gut mucosal barrier: successful strains and future challenges, *Antonie van Leeuwenhoek*, *70*, 347-358.

Schleifer, K.H., (1987) Recent changes in the taxonomy of lactic acid bacteria, *FEMS Microbiol. Rev. 46*, 201-203

Schleifer, K.H., Ehrmann, M., Beimfohr, Brockmann, E., Ludwig, W. and Amann, R. (1995) Application of molecular methods for the classification and identification of lactic acid bacteria, *Int. Dairy J. 5*, 1081-1094

Snel, J., Blok, H.J., Kengen, H.M.P., Ludwig, W. Poelma, F.G.J., Koopman, J.P. and Akkermans, A.D.L. (1994) Phylogenetic characterisation of *Clostridium* related segmented filamentous bacteria in mice based on 16S ribosomal RNA analysis, *Syst. Appl. Microbiol. 17*, 172-179

Spencer, R.J. and Chesson, A. (1994) The effect of *Lactobacillus* spp. on the attachment of enterotoxigenic *Escherichia coli* to isolated porcine enterocytes, *J. Appl. Bacteriol. 77*, 215-220

Stiles, M.E. and Holzapfel, W.H. (1997) Lactic acid bacteria of foods and their current taxonomy, *Int. J. Food Microbiol. 36*, 1-29

Stingele, F. and Mollet, B. (1995) Homologous integration and transposition to identify genes involved in production of exopolysaccharides in *Streptococcus thermophilus*, in J.J. Ferretti, M.S. Gilmore, T.R. Klaenhammer and F. Brown (eds.) *Genetics of Streptococci, Enterococci and Lactococci*, vol. 85, Karger, Basel, Switzerland, pp. 487-493.

Stingele, F., Neeser, J.-R. and Mollet, B. (1996) Identification and characterisation of the *eps* (exopolysaccharide) gene cluster from *Streptococcus thermophilus* Sfi6, *J. Bacteriol. 178*,1680-1690.

Stingele, F., Vincent, S.J.F., Faber, E.J., Newell, J.W., Kamerling, J.P., and Neeser, J.-R. (1999) Introduction of the exopolysaccharide gene cluster from *Streptococcus thermophilus* Sfi6 into *Lactococcus lactis* MG1363: production and characterisation of an altered polysaccharide, *Mol. Microbiol. 32*, 1287-1295.

Tannock, G.W. (1997) Probiotic properties of lactic acid bacteria: plenty of scope for fundamental R&D, *Trends in Biotechnology 15*, 270-274

Umesaki, Y., Okada, Y., Imaoka, A., Setoyama, and Matsumoto, S. (1997) Interactions between epithelial cells and bacteria, normal and pathogenic, *Science 276*, 964-965

Vaughan, E.E. and Mollet, B. (1999) Probiotics in the new millennium, *Food, 43*,148-153.

Wang, J., Whetsell, M. and Klein, J.R. (1997) Local hormone networks and intestinal T cell homeostasis, *Science 275*, 1937-1939

Wiese, B.G., Strohmar, W., Rainey, F.A., Diekmannm H. (1996) *Lactobacillus panis* sp. nov., from sourdough with a long fermentation period, *Int. J. System. Bacteriol. 46*, 449-453.

Wilson, K.H., and Blitchington, R.B. (1996) Human colonic biota studied by ribosomal DNA sequence analysis, *Appl. Env. Microbiol. 62*, 2273-2278

Yamamoto, Y., Murosaki, S., Yamauchi, R., Kato, K. and Sone, Y. (1994) Structural study on an exocellular polysaccharide produced by *Lactobacillus helveticus* TY1-2, *Carbohyd. Res. 261*, 67-78.

Zoetendal, E.G., Akkermans, A.D.L., and DeVos, W.M. (1998) Temperature gradient gel electrophoresis analysis of 16S rRNA from human faecal samples reveals stable and host-specific communities of active bacteria, *Appl. Env. Microbiol. 64*, 3854-3859

IMMOBILISED CELL TECHNOLOGY IN WINERY AND FRUIT WINE PRODUCTION.

REMY CACHON AND CHARLES DIVIES.
Laboratoire de Microbiologie UMR INRA, ENSBANA, Université de Bourgogne, 1-esplanade Erasme, 21000 Dijon, France. E-mail: cachon@u-bourgogne.fr

Summary

Winemaking is largely concerned with the progress of biotechnology and especially with the use of high cell density reactors. Entrapment is the most widely method to immobilise cells; several matrix can be used (alginate, carrageenan, agar) with different geometry (beads, fibres, plates). Alcoholic fermentation of wine, malolactic fermentation, bottle fermentation known as the "Methode champenoise" and sparkling wines are among the industrial applications. Whereas prospects for this technology appear encouraging, further research is needed to optimise reaction variables, improve the long-term stability of the reactors, and understand more about secondary metabolite production by yeasts under these conditions. Nevertheless, several industrial trials have shown that fermented products with good flavour could be produced, and about 20 patents have been published which underlines the potential interest of this technology.

1. Introduction

There is considerable diversity in technology available to winemakers, leading to a wide variety of fermented wines (Flanzy, 1998). Over the centuries, oenology has accumulated numerous pragmatic acts which, in fact, do not correspond to an optimisation of the winemaking process. This is why countries only recently engaged in this activity have been able to develop their wine industry with the aid of a more scientific approach, and apply some of the newer advances in fermentation technology.
Winemaking involves two principal operations; first, preparation of the grape must to tailor its composition and maintain the qualities of the grape at harvest: and second, to conduct microbial fermentation through rational exploitation of the biochemical activities of yeast and lactic acid bacteria. White wine is produced by fermenting juice, which has been extracted without macerating the solid parts of the grape cluster. Lack of maceration is not an absolute factor and, in some cases, short maceration of the skin

M. Hofman and P. Thonart (eds.), Engineering and Manufacturing for Biotechnology, 413–421.

is carried out (Lefur, 1990). Red wine is a maceration wine, using traditional method (maceration and fermentation are practically simultaneously), carbonic maceration (under carbon dioxide atmosphere without grape crushing), or thermovinification (grapes are crushed and the macerate is heated before pressing). Fermentation is done conventionally in a batch system. The volume of the vessels in which fermentations are conducted has increased considerably during the last 20 years. Previously, fermentation was done in 225-225 l barrels or in 6-12 hl vats. They have now been replaced by well-designed stainless steel fermenters with volumes of 1-340 m^3. These tanks are easy to clean and maintain, and may be constructed with a jacket in which cooling liquid is circulated for temperature control (Moresi, 1989a,b).

In microbial technology grape juices are inoculated with pure cultures of *Saccharomyces cerevisiae* at 1.10^6-5.10^6 cells/ml (ITV, 1994). Problems are occasionally encountered with slow or incomplete fermentations and, generally, these are related to the high concentration of sugar in grape juice at the onset of fermentation and, above all, the high alcohol concentration at the end. In the course of alcoholic fermentation, the yeast remains under a double dependence: the glucose effect (Fiechter *et al.*, 1981) and the action of oxygen required to ensure growth and survival (Lafon-Lafourcade, 1983; Ohno and Takahashi, 1986). Aeration of the must is recommended at the beginning (day 2-day 3) of the alcoholic fermentation (ITV, 1994), and precisely at the end of the growth phase - beginning of the stationary phase (Barre *et al.*, 1998).

2. Immobilised cell concept

The main techniques that enable biomass confinement are absorption on a support, autoflocculation and entrapment in gels. In contrast to lactic acid bacteria, yeasts readily form biofilms on the surface of supports (Navarro, 1978). Rapid desorption of the biofilm can result from an increase in the velocity of the liquid passing over its surface or by autolysis of underlying cells during operation. The release of large quantities of gases during alcoholic fermentation considerably disturbs the biofilm stability. Finally, access of oxygen to the underlying cells of the biofilm is a specific problem that is difficult to resolve. The flocculation of microbial cells to form a dense concentration of biomass is encountered naturally in a number of yeast strains (Stewart and Russell, 1986; Calleja, 1987). It is a very attractive method of biomass retention, since it involves decanting, the most basic method of liquid-solid distribution (Ghommidh and Navarro, 1986; Salou *et al.*, 1988). In the case of yeasts, a biomass density of 60-110 $g.l^{-1}$ can be maintained with little or no physical constraints for the microorganisms except for the reactor vessel. The fermentation process requires a reactor configuration in the form of towers that house the biomass through which the liquid is passed. Industrial installations have been described in breweries (Maule, 1986). Entrapment involves imprisoning living cells within a rigid network which permits the diffusion of substrates and products, thereby making possible the growth and maintenance of active cells (Diviès, 1975). Natural polymers such as alginate, carrageenan, chitosan and agar, enable polymerisation in very mild conditions and leave the cell integrity intact (Groboillot *et al.*, 1994). Entrapment in alginate, for example, is a very simple process;

a uniform suspension of the cells is prepared in a 2% sodium alginate solution. The suspension is the added to a solution of calcium chloride which catalyses polymerisation of the alginate into a gel in the form of beads of fibres with diameters between 0.2 and 3 mm.

The use of immobilised cells offers several advantages:
- Improved productivity of fermentations,
- Adaptation to continuous processes that can be better optimised andcontrolled,
- Simplified systems for removing microbial cells from batch processes,
- Greater tolerance to inhibitory substances,
- Smaller scale fermentation facilities (reduced capital and running costs),
- Possibilities of using a variety of microbial strains including genetically modified organisms,

and some potential disadvantages:
- Cell overgrowth what increases turbidity of the fermented beverage,
- Mechanical stability of the matrix used to immobilise microbial cells,
- Loss of activity on prolonged operation,

To be attractive in commercial practice the method must be (Janssen, 1993):
- Cheap,
- Easily performed in an industrial situation,
- Not liable to cause oxidation of the wine,
- Robust,
- Not susceptible to contamination,
- Able to impart correct flavour changes to the wine,
- Must use commercially acceptable supports and organisms,

Many applications have been studied and published in the last 25 years. The publication of many patents on the use of immobilised microbial cells in the production of fermented beverages has shown the potential industrial interest for this technology.

3. Possible applications in winery and fruit wine production

3.1. ALCOHOLIC FERMENTATION

3.1.1. Alcoholic fermentation without CO_2 pressure

The increase of productivity with saving in invested capital and labour costs, is of interest for alcoholic fermentation in oenology. Alginate is a good matrix for yeast immobilisation (Bertuccioli *et al.*, 1988; Cantarelli, 1989). Our own laboratory and pilot-scale studies with entrapped cells of yeasts demonstrated the potential value of immobilised cells for oenology. In the case of white grape juice fermentation (180 $g.l^{-1}$), complete fermentation can be obtained in 6 h at 23°C. Such a process is also used by Kyowa Hakko in Tokyo (Japan) to produce feed ethanol. A pilot scale ethanol production unit was built around five reactors with a total volume of 4 m^3 and produced 2400 l/day of pure alcohol. The alcoholic fermentation by yeasts immobilised in alginate gel beads is accelerated which has been related with changes in cell composition and function (Galazzo and Bailey, 1990). In batch culture immobilised cells do not show lag or exponential phase but rather a linear fermentation kinetic (Cantarelli, 1989). A first continuous process was proposed by Diviès (1975) and used *Saccharomyces cerevisiae* immobilised in poly-acrylamide gel beads for the alcoholic fermentation of must. A productivity ten fold higher than the corresponding free cell reactor was obtained. An alcoholic fermentation using *Schizosaccharomyces pombe* immobilised in alginate gel beads also showed a high productivity with the advantage of simultaneous transformation of malic acid to ethanol. Such a process has been proposed by Yokotsuka *et al.* (1993) with *Saccharomyces cerevisiae* and *Schizosaccharomyces pombe* separately immobilised in double-layer Ca-alginate fibres. Hsu (1987) proposed a batch process with a porous retaining screen inside the reactor for maintaining efficient contact of substrate and yeast-containing particles. Stepwise processes can also be used (Diviès and Deschamps, 1986). For example, in the first reactor the sugar concentration of the fruit juice is lowered by using an obligate aerobic yeast (for example *Rhodotorula glutinis*). In the second reactor, a fermentation yeast is used to carry out fermentation, so that the low alcohol product is obtained. Thus, the undesirable sweetness of low alcohol beverages can be overcome. A column reactor containing yeast cells immobilised on the surfaces of a substantially non-compressible carrier (DEAE cellulose) having anion exchange properties has been used by Lommi and Ahvenainen (1990) for apple juice fermentation. The electric forces established between the positively charged resin and the negatively charged yeasts cells are primarily responsible for the binding of yeast cells to the surface of the resin. The greater contact area in the column reactor results in a faster fermentation than free cell fermentation. In pilot scale set-ups, this process has been utilised for at least 13 weeks without need for regeneration. Ogbonna *et al.* (1989) designed a horizontal reactor for wine fermentations that exploited the successive activities of cells of *Schizosaccharomycesi* and *Saccharomyces*, which were immobilised on plates coated with alginate.

3.1.2. Alcoholic fermentation with CO_2 pressure : elaboration of sparkling wines

3.1.2.1. Bottle-fermented sparkling wines ("méthode Champenoise"). In the conventional process a blend of dry stabilised wine is mixed with sugar (about 25 g) and the secondary fermentation is conducted in the bottle by the inoculation of yeast in liquid suspension. The secondary fermentation produces carbon dioxide up to 5 bars while yeast metabolism and yeast autolysis participate to typical aroma and flavour. Subsequent sedimentation and removal of the yeast cells requires the lengthy and expensive procedure of "remuage". This operation occupies 35-40% of the space in cellars during 6 weeks. The sediment is removed by freezing the neck of the bottle (-25°C), which is then manually opened and the ice plug is squeezed out of the bottle by the internal pressure, it is the procedure of "degorgement". This conventional method can be advantageously shortened using immobilised yeasts in place of free cells, and using this method, "remuage" requires only 20 sec/bottle.

Different methods have been proposed such as the use of entrapped yeasts (Diviès, 1978; Fumi *et al.*, 1988; Godia *et al.*, 1991; Hill, 1990; Busova *et al.*, 1994; Yokotsuka *et al.*, 1997) and bottle cap with membrane cartridge like "Millispark" of Millipore S.A. (Lemonnier, 1986, 1990, 1992) or other origin (Spooner, 1973; Quetsch, 1990; Poirat, 1997). The kinetics of the "prise de mousse" (formation of champagne bubbles in the bottle) differentiate the processes. For the same initial population of yeast, the "prise de mousse" is terminated in 40 days for free and entrapped cells, while the membrane method takes longer, 90 to 120 days, which might lead to possible contamination by acidophilic bacteria. Nevertheless, the membrane system proposed by Spooner (1973) enables the opening of the bottle without cooling by slowly reducing the pressure prior to removing the closing of the bottle, which is performed by piercing the bottle cap. This method allows a slow pressure drop to obviate excess foaming and loss of material. The system of Quetsch (1990) consists in immobilised biocatalyst stoppers, made of polyethylene or cork. A string passes through the top of the stopper to the cover of a housing with the immobilised yeast cells, enclosed in a micro-filter. The housing can be pulled into a stopper cavity where an elastic seal provides a tight seat.

In such systems the critical parameter is mass transfer, which modifies the reaction rate for immobilised cells only at the end of fermentation, because of the increase of cell density inside the matrix which modifies the effective diffusion coefficients. Using cells immobilised in gel beads, the fermentation delay can be adjusted by the choice of the number of beads and of their specific area. When the specific area is doubled, the time required for finishing can be halved. To avoid cell leakage from beads, beads can be coated with a sterile calcium alginate layer (Fumi *et al.*, 1988; Hill, 1990; Busova *et al.*, 1994). Yokotsuka *et al.* (1997) used also double-layer calcium alginate fibres. Godia *et al.* (1991) have compared alginate and carrageenan gel beads, they observed that alginate showed a better structure to retain cells, 10-12 g/bottle was optimum to guarantee a clean wine free of cells. In the case of coated beads, the beads needed per bottle are reduced by half. The membrane method stills present two major faults: it has insufficient surface area for reaction exchange and it creates local super-saturation of carbon dioxide, which further limits exchanges.

It has been shown that organoleptic properties of the wine was equal or better than the same wine elaborated by the conventional technology (Fumi *et al.*, 1988; Godia *et al.*, 1991; Busova *et al.*, 1994).

Industrial utilisation of immobilised cells has been studied by Champagne Moët et Chandon which have proposed a process for large-scale production of immobilised cells (Duteurtre *et al.*, 1984a). For automation, the process using entrapped cells in gel beads requires the use of drying beads (dry matter 80-95%). Methods have been proposed by Diviès *et al.* (1989) with the Champagne Moët et Chandon (for normal and coated beads) and by Hill (1990), which also used a special system for transfer and dosing of beads in bottles (Duteurtre *et al.*, 1984b). This company has developed an industrial machine for the delivery of beads at the rhythm of 20,000 bottles per hour (and needs 1 m^3 of beads per day). An economic study undertaken for a plant producing 3,000,000

bottles yearly has demonstrated the competitiveness of this new process (Valade and Rinville, 1990). In 1992, about 500,000 bottles were produced using this technique.
The technology developed for champagne production can be transposed to other bottle-fermented sparkling beverages. Particularly interesting possibilities are the re-fermentation of wine supplemented before "prise de mousse" with an infusion of fruits obtained by hydro-alcoholic maceration (Lenzi and Cavin, 1985), and second fermentation of fruit wine such as cider and pineapple wine (Diviès and Deschamps, 1986). These new products can thus be obtained under more thoroughly controlled conditions.

3.1.2.2. Elaboration of sparkling wines in closed reactors. Diviès and Deschamps (1986) used a pressurised batch reactor to produce cider, sparkling wine or semi-sparkling grape juice using yeasts immobilised in alginate gel beads. For example, 1,500,000 bottles of sparkling wine per year could be produced by a reactor (1 m^3) operating for 220 days. A continuous system has been experimented by Fumi *et al.* (1989), a sparkling wine with composition and sensorial properties comparable to a product obtained conventionally was produced, but with a productivity greatly enhanced. In the fermentation process of Lommi and Ahvenainen (1990) previously described (*3.1.1*) the reactor can be pressurised up to 14 bars to obtain sparkling product. Similar continuous process have been proposed by Sarishvili *et al.* (1987) to produce "champagne-like" sparkling wines.

3.2. MALOLACTIC FERMENTATION OF WINE

The malolactic fermentation (MLF) of wine allows a reduction in acidity of wines and contributes to the development of subtle flavours that contribute to sensorial quality of wines. It also stabilises wines and lowers the risk of fermentation in bottles. MLF is the degradation of malic acid in lactic acid and carbon dioxide. The main microbial strain involved in malolactic fermentation is *Oenococcus oeni* (its old name is *Leuconostoc oenos*). MLF can occur several weeks after alcoholic fermentation, and even in the case of wine inoculation by selected starters, there is no guarantee that the fermentation will occur. This is because wine is unfavourable for growth of microorganism (pH<3.0-3.5, ethanol>10%, T<15°C in northern countries, high SO_2 concentration, lack of nutrients...). In 1976, Diviès and Siess proposed an immobilised cell process using *Lactobacillus casei* entrapped in poly-acrylamide gel lattice, this process operated for 12 months without loss of activity. Several reactor configurations have been tested (Crapisi *et al.*, 1987a, 1987b; Spettoli *et al.*, 1982; Cuenat and Villetaz, 1984; Rossi and Clementi, 1984; Naouri *et al.*, 1991) Some problems have been related, associated with microbial contamination of the reactors, transfer of flavour taints to the wine, loss of activity on prolonged operation and leakage of cells from the solid support. This is in contradiction with Fleet and Costello (1991) which published a patent on MLF of wine, the authors found that the bacteria can remain biocatalytically active for an indefinite time, thus allowing the process of the invention to operate continuously, or with interruptions. For a working winery, the simpler method of adhesion (on oak chips) might be recommended to adapt the method to industrial practice (Janssen, 1993).

4. Conclusion

Immobilised cells have shown several possibilities to facilitate the conduct of fermentation, especially in the field of sparkling wines production. The use of this technology on an industrial scale needs furthers scale-up studies, and a good scientific knowledge of the effect of immobilisation on physiology of industrial strains (metabolic fluxes distribution, kinetic of autolysis).

In the case of malolactic fermentation, opposite results have been published, and a better knowledge of the physiology of *Oenococcus oeni* is required, taking into account the physico-chemistry of the culture medium in which the FML is studied. In addition, there are recent advances on the physiological response of bacteria to environmental stress, and it can be expected that new strains well adapted to wine medium will be proposed in the next years.

References

Barre, P., Blondin, B., Dequin, S., Feuillat, M., Sablayrolles, J.M. and Salmon, J.M. (1998) La levure de fermentation alcoolique. In *Oenologie*, pp. 414-495, Technique et Documentation Lavoisier, Paris.

Bertuccioli, M., Rosi, I. and Costamagna, L. (1988) Recent progress in the use of immobilised yeasts in wine fermentation. In *Proceedings of the second international cool climate viticulture and oenology symposium*, pp. 304-307, New-Zealand society of viticulture and oenology, Auckland.

Bidan, P., Diviès, C. and Dupuy, P. (1978) Procédé perfectionné de préparation de vins mousseux.

Busova, K., Magyar, I. and Janky, F. (1994) Effect of immobilised yeasts on the quality of bottle-fermented sparkling wine. *Acta Alimantaria 23*, 9-23.

Calleja, G.B. (1987) Cell aggregation. In *The yeasts*, Vol. 2, pp. 164-298, Academic Press, London.

Cantarelli, C. (1989) Factors affecting the behaviour of yeasts in wine fermentation. In *Biotechnology applications in beverage production*, pp. 127-151, Elsevier Applied Science, London.

Crapisi, A., Nuti, M.P., Zamorani, A. and Spettoli, P. (1987a) Improved stability of immobilised *Lactobacillus* sp. Cells for the control of malolactic fermentation in wine. *Am. J. Enol. Vitic. 38*, 310-312.

Crapisi, A., Spettoli, P., Nuti, M.P. and Zamorani, A. Comparative traits of *Lactobacillus brevis, Lactobacillus fructivorans* and *Leuconostoc oenos* immobilised cells for the control of malolactic fermentation in wine. *J. Appl. Bacteriol. 63*, 513-521.

Cuenat, P. and Villetaz, J.C. (1984) Essais de fermentation malolactique des vins par des bactéries lactiques immobilisées du genre *Leuconostoc*. *Revue suisse viticulture arboriculture horticulture 16*, 145-151.

Diviès, C. (1975) Procédé enzymatique utilisant des microorganismes inclus. *French Patent #2320349.*

Diviès, C. and Siess, M.H. (1976) Etude du catabolisme de l'acide L-malique par *Lactobacillus casei* emprisonné dans un gel de polyacrylamide. *Ann. Microbiol. 127 B*, 525-539.

Diviès, C. and Deschamps, P. (1986) Procédé et appareillage pour la mise en œuvre de réactions enzymatiques et application à la préparation de boissons fermentées. *French Patent #2601687.*

Diviès, C., Lenzi, P., Beaujeu, J. and Herault, F. (1989) Process for preparing microorganisms incorporated within substantially dehydrated gels, gels obtained and their use in the preparation of fermented beverages. *World Patent #9000602.*

Duteurtre B.H.J., Ors, P., Charpentier, M.S. and Midoux, N. (1984a) Procédé et installation de fabrication de goutelettes essentiellement sphériques de dimension approximativement uniforme et de diamètre supérieur à au moins environ deux millimètres à partir d'un matériau susceptible de s'écouler, solidifiable, coagulable ou gélifiable contenant de préférence des particules en particulier formées par les levures ou microorganismes de fermentation. *French Patent #2570959.*

Duteurtre, B.H.J., Coulon, P.A. and Goutard, R.M. (1984b) Procédé et appareil de transfert et/ou de dosage de particules fragiles, à cadence élevée, en vue, par exemple, de l'insertion dans un récipient. *French Patent #2571135.*

Fiechter, A, Fuhrmann, G.F. and Kappeli, O. (1981) Regulation of glucose metabolism in growing yeast cell. *Adv. Microbiol. Physiol. 22*, 123-183.

Flanzy C. (1998) *Oenologie*, Lavoisier Techniques et Documentation, Paris.

Fleet, G.H. and Costello, P.J. (1991) Malolactic fermentation of wine, *U.S. Patent #5104665*.

Fumi, M.D., Trioli, G., Colombi, M.G. and Colagrande, O. (1988) Immobilisation of *Saccharomyces cerevisiae* in calcium alginate gel and its application to bottle-fermented sparkling wine production. *Am. J. Enol. Vitic. 39*, 267-272.

Fumi, M.D., Bufo, M., Trioli, G. and Colagrande, O. (1989) Bulk sparkling wine production by external encapsulated yeast bioreactor. *Biotechnol. Letters 11*, 821-824.

Galazzo, J.L. and Bailey, J.E. (1990) Growing *Saccharomyces cerevisiae* in calcium-alginate beads induces cell alterations which accelerate glucose conversion to ethanol. *Biotechnol. Bioeng. 36*, 417-426.

Ghommidh, C. and Navarro, J.M. (1986) Flocculation-fermentation. In *Bioreacteurs*, pp. 89-112, Société Française de Microbiologie, Paris.

Godia, F., Casas, C. and Sola, C. (1991) Application of immobilised yeast cells to sparkling wine fermentation. *Biotechnol. Prog. 7*, 468-470.

Groboillot, A., Boadi, D.K., Poncelet, D. and Neufeld, R.J. (1994) Immobilisation of cells for application in the food industry. *Crit. Rev. Biotechnol. 14*, 75-107.

Hill, F. (1990) Immobilisation of yeast in alginate beads for production of alcoholic beverages. *U.S. Patent #5070019*.

Hsu, W-P. (1987) Apparatus for the production of ethanol and fermented beverages. *U.S. Patent #4790238*.

ITV (1994) *Choix et emploi des micro-organismes en oenologie*, ITV - Communication, Paris.

Janssen, D. (1993) Immobilised-cell malolactic fermentation. *Vitic. Enol. Sci. 48*, 219.

Lafon-Lafourcade, S. (1983) Wine and brandy. In *Biotechnology*, vol 5, pp. 81-163, VCH Verlagsgesellschaft, Weinheim.

Lefur, Y. (1990) Typicité et macération pelliculaire. Application au cépage Chardonnay en Bourgogne, *Revue des Oenologues 57*, 23-25.

Lenzi, P. and Cavin, J.F. (1985) Procédé de fabrication de boissons fermentées aromatisées. *French Patent #2580665*.

Lemonnier, J. (1986) Cartouche tubulaire filtrante et son application à la fabrication de vin mousseux en bouteille. *European Patent #0210915*.

Lemonnier, J. (1990) Insert for second fermentation of champagne or wines produced by the traditional method. *European Patent #0428428*.

Lemmonier, J. (1992) Cartouche de fibres creuses microporeuses pour la fermentation de boissons sucrées. *European Patent #0555603*.

Lommi, H. and Ahvenaimen, J. (1990) Method using immobilised yeast to produce ethanol and alcoholic beverages, *European Patent # 0361165*.

Maule, D.R. (1986) A century of fermenter design. *J. Inst. Brew. 92*, 137-145.

Moresi, M. (1989a) Fermenter design for alcoholic beverage production. In *Biotechnology applications in beverage production*, pp. 93-106, Elsevier Applied Science, London.

Moresi, M. (1989b) Optimal fermenter design for white wine production. In *Biotechnology applications in beverage production*, pp. 107-126, Elsevier Applied Science, London.

Naouri, P., Chagnaud, P., Arnaud, A., Galzy, P. and Mathieu, J. (1991) A new technology for malolactic bioconversion in wine. *J. Wine Res. 2*, 5-20.

Navarro, J.M. (1978) Production d'éthanol par des cellules adsorbées. In *Utilisation industrielle du carbone d'origine végétale par voie microbienne*, pp. 211-217, Société Française de Microbiologie, Paris.

Ogbonna, J.M., Amano, Y., Nakamura, K., Yokotsuka, K., Shimaza, Y., Wtanabe, M. and Hara, S. (1989) A multistage bioreactor with replaceable bioplates for continuous wine fermentation. *Am. J. Enol. Vitic. 40*, 292-298.

Ohno, T. and Takahashi, R. (1986) Rôle of wort aeration in the brewing process. Part 1: oxygen uptake and biosynthesis of lipid by the final yeast. *J. Inst. Brew. 92*, 467-470.

Poirat D. (1996) Procédé pour faire fermenter un liquide en bouteille, et cage amovible pour sa mise en œuvre. *French Patent #2745586*.

Quetsch, K-H. (1990) Immobilised biocatalyst bottle stopper - for sparkling wine with string to retract the filter housing inside the stopper, *German Patent #3931906*.

Rossi, J. and Clementi, F. (1984) L-malic acid catabolism by polyacrylamide gel entrapped *Leuconostoc oenos*. *Am. J. Enol. Vitic. 35*, 100-102.

Salou, P., Sablayrolles, J.M. and Barre, J.M. (1988) Production de levures d'intérêt oenologique. *Revue Française d'Oenologie 114*, 29-34.

Sarishvili, N.G., Storchevoi, N., Vaganov, V.M. and Reiblat, B.B. (1987) Process for producing sparkling wines, *U.S. Patent #4981700.*

Spettoli, P., Bottacin, A., Nuti, M.P. and Zamorani, A. (1982) Immobilisation of *Leuconostoc oenos* ML34 in calcium alginate gels and its application to wine technology. *Am. J. Enol. Vitic. 33*, 1-5.

Spooner, J.E. (1973) Method for producing champagne. *U.S. Patent #4009285.*

Steewart, G.G. and Russell, I. (1986) One hundred years of yeast research and development in the brewing industry. *J. Inst. Brew. 92*, 537-558.

Valade, M. and Rinville, X. (1990) Approches techniques et économiques des différents procédés de prise de mousse en bouteille. In *Les nouvelles techniques de tirage*, pp. 34-39, Association des techniciens supérieurs en viticulture et oenologie, Epernay.

Yokotsuka, K., Otaki, A., Naitoh, A. and Tanaka, H. (1993) Controlled simultaneous deacidification and alcohol fermentation of a high-acid grape must using two immobilised yeasts, *Schizosaccharomyces pombe* and *Saccharomyces cerevisiae*. *Am. J. Enol. Vitic. 44*, 371-377.

Yokotsuka, K., Yajima, M. and Matsudo, T. (1997) Production of bottle-fermented sparkling wine using yeast immobilised in double-layer gel beads or strands. *Am. J. Enol. Vitic. 48*, 471-481.

A NEW POLYSACCHARIDE DERIVED FROM PLANT RHIZOSPHERE : PRODUCTION, PURIFICATION AND PHYSICO-CHEMICAL PROPERTIES

CROMPIN J.M., GARNIER T., PAYOT T., DE BAYNAST R.
ARD (Agro-Industry Research and Development), F-51110 Pomacle
E-mail : ard@wanadoo.fr Fax : (33) 03.26.05.42.88

Summary

Microbial polysaccharides are of great interest on the market of hydrocolloïds. The development of a new one is to be considered only if its properties are different and complementary of those already existing. It is the case of "Soligel", a new exopolysaccharide produced by *Rhizobium sp.*

1. Introduction

The most part of industrial polysaccharides are originated from plants (starch, cellulose, gum arabic, pectins, guar gum, locust bean gum ...) and seaweed (agar, carrageenans, alginates). However, their production may be subjected to several hazards (drought, crop failure, war, famine...), causing lack of an assured supply and variations in quality.

Comparatively, microbial polysaccharides offer numerous advantages : their production in bioreactor is controlled and reproducible. They are synthesised by a great diversity of microorganisms, with different compositions and/or structures. These macromolecules still have higher production costs than traditional polysaccharides. In order to compete with the latter, they must have better rheological properties and/or new functions.

For instance, microbial polysaccharides are used as thickening, gelling, flocculating or moisturising agents, in a wide domain of applications : food, agriculture, cosmetic, pharmacy, environment... (Paul *et al.*, 1986 ; Sutherland, 1998).

The strategy of ARD is to valorise European crops and associated by-products. ARD has found potential substrates for production of polysaccharides by using these renewable raw materials. Consequently, the research of a new polysaccharide with original properties has been initiated through an AGRICE program (AGRIculture for Chemistry and Energy), with specific tasks in the areas of microbiology, chemistry and processing.

M. Hofman and P. Thonart (eds.), Engineering and Manufacturing for Biotechnology, 423–428.

2. Materials and methods

2.1. BACTERIAL STRAIN.

The strain YAS 34 was isolated from the rhizosphere of a sunflower plant. The selection of exopolysaccharide-producing isolates was carried out on high carbon : nitrogen ratio (C/N) liquid media (Hebbar *et al.*, 1992). The screening procedure was performed by the LEMIR-CNRS (Alami, 1997). Further genotypic studies indicated that the strain was belonging to the *Rhizobium* genus.

2.2. INOCULUM PREPARATION AND CULTURAL CONDITIONS

In order to produce the exopolysaccharide (EPS) on a large scale (10 m^3), the following inoculation procedure was achieved : starting from the inoculation of cryotubes into a 3-liter shake flask, the culture broth was progressively inoculated into larger scale reactors, - 20 and 450 litres - , before the final stage into the 10 m^3 fermentor.

Specific media were developed by ARD (Alami et al., 1998), respectively to ensure sufficient growth of the microorganism and to improve the production of the exopolysaccharide. The temperature and pH were controlled, respectively at 30°C and 7.0. Oxygen and mass transfer were monitored by controlling agitation, as well as aeration and overlay pressure.

2.3. RECOVERY AND PURIFICATION OF THE EXOPOLYSACCHARIDE

The exopolysaccharide was recovered after a multi-step downstream processing. The fermentation broth was first heated to a temperature between 80 and 100°C before being clarified by continuous centrifugation at 10,000 g. As far as it was possible because of the gelling properties of the exopolysaccharide, an ultrafiltration was carried out on the supernatant at 25°C (membrane molecular weight cut-off : 200 kDa). A step of diafiltration was made in order to eliminate fermentation residues. The retentate was further purified by alcoholic precipitation (v : v) after the addition of 0.5 M NaCl. The precipitate was washed with increasing proportions of alcohol (v : v, 70/30, 85/15, 100/0) and was dried at 40°C in a vacuum dryer.

2.4. RHEOLOGICAL ANALYSIS

The final product was evaluated by viscosimetric methods. The viscosity of the different polysaccharide samples was measured with a Searle viscosimeter (Haake, VT500) equipped with coaxial cylinders for the studies of solutions at low concentrations. The pH was adjusted by the addition of HNO_3 or NaOH. The measurements of the viscoelastic properties of the gels were performed using a rotary rheometer with cone-plate measuring heads (TA Instruments, AR1000).

Most part of the rheological studies was ensured by the CERMAV-CNRS for its contribution to the process development (Villain-Simonnet, 1999 ; Villain-Simonnet *et al.*, 2000) and by the Biophysics Laboratory of ENSIA for the formulation studies.

3. Results and discussion

3.1. FERMENTATION DATA

Production of 10 to 25 g EPS . kg^{-1} in the broth was observed from the laboratory to the pilot scale, with good conversion coefficient of substrate into exopolysaccharide (50 to 60%). Recent data indicated the possibility of improving the productivity and the conversion coefficient by feeding continuously the fermentor with carbon source. The kinetic data seem to be the same as those of xanthan (Amanullah *et al.*, 1998), showing a first phase where bacterial cells are produced in a low C/N based medium, whereas the EPS production is favoured by a limitation of nitrogen sources (high C/N ratio) in a second phase (Crompin, 2000).

3.2. DOWNSTREAM PROCESSING

The multiple effects of the heat treatment on further purification of the exopolysaccharide were previously reported (de Baynast *et al.*, 1998). The heat treatment induce the conformational transition of the EPS, allowing a better clarification by the sharp decrease in broth viscosity, and conferring enhanced gelling properties to the EPS (figure 1). This phenomenon could be explained by the restoration of an ordered conformation with the formation of junction zones.

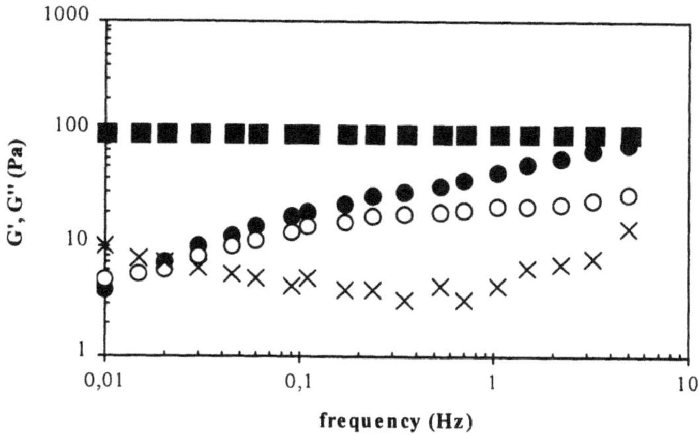

Figure 1 : Mechanical spectra of the polymer at 10 g.l^{-1} in NaCl 0.1 M under its sodium salt (a) in the native form : • G', ○G'', (b) after heat treatment : ■G', x G'' (10 % strain).

It is also possible to get a good clarification by combining heat and acid treatments. In addition to the previous results, we can suppose that the additional hydrolysis of the EPS leads to new physico-chemical properties, due to partial depolymerisation.

3.3. GELLING PROPERTIES

By changing the duration of the heat treatment and/or the concentration of the polysaccharide, it is possible to adjust the melting point of the gel for values ranging from 37 to 70°C. The study of the effect of monovalent and divalent ions have shown that the storage modulus is independent from the nature and concentration of salts (for [salts] > 0.04N).

This behaviour is original in comparison to others gelling polysaccharides. The strength of the gels is very dependent on the length of the heat treatment (figure 2), on concentration of polysaccharide (figure 3), and gel formation is possible without addition of salts.

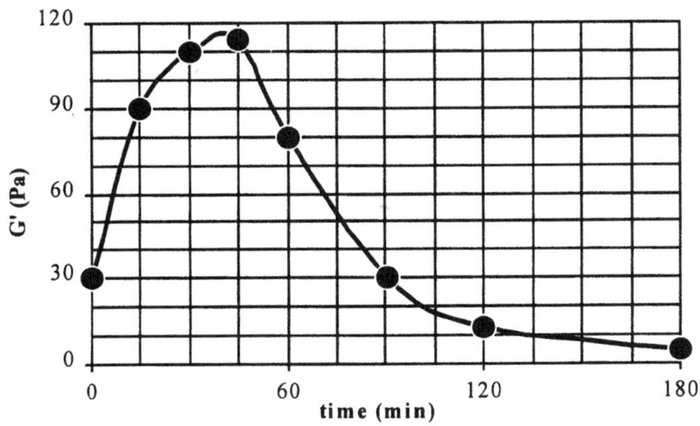

Figure 2 : Evolution of gel strength (G' at 25°C, 1 Hz) with duration of heat treatment (polymer at 10 g.l^{-1} in NaCl 0.1 M under its sodium salt).

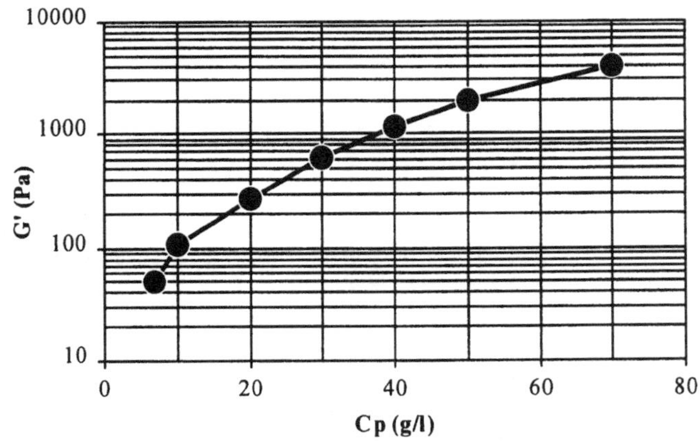

Figure 3 : Evolution of gel strength (G' at 25°C, 1 Hz) with concentration of polymer (Cp).

Tests for food formulation indicated that the EPS has a good stability in a broad range of pH (figure 4), show a very important cohesiveness and very good healing after the applications of important strains (figure 5). During storage, the gelling properties are reinforced, indicating a renaturation phenomenon ; no syneresis is observed.

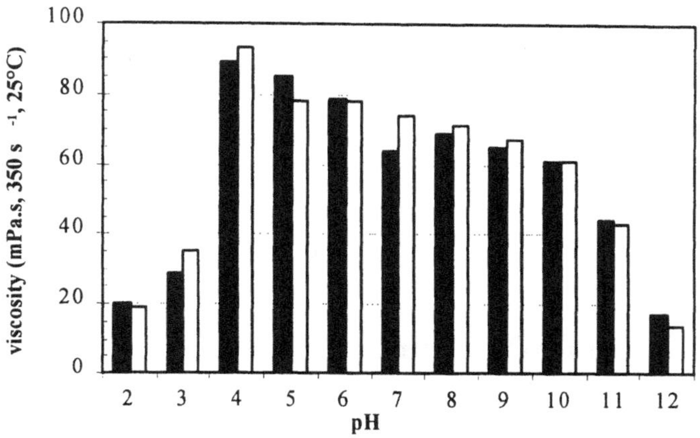

Figure 4 : Influence of pH on viscosity, ■ after preparation of solution,□ after 1 week (polysaccharide at 10 g.l^{-1} in water).

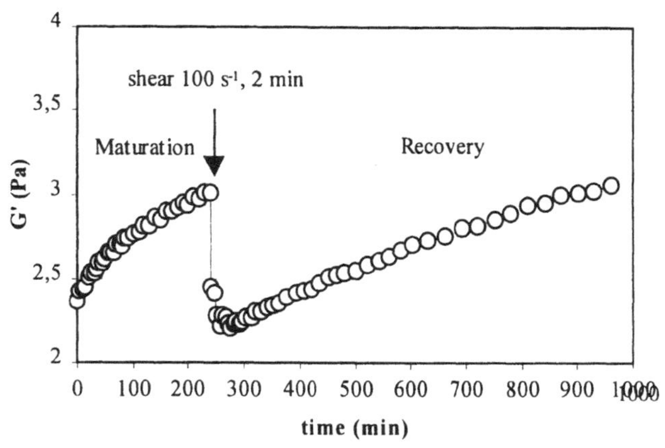

Figure 5 : Evolution of gel strength (G' at 20°C, 1 Hz, strain 10%) after shear (polymer at 5 g.l^{-1} in NaCl 0.1 M under its sodium salt).

4. Conclusion

Original properties of the EPS in terms of gelling effect have been underlined. Numerous applications are to be considered in food and non-food areas. First sampling and evaluation have been made towards cosmetic, pharmaceutical and food industries. Collaborations in these domains are in progress with industrial partners. The behaviour of "Soligel" joined to a good yield of conversion and large availability of raw materials authorise optimistic views for this new polysaccharide.

Acknowledgements

The authors are grateful to Pr M. Rinaudo (CERMAV-CNRS), Dr T. Heulin (LEMIR-CNRS), Pr G. Cuvelier (ENSIA) and Pr F. Duchiron (Europol'Agro) for their contribution to this work. This research was in part supported by the ADEME and a doctoral fellowship (Europol'Agro, Reims).

References

Alami, Y. (1997) Role of an exopolysaccharide-producing bacterium in the aggregation of rhizospheric soil of sunflower : consequence of inoculation on the structuration of the soil and on the mineral nutrition of the plant, PhD thesis, Université H. Poincaré, Nancy, 139 pp.

Alami, Y., Heulin, T., Milas, M., de Baynast, R., Heyraud, A.and Villain, A. (1998) Polysaccharide, microorganism and method for obtaining same, composition containing it and application, Patent WO 985993.

Amanullah, A., Satti, S. and Nienow, A.W. (1998) Enhancing xanthan fermentations by different modes of glucose feeding, *Biotechnol. Prog. 4*, 265-269.

Crompin, J.M. (2000) Industrial scaling up of the production and purification of a new bacterial polysaccharide derived from plant rhizosphere, PhD thesis, Université Reims Champagne Ardenne, 240 pp.

De Baynast, R., Crompin, J.M., Garnier, T., Wintrebert, A., Heulin, T., Alami, Y., Achouak, W., Rinaudo, M., Milas, M. and Villain, A. (1998) Characterisation of new microbial polysaccharides isolated from plant rhizosphere, internal report.

Hebbar, K.P., Gueniot, B., Heyraud, A., Colin-Morel, P., Heulin, T., Balandreau, J. and Rinaudo, M. (1992) Characterisation of exopolysaccharides produced by Rhizobacteria, *Appl. Microbiol. Biotechnol. 38*, 248-253.

Paul, F., Morin, A. and Monsan, P. (1986) Microbial polysaccharides with actual potential industrial applications, *Biotech. Adv. 4*, 245-259.

Sutherland, I.W. (1998) Novel and established applications of microbial polysaccharides, *TIBTECH 16*, 41-46.

Villain-Simonnet, A. (1999) New polysaccharides of bacterial origin : structure and properties, PhD thesis, Université J. Fourier, Grenoble , 242 pp.

Villain-Simonnet, A., Milas, M. and Rinaudo, M. (2000) A new bacterial exopolysaccharide (YAS 34). II. Influence of thermal treatments on conformation and structure. Relation with gelation ability, *Int. J. Biol. Macromol 27*, 77-87.

INITIATION, GROWTH AND IMMOBILISATION OF CELL CULTURES OF *TAXUS* SPP. FOR PACLITAXEL PRODUCTION

CHI WAI TANG, EMAN ZALAT AND FERDA MAVITUNA
Department of Chemical Engineering, UMIST, PO Box 88, Manchester M60 1QD, UK.
E-mail: f.mavituna@umist.ac.uk Fax No: 44 161 200 4399

Summary

Paclitaxel (Taxol), a cytotoxic diterpene initially isolated from the bark of *Taxus brevifolia* (Pacific yew), has been approved for cancer treatment. Since total chemical synthesis is not economical, plant biotechnology can offer an alternative route for the production of this drug. Callus cultures were initiated from different explants of *Taxus* × *media* and *Taxus cuspidata* on various media using different plant growth regulators in different photo-regimes. Several fast growing, white, friable calli lines were established. Suspension cultures were initiated from these calli and the activities of suspension and immobilised cultures were monitored over 40 days in shake flasks and bioreactors. Cell immobilisation in reticulated polyurethane foam particles showed improved growth. Since the specific paclitaxel yield were the same for both suspended and immobilised cultures, higher biomass concentration in the immobilised cultures led to higher paclitaxel concentration in the medium.

1. Introduction

1.1. PHARMACEUTICALS FROM PLANTS

Plants remain to be a significant source of pharmaceuticals for the treatment of a wide range of human ailments. Natural products isolated from higher plants account for approximately 25% of the prescribed drugs used in "Western" civilisation (Balandrin *et al.*, 1993; Farnsworth and Morris, 1976). In the United States alone, in 1990, this corresponded to a market value of approximately $ 15.5 billion (Principe, 1996). In the developing countries, 85% of the drugs used for healthcare is based on traditional medicine mainly derived from plants (Pezzuto, 1996). This implies that about 80% of the world's population rely on plant-derived natural products for their primary

M. Hofman and P. Thonart (eds.), Engineering and Manufacturing for Biotechnology, 429–448.

healthcare (Akerele, 1993). A list of medicinal plants used worldwide contains 21 000 species (Penso, 1983). Some believe that this is a conservative estimate and the number of plant species used for medicinal purposes is between 35 000 and 70 000 (Farnsworth and Soejarto, 1991).

Some examples of prescribed drugs that are obtained from plants include steroids, cardiotonic glycosides (*Digitalis* glycosides), antimalarials (*Cinchona* alkaloids, artemisinin-based drugs from *Artemisia annua*), analgesics and antitussives (opium alkaloids), anaesthetic (cocaine), anticholinergics (belladonna-type tropane alkaloids), antihypertensive (reserpine), cholinergics (physostigmine, pilocarpine), antigout (colchicine), skeletal muscle relaxant (tubocurarine), antiviral drugs (michellamine B from *Ancistrocladus abbreviatus*) and anticancer drugs (Balandrin *et al.*, 1993; Cardellina *et al.*, 1993; Wright, 1995; Pezzuto, 1997; Phillipson *et al.*, 1997).

Anticancer drugs from plants include paclitaxel (Taxol®) from *Taxus brevifolia* L., vincristine (Oncovin®) from *Catharanthus roseus*, podophyllotoxin from *Podophyllum peltatum* L. and camptothecin from *Camptotheca acuminata* Decne. In addition to these natural compounds extracted from plants, scientists have also produced semi-synthetic compounds based on these with improved properties, such as water-solubility and different or enhanced pharmacokinetic profiles. Examples of these are vinorelbine (Navelbine®) which is closely related to vincristine, teniposide (Vumon®) which is a podophyllotoxin analog, topotecan hydrochloride (Hycamtin®) which is an analog of camptothecin, artemether which is based on artemisinin, and taxotere which is a semi-synthetic derivative of taxol (Pezzuto, 1996; Phillipson, 1997).

There are approximately 250 000 species of higher plants in the world and the majority of these have not been examined yet for their pharmocological properties. The ethnomedical information and epidemiological data often play important roles in the screening of plants for pharmaceuticals. The rational plant selection also benefits from using a literature-based correlative approach such as the NAPRALERT database (Loub *et al.*, 1985) coupled with advanced analytical techniques (Constant and Beecher, 1995) and bioassays. Although this approach does not lead to the identification of structurally novel compounds with pharmacological properties, it nevertheless enhances our understanding of molecular recognition sites and structure-activity relationships as well as leading to new uses of known compounds. The approach for novel drug discovery is usually based on bioassay-directed fractionation (Cordell *et al.*, 1991; Suffness and Pezzuto, 1991). There are exciting developments in computational chemistry, molecular recognition, combinatorial chemistry and high throughput screening that will have significant effects on the drug discovery process.

With advances in extraction, purification and isolation techniques, it has become possible to produce single active ingredients in standardised tablet or capsule form instead of the "old" medicinal plant extracts. These advances have also led to the chemical synthesis of the active compounds or the moiety of these natural plant compounds. Although the total chemical synthesis of most of these compounds has been generally successful (Nicolaou *et al.*, 1994a; Nicolaou and Sorensen, 1996), it has proved to be uneconomic on an industrial scale in most cases compared to extraction from plant material. The current production based on extraction from plant material however, has its own long-term problems in the case of compounds extracted from wild plants. The international trade in the medicinal plants in Europe is reported to involve mainly wild-collection species (Cunningham and Schipman, 1995). There is a serious threat of the loss of medicinal plant species through over-harvesting and habitat destruction. Considering the estimated 25% extinction of higher plants, corresponding to about 60 000 species, by the year 2050 (Akerele *et al.*, 1991), the increasing demand for medicinal plants should be met either from cultivated species or other methods of production should be found. It is in this respect that plant biotechnology can offer a potentially attractive alternative.

1.2. PLANT BIOTECHNOLOGY

The technical aspects of plant cell, tissue and organ culture and experimental protocols can be found in a number of publications (Street, 1977; Pierik, 1987; Fowler and Warren, 1992; Dixon and Gonzales, 1994). Although the initial expectation from the application of plant biotechnology was in the *in vitro* production of fine chemicals commercially using plant cell, tissue or organ cultures in the bioreactors (Bajaj, 1999), this has not materialised except for a few cases. Instead, the main application of plant biotechnology has been in the area of horticulture, agriculture and forestry (Lindsey and Jones, 1989; Vasil, 1994; Davey *et al.*, 1998).

The main reasons for the failure to produce fine chemicals by plant cell cultures commercially were our initial lack of understanding of the complex metabolism of plants, especially the biochemical routes for the synthesis of these compounds and the mechanisms for their regulation and control, and of plant molecular biology and genetic engineering. Further complications arose from the instability of plant cell lines, cell line storage problems and extreme susceptibility of plant cell and tissue cultures to microbial contamination. Although there is still a great deal to discover and elucidate in plant metabolism and genetics, with our current knowledge and technology in these fields, and the numerous strategies developed over the years for the manipulation of plant cell, tissue and organ cultures, the exciting scope for the production of fine chemicals *in vitro* by genetically engineered plant cell, tissue and organ cultures is still viable (Goodwin and Mercer, 1983; Herbert, 1989; Burbridge, 1993; Dennis *et al.*, 1997).

1.3. ANTITUMOR COMPOUNDS FROM *TAXUS* SPP.

Paclitaxel (Taxol) is a relatively new anticancer drug which was first isolated from *T. brevifolia*. Paclitaxel and the related taxane compounds mainly occur in the plants of *Taxus spp*. (Strobel *et al.*, 1993). The bark of *T. brevifolia* was reported to have the highest content of Taxol (Vidensek *et al.*, 1990; Witherup *et al.*, 1990; Fang *et al.*, 1993), but the concentration was very low, only about 0.007-0.04% of the dry weight (Vidensek *et al.*, 1990; Wheeler *et al.*, 1992; Elsohly *et al.*, 1994). Paclitaxel and related taxane compounds act on cells by stabilising microtubules, preventing their depolymerisation and therefore blocking mitosis at the transition between metaphase and anaphase. The inability of the cell to depolymerise microtubules induces cell death. The structure and biological activities of Taxol were first published in 1971 by Wani *et al.* Since then the total synthesis of Taxol has been achieved but on an industrial scale it has been considered both unfeasible and uneconomic (Holton *et al.*, 1994; Nicolaou *et al.*, 1994b). The current method of commecial production involves the extraction of intermediary compounds from the needles and then the chemical transformation of these compounds (Nicolaou *et al.*, 1994a; Hezari and Croteau, 1997). Extraction from the bark of the old trees is not a viable option because of the obvious environmental concerns. Therefore, plant biotechnology can provide an alternative source for this drug.

Several research groups have reported success in inducing suspension cultures of *T. brevifolia*, *T. baccata*, *T. canadensis*, *T. cuspidata*, *T.× media* and *T. chinensis* (Christen *et al.*, 1989; Jaziri *et al.*, 1991; Wickremesinhe and Arteca, 1991, 1993; Fett-Neto *et al.*, 1992, 1993, 1995; Durzan and Ventimiglia, 1994; Chee, 1995; Kim *et al.*, 1995) and production of taxane compounds (Christen *et al.*, 1991; Fett-Neto *et al.*, 1992, 1994a, 1994b; Zhiri *et al.*, 1994; Mirjalili and Linden, 1995). However, there are not many reports on the immobilised cultures of *Taxus spp* (Seki and Furusaki, 1996; Seki *et al.*, 1997).

Although secondary metabolites are mainly produced by slow growing or non-growing cultures, it is a good process strategy to grow cultures fast to reach the desired biomass concentration in the growth stage and then switch to a production strategy in the second stage. When considering such a production system with two stages, cell immobilisation is one of the most effective ways to maintain higher productivity of target metabolites. Therefore, this work provides data for initiating and growing cultures rapidly for the first stage and then immobilisation for the second stage in a production process.

In the immobilised plant cell cultures, the actual productivity often changes due to the effects of intraparticle mass transfer (Furusaki *et al.*, 1988), redifferentiation and physicochemical interaction between the immobilising materials and cells (Haldimann and Brodelius, 1987). Immobilisation of plant cells in reticulated polyurethane foam matrices *in situ* in the bioreactors is easy, fast and economical (Lindsey *et al.*, 1983; Mavituna and Park, 1985; Mavituna *et al.*, 1987; Williams and Mavituna, 1992). This paper describes successful initiation, growth and immobilisation of *Taxus* cultures for paclitaxel production. The immobilised cultures showed improvement of cell growth and paclitaxel production compared to the suspension cultures.

2. Materials and methods

2.1. PLANT MATERIAL AND CHEMICALS

Seedlings, stems and needles of *T.× media* cv. Hicksii and *T. cuspidata* var. nana were used as explants. These plants were obtained from a commercial nursery and grown in our department.

All reagents were purchased from Sigma Chemical Company (Poole, UK), except for the basal media which were purchased in a pre-made form excluding the plant growth regulators and sucrose from Imperial Laboratories Ltd. (Salisbury, UK). Vitamins used were in a pre-mixed form, while coconut water was prepared in the laboratory by draining the juice from a large number of coconuts, deproteinising by boiling for 10 minutes and then filtering before storing it in small batches in the freezer until required.

2.2. CULTURE INITIATION AND MAINTENANCE

2.2.1. Callus initiation

2.2.1.1 Explant Preparation. Seedlings, stems and needles of *T.× media* and *T. cuspidata* were used as explants in this study. Seedlings were established using the same protocol of Zhiri *et al.* (1994) for the in vitro germination of *T. baccata* seeds. Stems and needles from pot-grown plants were surface sterilised before placing on the media. For surface sterilisation, plant materials were washed and immersed in 70% ethanol for 2 min. They were then immersed in 2% sodium hypochlorite solution for 20 min. and rinsed with sterile water. Finally, the surface sterilised explants were aseptically cut into pieces of ca. 10 mm in length, placed on solid media, and incubated at 23 ±2°C either in the dark or using a 16 h light / 8 h dark photo-regime. The light intensity was 20-30 $\mu mol\ m^{-2}\ s^{-1}$.

2.2.1.2 Media Composition. Different basal media, MS (Murashige and Skoog, 1962), Gamborg's B5 (Gamborg, 1968) and woody plant (WP) (Lloyd and McCown, 1980) supplemented with different plant growth regulators (PGR), all at 1 mg/l, were used for the initiation of callus from the needles of T.× media (Table 1). In addition to the growth regulators, these basal media were supplemented with 20g/l sucrose and 3.5g/l Phytagel. The pH of the media was adjusted to 5.5 before autoclaving at 121 °C for 20 min.

After the first set of experiments on callus initiation and growth using different basal media and PGR, B5 (Gamborg *et al.*, 1968) and NAA were chosen to investigate further the effect of the plant species, explant type and coconut water on callus initiation.

For the maintenance of the calli, seven pieces of callus were placed on either CWT, a new medium formulated in this research (Table 2), or TM5 medium (Ketchum *et al.*, 1995) which was modified (Zalat & Mavituna, unpublished) by solidifying with 0.3g/l gelrite instead of agar and adding ascorbic acid (100 mg/l) aseptically by filtering after autoclaving and cooling.

2.2.2. Suspension culture

Cell suspension cultures were initiated by transferring callus from *T.× media* needles and seedlings and *T. cuspidata* seedling to Gamborg's B5 medium supplemented with 20g/l sucrose and 1mg/l NAA and 0.0025 mg/l BA. 50mg/l ascorbic acid and 292 mg/l glutamine were filtered sterilised through 0.2 μm cellulose acetate filter (Gelman, Northampton, UK) and added to the medium after autoclaving. The suspension cultures were subcultured every two weeks by transferring about 4g cells (fresh weight) into 100ml of fresh medium in a 250ml Erlenmeyer flasks.

Table 1. The basal media and different plant growth regulators (PGR) used in callus initiation.

Basal salt	2,4-D mg/L	Kinetin mg/L	NAA mg/L
	1	1	0
	1	0	0
MS	0	1	0
	1	0	1
	0	1	1
	0	0	1
	1	1	0
	1	0	0
B5	0	1	0
	1	0	1
	0	1	1
	0	0	1
	1	1	0
	1	0	0
WP	0	1	0
	1	0	1
	0	1	1
	0	0	1

Table 2. The composition of the new medium (CWT) obtained by various additions to B5 basal salt medium.

	mg/L		μg/L
Inositol	300	Thiamine HCl	6900
Glycine	75	Nicotinic acid	6250
Proline	115	Pyridoxine HCl	1750
Aspartic acid	133	Folic acid	500
Arginine	175	Biotin	50
Ascorbic acid*	5		
Kinetin	0.1		g/L
ABA	0.1	Sucrose	7.5
GA3	0.1	Fructose	7.5

* *Filter-sterilised after autoclaving.*

2.3. CELL IMMOBILISATION

Cells from suspension cultures were immobilised in reticulated polyurethane foam (Declon, UK) particles in shake flasks and sheets in the bioreactors. The pore size of this foam matrix was 45 ppi (pores per inch). Five empty $1 cm^3$ foam particles were threaded onto an L shaped, stiff, stainless steel wire, which was held stationary and submerged in the liquid medium in the 250 ml Erlenmeyer flask. After inoculation the plant cells and cell aggregates were self-immobilised in these empty foam particles and grew within the foam matrix. The details of immobilisation in shake flasks and bioreactors can be found in Mavituna *et al.* (1987).

2.4. BIOREACTORS

Four different types of reactors were used in this study; airlift, bubble, stirred tank and immobilised cell reactor. One 4L reactor, with 3.5L working volume and dimensions of 24cm height x 15cm diameter, was modified to work as airlift, stirred tank and immobilised cell reactor. Air was introduced to the bioreactor through an L-shaped sparger and the airflow was maintained during the whole experiment at 0.25 v.v.m. In order to use it as an airlift bioreactor, an additional stainless steel tube draft with dimensions of 17cm length x 13cm diameter was placed in the centre of the bioreactor. In the stirred tank operation, two impellers of 9cm diameter and 12cm apart, rotating at 90 rpm were used. In order to use it as an immobilised cell reactor, 8 sheets of 3cm x 15cm X 0.5cm reticulated polyurethane foam were arranged vertically like baffles around the impeller shaft in the stirred tank bioreactor (Figure 7). Only one impeller was used with 90 rpm in the initial stage of immobilisation what took about 7 to 10 days. After this, the bulk medium was mixed by sparged air only.

A glass jar of 2L capacity with 1.5L working volume and dimensions of 18cm height x 12cm diameter was used as the bubble reactor. A sintered glass sparger placed in the bottom centre of the jar was used for the aeration and mixing. In all the bioreactor experiments, an inoculum size of 10% v/v was used in CWT liquid medium.

2.5. ANALYTICAL MEASUREMENTS

2.5.1. Growth

0.8g Callus was aseptically transferred to 50ml capped jars (Fisher Scientific) containing 10ml of either CWT or modified TM5 medium. On TM5, callus growth was determined by both the fresh and dry weight measurements at close intervals. On CWT, callus growth was monitored by measuring the increase in fresh weight which was then used to calculate the growth index as [(final wt - initial wt)/initial wt]. All samples were in triplicates.

In order to study the growth of suspension cultures, one gram fresh weight of cell suspension was inoculated into 50 ml medium in 125-ml Erlenmeyer flasks that were placed on shakers with 100 rpm. Three flasks were harvested at each point for the determination of fresh and dry weight, pH of the medium, viability of cells as well as sucrose, glucose and fructose concentration in the medium.

2.5.2. Viability

Cell viability of the various cultures was determined by fluorescein diacetate (FDA) staining according to the method of Widholm (1972).

2.5.3. Sugar analysis

The residual sugars in the suspension medium were analysed by HPLC using a Spherisorb5 NH2 column (Phenomenex, Macclesfield,UK). Acetonitrile:water ratio of 80:20 v/v was used as the mobile phase at a flow rate of 1ml/min. Sugars were detected by a differential refractometer (Waters 110). The concentration of glucose, fructose and sucrose were calculated using a standard curve for each sugar.

2.5.4. Taxane analysis

Paclitaxel and related taxane compounds were analysed by extracting the cell free medium with equal volume of dichloromethane. The organic layer was evaporated completely, and the residue was resuspended in HPLC grade methanol and filtered through 0.45μm filter for HPLC analysis (Waters). Analyses were performed using Curosil G column (Phenomenex, Macclesfield, UK), 250mm x 4.6mm, and acetonitrile:water ratio of 45:55 v/v as the mobile phase at a flow rate of 0.8ml/min. Taxol was detected using photo-diode array detector and identified by its retention time and UV spectrum.

3. Results and discussion

3.1. CALLUS INITIATION

3.1.1 Effect of media and plant growth regulators

Table 3 shows the result of callus initiation from the needles of *T.× media*. It was observed that callus initiation started from the cut surface of the needles and anywhere else on the leaf surface where the epidermal layer was stripped off. The initiated callus was fine, friable and light yellow.

Although there was not a significant difference between the different types of basal salt media in terms of their effect on callus initiation, in general, MS and B5 gave better results than WP as the callus yields were higher.

Hormone regimes played a key factor in callus formation. There was no significant difference between 2,4-D and NAA in terms of their effect on callus initiation. When the media contained both 2,4-D and NAA, this would slightly increase the percentage of callus formation. The media containing 1mg/L kinetin did not produce any callus or produced significantly smaller amounts of callus compared to the other treatments. This was because of the detrimental effect of kinetin on the explants (the explants turned brown). Kinetin also gave the lowest percentage of callus initiation in all groups. If either 2,4-D or NAA was added to kinetin in the medium, the percentage of callus initiation would increase significantly compared to the case when only kinetin was present.

The effect of increasing the NAA concentration on callus initiation was also evaluated. It was found that increasing the concentration of NAA to 5mg/l was better for callus initiation as it increased both the percentage of callus induction from the needles and the amount of callus produced per explant (data not shown).

Table 3. Percentage of callus initiation from the needles of T. media on different media and photo-regimes after 30 days. The medium specification is given in Table 1.

PGR	MS		B5		WP	
	Dark	Cycled light	Dark	Cycled light	Dark	Cycled light
2,4-D + K	54	36	68	38	14	6
2,4-D	68	62	88	62	64	32
K	12	0	32	4	24	10
2,4-D + N	72	76	72	56	66	52
K + N	64	22	28	4	18	22
N	28	54	78	50	56	36

3.1.2 Effect of light on callus initiation

Callus initiation was reduced when explants were kept under cycled light. Callus induced under cycled light was hard, green and grew very slowly compared to the yellow-white callus induced in the dark. Most of the calli induced under cycled light photo-regime did not proliferate when subcultured and eventually turned brown.

3.1.3 Effect of plant species and explant type on callus initiation

The effects of plant species and explant type on callus initiation were evaluated using B5 medium supplemented with 5 mg/l NAA and 20g/l sucrose. Seedlings, stems and needles of *T.*× *media* and *T. cuspidata* were used as explants.

Because mature *Taxus* seeds have a dormancy requirement of up to 2 years, a method for rapid germination of *Taxus* embryos was required. The same protocol for the rapid *in vitro* germination of *T. baccata* L. cv. Stricta embryos (Zhiri *et al.*, 1994) was used for *T.*× *media* and *T.cuspidata* seeds. A 100% germination frequency was observed after 7 days of culture of the excised zygotic embryos on modified MS medium and incubated in the dark. These embryos were obtained by dissecting the seeds after soaking in tap water for 7 days as this helped break the seed dormancy by leaching of the endogenous abscisic acid. This result is consistent with that of Zhiri *et al.* (1994) working with *T. baccata* and *T. canadensis*.

It was observed that 16 hours photoperiod was required for the optimal growth of the germinated embryos. However, dark incubation stimulated callus formation on the embryo axis. This result is consistent with that of Flores *et al.* (1993) working with *T. brevifolia* embryos, who found that light improved embryo germination and growth into seedlings.

The best source of explant for callus induction was the seedling since they exhibited a shorter induction time (7 days) and produced more callus. In addition, this callus had a relatively higher growth rate during the next few subcultures. As shown in Table 4, callus induction on needles started within 10 days compared 20 days on stems.

However, smaller calli were produced on needles and their growth rate during the following subcultures was too low. No major difference in either the induction time or the rate of callus induction was observed between the two species studied.

Table 4. The effect of coconut water on callus initiation from Taxus spp. placed on Gamborg's B5 salt supplemented with 2xB5 vitamin, 20g/l sucrose, 5mg/l NAA and incubated in the dark. +++ = more than 90% of the explants produced callus. ++ = 50-90% of the explants produced callus, + = less than 50% of the explants produced callus, 0= no callus was produced.

Plant Species	Explant Types	With coconut water			Without coconut water		
		Incubation Period (days)			Incubation Period (days)		
		10	20	30	10	20	30
T.xmedia	Needles	+	++	+++	0	+	++
	Stems	0	+	+++	0	0	++
T.cuspidata	Needles	+	++	+++	0	0	++
	Stems	0	+	+++	0	0	+

3.1.4 Effect of coconut water on callus initiation

The effect of coconut water on callus induction from the needles and stems of *T.x media* and *T. cuspidata* was studied using the optimum concentration of plant growth regulators (NAA 5mg/l) in the dark. Addition of coconut water (10% v/v) to the medium not only increased the percentage of induction but also shortened the induction time (Table 4).

3.2. CALLUS GROWTH AND MAINTENANCE

Although the previous B5 medium supplemented with the optimum plant growth regulator (5mg/l NAA) was found to be suitable for callus induction from different explants, the rate of callus growth on this medium was slow. In order to improve the callus growth, the calli of the different cell lines were subcultured on two different media, CWT and TM5. A substantial improvement in the callus growth was observed, however a red-brown exudate leached into the culture medium and this yielded a relatively hard, clumpy and brown callus. This browning problem is reported by many researchers working with *Taxus* species (e.g. Gibson *et al.*, 1995). After modifications of TM5 medium (Zalat and Mavituna, unpublished), the production of this brown exudate by the calli was completely eliminated. Within a few subcultures (4-6), we were able to obtain soft, friable, white or pale yellow and fast growing cell lines from the seedlings of the two *Taxus* species.

3.2.1 Effect of explant type

It was observed that the type of explant affected the appearance and the rate of growth of the newly initiated callus. The calli derived from seedlings showed a faster improvement in their growth during the following subcultures compared to those initiated from stems and needles. The two cell lines initiated from *T. × media* and *T. cuspidata* seedlings were designated as TMSD and TCSD and they were used in this study in addition to another cell line (TMNO) which was already initiated in our laboratory two years earlier from *T.× media* needles. It was observed that the growth of the different cell lines improved significantly with time during the subsequent subcultures. This could explain the reason for the fast growth of TMNO cell line, compared to the other two newly initiated cell lines of TMSD & TCSD (Figure 1).

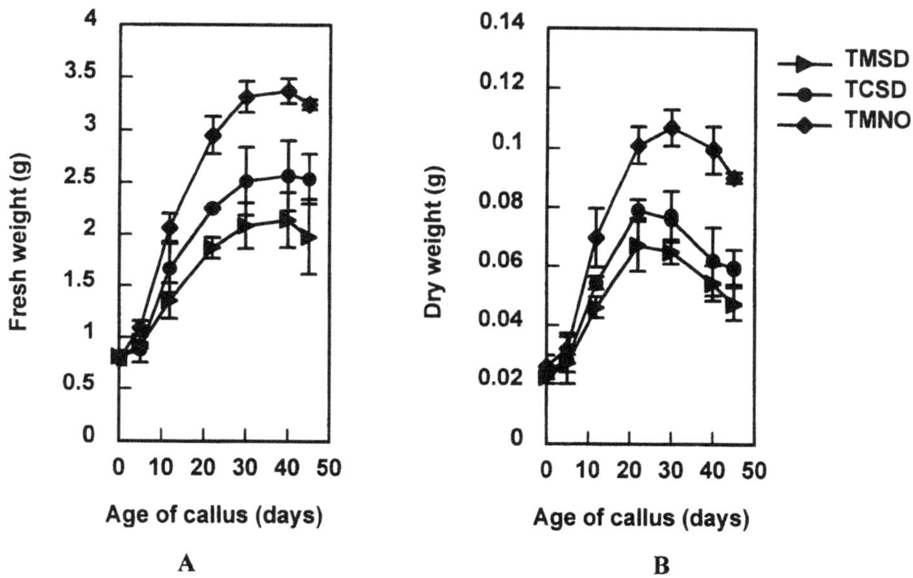

Figure 1. Growth curve of callus culture of T. species cell lines grown on modified TM5 medium in the dark. (A) Fresh weight (B) Dry weight

3.2.2 Effect of light on callus growth

Figure 2 illustrates the growth index of TMNO callus cultured on CWT medium under either cycled light photo-regime or in the dark. It was observed that callus cultured in the dark had a better growth rate than those growing under cycled light photo-regime.

After subculturing for 10 days, the callus growth index started to show a significant difference between the two photo-regimes. At the end of 30 days, the growth index of callus grown in the dark was nearly four times that of callus under cycled light treatment. The cultures incubated in cycled light showed a long lag phase for 16 days compared with just 10 days in the dark. Callus grown in the cycled light also seemed to enter a stationary phase after 20 days.

In general, the callus cultures grown under cycled light showed a decline in growth rate over several subcultures. By the end of the 5th subculturing, only callus induced from WP medium could be maintained with steady growth; all other calli turned brown and their viability decreased under cycled light photo regime.

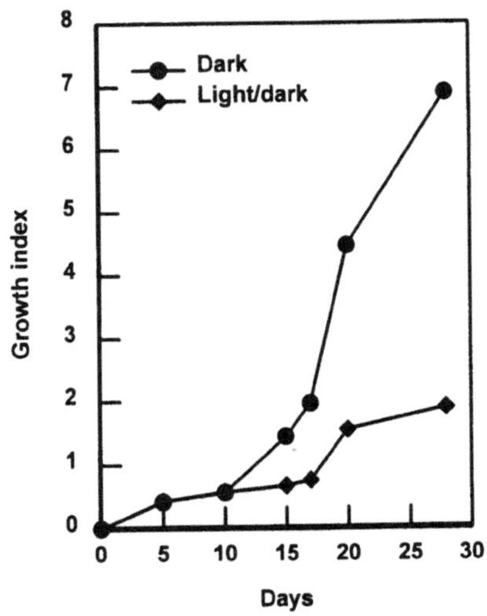

Figure 2. The growth index of Taxus x media callus on CWT medium.

3.3. SUSPENSION CULTURES

The growth kinetics of the suspension cultures of the three cell lines from *Taxus* species was studied in the dark. Cultures displayed a characteristic, slightly bi-phasic pattern for the increase in fresh weight over time in all the cell lines studied.

The time course of the biomass increase (fresh and dry weight), viability of cells and medium pH of TMNO cell suspension cultured in modified B5 medium, in the dark is illustrated in Figure 3. The medium pH of the different cell lines growing in the dark dropped slightly after the first five days of culture and then increased gradually. The same pattern of the pH change with time was reported for *T. cuspidata* cell suspension in shake flasks (Fett-Neto *et al.*, 1994b) and *T. baccata* cell suspension in bioreactor (Srinivasan, *et al.*, 1995). They suggested that the changes in the medium pH may be related to the uptake and utilisation of nitrogen sources from the medium by the cells.

It was observed that the viability of the cells of TMNO suspension culture was higher than 95% during the first 25 days of incubation, then it began to decrease and dropped to about 75% on day 45 of culturing. Retention of the high viability of cells in suspension culture during the growth cycle is one of the most important requirements for increasing the subsequent culture productivity.

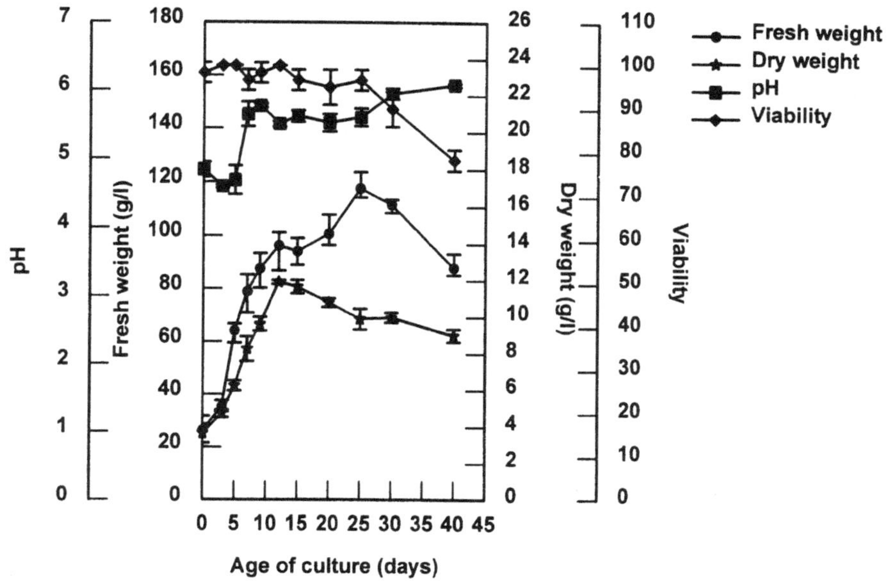

Figure 3. Batch culture of TMNO cell suspension in the dark.

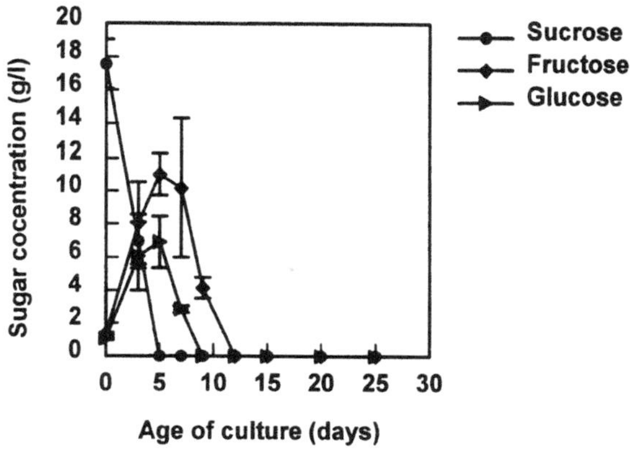

Figure 4. Sugar consumption by TMNO cell suspension in the dark.

Rapid uptake of sucrose and its hydrolysis to glucose and fructose followed by preferential uptake of glucose was observed in all cell lines (Figure 4). This has also been observed in other suspension cultures of *Taxus* species (Fett-Neto *et al.*, 1994b, Wickremesinhe and Arteca, 1994; and Pestchanker *et al.*, 1996). The accumulation of biomass was closely linked to the consumption of sugars in the medium. The onset of reduction in dry weight concentration coincided with the exhaustion of sugars from the medium on day 12.

3.4. IMMOBILISATION

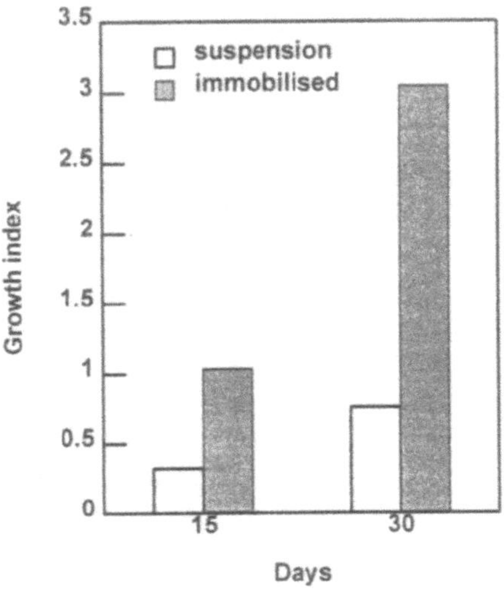

Figure 5. The growth index of free and immobilised cell cultures of T.× media.

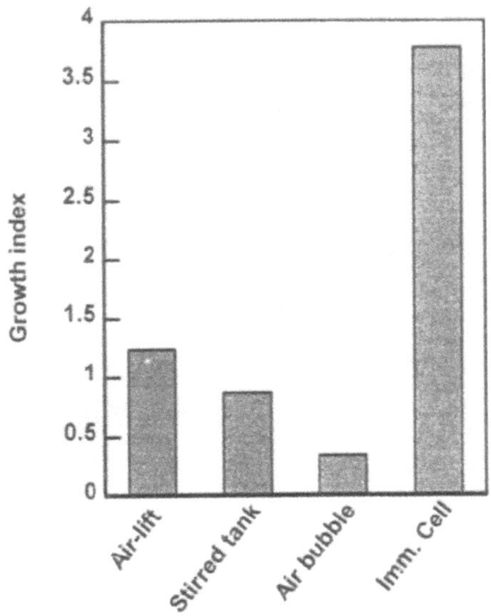

Figure 6. The growth index of T. × media cultures in different bioreactors after 30 days.

T.× media cells were immobilised in the open (continuous) pore network of the reticulated polyurethane foam matrices initially by the process of filtration brought about by agitation in the bulk liquid (Mavituna *et al.*, 1987). The initial entrapment through filtration led to the adhesion of cells and cell aggregates to the polyurethane fibres within the pores of the reticulated foam. As the pores of foam matrix were filled with the subsequent growth of the cell aggregates, the cells started to grow on the exterior surface of the foam matrix. Figure 5 shows a comparison of growth of suspension and immobilised *T.× media* cells in 250 ml shake flasks containing 100ml CWT medium. From this figure it is clear that immobilisation promoted cell growth. This observation, which was a repeat of our previous experience with other plant cells immobilised in reticulated foam matrices (Mavituna *et al.*, 1987), indicates that plant cells seem to benefit physiologically from being immobilised (Haldimann and Brodelius, 1987). The biomass yield per flask was about 4 times that of the freely suspended culture, starting with the same inoculum concentration, after 30 days of cultivation.

3.5. GROWTH IN BIOREACTORS

The cells in suspension culture were small and tended to float to the top of the liquid medium. Therefore, cells aggregated easily to form a "meringue" on the liquid surface in the headspace in the reactors. There was not much indication of growth in the bubble reactor. The best growth was obtained in the airlift reactor, but it formed a "meringue" of cells too. The stirred tank reactor had two impellers and the cells settled on the top impeller and again formed a "meringue". The immobilised cell reactor provided the best result in terms of increased growth yields and cell viability (Figure 7). Figure 6 gives the growth index after 30 days of cultivation in different bioreactors.

In terms of operation, the immobilised cell bioreactor was easier to handle. It was very easy to drain off the old medium and supply fresh medium aseptically keeping the immobilised cells in the bioreactor. This system could be run in an extended period of drain-and-fill repeated batches or in continuous mode of operation.

3.6. PACLITAXEL PRODUCTION

Suspension cultures of *T.× media* yielded 159 μg/l paclitaxel in shake flasks on the 15th day of culturing in CWT medium corresponding to 12.4 μg paclitaxel per g dry weight cells. The immobilised cells under the same conditions yielded 294 μg/l paclitaxel what corresponds to 11.0 μg paclitaxel per g dry weight cells.

Figure 7. The photographs of immobilised cell bioreactor at an early and later stage.

4. Conclusions

Calli of *T.×. media* and *T. cuspidata* were efficiently induced from needles, stems and seedlings on Gamborg B5 medium supplemented with 20g/l sucrose, 5mg/l NAA, 10% v/v coconut water. Addition of coconut water to the culture medium was found to have an enhancing effect on callus initiation. Kinetin and cycled light photo regime had detrimental effects on callus initiation. The CWT medium formulated in this work yielded fast growing cultures.

One of the major problems in plant cell culture of *Taxus* species is the browning of the culture which results in very slow growth rates and often cell death. We succeeded in preventing cell browning completely in the callus by modification of TM5 medium and within 3-5 subcultures, we were able to obtain a white, soft and friable callus. *T.× media* cells were immobilised successfully in the reticulated polyurethane foam matrices both in shake flasks and a 4L bioreactor. Immobilisation affected the culture performance positively. Compared to the suspension cultures, the biomass yield of the immobilised cultures were higher and the specific paclitaxel yield was almost the same. This means that our immobilised cultures lead to increased product concentration in the bulk liquid. It is also very easy to change the culture medium in our immobilised cell bioreactor allowing drain-and-fill repeated batch cycles and integration of product separation with bioreactor operation.

Strategies involving cell line selection and storage, manipulation of culture morphology and redifferentiation, control of the physico-chemical environment of the cultures through bioreactor design and mode of operation, cell immobilisation, continuous product removal, genetic manipulation of plant cells, integration of plant cell culture activity with biotransformations and combinatorial chemistry techniques should bring the potential use of plant cell, tissue and organ cultures for *in vitro* production of pharmaceuticals and other fine chemicals a step closer to realisation.

Advances in metabolic engineering (Stephanopoulos *et al.*, 1998), plant genetic engineering (Collins and Shepherd, 1997; Hall, 1999), combinatorial chemistry, instrumentation and analytical methods, computers and IT should accelerate the progress in this field. There are already established techniques for the production of various pharmaceutical compounds via transgenic plants as a consequence of these developments in other fields of science and engineering (Owen and Pen, 1996; Shahidi *et al.*, 1999). As our need for new and affordable medicine continues, so will the interest in the use plants as a valuable source of new therapeutic and chemo-preventive drugs.

Acknowledgement

We would like to thank the Egyptian Government for their financial support for E Zalat and UK CVCP, HEFCE for the ORS award and UMIST for the Graduate Research Scholarship for C W Tang during this project. We are also grateful for the taxane standards from the Drug Synthesis and Chemistry Branch, Developmental Therapeutics Program, Division of Cancer Treatment of the National Cancer Institute, USA.

References

Akerele, O. (1993) Nature's medicinal bounty: don't throw it away, *World Health Forum* 14, 390-395.

Akerele, O., Heywood, V. and Synge, H. (eds) (1991) *The conservation of medicinal plants*, Cambridge University Press, Cambridge.

Bajaj, Y.P.S. (1999) *Transgenic Medicinal Plants (Biotechnology in Agriculture and Forestry, 45)*, Springer Verlag, Heidelberg, Germany.

Balandrin, M.F., Kinghorn, A. D. and Farnsworth, N. R. (1993) Plant derived natural products in drug discovery and development: an overview, in A. D. Kinghorn and M. F. Balandrin (eds.), *Human medicinal agents from plants*, ACS Symp. Ser. No. 534, American Chemical Society, Washington D.C., USA, pp. 2-12.

Burbridge, A. (1993) Secondary plant metabolites from tissue culture, in BIOTOL, *In Vitro Cultivation of Plant Cells*, Butterworth-Heinemann Ltd., pp. 131-150.

Cardellina, J. H., Gustafson, K. R., Beutler, J. A., McKee, T. C., Hallock, Y. F., Fuller, R. W. and Boyd, M. R. (1993) National Cancer Institute intramural research on human immunodeficiency virus inhibitory and antitumor plant natural products, in A. D. Kinghorn and M. F. Balandrin (eds.), *Human medicinal agents from plants*, ACS Symp. Ser. No. 534, American Chemical Society, Washington D.C., USA, pp. 218-227.

Chee, P.P. (1995) Organogenesis in *Taxus brevifolia* tissue cultures, *Plant Cell Report* 14, 560-565.

Christen, A.A., Bland, J. and Gibson, D.M. (1989) Cell culture as a means to produce Taxol, *Proceeding of America Association of Cancer Research* 30, 566.

Christen, A.A., Gibson, D.M. and Bland, J. (1991) Production of Taxol or Taxol-like compounds in cell culture, *US Patent* 5,019,504.

Collins, G.B. and Shepherd, R.J. (1997) *Engineering Plants for Commercial Products and Applications (Annals of the New York Academy of Sciences, Vol 792)*, New York Academy of Sciences, New York.

Constant, H. L. and Beecher, C. W. W. (1995) A method for dereliction of natural product extracts using electrospray HPLC/MS, *Natural Product Letters* 6, 193-196.

Cordell, G. A., Beecher, C. C. W. and Pezzuto, J. M. (1991) Can ethnopharmacology contribute to the development of new anticancer drugs? *J Ethnopharmacol* 32, 117-133.

Cunningham, T. and Schippman, U. (1995) Concept and working programme of IUCN medicinal plant specialist group, *ICMAP Newsletter* 1, 16-18.

Davey, M. R., Alderson, P. G., Lowe, K. C. and Power, J. B. (1998) *Tree Biotechnology: Towards the Millennium*, Nottingham University Press, Nottingham.

Dennis, D. T., Turpin, D. H., Lefebvre, D. D. and Layzell, D. B. (eds.) (1997) Plant Metabolism, 2nd Edn., Addison Wesley Longman Ltd., Singapore.

Dixon, R. A. and Gonzales, R. A. (eds.) (1994) *Plant Cell Culture: A Practical Approach*, IRL Press at Oxford University Press, Oxford.

Durzan, D.J. and Ventimiglia, F. (1994) Free taxanes and the release of bound compounds having taxane antibody reactivity by xylanase in female, haploid-derived cell suspension cultures of *Taxus brevifolia*, *In Vitro Cell Developmental Biology* 30, 219-227.

Elsohly, H.N., Croom, E.M., El-Kashoury, E, Elsohly, M.A. and McChesney, J.D. (1994) Taxol content of stored fresh and dried *Taxus* chippings, *Journal of Natural Products* 57, 1025-1028.

Fang, W.H., Wu, Y., Zhou, J.Y., Chen, W.M. and Fang, Q.C. (1993) Qualitative and qualitative-determination of Taxol and related – compounds *in Taxus cuspidata* sieb *et* Zucc, *Phutochemical Analysis* 4, 115-119.

Farnsworth N. R. and Soejarto, D. D. (1991) Potential consequences of plant extinction in the United States in the current and future availability of prescription drugs, *Economic Botany* 39, 231-240.

Farnsworth, N. R. and Morris, R. W. (1976) Higher plants: the sleeping giants for drug development. *Am J Pharm* 147, 46-52.

Fett-Neto, A.G., DiCosmo, F., Reynolds, W.F. and Sakata, K. (1992) Cell culture of *Taxus* as a source of the antineoplastic drug Taxol and related taxanes. *Biochemistry and Biotechnology* 10, 1572-1575.

Fett-Neto, A.G., Melanson, S.J., Nicholson, S.A., Pennington, J.J. and DiCosmo, F. (1994a) Improved Taxol yield by aromatic carboxylic acid and amino acid feeding to cell cultures of *Taxus cuspidata*, *Biotechnology and Bioengineering* 44, 967-971.

Fett-Neto, A.G., Melanson, S.J., Sakata, K. and DiCosmo, F. (1993) Improved growth and Taxol yield in developing calli of *Taxus cuspidata* by medium composition modification, *Biochemistry and Biotechnology* 11, 731-734.

Fett-Neto, A.G., Pennington, J.J. and DiCosmo, F. (1995) Effect of white light on Taxol and baccatin III accumulation in cell cultures of *Taxus cuspidata* Sieb and Zucc., *Journal of Plant Physiology* 146, 584-590.

Fett-Neto, A.G., Zhang, W.Y. and DiCosmo, F. (1994b) Kinetics of Taxol production, growth, and nutrient uptake in cell suspension of *Taxus cuspidata, Biotechnology and Bioengineering* 44, 205-210.

Flores, T., Wagner, L.J. and Flores, H.E. (1993) Embryo culture and taxane production in *Taxus spp.*, *In Vitro Cellular & Developmental Biology-Plant* 29, 160-165.

Fowler, W. F. and Warren, G. S. (eds.) (1992) M. M. Moo-Young (Ed. in Chief) *Plant Biotechnology: Comprehensive Biotechnology Second Supplement*, Pergamon Press, Oxford.

Furusaki, S., Nozawa, T., Isohara, T. and Furuya, T. (1988) Influence of substrate transport on the activity of immobilised *Papaver somniferum* cell, *Applied Microbiology and Biotechnology* 29, 437-441.

Gamborg, O.L., Miller, R.A. and Ojima, K. (1968) Nutrient requirements of suspension cultures of soybean root cell, *Experimental Cell Research* 50, 148-151.

Gibson, D.M. and Ketchum, R.E.B., Hirasuna, T.J. and Shuler, M.L. (1995) Potential of Plant Cell Culture for Taxane Production, in M. Suffness (ed.). *Taxol Science and Applications*. CRC Press Boca Raton, Florida, pp 71-95.

Goodwin, T. W. and Mercer, E. I. (1983) *Introduction to Plant Biochemistry*, 2nd Edn., Pergamon Press, Oxford.

Haldimann, D. and Brodelius, P. (1987) Redirecting cellular – metabolism by immobilisation of cultured plant-cell – a model study with *Coffea arabica, Phytochemistry* 26, 1431-1434.

Hall, R.D. (1999) *Plant Cell Culture Protocols (Methods in Molecular Biology , 111)*, Humana Press, Totowa.

Herbert, R. B. (1989) *The Biosynthesis of Secondary Metabolites*, Chapman and Hall, London.

Hezari, M. and Croteau, R. (1997) Taxol biosynthesis: an update, *Planta Medica* 63, 291-295.

Holton, R.A., Somoza, C., Kim, H.B., Liang, F., Biediger, J., Boatman, P.D., Shindu, M., Smith, C.C., Kim, S., Nadizadeh, H., Suzuki, Y., Tao, C., Vu, P., Tang, S., Zhang, P., Murthi, K.K., Gentile, L.N. and Liu, J.H. (1994) First total synthesis of Taxol, *Journal of America Chemistry Society* 116,1597-1600.

Jaziri, M., Diallo, B.M., Vanhaelen, M.H., Vanhaelen-Fastre, R.J., Zhiri, A., Becu, A.G. and Homes J. (1991) Enzyme-linked immunosorbent assay for the detection and semi-quantitative determination of taxane diterpenoids related to Taxol in *Taxus sp.* and tissue cultures, *Journal of Pharm Belg* 46, 93-96.

Ketchum, R.E.B. and Gibson, D.M. (1996) Paclitaxel production in suspension cell cultures of Taxus, *Plant Cell Tissue and Organ Culture* 46, 9-16.

Ketchum, R.E.B., Gibson, D.M. and Gallo, L.G. (1995) Media optimisation for maximum biomass production in cell cultures of Pacific yew, *Plant Cell Tissue and Organ Culture* 42, 185-193.

Kim, J.H., Yun, J.H., Hwang, Y.S., Byun, S.Y. and Kin, D.I. (1995) Production Taxol and related taxanes in *Taxus brevifolia* cell cultures: effect of sugars, *Biotechnology Letters* 17, 101-106.

Lindsey, K. and Jones, M. G. K. (1989) *Plant Biotechnology in Agriculture*, Open University Press, Milton Keynes.

Lindsey, K., Yeoman, M.M., Black, G.M. and Mavituna, F. (1983) A novel method for the immobilisation and culture of plant-cells, *FEBS Letters* 155, 143-149.

Lloyd, G. and McCown, B. (1980) Commercially feasible micropropagation of mountain laurel, *Kalmia latifolia*, by use of shoot-tip culture, *Proceeding of International Plant Propagation Society*, 30, 421-427.

Loub, W. D., Farnsworth, N. R., Soejarto, D. D. and Quinn, M. L. (1985) NAPRALERT: Computer-handling of natural products research data, *J Chem Inf Comput Sci* 25, 99-103.

Mavituna, F. and Park, J.M. (1985) Growth of immobilised plant-cells in reticulate polyurethane foam matrices, *Biotechnology Letters* 7, 637-640.

Mavituna, F., Park, J.M., Williams, P.D. and Wilkinson, A.K. (1987) Characteristics of immobilised plant cell reactors, in C. Webb and F. Mavituna (eds.), *Plant and Animal Cell Process Possibilities*, Ellis Horwood Publ.Chichester, UK, Chp.9, pp.92-115.

Mirjalili, N. and Linden, J.C. (1995) Gas phase composition effects on suspension cultures of *Taxus cuspidata*, *Biotechnology and Bioengineering* 48, 123-132.

Murashige, T. and Skoog, F. (1962) A revised medium for rapid growth and bioassays with tobacco tissue cultures, *Physiologia Plantarium* 15, 473-497.

Nicolaou, H. C., Yang, Z., Liu, J. J., Ueno, H., Nantermet, P. G., Guy, R. K., Claiborne, C. F., Renaud, J., Couladouros, E. A., Paulvannan, K. and Sorensen, E. J. (1994b) Total synthesis of Taxol, *Nature* 367, 630-634.

Nicolaou, K. C. and Sorensen, E. J. (1996) *Classics in Total Synthesis: Targets, Strategies, Methods*, VCH, Weinheim.

Nicolaou, K.C., Dai, W.M. and Guy, R.K. (1994a) Chemistry and biology of Taxol, *Angew Chem Int Ed Engl* 33, 15-44.

Owen, M.R.L. and Pen, J. (1996) *Transgenic Plants: A Production System for Industrial and Pharmaceutical Proteins,* John Wiley & Son Ltd, New York.

Penso, G. (1983) *Index plantarum medicinalium totius mundi eurumque synonymorum*, Organizzazione Editorale Medico-Farmaceutica, Milan.

Pestchanker, L.J., Roberts, S.C. and Shuler, M.L. (1996) Kinetics of Taxol production and nutrient use in suspension cultures of *Taxus cuspidata* in shake flasks and a Wilson-type bioreactor, *Enzyme and Microbial Technology* 19, 256-260.

Pezzuto, J. M. (1996) Commentary: Plant derived anticancer agents, *Biochem. Pharmacol.* 53, 121-133.

Phillipson, J. D. (1997) Medicinal plants, *J. Biol. Edu.* 31 (2), 109- 115.

Pierik, R. L. M. (1987) *In Vitro Culture of Higher Plants*, Martinus Nijhoff Publishers, Dordrecht.

Principe, P. P. (1996) Monetizing the pharmacological benefits of plants, in M. J. Balick, E. Elisabetsky and S. A. Laird (eds.) *Medicinal Plant Resources of the Tropical Forest*, Columbia University Press, New York, pp. 191-218.

Seki, M. and Furusaki, S. (1996) An immobilised cell system for Taxol production, *Chemitech* 26, 41-45.

Seki, M., Ohzora, C., Takeda, M. and Furusaki, S. (1997) Taxol (Paclitaxel) production using free and immobilised cells of *Taxus cuspidata, Biotechnology and Bioengineering* 53, 214-219.

Shahidi, F., Kolodziejczyk, P. and Whitaker, J.R. (1999) *Chemicals Via Higher Plant Bioengineering (Advances in Experimental Medicine and Biology, 464),* Kluwer Academic Publishers, Dordrecht, The Netherlands.

Srinivasan, V., Pestchanker, L., Moser, S., Hirasuna, T.J., Taticek, R.A. and Shulet, M.L. (1995) Taxol production in bioreactors: kinetics of biomass accumulation, nutrient uptake, and Taxol production by cell suspensions of Taxus baccata, *Biotechnology and Bioengineering* 47, 666-676.

Stephanopoulos, G. N., Aristidou, A. A., and Nielsen, J. (1998) *Metabolic Engineering: Principles and Methodologies*, Academic Press, San Diego.

Street, H. E. (ed.) (1977) *Plant Tissue and Cell Culture*, 2nd edition, Blackwell, Oxford.

Strobel, G.A., Stierle, A., and Hess, W.M. (1993) Taxol formation in Yew–*Taxus*, *Plant Science* 92, 1-12.

Suffness M (ed.) (1995) *Taxol Science and Applications,* CRC Press Inc. Boca Raton, Florida, USA.

Suffness, M. and Pezzuto, J. M. (1991) Assays for cytotoxicity and antitumor activity, in K. Hostettmann (ed), *Methods in Plant Biochemistry*, Academic Press, London, Vol. 6, Chap. 4, pp. 71-133.

Vasil, I. K. (1994) Automation of plant propagation, *Plant Cell, Tissue and Organ Culture* 39, 105-108.

Vidensek, N., Lim, P., Campbell A. and Carlson C. (1990) Taxol content in bark, wood, root, leaf, twig and seedling from several *Taxus* species, *Journal of Natural Products* 53, 1609-1610.

Wani, M.C., Taylor, H.L., Wall, M.E., Coggon, P. and McPhail, A.T. (1971) Plant antitumor agents. VI. The isolation and structure of Taxol, a novel antileukemic and antitumor agent from *Taxus brevifolia, Journal of America Chemistry Society* 93, 2325-2327.

Wheeler, N.C., Jech, K., Master's S., Bronst, S.W., Alvarado, A.B., Hoover, A.J. and Sander, K.M. (1992) Effects of genetic, epigenetic, and environmental factors on Taxol content in Taxus brevifolia and related species, *Journal of Natural Products* 55, 432-440.

Wickremesinhe, E.R.M. and Arteca, R.N. (1991) Production of Taxol in callus and cell suspension cultures of *Taxus* x *media* Hicksii, *Journal of Plant Physiology* 96, 96.

Wickremesinhe, E.R.M. and Arteca, R.N. (1993) Establishment of *Taxus* callus cultures, optimisation of growth and confirmation of Taxol production, *Plant Cell Tissue and Organ Culture* 35, 181-193.

Wickremesinhe, E.R.M. and Arteca, R.N. (1994) *Taxus* cell suspension cultures – optimising growth and production of Taxol, *Journal of Plant Physiology* 144, 183-188.

Widholm, J.M. (1972) The use of fluorescein diacetate and phenosofranine for determining viability of cultured plant cell, *Stain Technology* 474, 189-194.

Williams, P. D. and Mavituna, F. (1992) Immobilised Plant Cells, in W. F. Fowler and G. S. Warren (eds.) M. M. Moo-Young (Ed. in Chief) *Plant Biotechnology*: *Comprehensive Biotechnology Second Supplement*, Pergamon Press, Oxford, Chap. 5, pp. 63- 78.

Witherup, K.M., Look, S.A., Stasko, M.W., Ghiorzi, T.J., Muschik, G.M. and Cragg, G.M. (1990) Taxus spp. Needles contain amounts of Taxol comparable to the bark of Taxus brevifolia – analysis and isolation, *Journal of Natural Products* 53, 1249-1255.

Wright, C. W. (1995) Natural products in the fight against AIDS, *Pharmaceutical Journal* 254, 583-587.

Zhiri, A., Jaziri, M., Homes, J., Vanhaelen, M. and Shimomura, K. (1994) Factors affecting the in *vitro*-rapid germination of *Taxus* embryos and the evaluation of Taxol content in the plantlets, *Plant Cell Tissue and Organ Culture* 39, 261- 263.

EFFECTIVE BIOFUEL PRODUCTION BY AN INTELLIGENT BIOREACTOR

HIDEKI FUKUDA[1], AKIHIKO KONDO[2], AND HIDEO NODA[3]
[1] Division of Molecular Science, Graduate School of Science and Technology, Kobe University, Japan
[2] Department of Chemical Science and Engineering, Faculty of Engineering, Kobe University, Japan
[3] Kansai Chemical Engineering Co., Ltd., Japan

Abstract

With the aim of contributing to efforts to solve global energy and environmental problems, a joint research project— Effective Biofuel Production by an Intelligent Bioreactor — has been set up with participants representing several universities, research institute, and industrial companies. Biofuel obtained from biomass resources is seen as an important source of 'clean energy' by virtue of features such as its biodegradability and low carbon and sulphur dioxide contents. By utilising an 'intelligent' bioreactor containing immobilised 'arming cells', it is expected that practical biofuel production can be achieved at a considerably lower cost than with conventional processes.

1. Introduction

The growing seriousness of the global energy problem and environmental pollution are substantially increasing the importance of using value-added products from biomass resources as biofuel. Biofuel produced from biomass, such as biodiesel or ethanol, have two significant advantages:

- biodegradability,
- better-quality exhaust gas emissions.

In addition, the atmospheric levels of carbon and sulphur dioxide are not raised because the organic carbon of biofuel is produced by photosynthesis in plants.

In current research on biodiesel production, rapeseed esters are being investigated in Europe [1] and palm oil esters in Malaysia [2]. Soybean oil esters also feature prominently as a potential diesel fuel alternative [3], and there is a wide range of ongoing research in this area. In fact, in recent years biodiesel has been produced from waste edible oil on a pilot scale in Japan.

M. Hofman and P. Thonart (eds.), Engineering and Manufacturing for Biotechnology, 449–455.

Though efficient in terms of reaction yield and time, the chemical approach to synthesising alkyl esters [4-6] from triglycerides has several drawbacks, including difficulties in the recovery of glycerol, the need for removal of salt residue, and the energy-intensive nature of the process. On the other hand, the use of biocatalysts allows for the synthesis of specific alkyl esters, easy recovery of glycerol, and trans-esterification of glycerides with high amounts of free fatty acids. In addition, this process can further be used to synthesise other value-added products, including biodegradable lubricants and additives for fuels and lubricants. However, it has not thus far been adopted industrially because of its high cost.

Over the past two decades, there has been considerable interest and activity in the production of ethanol for use as a fuel by fermentation. Sugar materials such as molasses, sugar cane, and sulphite waste liquor have been mainly utilised for ethanol production, since complicated saccharifying or lignin degrading pretreatment processes are required when starch or cellulosic materials are used. There is thus a need for a novel bioprocess by which ethanol can be produced directly from starch or cellulolytic materials without the necessity for any pretreatment.

The purpose of the project described here, which has been realised with the support of the New Energy and Industrial Technology Development Organisation (NEDO) of Japan, is to establish a practical bioprocess for biodiesel and/or ethanol production from biomass resources based on two key technologies as elucidated below.

2. Key technologies for biofuel production

2.1 INTELLIGENT BIOREACTOR USING IMMOBILIZED YEAST CELLS

The past two decades have seen rapid developments in the use of enzymes as catalysts for industrial, analytical, and medical purposes, leading to the appearance of a new field of research known as enzyme technology. With the aim of making their use more convenient, there is now considerable interest in the direct utilisation of immobilised cells as a means of catalysis.

Passive immobilisation using porous biomass support particles (BSPs) has been successfully applied in fundamental research involving a wide variety of microbial, animal, and plant cell systems [7].

Recently, the author's laboratory succeeded in effective enzyme production using cells of a flocculent yeast immobilised within BSPs [8], as well as in the development of a novel bioconversion process using immobilised recombinant flocculent yeast cells carrying an enzyme fusion gene between rat P4501A1 and yeast NADPH-P450 reductase [9]. Of particular interest was the finding that recombinant cells immobilised within the BSPs not only exhibited significant expression of the fused enzyme, but a high proportion of plasmid-carrying cells was maintained. This contrasted with a much lower proportion among freely suspended cells released from the BSPs, in which no expression of the enzyme could be detected. It was thus apparent, as illustrated in Figure 1, that only highly expressing yeast cells were spontaneously immobilised within the BSPs. A bioreactor packed with such cells, which we have termed an 'intelligent' bioreactor, possess excellent potential either for the production of a wide variety of useful compounds or for use in various bioconversion reactions.

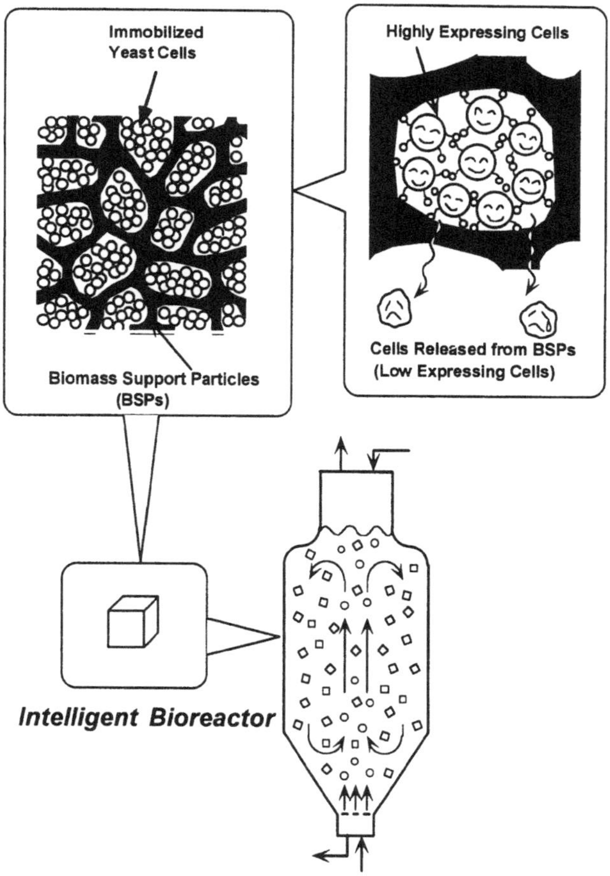

Figure 1. Illustration of an "intelligent" bioreactor

2.2 IMMOBILIZING PROTEINS ON THE SURFACE OF YEAST CELLS

The display of heterologous proteins on the cell surface of microorganisms is an important objective for many applications in micro- and molecular biology. In contrast to existing examples of surface expression on cells of eukaryotes, in the case of yeast cells, i.e. *Saccharomyces* sp., surface-expressed proteins become covalently linked to glucan in the cell wall instead of being linked to the plasma membrane, thereby rendering them resistant to extraction.

Recently, one group involved in the current project succeeded in constructing of a novel starch-utilising 'cell surface-engineered' yeast by displaying an amylolytic enzyme on the cell wall of *S. cerevisiae* [10]. Thus, the cell surface can be regarded as a new target for bestowing additional characteristics of metabolic reactions, and effective multipurpose yeast cells suitable for biofuel production, which have been termed 'arming" cells', can be created (illustrated in Figure 2).

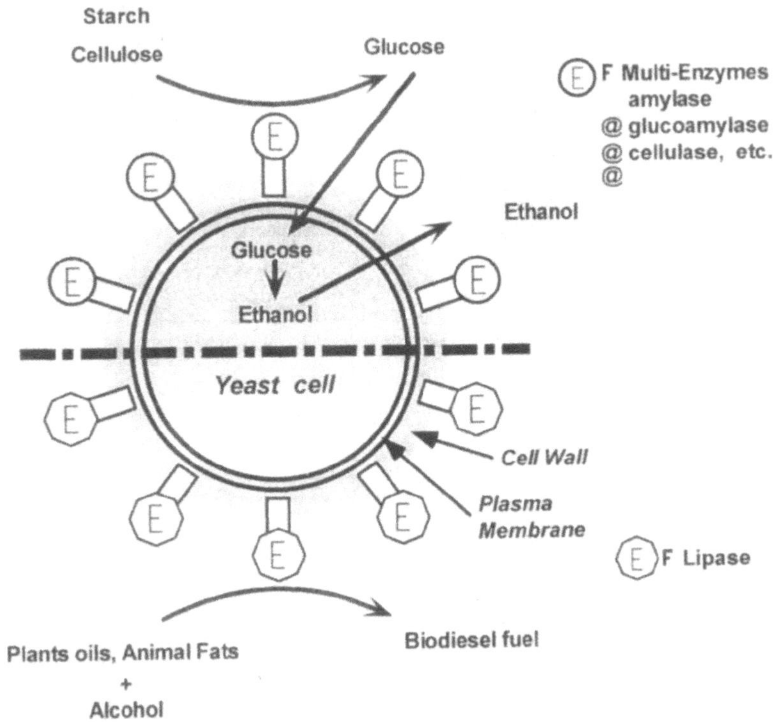

Figure 2. Illustration of "arming cells

3. Outline of ongoing research

To establish a practical process for biodiesel or ethanol production, research efforts will mainly focus on the development of (i) highly functional yeast cells, (ii) an intelligent immobilised bioreactor system, and (iii) an optimal control system in conjunction with an efficient monitoring system.

3.1 DEVELOPMENT OF HIGHLY FUNCTIONAL YEAST CELLS

The development of highly functional yeast cells that are effective for the production of biofuel should be able to realised using genetic engineering. Possible techniques include the introduction of enzyme-producing genes into flocculent yeast strains and/or of flocculation genes cloned from *Saccharomyces* strains [11] into enzyme-producing non-flocculent yeast strains both with and without cell-surface engineering.

In addition, techniques suitable for the screening and cloning of effective enzyme genes and protein conformation stabilisation are required. To stabilise the conformation of biofuel-producing enzymes and to generate strains that have higher activities than native enzymes, our laboratory will try to reveal the factors that affect the stability of the secondary structure, such as the alpha-helix, by structural analysis of synthetic model peptides and theoretical calculation. We will also endeavour to isolate thermostable proteins. The following researches are currently in progress.

- Construction of arming yeast cells
- Screening and cloning of lipase enzyme for biodiesel production [12-13]
- Stabilisation of protein conformation

3.2 DEVELOPMENT OF AN INTELLIGENT BIOREACTOR SYSTEM

Optimising the design of bioreactors for use in immobilised cell processes depends on several factors. The nature of the mass transfer requirement plays an important part, while the immobilisation method and particle characteristics must also be considered. In addition, the type of substrate, operational requirements, and hydraulic and economic considerations will also influence the design. Thus, the application of an intelligent bioreactor system suitable for biofuel production in industrial processes requires—mainly involving cell physiology, BSP materials and mass transfer within the bioreactor—to elucidate such properties. Researches in the following areas are now ongoing.

- Development of novel BSPs for engineered yeast cells
- Analysis of the suitability of bioreactors
- Construction of a novel intelligent bioreactor

3.3 DEVELOPMENT OF AN OPTIMAL CONTROL SYSTEM I N CONJUNCTION WITH EFFICIENT MONITORING

To achieve efficient online monitoring, a micro-sampling device, advanced measurement methods based on immunotechnology, and a novel flow injection analysis (FIA) system will be developed. We will also endeavour to formulate a model of mass transfer in immobilised cells and of the related biochemical reaction in order to develop an optimal control system for a bioreactor with immobilised cells. A simulator will be constructed to examine the responses of state variables to several kinds of control action. Researches are at present under way on the following topics.

- Modelling of the mass transfer and biochemical reaction in a bioreactor
- Construction of a simulator for use in the study of responses to control actions
- Construction of a novel intelligent control system as illustrated in Figure 3.

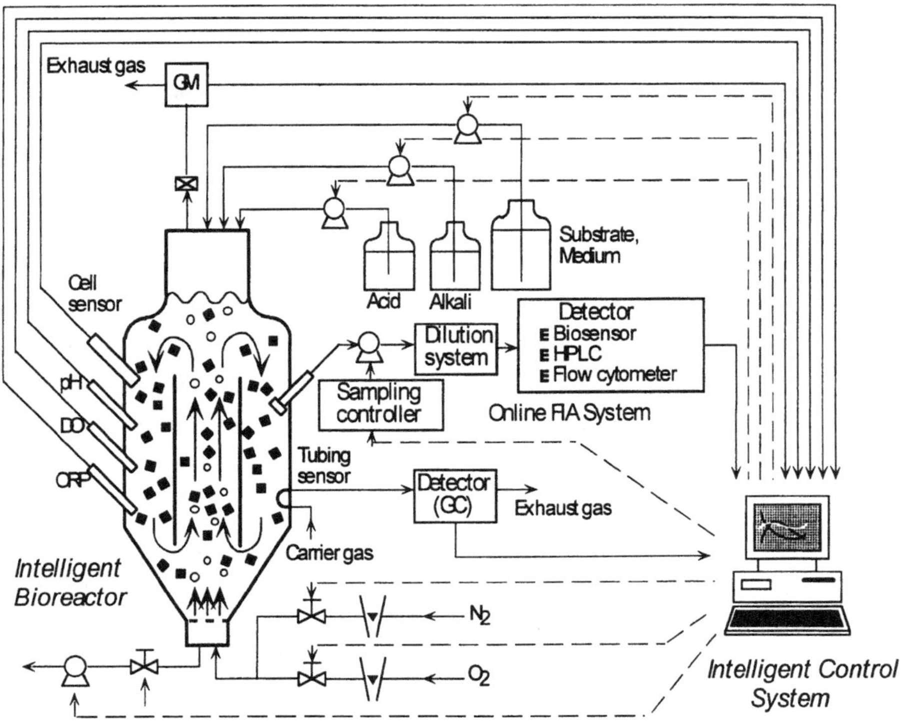

Figure 3. Illustration of a novel "intelligent" control system.

4. Conclusion

The joint research project—*Effective Biofuel Production by an Intelligent Bio-reactor*—which has been set up with the support of the New Energy and Industrial Technology Development Organisation (NEDO) of Japan, aims to establish a practical bioprocess for biofuel production based on two key technologies: an intelligent bioreactor and arming cells. Twenty-two scientists with backgrounds in biochemical engineering, enzyme engineering, biology, chemistry, and other relevant fields, working in three universities, two research institutes, and four industrial companies, are participating in the research. It is expected that the project will run for three years from 1998 to 2001, and if successful could significantly contribute to alleviating global energy and environmental problems.

References

1. Cvengros, J., and Cvengrosova, Z. (1994), *J. Am. Oil Chem. Soc. 71*, 1349-1352
2. Masjuki, H. H, and Sapuan, A.M. (1995) *J. Am. Oil Chem. Soc.72*, 609-612
3. Goering, C.E., and Fry, B.G. (1984) *J. Am. Oil Chem. Soc.61*, 1627-1632

4 Ali, Y, and Hanna, M.A. (1994) *Bioresource Technol. 50*, 153-163
5. Freedmann, B., Pryde, E.H., and Mounts, T.L. (1984) *J. Am. Oil Chem. Soc. 61*, 1638-1643
6. Lee, I., Johnson, L.A., and Hammound, E.G. (1995) *J. Am. Oil Chem. Soc. 72*, 1155-1160
7. Fukuda, H. (1995) *BIOREACTOR SYSTEM DESIGN*, Marcel Dekker, Inc.,New York, pp.339-375
8. Furuta, H., Arai, T., Hama, N., Shiomi, N., Kondo, A., and Fukuda, H. (1997) *J. Ferment. Bioeng. 84*, 169-171
9. Liu, Y., Kondo, A., Ohkawa, H., Shiota, N., and Fukuda, H. (1998) *Biochem. Eng. J. 2*, 229-235
10. Murai, T., Ueda, M., Yamamura, M., Atomi, H., Shibasaki, Y., Kamasawa, N., Osumi, M., Amachi, T., and Tanaka, A. (1997) *Appl. Microbiol. Biotechnol., 63*,1362-1366
11. Watari, J., Takata, Y., Ogawa, M., Murakami, J., and Koshino, S. (1991) *Agric. Biol. Chem. 55*, 1547-1552
12. Shimada, Y., Watanabe, Y., Samukawa, T., Sugihara, A., Noda, H., Fukuda, H., Tominaga, Y. (1999) *J. Am. Oil Chem. Soc. 76*, 789-793
13. Kaieda, M., Samukawa, T., Matsumoto, T., Ban, K., Kondo, A., Shimada, Y., Noda, H., Nomoto, F., Ohtsuka, K., Izumoto, E., and Fukuda, H. (1999) *J. Biosci. Bioeng. 88*, 627-631

PART IX
PATENTS AND LICENSES

TRANSLATING EUROPEAN BIOTECH INTO US PATENTS DO'S, DON'TS, & COSTS

THOMAS M. SAUNDERS
Lorusso & Loud
Boston, MA, USA Fax: +(617)723-4609; tmsaunders@aol.com

Introduction

Start-up biotechnology begins with an idea. European start-up biotechnology must consider how these ideas for novel drugs and methods will make the trans-Atlantic crossing to access the US market. Given the huge costs of US Food & Drug Administration (FDA) approval, exclusivity is a threshold requirement of new drug development. US patent portfolios are a means to exclusivity. Without the potential for exclusivity, it may be impossible to justify the cost of regulatory approval. Without potential exclusivity, intellectual property licensing income or venture capital for product development will not be available. This report considers significant differences between European and US patent systems, and offers suggestion patent strategy to develop a biotech patent portfolio.

1. Five important patent differences between Europe and the US

1.1. ONE-YEAR US GRACE PERIOD FROM FIRST USE OR SALE

One European rule for patentability is "absolute novelty." Absolute novelty means that disclosing an invention -- such as by delivering a talk or selling the thing invented -- is fatal to patentability. In contrast, the US offers a one-year grace period from first disclosure. Thus, in those instances in which the European rights were destroyed by untimely disclosure, that disclosure will not prevent applying for a patent in the US.

1.2. GRACE PERIOD (CONTINUED): *TEMPUS FUGIT*

The US definition of "sale" differs from the European definition. Under US law, outsourcing of a product or process may constitute a "sale" back to the hiring party. This "sale" will (perhaps) start the one-year clock. It also is possible to have an invention "on-sale" which has been designed but not built. For example, purchasing a

M. Hofman and P. Thonart (eds.), Engineering and Manufacturing for Biotechnology, 459–486.

novel nucleotide from an outsource can place the nucleotide "on sale." Soliciting business under a prepared protocol for a series of gene manipulations to yield a particular therapeutic or diagnostic result, if complete to the point at which a post-doctoral fellow or a clinician ("a person skilled in the art") could follow the protocol and obtain the result at least some of the time, could be deemed placing it "on-sale," thus starting the one-year clock.

1.3. DUTY OF DISCLOSURE

A European patent applicant has no specific requirement to inform the European Patent Office of significant prior art. In the US there is a duty of absolute candour. While no search need be made, an applicant in the US is responsible for informing the Patent Office of any art of which it is aware.

1.4. COMPUTER ALGORITHMS NOW PATENTABLE

Another recent court decision confirmed that computer algorithms are now patentable. This reversed a line of decisions that held algorithms to be "mere mathematical formulae" and unsuitable for patent protection.

1.5. FIRST TO INVENT VERSUS FIRST TO FILE

In the US, a patent is awarded to the "first to invent." In Europe and most of the world, a patent is awarded to the "first to file." This means that a later-filing patent applicant (the "junior party") can obtain a patent over an earlier-filing patent applicant (senior party) if the junior party can establish earlier "conception," "reduction to practice," and (usually) some level of "diligence." These elements are determined in the context of an intra-Patent Office proceeding termed a "patent interference."

Drug discovery, drug therapy, drug screening, and drug manufacture have become entwined with computer algorithms. High throughput and combinatorial chemistry are creatures of computer methodology. With some ingenuity, it is possible to delimit pharmaceutical methods, in part, by reference to algorithms. Now, algorithms can form an additional element of patent protection not to be overlooked -- whether as an offensive or as a defensive patent strategy

2. Basic patent game theory

- Only market exclusivity can justify the huge cost of drug regulatory approval.
- Market exclusivity can arise from either the patent system or (occasionally, in the US) from the FDA.
- Patent protection holds out the promise of market exclusivity, augmented profit margins, and great wealth.
- As a consequence, patents hold out the promise of wildly expensive litigation (the only kind).

- Applying for and obtaining a biotechnology patent in the US costs about $20-30,000, and about another $70,000 for all of Europe.
- For start-up biotechnology, tens of patent applications will be filed now, and the money paid now, for (maybe) one marketable patented product years down the road. IP applicants must plan on patent expenses of no less than $250,000 per year.
- The object of the game is to have patent protection for what it is you sell, and still avoid bankruptcy.

No one can pay for everything.
Miss a payment, lose a patent.
Choices must be made.
Not all choices will be correct.
Even if you have it, it may not be sufficient.

3. Invention germination

A technology venture must take pains to encourage disclosure of inventions. Inventive ideas are not limited to the senior scientists. One ignores post-doctoral students and technical staff at the company's peril. One way to obtain the disclosure of inventions is by invention disclosure forms.

3.1. INVENTION DISCLOSURE FORMS

3.1.1. Short forms only

While any number of forms can be devised, our view is that a form should be as brief as possible. The function of a disclosure form is merely that of prompting a corporate representative to meet face-to-face with the inventor. At that meeting, a skilled representative can inquire more fully into the invention and document the necessary information. Nothing is less conducive to unearthing inventions than requiring completion of a multi-page disclosure form with no feedback. From a legal perspective, it is foolhardy to offer an inventor a detailed questionnaire and then base corporate patent actions (or inaction) on those answers without further inquiry by the patent staff.

3.1.2. Who gets the forms?

Submitting invention disclosures directly to an invention committee or its representative (even if copied to supervisors) causes some problems and avoids others. From experience, we assure you that there is a real or imagined concern among inventors that a supervisor will suppress an invention or at least ask to "approve" a submission and try to claim inventorship. To avoid this concern, direct submission to a committee is useful. With direct submission, less-secure supervisors may interpret such submissions as theft of their own ideas. Periodic group idea sessions with minutes may be a partial solution. Another method is by simultaneous submissions to supervisor and committee. Mandatory lithium is a possibility. Call us if you come up with a better idea.

3.2. NO FORMS

A very useful approach for unearthing inventions is direct and frequent contact between all scientific personnel and the patent point person. The point person walks into all the labs at least every other week and asks what people are up to. Regular visits and interested inquiry from a source that is not "checking to see if you are working" is superior to invention disclosure forms, productivity reports, project summaries, E-mail, and everything else. Even a follow-up visit to an invention disclosure filing should prompt the patent point person to inquire beyond the basic disclosure. A comprehensive inquiry should include attention to a scientist's guesses, hunches, and views of what data are or will prove significant.

4. Invention selection

In the non-pharmaceutical world, there are guidelines for selecting which inventions or intellectual property (IP) to pursue as patents. "Almost as good and a lot cheaper" always sells. Also, things that work better but cost more can be sold. In the pharmaceutical world, marketing "almost as good" is close to impossible. In the pharmaceutical world, at the time the patent money is being spent, you will have no idea if it works at all, let alone better. In the pharmaceutical world, most things don't work.

4.1. IP FOCUS

What business are we in? Read your business plan or stock offering. If the technology under consideration is not at the heart of your business, expending time and money on a patent application is ill advised. First, technological judgement and market savvy drop off logarithmically as the distance between the discovery and the main business increases. Second, the ability to enjoy the profit potential of a patent is substantially based on being the source of the product or process employed. If the idea is a great one for some other business, that is a bad sign.

The "our business" test should be an expansive test considering the vertical aspects of your business. Will the discovery be the subject of further inquiry in the 12 months following conception? If the answer is "no," the technology is probably not central.

Be realistic. Breakthrough compounds emerge only rarely. In contrast, advantageous process and delivery technology improvements emerge more often and with a higher likelihood of commercial usefulness. Thus, super-producer clones, essential filtration steps, and dosage form architecture should not be overlooked as sources of IP. A patent at a process bottleneck (read "choke point") can offer market position, trade goods for other essential technology, an entree to cooperative relationships, and a royalty stream.[1] In fact, in a crowded field, it is often reasonable to maintain exclusivity by assembling a collection of tiny impediments to duplication and

1 Rats form the major component of the tiger's diet in the wild. They are easy to catch and, in quantity, provide for most nutritional needs.

encroachment in the form of "picket patents." With enough patents in an area, "me-too" producers may be dissuaded from entering a market unless the total market rises above perhaps $50,000,000. There is a patent litigation adage that say if one is being sued under multiple patents it is most unlikely to avoid all the claims of infringement. And a single claim of infringement will carry the day. Thus, a multiple patent arsenal is particularly protective of a commercial position.

4.2. THE LEARNING CURVE

Biotech patent applications are often filed at very early stages following conception. This can mean filing even before the first mouse has been dosed. In the US, a patent is typically filed on limited early data and then repeatedly refiled as continuing data arise. There is a one-year period between the US filing[2] and worldwide filing.

4.3. THE STAR WARS TEST

If the invention works as planned, do you anticipate that multiple Nobel Prizes would be awarded? This is nature's way of telling you that the likelihood of success is limited in a field accustomed to limited success.

4.4. IS THERE A MARKET?

4.4.1. Money

If the primary indication is a small population, a patent might not be justified or required. Orphan drug designation may be a cheaper alternative. Also, if the market is small enough, how attractive will such a market be to the generic houses? Below about $25,000,000 annual sales there may be no reason to seek a patent.

4.4.2. Perceived need

Does the world know that this technology is required? Such a threshold question was answered affirmatively as to NMR and liposome drugs. Betaseron for the symptomatic treatment of MS has run into marketing problems because MS patients are not prepared to undergo early severe side effects, whatever the benefit. In contrast, side-effect resistance does not limit the market for cancer chemotherapeutic agents.

[2]Under some circumstances, a second European application can be filed prior to a point 18 months from the US filing date. As this is not a treatise on patent prosecution, some nuances of timing and law will be simplified. For more specific answers, contact patent counsel.

Patentable in the US

Treatment of Humans

In general, and while not without an undercurrent of objection, there are few public policy restrictions on the patentability of genomic inventions. US courts have, for some time, accepted the patentability of life forms. This past year however, the Patent Office felt the need to prospectively proclaim the unpatentability of human/non-human chimeras. As a basis for proscription, the Patent Office cited public policy and morality.

In European practice, it is not possible to claim a method of treating humans -- "*en clair*". In the US, such claims are possible. A typical US method of treatment claims might be:

> *A method of treating a human subject in need of such treatment by administering to said human a therapeutically effective amount of Drug X.*

This may be a distinction without a difference. In European practice the same result is achieved in the form:

> *A use of Drug X for the manufacture of a medicament for therapeutic application Z.*

5. Points of decision

5.1. PATENT COMMITTEE

It is useful to establish a patent committee that meets, perhaps, quarterly (but certainly not more often than monthly). Representation on the committee should include Research, Development, Business Development, Marketing, and Patent Counsel. Each inventor should be invited to plead his or her case.

5.2. RATINGS

*5.5.1. **A** = File immediately*

The invention is at the core of our technology. We intend to make a product embodying this technology. Work is being done on this technology now or will begin immediately.

A' (a rare ancillary consideration) The invention covers a technological cusp which we are *sure* that a competitor must traverse, but which the competitor may have overlooked. [This sounds plausible but never pans out.]

*5.5.2. **B** = Review in six months*

This could prove interesting if more data were available. It is not at the heart of our business. At the review point there will only be more data if the proponent has been able to find the time, money, and personnel. If this has not occurred, the corporation has voted. The disclosure can be rolled over for additional periods or await other events such as the failure of a preferred line of inquiry.

*5.5.3. **C** = Indefinite hold*

The idea appears clever to non-experts. It is not germane to our business or to our competitors. The invention can be revisited if circumstances change.

It is important not to inhibit the willingness of the scientists to offer ideas by rating the ideas as "unusable, forever, under any circumstances" (though a smoke detector with a snooze alarm comes close).

Patentable in the US

Computer Algorithms

Algorithms are patentable in the US. Drug discovery, drug therapy, drug screening, and drug manufacture have become entwined with computer algorithms. High throughput and combinatorial chemistry are creatures of computer methodology. With some ingenuity, it is possible to delimit pharmaceutical methods, in part, by reference to algorithms. Now, algorithms can form an additional element of patent protection not to be overlooked -- whether as an offensive or as a defensive patent strategy.

Almost any transformation of data can be cast into the patentability requirement of a "useful, concrete, and tangible result." For example, a gene chip being read for hybridisation according to a particular mathematical protocol would offer a patent opportunity. It is likely that a dosage regimen of some complexity offers an opportunity for patent protection as well. A responsive approach to this development is to make note of any computation entailed in the manufacture and use of a drug or diagnostic for patentability consideration.

5.2.3. Hard financial facts

Basic exemplary calculations include attributing about $40,000 to $50,000 in the first 4 years for each maintained application that is filed internationally.

Presuming 10 patent applications are filed yearly:

First year costs: $100,000 -- ($90,000 legal, $10,000 patent fees)

This covers $10,000 for a totally new application, and rather less for a similar application, and some costs. For an accelerated delineation of the likely patent position, filing in the US from the outset may be indicated. This can be accomplished as a US filing without filing initially in the home country, or as parallel initial filings.

Second year costs: $165,000 -- ($110,000 legal, $55,000 patent fees)

In the second year, filing 10 new patent applications as before will cost another $100,000. Perhaps 5 of the earlier patent applications will require responses to the Patent Office adding another $20,000

Entering the PCT at the one year mark on the first 10 applications will add about another $30,000. Seven months thereafter (at 19 months) one must file a demand for preliminary examination at about $1,500 each, or $15,000.

Third Year Costs: $405,000 -- ($150,000 legal, $255,000 patent fees)

In the third year, filing 10 new patent applications as before will cost another $100,000. Perhaps 15 of the earlier patent applications will require responses to the Patent Office adding another $60,000
As before, entering the PCT at the one year mark on the next 10 applications will add another $30,000. Again, at the 19-month point the demands for preliminary examination add $15,000. And now at 30 months, entering the national stage on will cost perhaps $200,000.

In the fourth year it is reasonable to expect some patent applications to be supplanted by later applications, while others are abandoned outright. There may be a levelling off of the necessary responses to the Patent Office, but the responses may be more difficult, and hence more expensive. Offsetting any reduction, however, will be an increase in the need to respond to patent actions in the EPO. These tend to mirror responses in the US, but are not without cost. Adding up the foregoing and assuming $330,00 for the fourth year, the 4-year total is $1,000,000.

From experience there is an initial wave of filings in the first six month of corporate life. Thereafter -- and with considerable variability -- patent applications begin to emerge from research projects about 18 months after wet lab begins. At that point five or six process and product applications will be suitable for filing. Surely, the technology of your corporation will offer its own peculiar maturation time. However, once this time is known it will generally repeat forming a useful guideline for budgeting purposes.

6. Its just business

In a proprietary-obligate industry, the patent position being sought is, in many respects, no more than a reflection of corporate direction. If a start-up biotechnology or pharmaceutical technology company is sufficiently confident in the potential for success of a technology, and sufficiently committed to devote the resources to develop this technology, the patent position will follow.

6.1. WHAT IS IT REALLY WORTH TO DEVELOP AND MAINTAIN YOUR PATENT PORTFOLIO?

- Question. How do you exploit its value?
- Answer. It depends.

6.2. THE ONE TRUE ANSWER.

- Nothing in life is simple.
- Decisions must be made (really).
- The points of decision are subjective.
- Making the best decision is a matter of luck and don't let anyone tell you different.
- Making a reasonable decision that maximises the potential for a favourable outcome is a matter of skill.

Based on the factors above, a few helpful hints are offered (and life is no more certain than that).

If you really believe you're an extraordinarily high-tech company, then you're almost certainly spending a lot of money and taking risks. So, the conclusion is that you shouldn't take such risks without high likelihood of broad patent protection, right? Well, yes and no. In pharmaceuticals, yes! Yes, because (1) the product will probably be on the market for many, many years; (2) your FDA filings make it fairly easy for people to copy you with all the time they'll have; and (3) the markets are big enough that people will make money even with a small share if they've piggybacked on your favourable outcome without the risks.

Interestingly, while broad patents are generally better than narrow ones, a narrow patent can be valuable in this field.[3] For example, while even a slight change can avoid a narrow patent, e.g. replacing ethyl with butyl, it will still send the competitor back to the beginning in the development, clinical and US regulatory maze, and add to the development cost and development risk.

Contrast this with a high (but not quite as high) tech field like diagnostics. The markets are smaller, and the life cycles shorter. Odds are that a narrow patent doesn't keep out much competition. In addition, the years it takes to complete a major patent dispute, the products in question will be obsolete well before your lawyer has finished piling paper on the judge. It often makes more sense to spend the money on the next innovation instead of defending a patent. In businesses like diagnostics or software where the product/technology turnover is very rapid, speed, flexibility, and always getting there first with the next generation can be worth much more than a patent. Of course, there are exceptions. Some very broad patents exist in these fields too, and often available for license.

NB #1: Avoid patent directed experiments. Perform only those syntheses and experiments required within a development project. It is generally a waste of time, money, and effort to perform additional syntheses and experiments with the sole intention of generating data for expanded patent protection. In pharmaceuticals, most primary compounds fail. Thus, casting a net to cover the mere possibility of a secondary compound is ill considered.

Realise that in deciding what to patent and whether to go forward without one, the competition goes through the same thought process. If you don't want to go forward without a patent base, others won't either. That's why there are some great ideas that companies don't pioneer. Aspirin for heart attack prevention, for example. A small company would be unlikely to spend money on a claim to that cardiac indication when the customer could use any of the many generic substitutes already available. Of course, in the US a large producer has expended the effort, but backed it up with a huge marketing push. In most instances "method of treatment" patent claims are only valuable when you can police use of the necessary drug.

---

[3]A "broad" patent has a broad main claim. This does not exclude quite specific "picture claims" exactly coinciding with the product to be marketed.

NB #2: Policing uses. A method of using an OTC like aspirin can't be controlled. If, however, the method were to a novel use of a drug with only a hospital application such as doxorubicin, policing would be an easy matter.

This leads to the "under the radar" approach. Without a patent, go for a market that's big enough for you, but not big enough to attract the sort of big competition that needs a patent. Circular, but profitable. Example: A $10 million pharmaceutical market. Who's going to bother going for it with no patent anyway? Get there first, and secure a niche market.

6.3. NICE PACKAGE

A first objective of any patent strategy is turning what you have into a licensable package you can market to a partner.

6.3.1. Human pharmaceuticals

No major pharmaceutical house puts serious money into anything unless they see the likelihood of a strong patent position. In large part, their analysis leads them there, and their huge fixed cost reinforces this position, but this is also a deep-rooted cultural practice that is unlikely to change. It would be a daring executive indeed who risked a clinical failure when the up side was only $50,000,000.

The elements of a licensable package are

- Good basic underpinnings of science
- Data -- Well designed experiments that truly illustrate your point *in vitro* and, if possible, in animals
- A clear articulation of how the science translates into products
- A simple story on why the all-critical technology is novel and why you will have protection in the market, and not just a patent.
- Patent applications rather than issued patents have a potential that is yet to be defined (and cannot be dismissed). Issued patents are often less saleable.

6.3.2. Windage

Faster moving technological fields such as gene therapy have less defined patent requirements. This is not to say it does not have patent requirements. Gene therapy has pharmaceutical aspects in long development times, but vector and tool technology is changing daily. As a result, the potential value of the broader gene sequence patents and patented use of specific genes is undefined. If one is committed to such a technology, portfolio bulk is an important consideration. The hope is that some aspect of your proprietary technology will be useful for leverage in reaching any required accommodation with a competitor.

6.3.3. Exclusivity

Pharmaceutical players always want it. But diagnostics companies and others often settle for non-exclusivity. Ask yourself this: Does having all of the market for the few uses you can market on your own translate into more or less than a small piece of a truly vast market? Many, perhaps most, inventors want control and exclusivity when, in fact, reasonably cheap non-exclusivity to any and all comers is most profitable. One favourite example is the Dolby noise reduction system. Dolby could never have overcome the Sony's and Panasonic's making cassette player-recorders. But at about a 1% royalty on a non-exclusive basis, even the sleaziest knock-off artists in the world will take a license and pay the royalties. Following this patent strategy, Mr. Dolby is now richer than most biotech executives are.

6.3.4. More patent strategy, packages II

It's not rocket science to identify and assemble a licensable patent/patent application package. But it is sometimes cosmology, astrology, pure guesswork, or intangible management skill as to what a BigPharma executive will buy from you and when. No matter what you have, it must appear to be something on their list, and it must arrive when they are in the mood to make a deal. Timing counts and luck counts double.

It helps to have that certain "je ne sais quoi" which is French for how is it that Ligand, Onyx, Arris, and Millenium can do this again and again and again and my product is years closer to the market and I can't get a partner? Other companies have great science and great people too, but the aforementioned seem to have an exceptional ability to turn it into the big deal. Take heart: simple products ready for market are rarely bought by big licensors, but they can make you a very nice living.

Discussion question
What if you had a truly broad, valuable, and needed invention in an oligopoly industry? For example, dog and cat vaccines is about a 4 or 5 player business worldwide. You need a broad line and you need volume to be a player. One great vaccine can help sell the rest of your more ordinary line. Suppose you invent something really great, like a vaccine to prevent a big, common, serious disease. Consider selling non-exclusive licenses: the first one to pay up gets it for $100,000 and 2%. The second one to pay up is $500,000 and 4%. You overslept and last to buy it? Your price is $1 million and 10%. It may be worth it, because you can't afford to be the only one left out of the pool.

7. Filing a patent application

7.1. INPUT FROM THE INVENTOR

Filing a patent application requires providing the patent attorney with information. This is not a simple task, and an inventor must be prepared to commit the necessary time to adequately collect and supply information. Preparing a patent application is not like stopping by the photo shop to pick up developed pictures from film you dropped off earlier.

All of the available data should be provided. A full description of what the invention does, how this is accomplished, everything it might do, and the significance of the discovery is necessary. Often, this requires that the inventor perform calculations of ranges or working conditions, or calculated distinctions over the prior art.

> Housekeeping Tip #8
>
> Every possible source of invention -- researcher, post doc, technical staff, regulatory supervisor, university consultant, etc. -- must be obligated in writing to assign all rights in any inventions. They must also be obligated to assist the company in obtaining and perfecting patents. It is not unheard of to later decide that one contributor to an invention is not properly a co-inventor. Without such agreement (and often with such agreement) requiring them to sign off on the removal of their name from a patent application can be problematic.
>
> **Rigorously obtain and maintain agreements to assign and co-operate.**
>
> A patent application is NOT a scientific paper. This is noted because the more cautious inventors -- in applying the standards associated with scientific publications -- are reluctant to speculate. In patent filings the inventor is encouraged to make suppositions and extrapolations which include elements not actually established. Imagined experiments may be presented in the present tense, while actual experiments are presented in past tense. The first draft of an application is generally prepared in conjunction with an interview by the patent attorney or patent liaison. Thereafter, the inventor should review and comment upon the application prior to the final draft.

> Housekeeping Tip #8.1
>
> "Every possible source of invention" further includes cell lines, genomic material and pathogens identified in clinical subjects. The Informed Consent should contain ownership provisions for unique tissue so obtained. Agreement at the outset can avoid later dispute. European regulations are not yet clear, and this may not be possible for EU trials.

7.2. MORE INPUT FROM THE INVENTOR

Failed experiments and contrary information must be provided. Hiding such information merely provides grounds for invalidating a patent should one issue and the product succeed. The best mode of practising the invention must also be provided.

Patent sophisticated inventors know the advantages of being forthcoming (and make the patent attorney appear infinitely smarter). Unsophisticated inventors are hazardous to patent and patent attorney health. You've seen it on the crime shows, and believe it here: It is a mistake to lie to or mislead your attorney.[4]

[4]Of course, having been down this road, we rarely believe anything without extensive grilling.

7.3. COMPILE ALL RELEVANT ART

7.3.1. Why

A European patent applicant has no specific requirement to inform the European Patent Office of significant prior art. In the US there is a duty of absolute candour. In US practice, patent applicant -- inventor and company -- as well as patent counsel have a clear obligation to inform the Patent Office of any relevant art which is material to an invention of which they are aware. This is done in the context of providing an Information Disclosure Statement and, as necessary, supplementary Information Disclosure Statements. Technically (and isn't that why we are here) the required information is information published prior to the filing date of the application. However, as noted before, be forthcoming with patent counsel.

Wholeheartedly fulfilling the disclosure obligation is a particular responsibility of the inventor. Failing to fulfil this obligation can result in an invalid or unenforceable patent. The two best reasons to rigorously meet the requirement are (a) if there is *any* way to distinguish your invention over the prior art causing concern it can most easily be established in argument during patent examination[5], and (b) once the PTO has passed on a reference it is dead as a validity problem. A corollary is that, recognising the litigation prone nature of US biotechnology, any reference not disclosed will be characterised by an opponent as a clear effort to mislead the Patent Office and defraud the public. As Europe permits patent oppositions prior to patent issue, significant references will likely to be considered in Europe, even without a Duty of Disclosure.

Finally, as to housekeeping matters, make at least five notebooks of the art you compile. Additional sets are always required somewhere down the line. This task is made easier by maintaining a current company-wide filing system/database of references. Computer programs such as "Endnote" are available for this purpose. The alternative is searching each office for references with the filing of each patent application.

7.3.2. Yes, everything

In fact, it can be quite a substantial job to thoroughly present all relevant art of which one is aware. Consider art

7.3.2.1. Cited in related applications. Always present all art cited the application being filed and in related applications (including international versions).

7.3.2.2. Materials consulted in application preparation. Any journal articles or patents specific to the new application and consulted during its preparation must be considered for presentation to the PTO.

7.3.2.3. Reading files. Most scientists keep a rather extensive reading file in their area of research. These articles have typically been gathered over years and occupy several

[5]If you can't beat up the PTO, who can you beat up?

file drawers in the inventor's office. The files must be consulted, and relevant papers selected. It is a good practice to stand there while the scientist/inventor goes through the files. When you do, listen closely for those "*sotto voce*" remarks like "I don't know about this one?" React, seize the reference, and over include. [☞ It really does happen just this way.]

7.3.2.4. Computer runs. If a database was consulted and a number of abstracts reviewed, a copy of the computer run must be presented.

7.3.2.5. Company library. If there is a central patent and articles file for the entire company to which the inventor has access, some consideration must be given to what the library contains.

7.3.2.6. Don't under present. An inventor must be cautious and over present to patent counsel. Failure to do so looks like concealment, while complying effortlessly shifts responsibility and later blame to counsel. Its easy, it's fun.

7.3.3. When

The prior art need not be submitted with the application. It can be submitted up to about 3 months later, and is presented as an Information Disclosure Statement. Somewhat later and for a few months, it can be submitted along with a $200 fee. But who's kidding whom? If it's not done immediately, when do you think anyone will get to it?

7.3.4. Searches

It is not a usual practice to search the literature in advance of filing a patent application for the purpose of determining if subject matter is patentable. It is simply not cost effective. First, your researchers in the field are aware of what is known in their own field. Second, as a company is filing multiple applications in a single technology, the usual obstacles are usually known. Thus, if the composition or method of interest is clinically effective, and the identical product or method is not in use, it is likely that there is a patentable aspect to the effective approach, and the search step will not alter this.

8. The prior of prior art

8.1. IN THE US

Prior art is the public use or sale or the publication of aspects of your invention before your earliest filing date for the relevant subject matter. Thus, if in Patent Application A you disclose Composition A, and the application is filed February 1, 1996, disclosure of Composition A prior to February 1, 1996 is prior art. If, on August 1, 1996 you file a continuation-in-part application ("CIP") of the original application adding new matter.

The addition of new matter is the defining characteristic of a CIP. In the present example the new matter of the CIP is the disclosure of Composition A'. The CIP is subject to prior art existing prior to August 1, 1996 as to Composition A', but for Composition A, the prior art date remains February 1, 1996. As explained more fully below, there is an opportunity to "swear behind" a reference that is within one year of your filing date.

The effective date of a publication is calculated from the date the publication first arrives in the hands of a member of the public, and not the date printed on the publication. For example, a newsmagazine like Time is published, perhaps, a week before the date listed on the magazine, but the effective date is the date it hits the newsstands.

The effective date of an issued patent acting is its earliest filing date; not its issue date. Of course, if the issued patent is a CIP of earlier applications, one must look to each parent application to determine the earliest date at which specific new matter is disclosed.

Watch out for the Petard

International patent applications -- including yours -- are published at 18 months. Twelve months after such publication (thus, 30 months out) your own published patent application will act as prior art against the new subject matter of any CIP you have filed after the 30 month point. Remember, you cannot "swear behind" a reference over 12 months prior to your effective filing date.

8.2. NON-US

There is no CIP practice outside of the US. Thus, in Europe and elsewhere one is regularly faced with distinguishing your early patent application from the later CIP applications. While this is frustrating for obtaining patents, it presents a cost saving opportunity. If the first application is allowed, and in truth the distinctions between Composition A and Composition A' are slight, drop the second application and save the money.

9. Making U.S. filings/incurring

9.1. APPLICATION PREPARATION

9.1.1. New applications

To properly prepare a patent application, a patent attorney must devote about 6 to 8 weeks to drafting, at minimum. This minimum time period presumes that the inventors make themselves available to respond to questions, supply additional information as requested, and promptly read and comment upon drafts of the application. Legal fees for preparation of a biotech application start at about $10,000. There is a basic filing fee of $385 for an individual, non-profit, or small entity and double that for a large entity. Additional fees are required when a basic number of claims is exceeded, but in the U.S. this is usually only a few hundred dollars at most. In Europe the excess claim fees are very large, but different claim provisions permit a reduction in claims for European filings.

Patent Access

Including drafting, a biotech patent for can be obtained for about $30-50,000. This covers 250,000,000 people and a rich market. In comparison, obtaining a patent for all 13 countries of the EPO can run about $70,000 without patent drafting. An EPO patent covers about 500,000,000 people. [A patent in Japan runs about $70,000 for 125 million people.]

9.1.2. Provisional applications

U.S. law now provides for the filing of a "provisional" patent application. A provisional application differs from the usual application in that there are virtually no formal requirements to obtain a filing date. Within one year from filing a provisional application, the provisional application can be converted to a regular patent application. Conversion of a provisional applications to regular applications must be done within one year, and cannot extend to the next business day if the anniversary date falls on a Saturday, Sunday or Federal Holiday. The filing fee is $75 for a small entity and $150 for a large entity. If there is neither a US inventor nor US corporation involved, Provisional Applications are uncertain because, when the PCT application is filed it cannot be initially filed through the US receiving Office. Then, as the PCT application is not derived from a regular application may have a questionable priority.

9.1.3. Later applications

U.S. patent applications are regularly refiled exactly as originally filed, or refiled only slightly amended. Refiling occurs when one has not obtained allowed claims during the limited periods in which one can argue the merits of the application, amend the claims, or offer additional evidence to the Patent Office. In many instances, the Patent Office finds that multiple separate inventions are being sought within a single application. In such a case, the Patent Office enters a restriction requirement forcing the applicant to elect one invention. The non-elected inventions are then pursued in separate applications.

To keep up with the jargon, a re-filed application is called a "continuation" or "straight continuation" application. If some claims were not examined in the original application, the continuation is called a "divisional" application. If, at the time of refiling a continuation application, one wishes to pick up prosecution where the parent application left off, the filing is called a "file wrapper continuation" or FWC." If any new matter is added, the continuation application is called a "continuation-in-part" application ("CIP").

9.2. PATENT PROSECUTION

9.2.1. First matters

After a patent is filed, the applicant awaits action from the Patent Office. There is often some early activity as to ministerial matters, which incur minor costs. Such matters include the filing of a declaration by the inventors that they invented the subject matter of the application. It is not unusual for a declaration to be omitted in the last minute rush to file. Similarly, an assignment of the invention to the company may come later, as will formal drawings. Formal drawings are figures prepared by a draftsperson according to Patent Office rules.

9.2.2. Information disclosure statement (cont.)

As noted above, one item early item is the Information Disclosure Statement, which is to be filed within about 3 months after the original filing, if not with the original filing.

9.2.3. First office action

Under normal circumstances, and depending on the backlog in the section of the Patent Office to which your application is assigned, a first Office Action on the merits for a new patent application will come in about 12 to 18 months. A continuation or divisional application may receive a faster first Office Action. There are provisions whereby an application can be "made special" and advanced more quickly through patent examination. Given the regulatory delays attendant to pharmaceutical production, this seldom makes sense. Nevertheless, an issued patent may be of interest to unsophisticated inventors, or if a competitor is about to market a product your patent could block.

Except in unusual circumstances, applications are presented with claims to insure an initial **rejection** of the broadest claims. The thinking is that if the applicant secured a first Office Action allowance, and did not "over-claim" then the invention had not been claimed broadly enough at the outset.

9.2.4. First response

A response to an Office Action is due three months from the mailing date by the Patent Office. A response can be extended in monthly increments up to six months from the mailing date of the Office Action (that is up to 3 months of extension). The Patent Office motto is "If you've got the money, we've got the time." While a one-month extension is only $55 for a small entity, the full three months costs $465 (and double for a large entity). Legal fees for responding will run several thousand dollars.

At this point it is possible to amend claims and to offer supporting data. Later in prosecution, this will not be possible. It is also possible to interview the Examiner and urge the merits of the application. Interviews can be via telephone or in person at the Patent Office. In some instances, an inventor may participate in the interview.

For the best results, your inventors should be available to assist counsel in responding. In fact, this should not prove particularly onerous. A portfolio of patent

applications involves substantially the same references. Thus, after the first few responses in an area, all the art applied by the Patent Office will be known and the distinctions over the claimed subject matter, well defined.
Occasionally, more will be required of an inventor or corporate technical staff.

9.2.4.1. Swearing behind. In some instances, to avoid a reference with an effective date of publication within one year or less from the filing date of the application, an inventor will be asked to "swear behind" such a reference. In practice, this means that if the patent were filed on April 1, 1996, and the reference were dated March 1, 1996, then the inventor is permitted to file a declaration, and likely, some pages of a lab book, establishing that the information in the reference was known by the inventor prior to March 1, 1996. If the reference is over one year prior, there is no swearing behind

9.2.4.2. Expert Opinion. A declaration can be filed under other circumstances as well. When a calculation can be performed that distinguishes claimed subject matter from a reference, a declaration is the better way to offer such calculations (as distinguished from "mere attorney's arguments"). For example, a reference may state a particular concentration of reagents, but not state the pH, while the rejected claims are limited by pH. An inventor or other expert could offer a calculation that establishes that the pH of the reference could not overlap the claimed subject matter.[6] Similarly, an expert could offer an opinion as to what "one skilled in the art" (the standard measure) would have understood from a reference or a term. If a relative term such as "high dosage" appeared in a reference, and the claims were drawn to a some μg/ml or ng/ml dosage, would that have been understood by one skilled in the art as of the filing date to include the claimed subject matter? An expert could offer the view that these amounts were orders of magnitude apart.

9.2.4.3. Specific Comparative Experiments. It is possible to provide comparative test results as a means to establish patentability. This is a less favoured course for several reasons. In general, comparative experimental evidence is highly peccable. While the Patent Office may be satisfied, it offers a substantial point of attack in litigation. Also, the result may be other than what was hoped for.

There is also the question of the need for any specific comparisons. In biotechnology, if the drug works, and it has not been produced before, and the FDA is going to be satisfied, then the data required to establish patentability is out there or will be prepared in the normal course of business. Await this data. Doing experiments to satisfy a patent attorney is usually counterproductive and an inappropriate use of corporate resources.

---

[6]One must be cautious, as an error in calculation could be viewed as an attempt to mislead the Patent Office.

9.2.3. Further office actions and responses

The next Office Action on the merits either allows some claims, or (and more usually) presents a Final Rejection. A second non-final Office Action may be sent. If the second response is not a final rejection, it usually means that the applicant's arguments prevailed, but the Examiner is citing new art as the basis for rejection. If final rejection is received, there are several options.

While six months to respond is available (with extension fees); the first option is to respond within two months offering further argument. A quick response requires the Examiner to promptly issue yet another Office Action. Otherwise, the Examiner need not respond in a timely fashion. At this point, the ability to amend the claims is rather limited. The only amendments usually accepted are those to bring the claims into accord with what the Examiner will allow. New claims and new data will not be considered. Also, the Examiner can always be contacted. While the Examiner is not required to discuss an application under final rejection, a refusal to discuss an application is rare.

Whenever the final rejection is truly final, the prosecution options are to refile or to appeal to the Board of Appeals. An appeal comprises filing a Notice of Appeal, and, two to six months later, an Appeal Brief. Thereafter, there is the opportunity for a Reply Brief, as well as the option of oral argument before the Board of Appeals. The Board of Appeals is comprised of three senior Examiners. Given the variability of patent Examiners, the technical expertise of the Board often leads to a reverse of illogical final rejections.

The briefing required for an appeal can run about $7,500-15,000, (if the briefs are actually filed). An appeal will not be decided for several years. In most instances, and with additional data usually in the offing, the more effective course is to refile the patent application. Selecting between appeal and refiling is fact specific. Assessment can only be made with attention to the patent application, the actual rejection, the data available, and the market situation at the time of decision. In any event, the decision to appeal can later be countered simply by dropping the appeal and refiling. As noted below, when data is expected in the near tem, extensions for responses and filing notices of appeal can be made for reasons of increasing the time available.

9.4. MAINTAINING PENDING APPLICATIONS

No matter how many patents have issued on your technology, it is important to keep an application pending in the US on subject matter that will become a product. If a patent has issued and there are no pending applications, the options to modify an issued patent are limited to either patent "re-examination" or "re-issue."

9.4.1. Re-examination

"Re-examination" is a process that permits the US Patent Office to review issued patents in view of newly presented prior publications. Previously, the law permitted anyone -- even anonymous "third parties" -- to request re-examination. Under the old system, after requesting re-examination, a third party had little voice. As a method of attacking a patent, the chief virtue of re-examination was that it was a great bargain as compared with patent litigation. Under the new law, a party opposing a patent by re-examination has a voice in the entire proceeding.

Europe has longs had a system for opposing patents before the European Patent Office. For nine months after a patent issue in Europe, anyone can oppose the patent. The change in US law brings something like the opposition system to the US

Under the new US re-examination system (i) anonymous third parties are excluded, (ii) the opposer can argue against the patent at every step, and (iii) the results of re-examination are binding on the parties in any later infringement suit. This provision effects patents filed after November 29, 1999.

Without regard to imminent litigation, a patent holder is permitted to request re-examination if the patent holder finds a significant reference not previously considered. Re-examination is, at best, cumbersome. If there is a pending application, it is a simple matter to cite the reference to the Patent Office and argue it in the context of that pending application. While this does not absolutely dispose of the reference as to prior issued patents, a successful distinguishing argument in a pending application will clearly defuse the reference as a point of weakness in the issued patents.

9.4.2. Reissue

In an issued patent claims cannot be easily expanded, but, insofar as permitted, this is accomplished by "reissue." Claim expansion is sought when the patentee discovers that, through inadvertence, less has been claimed than might have been claimed. Such inadvertence often arises right after seeing a competitor's product. Even the limited right to increase claim scope is absolutely ended two years from issue.

9.4.3. Pendency

Maintaining a pending application is largely a matter of refiling the application and including new subject matter, or, while waiting later developments, pursuing rather narrow (but not frivolous) claims. Given six months to respond to each Office Action, and the availability of an additional six months based on a notice of appeal, a follow on application can run at least two years with minimal expense.

> <u>Hiring in-house Patent Counsel</u>
>
> When your costs for patent attorney bills for prosecution exceed about $400,000 per year (excluding patent fees) you might start thinking about hiring in-house patent counsel. However, a single patent attorney will not likely lower your bills. The real saving will come with patent staff of, at minimum, two attorneys and a secretary/paralegal. With one attorney in a start-up, at least half of the time will be devoted to non-prosecution (and likely non-patent) matters. The maximum docket that a single patent attorney can maintain is about 30 pending U.S.

10. Timing

10.1. U.S. APPLICATION FILING

The United States offers a one-year grace period from first publication, use, or sale of an invention within which to file for a patent. Within that one year, an inventor can file a patent application with no loss of U.S. rights. After one year, no patent is possible. Other than Canada, the rest of the world does not offer any grace period. The world system is one of "absolute novelty." Publication, use, or sale of an invention prior to filing a patent application eliminates the possibility of international patent protection. Filing an U.S. patent application prior to first publication, use, or sale is sufficient to preserve international rights. In those instances in which the European rights were destroyed by untimely disclosure, it is still possible to apply for a patent in the US. The grace period is also useful with "iffy" inventions. Given the availability of one year in which to manufacturing and market an invention, it may be possible to determine if a real market advantage exists, prior to investing patenting costs. Such a scenario is more likely with near term products such as diagnostic and screening methodologies. Note that U.S. patent applications have previously been maintained (more or less) in confidence and not publicly disclosed until issued as a patent. US applications filed after November 29, 2000 will be published at 18 months from the earliest priority date unless the US applicant agrees not to file international counterparts. In the rest of the world, patent applications are published 18 months from the earliest priority date.

10.2. ON-SALE BAR TO PATENTABILITY

Europe limits patentability to an "absolute novelty" standard. The US offers a one-year grace period after disclosure use or sale of an invention in which to file a patent application. Europe and the US, however, apply different criteria for acts which place an invention "on sale" The surprising result of the differential is that an invention might be unpatentable in the US by reason of exceeding the one-year grace period for patent filing, but remain patentable in Europe under the seemingly more strict absolute novelty standard.

10.2.1. Out source disaster

In the United States, after an invention has been "on-sale" (or made public) for more than one year, it is not possible to obtain a patent. The "on-sale" bar is an absolute to patentability. Of late, the courts have been defining "on-sale" quite expansively. Under recent judicial constructions of "on-sale," the "sale" of a novel compound or compound made by a novel process can be found within the transaction between a contract manufacturer and the contracting party. This "sale" starts the one-year clock.

The sale/non-sale determination may turn on such housekeeping issues as invoicing and accounting of payments between the contracting party and the manufacturing party. If the contract manufacturer is being paid to research and develop a process, or by "time and materials," it may be arguable that there is no "sale" taking place. If invoicing is per unit (weight, batch, etc.), there may be a sale. This seems particularly problematic when ordering a nucleotide or protein from a supply service. Even when a transaction is "confidential," the contracting party places an order for a specific novel nucleotide or protein, and the exact composition is created and placed in commerce – often over night. This looks like a sale.

One response to the on sale worry is to labour over each out-sourcing agreement and have excellent trial counsel. Another is to file any likely patent application within the year. To effectively respond to the danger of an unrecognised premature "sale," keep patent counsel apprised of all out source agreements *and results*. Patent applications tend to be easier to draft if the patent attorney (i) knows of the invention, and (ii) has fixed an outside date for patent application filing.

10.2.2. Concept offered for sale

Under US law, an invention can be "on-sale," even if the invention is yet to be reduced to practice.

A case recently decided by the Supreme Court dealt with an inventor who designed, but did not build, a new computer chip socket[7]. Drawings of the socket were sent to a manufacturer and shown to Texas Instruments before the critical date one year prior to filing the patent application, but no manufacturing was begun until after the critical date. The Supreme Court granted review specifically to address the question of whether there can be a sale prior to a reduction to practice. The answer is yes.

The Court read the patent statutes as making no requirement for reduction to practice. Thus, it reasoned, an invention was actually invented when drawings or other descriptions had been prepared "that were sufficiently specific to enable a person skilled in the art to practice the invention[8]."

Consider the situation of a protocol for series of gene manipulations to yield a particular therapeutic or diagnostic result. If the protocol were complete to the point at which a post-doctoral fellow or a clinician ("a person skilled in the art") could follow the protocol and obtain the result at least some of the time, any commercial activity after that date could begin the one-year clock. Note, too, that even if the method by which the effect of the invention is to be achieved remains secret, a sale can be consummated. Thus, an offer to a laboratory to provide a diagnostic genomic screen, if offered at the time at which a written protocol existed to accomplish that particular screen, could well be on-sale activity under the current rule.

The safe course is to avoid any disclosure or commercial activity in advance of filing for a patent. Such "premature" patent application efforts are a necessary trade-off for obtaining secure patent protection, and may necessitate filing supplemental patent applications later as the imagined protocol encounters laboratory realities.

---

[7] Pfaff v. Wells Electronics, Inc., 48 U.S.P.Q. 2d 1644 (1998).

[8] Pfaff, 48 U.S.P.Q. 2d at 1642.

10.3. INTERNATIONAL APPLICATION FILING

10.3.1. Priority dates

By various treaties, the US will accept international filing dates and the rest of the world will accept the US filing date as the priority date for applications filed within one year of an original application. To file internationally, the basic choices are to file under the Patent Co-operation Treaty ("PCT"), or to file directly in a specific countries and/or regions. In particular, a US filing or a filing in any country of Europe it is possible to file directly in the European Patent Office ("EPO") designating 17 countries in Europe (including all the major countries).

10.3.2. Internationally file the CIP

The vanilla US based company patent strategy is to file a US application on work actively underway, and then, just prior to one year from US filing, update the US application in the form of a CIP. It is then the CIP which is simultaneously filed in the US (replacing the original US application), and worldwide. A US first procedure (or home country and US) is also available to international applicants.

10.3.3. Filing costs

Filing in the PCT currently costs about $164 for each designation up to 11 designations, and peaks at about $3000 total with additional fees to designate the entire world including transmittal fees, yearly fees, and other basic fees. There are even additional fees for applications in excess of 30 pages. Ultimately, designating the world is just a holding action pending getting a major partner. This is because at 30 months one is faced with paying to enter each country, which generally makes no sense.

In the PCT, by, at most, 30 months, an application must enter the national stage. At that point, the PCT application ends and the application exists only in each country or regional patent office (*e.g.*, EPO). At the 30 month point, the cost to enter the national stage in each country of Europe is about $15,000 total. Entering Canada is about $1,500, Japan about $4,000, and Australia about $2,000. In the end, anticipate paying $70,000 for Europe and $50,000 for Japan through to issue.

10.3.4. The EPO option

One option is to avoid the PCT and file directly by region or country. This makes sense when the countries of interest are the usual suspects. In our view, this means the EPO, and Canada, with Japan and Australia as likely additions. The distinction between PCT and direct filing is that, while expenses incurred at 30 months in the PCT are incurred at 12 months by direct filing, in most cases the initial expense is lower with direct filing. One clear advantage is that by avoiding the PCT one avoids a number of procedural pitfalls that can derail an application. On the other hand, using the PCT offers the seeming advantage of an option to drop an application before certain fees are required. This dropping advantage is only realised if the underlying technology is found inoperative in that period between 12 and 30 months when a PCT application must go to the national stage. In practice however, no new data arises and no applications are dropped in this window. It is sad but true that incurring the PCT expenses in the hope that BigPharma will eventually pay extra because rights in the former Soviet empire (or somewhere) never pays off. In the aggregate the costs are more at the front and not enough at the back on the one hit. Remember that these expenses will be incurred on every patent application for every project, while a deal will only apply to a few patents or patent applications at a later date.

10.3.5. Country selection

Patents exist on a country-by-country basis. A US patent can only be enforced in the US; a French patent, only in France. Do not pursue patents in countries without effective enforcement systems. This is a basis for espousal of limiting foreign filings to the EPO and Canada, and, maybe, Japan. While Australia has an efficient judicial system, an effected population may be too limited. Remember that the costs are incurred now, while revenue will not be generated until later, if ever.

One cost sensitive licensing point is agreeing to extensive international filings only if <u>fully</u> paid for by the licensee. The rub comes when licensor's corporate staff time is not reimbursed. Patent prosecution in the more exotic countries requires numerous notarisations by state and consular officials and represents a substantial time burden. Agreeing to such filings with BigPharma is not a deal breaker, but be quick to suggest that their patent counsel handle these matters at their expense.

10.3.6. Annuity fees

Many countries charge yearly fees on pending applications and on issued patents. These fees increase as the patent portfolio increases and ages, forming a constant and substantial drain on finances.

11. Patent Position in Action

"*We aren't worried about their patents, we have one of our own.*"

11.1. PATENT CLAIMS VERSUS PRODUCTS

Patent infringement arises from any claim of one party's patent covering another party's product or process. The issue is never Patents v Patents except in fighting it out within the US Patent Office to prove that your side is first to invent. This means that in the US, a patent applicant who is second in time of filing (the "junior party") can obtain a patent over an earlier filed patent application (senior party) if the junior party can establish earlier "conception," "reduction to practice" and, usually, some level of "diligence." These elements are determined in the context of an intra-Patent Office proceeding termed a "patent interference."

This report is not a legal treatise, and the exact meaning of these terms will not be developed here to a legal certainty. Should you find yourself in an interference, the terms will be explained to you for in excess of $1,500,000 in the first year of proceedings. And if you are junior by more than about three months, you are unlikely to be pleased with the explanation.

> **Good Patent Practice**
>
> **To be in a position to prevail in an interference, it is necessary to have documentation of conception, and the work performed as performed. In biotech, documentation is by way of laboratory notebooks. All researchers should maintain laboratory notebooks by date, which are regularly signed, witnessed, and securely archived**
>
> **It is, however, both possible and wise to settle an interference between the parties. The usual settlement is by means of cross licensing. Settlement is best completed before each party spends the first one million dollars. Settlement is particularly wise because it is possible for both parties to lose an interference and have no patents issue or to have the surviving patent so limited as to be of little value. And, given the Byzantine rules associated with interference practice, even a winning position can be defeated by the procedural maze.**
>
> **Interferences aside, if their patent covers your product, you have a problem. If, however, it is also true that your patent covers their product, you have the makings of a wonderful relationship. It is further true that if you have other patents that cover their other products or projects, reason may well prevail.**

11.2. LITIGATION VERSUS LICENSE

11.2.1. Litigation

The absolute minimum tab for a US patent suit is $1,000,000 -- to the courthouse steps. Of course, such bargains are rarely available. And forget finishing litigation. A few years of legal bills of several hundred thousand dollars per month brings most people to the table. The winner in litigation (other than the lawyers) cannot be predicted with certainty. Litigation is extremely distracting to that centre of the company that is responsible for advancing the corporate scientific front. In start-up biotechnology, this may be only five or ten people.

11.2.2. Licensing

If a license is taken, there is no infringement. If the problem is thought to be rather speculative, consider taking the license or option to a license at a lower price. The reduced price represents the tenacity of the position of the patent holder.

Taking a license is the only step that absolutely disposes of a potential infringement problem.

Unless counsel says that there is no conceivable infringement issue, take the license -- if available -- under even minimally acceptable terms. With the uncertainty and expense of litigation, a license now can represent a great saving.

11.2.3. More timing

Be clear on timing. There may be no real problem of infringement. U.S. patents have had a 17-year life span, though some drugs secure a Patent Term Extension of up to 5 years based on a patent having issued prior to FDA approval. Under the new GATT rules that became effective in June of 1995, U.S. pharmaceutical patents will generally have a considerably shorter life span, based on a maximum of 20 years from filing. If the blocking patent is some years post-issue, and the product horizon for sale after FDA approval is some years off, there may be no problem. If it is anticipated that, subsequent to FDA approval of your product, the remaining term of the blocking patent will be brief, one may look to international sales to bridge this restricted period. Of course, this presumes that there are countries other than the U.S. presenting reasonable commercial prospects, and lacking patent coverage. Raising these alternative strategies with a recalcitrant patent holder may also bring them into agreement.

11.3. SURVIVAL CLAIM READING

11.3.1. Look only to the words of the claims

Do not be distracted by what the patent is "really" directed at, or whether or not the underlying science is bogus.

11.3.2. Element-by-element comparison

If a claim recites elements A, B, and C, does the product in question have each element? Having A, B, C, and D usually does not avoid the fact that you also have A, B, and C. It follows that A, B and D is <u>not</u> A, B, and C and is clear of infringement. Also, note the distinctions between <u>**and**</u> and <u>**or**</u>.

11.3.3. Numerical claim limitations

A concentration or amount recited in a claim can provide a bright-line distinction to define infringement.

11.3.4. Ignore predicate phrases

Predicate or preamble phrases such as "A composition for growing hair" do not comprise a limitation of the claimed invention. If the potential blocking patent claims composition has A, B, and C, and your composition has A, B, and C, there is substantial cause for concern -- even if the intended use of your composition has nothing to do with hair growth. One bottle of A, B, and C looks pretty much like another bottle of A, B, and C.

11.3.5. Terms of art

In U.S. practice, certain terms have very specific meanings in patentees, not equalled in general usage: "comprising" v. "consisting of." In patentees "comprising" means the thing recited, and anything else. In contrast, "consisting of" means a rather strong limit to the recited elements. "Consisting essentially of" means almost exactly the recited elements.

11.3.6. Definitions

The issue of coverage, validity, or infringement may turn on the specific meaning attributed to a given term. If a claim recites "lower alkyl alcohol" how many carbons does "lower" mean? A hierarchy of definitions is applied. What the patentee stated in the patent or in the file history will determine this meaning first. If neither patent nor file history defines this term, then one may look to its general definition in the art.

Be sure to check the patent for a definition of all the critical terms before infringement is ruled in or out. In almost every patent, a unique definition is applied to some claim term that may rule infringement in or out.

11.3.7. File wrapper estopped

If the applicant retreated from one position to a narrower position to avoid some prior art, this retreat forms an outer boundary of claim scope. The patentee is "estopped" from attempting to regain in court that which was surrendered in the Patent Office. The corollary here is that if one practices what was given up in prosecution of the blocking patent, one is in a "safe haven" where infringement is not possible (as to that patent).

11.3.8. Things that won't help

11.3.8.1. Different product. The product the bad guys make does not define or reflect what their patent claims, or is totally distinct from your product. It is their patent claims against *your* product. Their product has nothing to do with it.

11.3.8.2. Junk Science. Bad underlying science generally will not remove a claim or destroy a patent. Patents are not journal papers written to prove and convince. Patents are merely recipes that are required to work much of the time, if followed. And even if the broad claims are wildly optimistic, specific narrow claims may fall within the range of reality.

12. Digging for patent dirt

12.1. FILE HISTORIES

Obtain full file histories of all licensed patents, patent applications, and all related applications and patents. The file histories of issued patents are publicly available documents. This is true in the U.S. and Europe. In the U.S., file histories of pending applications are confidential documents of the Patent Office, and not publicly available.

12. 2. COMPUTER SEARCHING

The INPADOC (**IN**ternational **PA**tent **DO**cumentation Center) database provides a means to generate a printout of all countries where a patent application has been filed and the current status as to each country.

Another database is the Derwent World Patent Index. A computer search of patents and patent publications on the Derwent data base (1963 to present) can uncover international patents and patent applications, whether or not issued. In the free-to-market context, the real point of a Derwent search is to uncover pending US patent applications. The search output identifies all countries in which the application is pending. The Derwent information is less than conclusive because the 18-month delay creates a window of uncertainty to be recognised. Confirm the identity of what you are buying as to related applications and patents, and which countries. As noted above an INPADOC search is a quick check on what is out there for the patents and patent applications under consideration.

A number of new on-line search options are available on the Web or by fax.

These include

- the US Patent Office (http://www.uspto.gov)
- an IBM patent site (http://patent.womplex.ibm.com),
- the European Patent Office (http://dips.patent.gov.uk/dips/gb/en/level1.htm),
- MicroPatent Services (http://www.micropat.com),
- Corporate Intelligence (http://www.first.com/).

Conclusion

The pharmaceutical industry lives on proprietary technology. Patents comprise the major proprietary element. Building a patent portfolio requires *timely* attention. Timely, under these circumstances, means before the validity or value of the invention are proven. Building a patent portfolio also requires large infusions of cash. Building a *marketable* patent portfolio requires foresight (which is often indistinguishable from luck.).

INDEX

MIX
Papier | Fördert
gute Waldnutzung
FSC® C083411

Kolibri 360 GmbH
Ferdinand-Jühlke-Straße 7
99095 Erfurt, Deutschland
produktsicherheit@kolibri360.de